Management, Control and Evolution of IP Networks

Management, Control and Evolution of IP Networks

Edited by
Guy Pujolle

Parts of this book adapted from "L'internet ambiant" (2004), "Les évolutions du monde IP" and "Contrôle dans les réseaux IP" (2005) published in France by Hermès Science/Lavoisier
First Published in Great Britain and the United States in 2007 by ISTE Ltd

ISTE Ltd
6 Fitzroy Square
London W1T 5DX
UK

ISTE USA
4308 Patrice Road
Newport Beach, CA 92663
USA

www.iste.co.uk

Library of Congress Cataloging-in-Publication Data

Management, control, and evolution of IP networks/edited by Guy Pujolle.
p. cm.
"Parts of this book adapted from "L'internet ambient" (2004), "Les évolutions du monde IP" and "Contrôle dans les réseaux IP" (2005) published in France by Hermes Science/Lavoisier."
Includes index.
ISBN-13: 978-1-905209-47-7
1. Computer networks. 2. TCP/IP (Computer network protocol) I. Pujolle, G., 1949-
TK5105.5.M35767 2006
004.6'2--dc22

2006033295

British Library Cataloguing-in-Publication Data
A CIP record for this book is available from the British Library
ISBN 13: 978-1-905209-47-7

Printed and bound in Great Britain by Antony Rowe Ltd, Chippenham, Wiltshire.

Table of Contents

Part 1

Control of IP Networks

Chapter 1

Introduction

1.1. Introduction

Packet-switched networks form a very complex and difficult to control world. With circuit switching networks, if all circuits are busy, the network cannot accept additional clients. With networks that move information in packets, the limit where they stop accepting new clients is vague. The primary objective of IP network control is to determine that limit. Other major objectives are: avoiding congestion when a node is completely blocked, putting in place security components, managing client mobility, etc.

This chapter is meant as an overview of some of the important control mechanisms in IP networks. We will start with flow control, which can be done in different ways, such as the opening of another node with generally important priorities on that node or the statistical utilization of resources, as will be shown with DiffServ technology.

The major requirement for efficient control is based on the presence of messages capable of transmitting control information. The system that will generate these messages is called a signaling network: events and decisions must be flagged. Signaling information transmission is a major component of network infrastructure. One can go so far as to say that the future of networks resides in our capacity to drive and automate their configuration. Signaling objective means flagging information, for example, the control and set-up activation of a new route or reserving a part of the infrastructure in order for a software application to run

---

Chapter written by Guy PUJOLLE.

efficiently. Signaling has long been studied by normalization groups, especially the ITU-T. It has greatly evolved in the last 10 years and must continue to adjust as the IP world changes. The Internet's normalization group, IETF, has partially taken over particularly the integration of telephony over IP environments.

Internet flows also require control. If we want to achieve QoS (Quality of Service), it is imperative that we control the flows and the network has to be capable of slowing down or accelerating them according to their importance. Another way of controlling a network is to implement rules according to users' requests. This solution has been developed a few years ago and is called Policy-Based Management (PBM).

Some network functionalities also require rigorous control, such as security and mobility. Let us start by introducing security control mechanisms and then move to mobility management in a network where terminal units can move while remaining connected to the network. In this book, we will detail these extremely important control mechanisms. Finally, we will go to the core of the networks as we will discuss optical networks.

These control mechanisms will be examined briefly in this chapter. The first section of this chapter is a quick overview of signaling. This section will introduce some basic notions with examples, then we will examine flow and congestion control mechanisms, followed by PBM and security and mobility management. We will finish with a discussion on the management of the core of the network.

1.2. Signaling

Signaling means the steps that need to be put in place in order for the information to be transmitted, such as the set-up or closing of a path. It is present in all networks, including those such as IP, that need signaling in its most basic form in order to preserve the system's simplicity. Signaling must therefore be able to function in all network environments, especially IP networks.

Signaling usually needs to function in routing mode. Indeed, it is essential to indicate to whom the signaling is addressed and, in order to do that, the complete address of the receiver must be indicated in the signaling packet. Therefore, all switched networks need a routing process in order to activate signaling.

Signaling functionality is capable of taking over services at different levels of the architecture. For example, it must be able to negotiate SLA (Service Level Agreement) in order to request user authentification, to collect information on available resources, etc. Signaling protocols must be expandable in order to easily

Signaling must be robust, effective and use the least amount of resources in the network. It must be able to function even when there is massive congestion.

The network must be able to give priority to signaling messages. This will reduce signaling transit delays for high priority applications. Attacks by denial of service are also a threat to be aware of, as they can overload the network with high priority signaling messages.

Signaling protocol must allow for grouping of signaling messages. This may include, for example, grouping of refresh messages, such as RSVP, thus avoiding individually refreshing soft-states.

Signaling must be scalable, meaning it has to be able to function within a small network as well as in a major network with millions of nodes. It must also be able to control and modify the multiple security mechanisms according to the applications' performance needs.

1.3. Flow control and management techniques

Flow control and management techniques are imperative in the networking world. Frame or packet-transfer networks are like highways: if there is too much traffic, nobody is going anywhere. It is therefore imperative to control both the network and the flow of traffic within that network. Flow control acts as preventive; it limits the amount of information transferred by the physical capacity of transmission. The objective of congestion control is to avoid congestion within the nodes and to resolve jams when they appear.

Both terms, flow control and congestion control, can be defined in more detail. Flow control is an agreement between two entities (source and destination) to limit service transmission flow by taking into account the available resources in the network. Congestion control is made up of all the actions undertaken to avoid and eliminate congestions caused by a lack of resources.

Under these definitions, flow control can be considered as a unique part of congestion control. Both help ensure QoS.

QoS is defined by the ITU-T's E.800 recommendation which states: "collective effect of the service of performance which determines the satisfaction degree of a system user". This very broad definition is explained in more detail in the I.350 recommendation, which defines QoS and network performance (NP).

NP is evaluated according to the parameters significant to the network operator and which are used to measure the system, its configuration, its behavior and maintenance. NP is defined independently of the machine and the user's actions. QoS is measured in variable conditions that can be monitored and measured wherever and whenever the user accesses the service.

Figure 1.1 illustrates how QoS and NP concepts can be applied in a networked environment. Table 1.1 establishes the distinctions between QoS and NP.

A 3×3 matrix has been developed by ITU-T in the appendix of recommendation I.350 to help determine the parameters to take into account when evaluating QoS and the NP.

This matrix is illustrated in Figure 1.2. It is composed of six zones that must be explicitly defined. For example, if we look at the first column, the access capacity, the user information transfer capacity and finally the maximum capacity that might need to be maintained when a user disengages must be determined. The second column corresponds to the parameters that ensure validity of access, transfer and disengagement actions. The last column takes into account the parameters that control the secure operation of the access, transfer and disengagement.

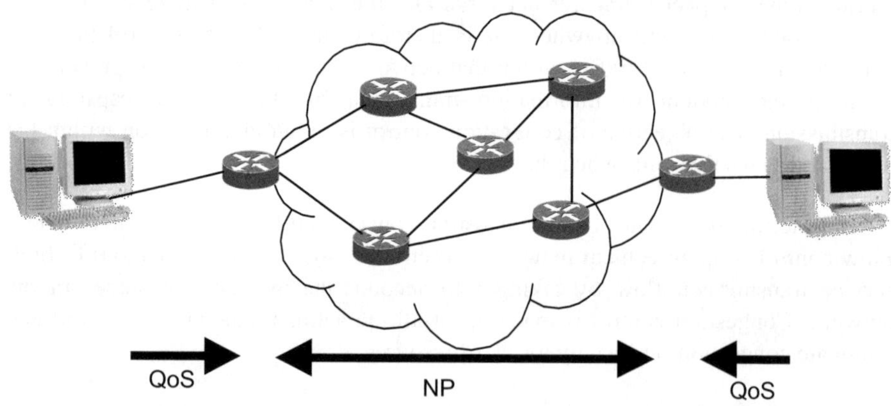

Figure 1.1. *Quality of Service (QoS) and network performance (NP)*

Quality of Service (QoS)	Network performance (NP)
Client oriented	Network oriented
Service attribute	Connection attribute
Observable by user	Controls planning, performance control maintenance
Between access nodes	Between the two network connections

Table 1.1. *Distinction between QoS and NP*

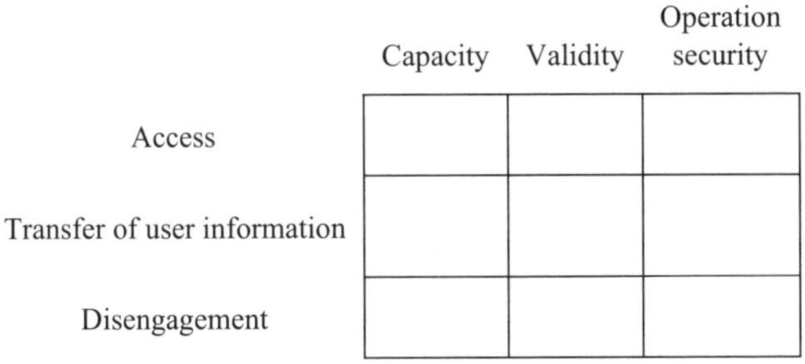

	Capacity	Validity	Operation security
Access			
Transfer of user information			
Disengagement			

Figure 1.2. *3 × 3 matrix defining QoS and NP*

1.3.1. *Flow control techniques*

The ITU-T and the IETF have defined a multitude of flow control techniques. Among the leading techniques, we find the following.

UPC/NPC (Usage Parameter Control/Network Parameter Control)

Usage Parameter Control/Network Parameter Control (UPC/NPC) consolidates all actions taken by the network to monitor and control user access traffic and compliance of the open connection to network access. The main objective of this technique is to protect the network against violations of traffic shaper that can lead to a degradation of the quality of service to other user connections.

Priority management

Priority management generally controls three service classes. The first high priority class is designed for real-time applications such as voice telephony, one average priority class maintains good packet transmission but offers no guarantee on transit times and a low priority class has no guarantee whatsoever.

NRM (Network Resource Management)

Network Resource Management (NRM) groups together the forecasts of network resource allocations to optimize traffic spacing according to the service properties.

Feedback technique

Feedback techniques are the set of actions taken by the users and the network to regulate traffic on its many connections. This solution is used in operator networks like ATM networks with procedures such as ABR (Available Bit Rate) as well as in IP networks with Slow Start and Congestion Avoidance technique, which lowers the value of the transmission window as soon as the return time goes beyond a certain limit.

Among traffic management methods making it possible to avoid network overload, we find traffic control mechanisms whose function is to ensure that traffic is compliant with the traffic shaper, Fast Reservation Protocol (FRP) and Explicit Forward Congestion Indication/Backward Congestion Notification (EFCI/BCN).

The biggest challenge is to constantly design flow control mechanisms that will enable efficient utilization of network resources and satisfy the required QoS. In traditional networks, control of flow by window mechanism is the most widely used. In ATM networks, on the other hand, “send and wait” type protocols do not perform adequately because the propagation delay is too long compared to transmission time. Many other adaptive flow control methods can also be implemented in the upper layers. In general, these control methods work on the size of the window or throughput layer and parameter values are decided by the destination node according to the state of the network.

This system’s implicit assumptions, such as information on the state of the network or the receipt on time of information on the state of the network, can also cause problems. Even if congestion is detected in the network, it is difficult to estimate duration, to locate the congested node in time, to measure the importance of the congestion and therefore to reduce the size of the window adequately.

ITU-T has defined numerous access and traffic control mechanisms. The role of the user parameter control and of the UPC/NPC network is to protect network resources from malicious users and from involuntary operations that have the ability to degrade the QoS of previously established connections. The UPC/NPC is used to detect shaper violations and to take appropriate actions.

In order to avoid cell loss at UPC/NPC level, the UPC/NPC emulation can be executed at sender level. This function is called Source Traffic Smoothing (STS), so that it can be distinguished from the UPC/NPC. From the user's standpoint, the STS function is a nuisance since it means an additional delay and needs more buffer space.

The Virtual Scheduling Algorithm (VSA) recommended in norm I.37 represents the first possibility to detect irregular situations and bring back an acceptable flow from the traffic shaper. Its role is to monitor the peak rate of a connection while at the same time guaranteeing a jitter limit. Simply put, if a frame arrives sooner than expected, it is put on hold until the moment where it should have arrived. Only at that moment is it transmitted on the network and it becomes once again compliant. If the frame arrives later than expected, either it arrives in a short enough time frame to stay compliant with the jitter – and it becomes compliant – or it arrives too late to be within the acceptable limit and it then becomes non-compliant.

The Leaky Bucket (LB) is another mechanism of the UPC/NPC and STS. It consists of a counter (c), a threshold (t) and a leaky rate (l). The counter is incremented of one frame each time it arrives in the buffer and decremented by the leaky rate. If a frame arrives when the counter value is equal to the threshold, it is not memorized in the buffer. In other words, when the buffer is full, the arriving frame is rejected.

1.3.2. *Congestion control methods*

Congestion control methods also vary and even more so whether we are referring to label switching (also called packet-switching) or routing transfer (also called packet-routed) networks. Packet-switched networks correspond to telecommunications operators' views: packets of the same flow have to follow a path. Packet-routed networks are symbolized by the IP world: packets of the same flow can be routed on different routes.

In a packet-switched network, even if each source respects its traffic shaper, congestions may happen from piggy backing of multiple traffics. Several recommendations have been to put in place a method to selectively reject frames in order to ease traffic on the network when there is congestion. For example, in an ATM environment, when the cell header CLP (Cell Loss Priority) bit is marked (CLP = 1), the cell is destroyed first when a congestion is detected. These methods can be useful to relieve the network without much degradation of the QoS.

However, they can result in a waste of network resources and of its intermediary nodes, especially if the congestion duration is overly long. The CLP can also be marked either by source terminals, indicating that the cell has inessential information, or by the UPC/NPC method, specifying that the cell violates the traffic limit negotiated with the CAC.

In the case of packet-routed networks, congestion control methods are handled by the packets themselves, independently from the network's structure. The most traditional solution is to put an interval timer in the packet which, when it expires, destroys the packet. This interval timer is positioned within the TTL (Time To Live) field of IP packets. In fact, in order to simplify time comparison procedure on timers that are rarely synchronous, the IP world prefers rounded values in the TTL field which decrements at each node pass, so that when the packet makes more than a certain number of hops, 16 for example, the packet is destroyed. This way may be somewhat removed from a congestion control solution but it favors the destruction of lost packets or the execution of loops on the network.

1.3.3. *Priority technique*

A flow control solution, that we have not examined yet, consists of associating a priority to a packet or a frame and to process these entities by their priority. This priority can be either fixed or variable in time. The latter are called variable priorities. Several publications have shown that the priority sequencing method in a transfer node could bring about a relatively high resource usage rate of the node. The fixed priority method is the simplest one.

In IP networks, "premium" (or platinum) clients always have a higher priority than the ones in the class just below, "Olympic" clients (Olympic clients are generally subdivided into three classes: gold, silver and bronze), who have themselves a higher priority than the lower class clients, the "best effort" clients.

As opposed to a fixed priority method, a variable priority method will change according to the control point. For example, delay sensitive services become a priority for frames coming out of the buffer. However, an operator can, if he wishes, put the packets back in the Olympic flow in order to transmit at a lower cost, if the transmission rate is sufficiently powerful. Services sensitive to loss have priority for frames entering the buffer: if a frame sensitive to loss requests entry in a memory during overflow, a delay sensitive frame will be rejected. There are several variable priority methods:

– In the Queue Length Threshold (QLT) method, priority is given to frames sensitive to loss if the number of frames in the queue crosses a threshold. Otherwise, delay sensitive frames have priority.

– In the Head Of the Line with Priority Jumps (HOL-PJ) method, several priority classes are taken into account. The highest priority is given to the traffic class that requires strict delays. Non-pre-emptive priority is given to high priority frames. Finally, low priority frames can pass through to a higher priority queue when the maximum delay has been reached.

– In the push-out method or partial buffer sharing, selective rejection is executed within the switching elements. An unmarked frame can enter a saturated buffer if marked cells are awaiting transmission. One of the marked cells is rejected and the unmarked cell enters the buffer. If the buffer only counts unmarked cells, the arriving unmarked cell is rejected. In the partial buffer sharing method, when the number of cells within the buffer reaches a predetermined threshold, only unmarked cells can now enter the buffer. The push-out method can be improved in many ways. For example, instead of destroying the oldest or most recent marked frame in the queue, it would be possible to destroy a larger number of marked frames that correspond to a unique message. Indeed, if we destroy one frame, all frames belonging to the same message will also be destroyed at arrival; it would make sense to destroy them directly within the network. That is the goal of the improved push-out method.

1.3.4. *Reactive congestion control*

Reactive congestion control is essential when simultaneous bursts generate instant overloads in the nodes. Congestion can happen following uncertainty on the traffic or during incorrect modeling of statistical behavior of traffic sources.

The EFCI/BCN mechanism was first introduced by the ITU-T among their recommendations. The role of the Explicit Forward Congestion Indication (EFCI) mechanism is to transmit congestion information along the path between the transmitter and the receiver. The frames or packets that go through an overloaded node are marked in the heading. In the ATM networks, the destination node, receipt of cells marked by congestion indicators (PTI = 010 or 011) indicate congestion in certain nodes of the path. The Backward Congestion Notification (BCN) mechanism enables the return of information throughout congestion to the transmission node. This node can then decrease its traffic. The notification to the transmitter is executed through a supervision flow. This method requires an efficient flow control mechanism, reactive to the internal congestion.

In traditional networks, window-based flow control has been the most widely used. Recent studies propose adaptive methods of window-based flow control. The window size is calculated by the recipient or automatically increased by the arrival of an acknowledgement. These methods were developed for data service and can be linked to error control. The very long propagation delay compared to the transmission time makes the use of a window-based flow control mechanism difficult. Furthermore, these methods use strong assumptions, such as the knowledge of the network's state or that the propagation time will be short enough to return control information in adequate time.

1.3.5. *Rapid resource management*

It is possible, by adapting resources reservation within the network in the entering traffic, to have better control. Obviously, this control would be complicated to implement because of the discrepancy between transmission speed and the propagation delay. The Fast Reservation Protocol (FRP) method has had strong support at the beginning of the 1990s to attain QoS. It is made up of two variables, FRP/DT (Fast Reservation Protocol/Delayed Transmission) and FRP/IT (Fast Reservation Protocol/Immediate Transmission). In the first case, the source transmits only after securing the necessary resources for the flow of frames at every intermediary node level. In the second version, the frames are preceded by a resource allocation request frame and are followed by a resource deallocation frame.

1.4. Policy-based management

Telecommunications operators and network administrators must automate their node configuration and network processes. The two goals of this automation are to control information flow being transmitted in these nodes and to manage networks more easily. These needs have been translated into a policy-based management system, to which we can include control, an integral part of any management system.

The goal of this section is to present this new paradigm, consisting of maintaining and controlling networks through policies. We start by introducing the policies themselves and then detaining the architecture associated to the signaling protocol used in this environment.

A policy takes the form "if condition then action". For example, "if application is voice over telephone type, then make all packets Premium priority". Chapter 6 reviews in detail the policies and their control utilization, as well as the signaling protocol responsible for deploying policy parameters and the different solutions available to put in place policy-based management. You will find here some of the basic elements of policy-based management.

A policy can be defined at multiple levels. The highest level corresponds to the user level, since the choice of a policy is determined by a consultation between the user and the operator. This consultation can be done using the natural language or rules put in place by the network operator. In this case, the user can only choose the policy he wants to see applied from the network operator's rules. This policy is based on the business level and it must be translated into a network language in order to determine the network protocol for quality of service management and its relevant parameters. Finally, this network language must be translated into a lower level language that will be used to program the network nodes or node configuration.

These different levels of language, business, network and configuration are maintained by an IETF workgroup called Policy. The final model comes from another workgroup, DMTF (Distributed Management Task Force) and is called CIM (Common Information Model). Nowadays the two workgroups work together to develop the extensions.

The goal of the standardization of information models for the different language levels process is to create a template that can be used as information models by domain, as well as an independent representation of equipment and implementations. Chapter 6 is dedicated to this solution.

1.5. Security

Security is at the heart of all networks. Since we do not directly see the person with whom we communicate, we must have a way to identify him. Since we do not know where all of our information goes, we need to encrypt it. Since we do not know if someone will modify our transmission, we must verify its integrity. We could go on and on about security issues which networks have to be able to handle all the time.

Globally, security can be divided into two parts: security when we open a session and security during the transmission of data. There are a great number of techniques used to run these two security modes and new ones are invented every day. Each time an attack is blocked, hackers find new ways to thwart systems. This game of pursuit does not make it easy for the presentation and implementation of security mechanisms. In this book, we will limit ourselves to the control of security in network environments without analyzing security of the equipment and software applications themselves.

This section offers a general overview of the security elements within a network, following ISO's security recommendations. These recommendations were done at the same time as the reference model. Then we will present the more traditional security control mechanisms, such as authorization, authentification, encryption, signature, etc.

1.5.1. *General overview of security elements*

The security of information transmission is a major concern in network environments. For many years, complete system security required the machine to be in total isolation from external communication. It is still the case in many instances today.

In IT, security means everything surrounding the protection of information. The ISO has researched and taken account of all necessary measures to secure data during transmission. These proceedings have helped put in place an international architecture standard, ISO 7498-2 (OSI Basic Reference Model-Part 2: Security Architecture). This architecture is very useful for anyone who wants to implement security elements in a network as it details the major capabilities and their positioning within the reference model.

Three major concepts have been defined:

– security functions, determined by the actions that can compromise the security of a company;

– security mechanisms, that define the algorithms to put in place;

– security services, which are the applications and hardware that hold the security mechanisms so that users can have the security functions that they need.

Figure 1.3 explains security services and the OSI architecture levels where they must be put in place.

Five security service types have been defined:

– *confidentiality*, which must ensure data protection from unauthorized attacks;

– *authentification*, which must make sure that the person trying to connect is the right one corresponding to the name entered;

– *integrity*, which guarantees that the information received is exactly the same as the one transmitted by the authorized sender;

– *non-repudiation*, which ensures that a message has really been sent by a known source and received by a known recipient;

– *access control*, which controls prevention/notification of access to resources under certain defined conditions and by specific users.

Within each one of these services, there can be special conditions, explained in Figure 1.3.

	Levels						
	1	**2**	**3**	**4**	**5**	**6**	**7**
Confidentiality							
with connection	yes	yes	yes	yes		yes	yes
without connection		yes	yes	yes		yes	yes
of a field						yes	yes
Authentification			yes	yes			yes
Integrity							
with recovery			yes	yes			yes
without recovery				yes			yes
of a field							yes
Non-repudiation							yes
Access control			yes	yes			yes

Figure 1.3. *Security and OSI architecture levels*

By using the five security services presented earlier and studying the needs of the sender and recipient, we obtain the following process:

1. The message must only get to its recipient.
2. The message must get to the correct recipient.
3. The message sender must be identified with certainty.
4. There must be identity between the received message and the sent message.
5. The recipient cannot contest the receipt of the message.
6. The sender cannot contest the sending of the message.
7. The sender can access certain resources only if authorized.

Number 1 corresponds to confidentiality, numbers 2 and 3 correspond to authentification, number 4 to data integrity, numbers 5 and 6 to non-repudiation, and number 7 corresponds to access control.

1.6. Mobile network control

Networks are becoming global: a client can connect at any moment, anywhere with large throughput. Access networks that allow Internet access are wireless networks or mobile networks. Resources are limited and control management is imperative in order to ensure quality of service. Furthermore, clients can be nomads or mobiles. Nomad clients can connect in different places and get back to whatever they were working on earlier. Mobile clients can continue to work while moving, staying connected. Handovers, i.e. changing receiving antennas, can happen without affecting communication. These environments need control management so that their features can be verified.

Together, nomadism and mobility are part of a bigger setting, which we call global mobility. Global mobility is a vast concept combining terminal mobility, personal mobility and services mobility. This global mobility has become the decisive advantage of third generation networks over today's mobile networks. Controlling this global mobility is an important concept and, in this chapter, we want to delve deeper into this issue.

Terminal mobility is the capacity of the terminal to access telecommunications services, wherever the terminal may be and regardless of its traveling speed. This mobility implies that the network is able to identify, locate and follow the users, regardless of their moves, and then to route the calls to their location. A precise mapping of the user and his terminal's location must be maintained. Roaming is linked to the terminal's mobility, since it allows a user to travel from one network to another.

Personal mobility corresponds to the capacity of a user to access inbound and outbound telecommunications services from any terminal, anywhere. On the basis of a unique personal number, the user can make and receive calls from any terminal. Personal mobility implies that the network is able to identify users when they travel in order to service them according to their services profile and to locate the user's terminal in order to address, route and bill the user's calls.

Services mobility, also called services portability, refers to the capacity of the network to supply subscribed services wherever the terminal and user are. The actual services that the user can request on his terminal depend on the terminal's capacity at this location and on the network which serves this terminal. Portability of services is ensured by regular updates of the user's profile and queries for this profile if necessary. Services mobility links services to a user and not to a particular network access. Services must follow the users when they travel.

Linked to services mobility, VHE (Virtual Home Environment) takes care of roaming users, enabling them to access the services supplied by their services providers in the same way, even when they are out of their area. Due to VHE, a user has access to his services in any network where he is located, in the same way and with the same features as when he is within his own subscriber network. He then has at his disposal a personalized services environment, which follows him everywhere he goes. VHE is offered as long as the networks visited by the user are able to offer the same capabilities as the user's subscriber network.

Within the user mobility concept, terminal mobility and personal mobility are often grouped.

The control of these mobilities is closely studied by normalization and promotion groups, from the ETSI and the 3GPP or the 3GPP2, to the IETF and the IEEE.

1.7. Optical network control

In the previous sections, we have mostly been interested in the control of the network edge and of the local loop, that is on telecommunication ways that provide user access to the heart of the operator.

We must also mention core networks, especially optical networks that today make up the central part of the interconnection networks.

Optical network control means optimization of bandwidth usage. The technique used until now is circuit switching. The concern with this comes from the power of the wavelengths that enable throughputs of 10, even 40 Mbps and soon 160 Gbps. No user alone is capable of utilizing a wavelength allowing end-to-end communication of this throughput. It is therefore important to control multiplexing of users. Other solutions are being studied, such as the opening and closing of optical circuits corresponding to very short times like Burst Switching techniques. Burst Switching is basically optical packet switching but of very long packets which can consist of hundreds of microseconds. Control of these bursts is tricky since it is not possible to memorize bytes of the packet within intermediary elements.

It is also important to control the reliability of core networks. For example, to ensure good telephony, the reliability of the network must reach the 5 "9", that is 99.999% of the time. Control will have something to say in this functionality.

We will examine all these controls within optical networks at the end of this book.

1.8. Conclusion

An uncontrolled IP network cannot work. A minimal control takes us to the Internet as we know it today. Trying to introduce QoS is a complex task but it is quickly becoming a necessity. In order to achieve this, an excellent knowledge of networks is essential. To achieve this knowledge, a high level of control is mandatory.

1.9. Bibliography

[ADA 99] ADAMS C., LLOYD S., KENT S., *Understanding the Public-Key Infrastructure: Concepts, Standards, and Deployment Considerations,* New Riders Publishing, 1999.

[AFI 03] AFIFI H., ZEGHLACHE D., *Applications & Services in Wireless Networks*, Stylus Pub, 2003.

[AUS 00] AUSTIN T., *PKI: A Wiley Tech Brief*, Wiley, 2000.

[BAT 02] BATES R. J., *Signaling System 7,* McGraw-Hill, 2002.

[BLA 99] BLACK D. P., *Building Switched Networks: Multilayer Switching, Qos, IP Multicast, Network Policy, and Service Level Agreements,* Addison Wesley, 1999.

[BOS 02] BOSWORTH S., KABAY M. E., *Computer Security Handbook*, Wiley, 2002.

[BRA 97] BRADEN B., ZHANG L., BERSON S., HERZOG S., JAMIN S., *Resource ReSerVation Protocol (RSVP)-Functional Specification, IETF RFC 2205*, September 1997.

[BRO 95] BRODSKY I., *Wireless: The Revolution in Personal Telecommunications,* Artech House, 1995.

[BUR 02] BURKHART J. *et al., Pervasive Computing: Technology and Architecture of Mobile Internet Applications,* Addison Wesley, 2002.

[COO 01] COOPER M., NORTHCUTT S., FEARNOW M., FREDERICK K., *Intrusion Signatures and Analysis*, New Riders Publishing, 2001.

[DOR 00] DORNAN A., *The Essential Guide to Wireless Communications Applications, from Cellular Systems to WAP and M-Commerce,* Prentice Hall, 2000.

[DRY 03] DRYBURGH L., HEWETT J., *Signaling System No. 7 (SS7/C7): Protocol, Architecture, and Applications,* Pearson Education, 2003.

[DUR 00] DURHAM D., BOYLE J., COHEN R., HERZOG S., RAJAN R., SASTRY A., "The COPS (Common Open Policy Service) Protocol", *IETF RFC 2748*, January 2000.

[DUR 02] DURKIN J. F., *Voice-Enabling the Data Network: H.323, MGCP, SIP, QoS, SLAs, and Security,* Pearson Education, 2002.

[GOL 04] GOLDING P., *Next Generation Wireless Applications*, John Wiley & Sons, 2004.

[HAR 02] HARTE L., *Telecom Basics: Signal Processing, Signaling Control, and Call Processing,* Althos, 2003.

[HAR 02] HARTE L., DREHER R., BOWLER D., *Signaling System 7 (SS7) Basics,* Althos, 2003.

[HAR 03] HARTE L., *Introduction to SS7 and IP: Call Signaling using SIGTRAN, SCTP, MGCP, SIP, and H.323,* Althos, 2003.

[HAR 04] HARTE L., BOWLER D., *Introduction to SIP IP Telephony Systems: Technology Basics, Services, Economics, and Installation,* Althos, 2004.

[HEI 03] HEINE G., *GPRS - Signaling and Protocol Analysis – Volume 2: The Core Network,* Artech House, 2003.

[HER 00] HERZOG S., BOYLE J., COHEN R., DURHAM D., RAJAN R., SASTRY A., *COPS Usage for RSVP*, IETF RFC 2749, January 2000.

[HOU 01] HOUSLEY R., POLK T., *Planning for PKI: Best Practices Guide for Deploying Public Key Infrastructure,* Wiley, 2001.

[JAN 04] JANCA T. R., *Principles & Applications of Wireless Communications*, Thomson Learning, 2004.

[JOH 04] JOHNSTON A. B., *SIP: Understanding the Session Initiation Protocol,* 2nd edition, Artech House, 2004.

[KAU 02] KAUFMAN C. *et al.*, *Network Security: Private Communication in a Public World,* Prentice Hall, 2002.

[KOS 01] KOSIUR D., *Understanding Policy-Based Networking,* Wiley, 2001.

[LIN 00] LIN Y. B., CHLAMTAC I., *Wireless and Mobile Network Architectures,* Wiley, 2000.

[MAC 97] MACARIO R. C. V., *Cellular Radio, Principles and Design,* 2nd edition, Macmillan, 1997.

[MAX 02] MAXIM M., POLLINO D., *Wireless Security*, McGraw-Hill, 2002.

[MCC 01] MCCLURE S., SCAMBRAY J., KURTZ G., *Hacking Exposed: Network Security Secrets & Solutions*, McGraw-Hill, 2001.

[MUL 95] MULLER N. J., TYKE L. L., *Wireless Data Networking,* Artech House, 1995.

[NIC 01] NICHOLS R. K., LEKKAS P. C., *Wireless Security: Models, Threats, and Solutions*, McGraw-Hill, 2001.

[NIC 98] NICHOLS K., BLAKE S., BAKER F., BLACK D., *Definition of the Differentiated Services Field (DS Field) in the IP4 and IP6 Headers,* IETF RFC 2474, December 1998.

[PRA 98] PRASAD R., *Universal Wireless Personal Communications,* Artech House, July 1998.

[STR 03] STRASSNER J., *Policy-Based Network Management: Solutions for the Next Generation,* Morgan Kaufmann, 2003.

[TOH 01] TOH C. K., *Ad-Hoc Mobile Wireless Networks: Protocols and Systems,* Prentice Hall, 2001.

[VAN 97] VAN BOSSE J., *Signaling in Telecommunication Networks*, Wiley-Interscience, 1997.

[VER 00] VERMA D., *Policy-Based Networking: Architecture and Algorithms,* Pearson Education, 2000.

[WEL 03] WELZL M., *Scalable Performance Signaling and Congestion Avoidance,* Kluwer Academic Publishing, 2003.

[YAR 00] YAVATKAR R., PENDARAKIS D., GUERIN R., "A Framework for Policy-Based Admission Control", *IETF RFC 2753*, January 2000.

Chapter 2

Quality of Service: The Basics

2.1. Introduction to Quality of Service

Quality of Service (QoS) is a widely used term. Hardware vendors sell equipment supplying "QoS solutions", operators offer "QoS guaranteed" services; QoS has been the subject of countless works, articles and journals over the years. It is not a new trend; on the contrary, QoS is a growing phenomenon. It is not the goal of this chapter to revisit QoS (there are books dedicated to the subject); we aim to present some useful concepts to understand the QoS problems in networks, and especially IP networks. We will present the currently used QoS parameters, as well as the basic mechanisms implemented at the heart of the hardware. Complete architectures and protocols using these mechanisms will be presented in the next chapter.

Before we go on to discuss technicalities, we should ask ourselves basic questions. Why are we talking about QoS? Why do we even need QoS? How do we define it? And what are the major concerns?

2.1.1. *Why QoS?*

The answer to this question could be as simple as: because there are services, but even more precisely, because there are users who pay for these services. Each time a new service is defined, it implies (or should imply) a definition of the expected

---

Chapter written by Benoît CAMPEDEL.

result. And on the other side, each new user of a service implies a perception (more or less subjective) of this result, of the QoS received.

QoS is often dealt with by itself, separately from services, because certain issues are common to all services. It is therefore important to have a global recognition, regardless of domains and applications, in order to find the most generally sound solutions.

It can also be said that QoS has become fashionable in the last few years. This is due to the Internet explosion, its uses and the rapid evolution of resulting demands.

2.1.2. *The needs*

The needs are generated by the ever-growing number of users and of the applications they use.

Within IP networks, we first find new applications, which are very sensitive to network dysfunctions. Among them are the following:

– telephony over IP and videoconferences: companies are moving toward internal IP solutions and it is not uncommon today for connected users to communicate via webcams;

– ASPs (Application Service Providers): with actual distributed infrastructures, the network is becoming more and more a critical resource for the accurate behavior of a lot of applications;

– network gaming: they have been there since the beginning of networks, but today's computing power enables the design of an ever-growing number of memory-intensive games. Furthermore, the ease of accessing the Internet now enables the use of this network to play. The main console makers even supply online gaming over the Internet.

On top of these new applications, some of the older critical applications are slowly transferred over to IP. For example:

– administrative functions (remote control);

– online banking, stock exchange;

– medical applications.

Speaking of medical applications, apart from information and management access, we can name a new application that could become global and which obviously needs a perfect communication quality: telesurgery. We may remember the first procedure on a human being in September 2001 during which Dr

Marescaux from New York performed an operation on a patient in a hospital in Strasbourg, France.

2.1.3. *Definition*

There is a diversity of applications, and it is not easy to give a global definition of QoS. In fact, it all depends on the point of view.

A user, for example, will want the system to work *correctly*. This term groups different criteria depending on the services used and may be subjective at times. For example, in the case of file transfers, the main judging criteria will be speed. On the other hand, for a videoconference, the user must be *audible* and *recognizable*. For movie viewing, there needs to be a good image definition and a good refresh speed (especially for an action movie). In the case of online banking, we will be mostly concerned with reliability and security for our transactions. Finally, if we go back to our network games example, there must be a good synchronization between gamers' machines. We could probably come up with still more criteria if we brought up other applications.

On the contrary, an operator will have a much more technical view, which enables more objectivity. We will then talk about bandwidth, error ratio, etc. Detailed examples of these parameters are presented later on.

2.1.4. *The concerns*

Once the service quality criteria are well defined, the means to guarantee them has to be implemented. In order to do that, we must proceed to an integration of several elements within a general end-to-end model. There are providers for turnkey solutions, but these are mainly responsible at the network level and are only concerned with the interconnection between two sites or the connection to another network. One of the major concerns is a successful integration at two levels: vertical and horizontal.

Vertical integration consists of traveling from user needs to the physical resources. Each layer has its own control mechanisms and it is important to ensure the correct transfer between them. The translation of needs must be completed and interoperability of mechanisms must be ensured.

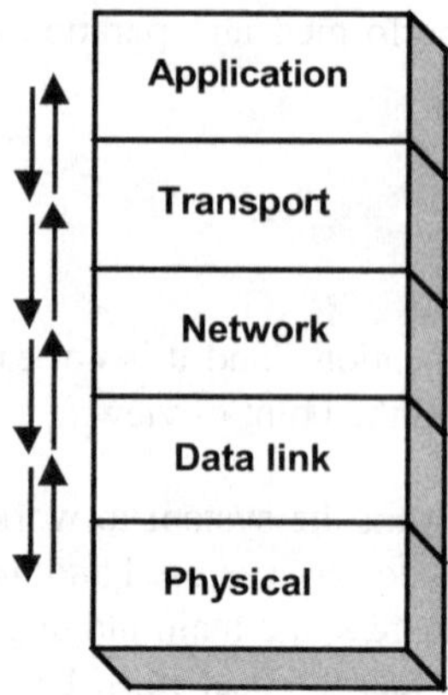

Figure 2.1. *Vertical integration*

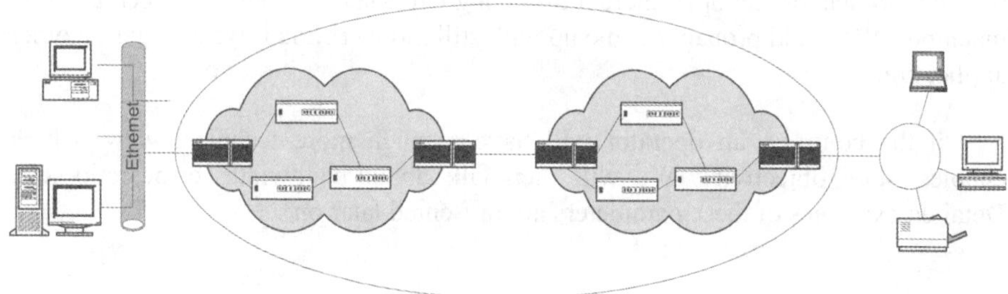

Figure 2.2. *Horizontal integration*

Horizontal integration is basically the hardware connecting two communicating extremities. We might have a combination of multiple operators, which can lead to negotiation problems, and we probably also have a variety of implemented QoS technologies. The crossing of the network requires interoperability.

The problems will also vary according to the context. In the case of closed environments, or proprietary, the operator has total control over his environment. Moreover, the environment is usually homogenous. It is then easier to put dedicated solutions in place (leased lines, ATM, MPLS, etc.).

On the other hand, an open environment will be harder to manage. It is probably heterogenous (horizontal view) and the more popular technologies (IP, Ethernet) are not adapted to QoS (IP best effort mode).

In order to overcome these problems it is necessary to:

– define the mechanisms (basic layer) to manage QoS within different layers;

– define the integration architectures/models of these mechanisms to supply requested services;

– define the interaction (vertical and horizontal) of these models for an end-to-end QoS.

This chapter will mostly deal with network considerations, staying within the basic layer, but we must remember that a viable solution requires a global model that goes back to the user of the service (be it an application or a human being).

2.2. Network parameters

In order to be able to discuss the requirements in terms of QoS and to be able to establish contracts and verify afterward that the clauses are respected, we need a clear set of criteria.

When we get into technical considerations, it is much easier to define the parameters and to be objective in their evaluation (which does not mean that the measurement is easy).

The relevant parameters can vary according to the environments and there is not really a universal criterion (except maybe for bandwidth). However, we can name five parameters that are traditionally found associated with networks and that have a direct impact on applications: availability, bandwidth, latency, jitter and loss ratio.

These parameters are not completely decorrelated, but allow for different needs.

In this section, we will try to give a precise definition, to show the impacts on applications and to identify the multiple elements responsible for their degradation.

2.2.1. *Availability*

2.2.1.1. *Definition*

Network availability can be defined as the ratio between the time when the connection to the network was available and the total time that the system should have been open. We then get a percentage that gives us a first glance at the offered service. For example, if there were three failures last week lasting a total of 5 hours, we could say that the network was available 97% of the time during the week.

We must be careful here because this parameter only takes into consideration the network connection, with no other measurement of quality. There is nothing to tell us that during the network availability periods, the global QoS was sufficient to execute the user's applications properly.

That is why we will also find another notion, which is *availability of service*. This more general notion is defined as the ratio between the time during which the network offered the expected QoS and the total time the service was available. This indicator will obviously be inferior to the previous one.

It is important here to clarify *subjective* perceptions, which we mentioned earlier. If a user uses the network twice during the week and he encounters an interruption of service his second connection, he will feel like he had a 50% availability rate, even if *objectively* the availability was 97%. It is therefore important to clearly define the criteria, the measuring tools and to clarify the reports.

In order to avoid redundancy with the other parameters, we generally take into account the availability of the network connection, which is defined separately from the other parameters.

2.2.1.2. *Impact on applications*

Several factors will come into play to determine the impact on applications. The main factor will be the length of time during which the connection is unavailable. However, this time will be more or less critical depending on the application used.

Simply put, we can say that if the disconnection time is long enough, the application will be alerted and the associated service will be interrupted. On the other side, if the disconnection time is short enough, mechanisms of the transport layer will mask this interruption and ensure continuity of service. An example of this continuity can be seen during a TCP session. If we unplug and replug the network cable, the session does not go down.

However, even if the connection with the application is held, thus masking the network outage, it will still have an incidence on QoS. In reality, all the parameters will be affected. The resent packets will have a more important routing delay, risking jitter. Furthermore, the available bandwidth is reduced to zero during the break, and possible congestion control mechanisms can be activated and reduce the throughput, even when the connection is back (slowstart TCP). This can be translated into degradations at the application level, without incurring loss of service. For example, telephony over IP session can be maintained, even though the user has experienced the break-up in the middle of a word.

2.2.1.3. *Degradation*

The causes for network connection loss vary and are mostly due to some kind of outage. There is another cause that is not so rare: maintenance operations.

It is indeed frequent to have intentional interruptions of service on certain networks. This can happen with operators during important migrations (at the time of a changeover). Certain suppliers of mass ADSL access, for example, disconnect their service every 24 hours. These interruptions are very short and are mentioned in their contract clauses. They are, as much as possible, planned for times when the least number of users will be affected, but some people take advantage of the night times to download and may have noticed a network outage (that might have lead to a change in their dynamic IP address).

2.2.2. *Bandwidth*

2.2.2.1. *Definition*

Bandwidth is probably the most widely known parameter. Most of the operator offers mention throughputs, commercially at least. We often illustrate bandwidth as the width of a pipe. We also show the image of a highway and imagine the bandwidth as the number of available lanes.

More precisely, the bandwidth is defined as the number of bits that can be transmitted per second, or today as the number of kilobits (Kbit) or megabits (Mbit) per second. When we speak of a 512 Kbit/s connection, we mean a bandwidth of 512 kilobits per second.

When we speak of bandwidth, we sometimes make a distinction between gross bandwidth and *useful* bandwidth. In WiFi networks, for example, we can say that 802.11b can supply a bandwidth of 11 Mbit/s, but a useful load of only about 6 Mbit/s. The difference is caused by the fact that control and signaling traffic does not represent the communication data. In the QoS network parameters, we refer to gross bandwidth because it is not always possible to calculate the precise useful load (which depends on the applications used).

Another fact to consider is the application domain of the bandwidth which looks at the possible throughput between two points. In the case of an operator that supplies an Internet access, we generally refer to the bandwidth between the customer's hardware and the operator's network. There is no guarantee concerning zones out of the operator's network. We could then have a 2 Mbit/s leased line and do a file download on another network that is limited to 10 KB/s. In the case where it would be important to guarantee bandwidth through multiple networks, we must have reservation mechanisms (whether they are static or dynamic like RSVP). We are now talking about end-to-end bandwidth.

In conclusion, we should mention that the bandwidth concerns one given direction, for example, in sending. It is entirely possible to have different values in both directions, whether the connection is symmetrical or asymmetrical.

2.2.2.2. *Impact on applications*

The impact will vary according to the nature of the applications, since some of them will be able to adapt to a smaller throughput, whereas others will not be able to function.

Among the applications that can adapt, we have file transfer. Even though the speed of transfer is the main QoS parameter for this application, it is still possible to continue to transfer data with a lower throughput. This is especially true if the transfer does not contain much data, as is the case with sending emails or during Web browsing (if we put aside slow loading graphics). The impact will then not be as much at the application level but more for the user who might find it slower (if he is waiting for the end of a file transfer, for example).

It is important not to minimize the impact of the bandwidth for transfers because it may be vital sometimes. Many corporations start backups and programmed synchronizations during the night or over the weekend, and start other automatic operations later. A long delay in the backup process can cancel the procedure and even, in the worst-case scenario, compromise the integrity of the data if another operation starts before the end of a synchronization.

Examples of applications that adapt poorly are those that require a constant or minimal throughput. Some of those include telephony over IP, streaming content (audio and video), etc. With telephony or videoconference, under a certain throughput threshold, correct encoding becomes impossible and communication may be interrupted. In the case of streaming content there are buffer mechanisms that absorb the temporary bandwidth decrease and synchronizing mechanisms in case data gets lost. But then again, after a certain threshold, the application cannot continue. At best, it can put the transmission on hold (if it is not live).

2.2.2.3. *Degradation*

The bandwidth depends on the physical supports used, but also on the processing capacity of the feedthrough network equipment.

In general, it decreases when congestion is detected (by the TCP flow control, for example). When no measure is taken to guarantee QoS, the network functions in best effort mode. If more than one application is using the resources, these are allocated more or less evenly, thus reducing available bandwidth for each application.

Congestion should not logically happen if the bandwidth is guaranteed and therefore reserved. There are always breakdown or deterioration risks of the supports and equipment. Material problems (cables, network cards) can generate important error rates and increase the resending of packets. Hardware breakdowns can force a changeover to emergency hardware, which makes it possible to maintain the network connection, though in low quality mode, their capacity being sometimes inferior.

It is also possible that an operator incurs a peak of significant traffic, which has not been anticipated and is then found at fault if he practices overbooking, offering more resources than he actually has (somewhat the same as what we see with airline reservations). This behavior is not aberrant as long as it is possible to perform statistical estimations and it is rare that all customers will use the complete bandwidth dedicated to them all of the time. We do not see this behavior much anymore, however, since the operators today apply an oversizing policy, which is easier to maintain and less expensive, due in particular to optical technologies. An interesting consequence is that we sometimes see today the extreme case reversed, where the available bandwidth is superior to the one negotiated (the bandwidth negotiated is then considered minimum).

2.2.3. *Delay*

2.2.3.1. *Definition*

The delay corresponds to the time a packet takes to cross the network, from sending the first bit to receiving the last bit. It depends on:

– the support: the *propagation time* varies according to the technologies used. Fiber optic communication, for example, is much faster than a twisted pair or satellite communication;

– the number of hardware crossed: each piece of equipment will add *processing time* to the packet. This delay depends on the equipment. Therefore, a switch will process faster than a router translating addresses;

– the size of the packets: *serialization time* is also taken into account, i.e. the time required to send the packet on the network link bit by bit (same thing for the reception).

This definition helps to understand that it is not enough to increase the bandwidth in order to ensure better delays, contrary to what we may sometimes think. We can recall here the analogy with highways. The bandwidth corresponds to the number of vehicles passing through a fixed mark per second. The speed being limited at 130 km/h and security distances respected, we must increase the number of lanes on the highway in order to increase throughput, going from 2 to 4 lanes, for example. We can then double the bandwidth. However, with the speed still limited at 130 km/h, a vehicle will still take the same amount of time to reach his destination and therefore the delay as not changed.

In the case of bandwidth decrease, particularly because of congestions, there can be a notable increase in the delay of certain packets that end up in queues or are simply deleted (or resent later).

2.2.3.2. *Impact on applications*

Some applications, by nature, require very short delays and therefore very low network latency. It is the case with interactive applications, but also with applications requiring a strong synchronization between equipment and networks.

Telephony is an example of an interactive application where delays must be low in order for the utilization to be comfortable (we estimate that 100 ms is the maximum). A delay that is too long disrupts communication and causes repetitions and interruptions. This problem can easily be seen when you look at TV news channels during conversations with correspondents on foreign soil. Communication often goes through satellite, where delays are very long (they may reach 500 ms). In almost every case, we see the announcer ask a question and when he does not get an answer within his *comfort interval*, he asks the question again. Immediately after, the correspondent starts to answer and stops (he is receiving the question the second time it was asked), apologizes and continues his response. The problem happens the same way if one of the speakers tries to talk over the other. Put in network terms, the delay surpasses the time-out for the receipt of the message and therefore it is resent.

The problem with strong synchronization is a little different since it is not possible to compensate by resending if the message takes too long to arrive at destination. Contemporary applications that illustrate this problem are network games, in particular real-time combat games. Most of the time, movements are extremely fast and a very short delay is imperative in order to synchronize the position of the adversaries.

In order to give a professional example of such an application, we should consider online banking, especially market operators. There are networks offering real-time information flow (Reuters, etc.) and there is also a network where orders are placed. These two networks are critical. If he does not have accurate on-time information, the operator can miss an important deal and if his transactions are not registered fast enough, he can easily lose enormous amounts of money. This is true for manually entered orders, but is even more critical in the context where we find multiple reactive automatons that can rapidly place orders.

Globally, we can say that any application with real-time constraints (or close to real time) is impacted by problems with delays.

2.2.3.3. *Degradation*

The different elements that can affect the value of latency have been described in the previous definition.

Except in very special cases, serialization and propagation times are not deteriorating factors because their variance is limited and is pretty well controlled. If we want to see the impact of serialization, we can play around by testing the UNIX ping command. By choosing a machine on the local network and by varying the size of the test packets (between 50 and 1,024, for example), we will often see a 2 factor in the result.

The most critical point is the network crossing equipment. Each component produces minimum processing, which is necessary in order to function properly. This defines its minimal delay, which is not only linked to the reception and sending of packets, but also to processes. In the case of a router, there needs to be an analysis of the destination and of the routing table. If this router integrates QoS control mechanisms, this complicates somewhat the processes (we will show some examples later on). In the case of a bridge, if we must fragment packets, encapsulate them into another technology, translate addresses, filter them, etc., latency will be increased.

As we can see, the more active and intelligent the network components, then the more the delay increases. Technologies and components have to be very powerful, but as simple as possible, in order to reduce the delay. This explains the importance of evolved switching techniques (MPLS and others) to lighten processes in the heart of networks.

2.2.4. *Jitter*

2.2.4.1. *Definition*

Jitter is the variance in the latency. It measures the distance between the transmission delays of the different packets in the same flow.

It is possible to have a clear idea of the minimum time that a packet will need to go through the network, but it is harder to know the exact time because of the multiple flows that are present on the network as well as all the components that come into play. Even in the absence of congestion, we see variations of latency. We can again use the ping command and see that on many occurrences we do not find the exact same value, whether it is on the local network or, more logically over the Internet.

Jitter is therefore non-zero, but this does not cause problems unless the gaps become important.

2.2.4.2. *Impact on applications*

We can again identify different consequences, depending on the applications. However, the ones more sensitive to jitter are those evolving in the multimedia world and that need synchronization.

We have seen that latency is critical and could compromise the operation of connections if it is too high, whether it is caused by time-outs or insufficient quality. There are, however, applications that can adapt to network constraints such as high latency (without exaggerating) and low bandwidth. Streaming applications, whether we are talking about listening to an audio flow (radio on the Web, for example) or watching a video (rebroadcast of a remote event), can in general offer multiple coding quality values, according to the available bandwidth. Since the communication is not interactive, the delay problem does not have the same impact, for it is possible to receive the flow with a delay of a few hundreds of ms, maybe even a few seconds. The constraint here is essentially having a constant flow and relatively stable inter-packet delays. In reality, even if the video is completely rebroadcast, it is not advisable to have pauses in the middle or even to slow down the action! An image arriving later than expected would be pointless, deleted and would include a blank space in the video rendering, generally visible (in the case of relative coding, where the transmission is not of an entire image but of the differences, the loss of only one packet can disrupt the display during many seconds if the action is slow). Thankfully, there are mechanisms within the applications (memory-buffers, for example) that can compensate for the inevitable jitter. However, these mechanisms have their limit; they can only handle a set maximum variation. If there is a serious congestion, they can be insufficient. In certain instances, as we can see here, a lower latency may not necessarily be a good thing. Packets arriving too early will be memorized in the buffer, but will be deleted if the buffer is full (although we can always ask for a rebroadcast).

We have recalled above a latency adaptation mechanism, especially to jitter. For other applications, one way to control jitter will be to size time-out values somewhat wider than necessary. However, we cannot exaggerate the values, as this would diminish its responsiveness in the case of a real problem (packet lost and not delayed, for example).

2.2.4.3. *Degradation*

The elements that will contribute to jitter are obviously linked to elements that initially define latency. Therefore, a piece of network hardware, during congestion, will take more time to treat packets, to control its queues and thus will increase its routing delay. One thing to consider is that the higher the number of feedthrough equipment, the greater the minimal feedthrough delay, but also the *risk* of jitter will increase. That is why, we prefer the shorter roads that enable a better latency control (minimum value and jitter).

Another factor to take into account is the dynamic changes of the elements involved in the calculation. During routing change, for example, the number of equipment, their nature and associated processing times can vary greatly.

For all these reasons, jitter is probably the hardest parameter to manage, especially on large networks and particularly when they are open (as with Internet).

2.2.5. *Loss ratio*

2.2.5.1. *Definition*

The loss ratio is defined as the ratio between the number of bytes transmitted and the number of bytes actually received. This calculation integrates the undetected losses (for example, UDP), as well as the retransmitted data.

This makes it possible to have an overview of the *useful* capacity of the transmission and to identify the percentage of traffic output due, not to the client activity, but to network failures (useful data for invoicing).

2.2.5.2. *Impact on applications*

There are mechanisms in place to detect and correct losses. TCP is an example of a transport layer enabling a connected mode and the retransmission management of lost packets.

Applications using this type of service are not generally concerned with loss ratio. The effects are usually due to the impact of these losses over other parameters. Thus, a very high ratio will result in a high bandwidth decrease, all the way to a network outage. In general, losses will mean retransmission of packets and therefore a consequent extension of their effective transmission delay. These are then the consequences linked to the parameters previously discussed. Such applications as file transfer or a Website inquiry will not be seriously impacted by a low loss ratio.

Applications that are more sensitive are those not using these guarantees and that rely on offline modes; multimedia applications are a good example. They can be using, for example, the RTP protocol (in order to sequence packets) which is then encapsulated into UDP. The network will not detect lost packets, as this protocol offers no guarantees. It is therefore important to keep this aspect in mind during application design in order to be able to function even with the absence of lost packets. In spite of these drawbacks, UDP is often preferred here because of its simplicity and speed, and especially because it does not implement TCP flow control, thus enabling a constant flow (amid lost packets), decreasing jitter.

2.2.5.3. *Degradation*

One possible cause for packet loss is an equipment or transmission support problem. However, the latter is getting more and more reliable and with today's optical fiber, the error margin reaches the 10^{-12} level. The level of trust for transmission support explains the fact that most new technologies execute less checks in the lower layers of the OSI model.

In practice, the major cause for loss of packets is network congestion. When a piece of equipment is saturated with respect to the processes that it can manage and its queuing capacity, it deletes the new packets.

2.3. Overview of the basic mechanisms on IP

As we have seen, QoS essentially depends on processes executed on network equipments. Whether static or dynamic architectures are put in place, or whether the solutions are based on resource reservation or flow clustering and differentiation of services, it will be imperative to be able to use a certain number of mechanisms or building bricks in order to manipulate packets at equipment level.

After analyzing the simplified case of a standard router, later on in this section, we will discuss the operating principles of a QoS router and we will detail the different mechanisms involved in order to understand their operation.

This section examines the case of an IP router, but the principles may apply to other network layers, as well as to some switching equipment.

2.3.1. *Standard router*

Most of today's available routers already integrate the mechanisms that will be discussed here, but in order to understand its usefulness, it may be interesting to go back to the basic router operation.

The architecture of a standard router is described in the figure below. Common components are:

– input/output interfaces, used to receive and transmit packets;

– the routing function, which is responsible for finding the destination interface of a packet from a routing table;

– management functions, which integrate equipment control functions as well as the management of routing protocols (to update its table).

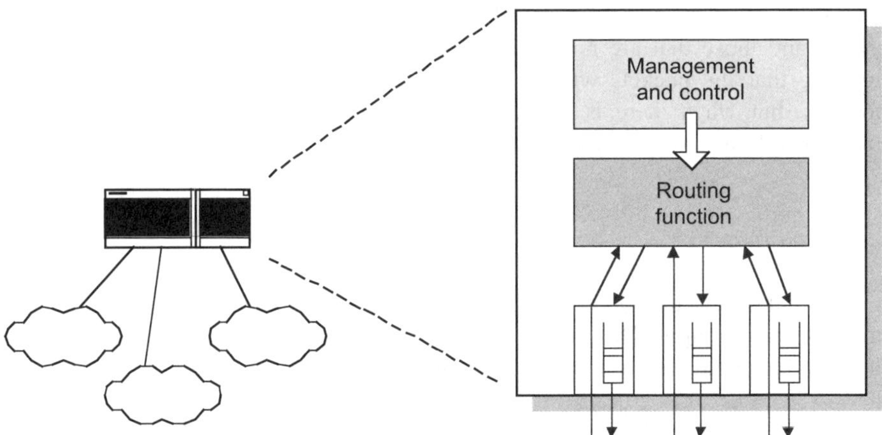

Figure 2.3. *Standard router*

The principles of operation are pretty simple:

– receiving of a packet on the router interface;

– consultation of routing table to determine the output interface to use;

– adding to the queue (FIFO – *First In First Out*) of the output interface. If the queue is full, the packet is deleted.

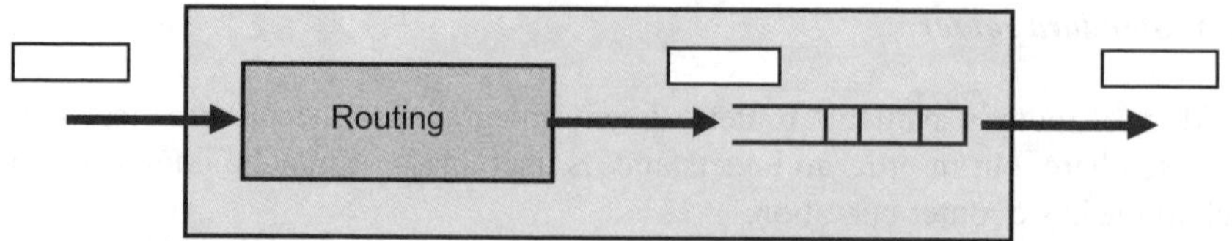

Figure 2.4. *QoS router operation*

This operation can be very fast. Congestion is in the routing function, which is used by all interfaces, but the newer equipment eliminates this problem by using multiple dedicated processors and internal buses at very high throughput.

The main problem here is in the output queue, which is often FIFO.

When there is congestion, the queue fills rapidly. Once full, there is no way to process new packets and they are deleted without distinction between those that are critical and those that are not. This illustrates what we call the best effort mode, meaning that the packets will go through the equipment as long as there is no problem, but when there is congestion, the operation cannot be foreseen. Some jokingly call this mode the non-effort mode.

2.3.2. *QoS router*

In order to discuss QoS, we must be able to differentiate the flows and apply processes accordingly. This means taking into account the specific needs for this or that application, as well as the requirements of the users with valid contracts.

For this, we integrate a number of mechanisms within the router:

– *classification*: packets are analyzed at input level in order to determine the process to apply;

– *policing* and *marking*: policing enables the verification of traffic according to predictions. If this is not the case, it is possible to mark packets as out of profile for further processing (they will, for example, be deleted first);

– *queuing*: a more detailed control than the simple FIFO mechanisms enables the optimization of packet processes and a better congestion management;

– *scheduling*: packets must be transmitted for output, according to the pre-executed classification.

2.3.3. *Classification*

The objective of classification is to determine what process to apply for a received packet and to associate it with an anticipated traffic profile. This will be based on information within the packet, particularly its header.

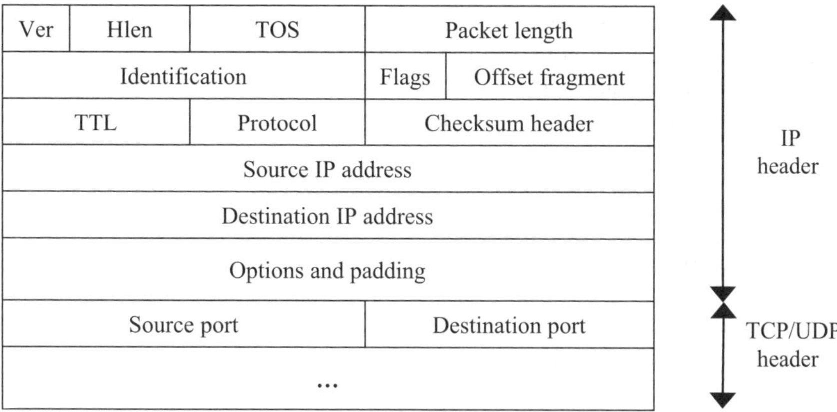

Figure 2.5. *IP header*

The information that is used will depend on the classification type we determine, knowing that we can rely on one or more fields.

2.3.3.1. *Simple classification* (*TOS, DSCP*)

In its simplest form, we use only one byte, initially designed to differentiate IP services, called the TOS (Type Of Service) field. Practically, this field has been diverted from its initial definition and is today called the DSCP (Differentiated Service Code Point) in reference to the QoS *DiffServ* model that redefines it (RFC2474).

There are many advantages to using a unique field to control classification. Filtering will be faster and, in the case of the DSCP field, it stays accessible even during secure transfers (IPSec, for example). However, it is then presumed that the packets have been adequately marked previously and this limits the number of possible cases to process (in the case of bytes, theoretically 256).

We will discuss in more detail the DSCP field and the flow process in the following chapter (which describes different protocols of QoS management).

2.3.3.2. *Multifield classification*

A more precise way to classify packets is to look in more detail at IP headers as well as transport (TCP and UDP). The goal being to identify a particular flow, we will obviously analyze the origin and destination of packets.

The fields more frequently used are TOS, IP source and destination addresses, as well as source and destination ports, which allow the identification of the appropriate applications in more detail. It is important to note that using the fields in the transport layer can cause problems if there is IP fragmentation. Complementary mechanisms must then be put in place in order to associate fragments in the same session.

2.3.3.3. *Implementation*

The choices of implementation will depend on the global model that we will want to use. There are, however, certain constraints to take into account. It is especially important to determine the classification for the speed of packet arrival. In particular, it is important to execute classification at the speed of packet arrival.

There are other issues to consider. Multifield classification is not adapted to secure networks. The useful load being encrypted, the router has no access to the transport header, for example. Within the global model, we must ensure the reliability of information used for classification. Without control mechanisms, a user could send all his packets with a DSCP field corresponding to the best service, even if he does not have that right.

2.3.4. *Policing and marking*

The objective of policing is to verify that the traffic flows according to predictions and, if it is not the case, to make sure it does not affect other flows.

Compliance is based on a flow definition previously done at the equipment level. This means making sure that the contract is respected by the transmitter. Avoiding that exceeding traffic disturbs other flows constitutes the equipment's respect of the contract.

This is done on two levels: traffic measuring tools and management techniques for non-compliant traffic.

2.3.4.1. *Non-compliant traffic*

As mentioned previously, it is important that traffic does not affect the other flows. A simple way to solve this problem is to delete exceeding packets, but that is not the only solution.

For example, it is possible to smooth traffic within the same flow. We memorize the exceeding packets in order to transmit them later, when the volume of traffic is back below allowed resources. Of course, we cannot memorize packets indefinitely and we thus increase latency. However, if we consider that the traffic exceeds the limits of the contract, it is not critical.

We could, on the contrary, let the traffic pass by marking non-compliant packets. This *marking* will consider these packets as non-priority and delete them directly if congestion happens.

The advantage of this solution is that it does not introduce an additional delay and it will use the available general resources in the best possible way.

In the case of smoothing, we would take advantage of a decrease in traffic from the same flow later on.

In the case of marking, we take immediate advantage of the possible decrease of the other flows.

Marking is used during queue management within the same equipment. If we want it to spread to other equipment (not always desirable), it is necessary to use a specific field to transport the information (modification of TOS for IP or the CLP byte in ATM, for example).

2.3.4.2. *Leaky bucket model*

This model is defined as a mechanism based on the operation of a leaky bucket. We can fill it all we can, until it is full, but its output will be limited by the size of the hole. An example of a *leaky bucket* is illustrated in Figure 2.6.

The parameter that defines the *hole* makes it possible to control precisely and limit the output bandwidth. If the input traffic varies, this mechanism enables a more regular output.

The bucket's capacity, although limited, will accept specific traffic and will smooth it. Once the limit is reached, exceeding packets are then considered non-compliant.

The major drawback for this mechanism is its output limit. We cannot take advantage of the fact that our network is not full, at some point, to empty the bucket and thus avoid possible saturation while decreasing latency for the queuing packets.

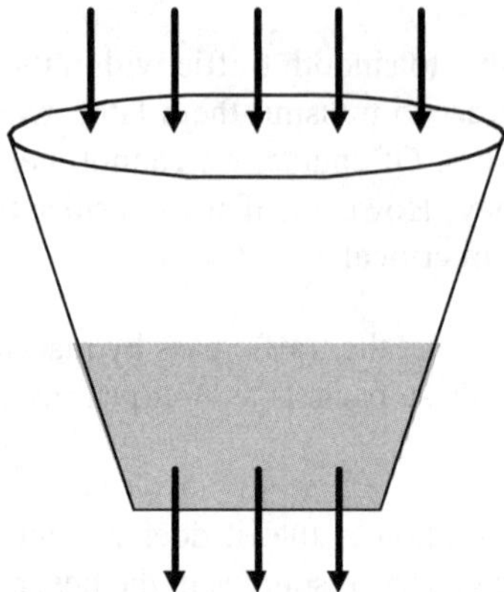

Figure 2.6. *Leaky bucket*

2.3.4.3. *Token bucket model*

This model is based on a token mechanism. These tokens are allocated regularly and indicate how many packets can be output.

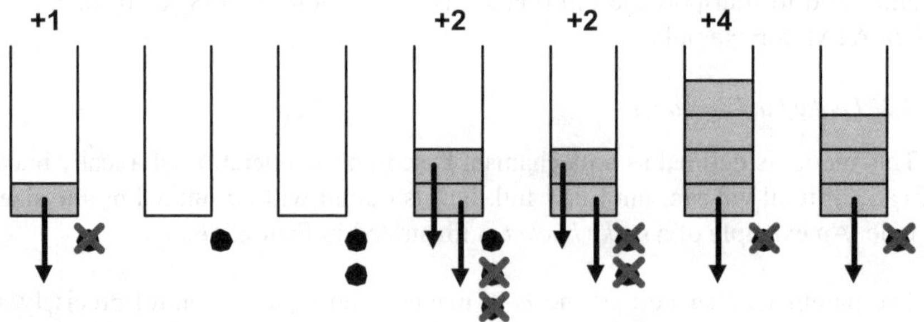

Figure 2.7. *Token bucket*

In the figure above, we see the evolution, over time, of the content of the bucket. Each moment, an additional token is allocated (black spot). The number on top (+ n) indicates packets arriving. The arrows show which packets are transmitted, with the crossed out tokens (consumed).

As you can see in this example, we minimize the leaky bucket model problem. Even though we only allocate one token at a time, we can recognize the output of two packets. Furthermore, if we look closer, without the management by token mechanism, a packet would have been deemed non-compliant at the sixth interval (4 packets arriving). Therefore, we can better control the output bandwidth and we avoid packet deletions.

This mechanism offers more flexibility than the previous mechanism but it must stay compliant with the initial output contract. Instead of being limited to a specific level, the output bandwidth is averaged. The peaks should not get too big and therefore we must limit the number of memorized tokens (this number defines the bucket's peak output).

We must note that this mechanism, even though it corresponds to a way of implementing flow control, is also used to specify the traffic models for negotiations (RSVP, for example), and this should be done independently from the actual coding of equipment control mechanisms.

2.3.5. *Queue management*

The objective for the queue management mechanism is to optimize traffic processing by taking into account peaks, for example, and it is also responsible for the minimization of the size and placement of queues. This last point is important in order to reduce delays and processing time adding to global latency. It is a matter of finding the correct compromises and having mechanisms in place to control congestion and avoid queue overflow.

In order to do this, we must ensure an active flow control, anticipate congestion and attempt to regulate faulty transmitters. This regulation has to go through TCP flow control, since UDP does not supply an appropriate mechanism. We normally presume that the IP traffic is essentially composed of TCP packets.

2.3.5.1. *TCP flow control*

The goal of flow control is to decrease the transmitter's throughput in order to resolve or anticipate congestion problems on the network.

This should be done in an explicit manner. The transmitter must be advised to decrease his transmission window. This makes it possible to target the guilty party but, at the moment, there is no standard defined for IP (there are, however, some propositions such as RFC2481).

Then, the following must be done in an implicit manner, i.e. ensuring that the remote TCP layer detects a problem and acts accordingly. To do this, it must delete exceeding packets. The TCP layer then automatically reduces its window. This solution, however, creates new problems. On the one hand, it is difficult to target the guilty source, but more importantly we risk affecting all sources and so have a quick return of congestion (all flows coming back at the same time to their maximum throughput, when bandwidth becomes available again).

To improve this mechanism, we use more complex packet deletion methods. The most commonly used method is RED (*Random Early Detection*).

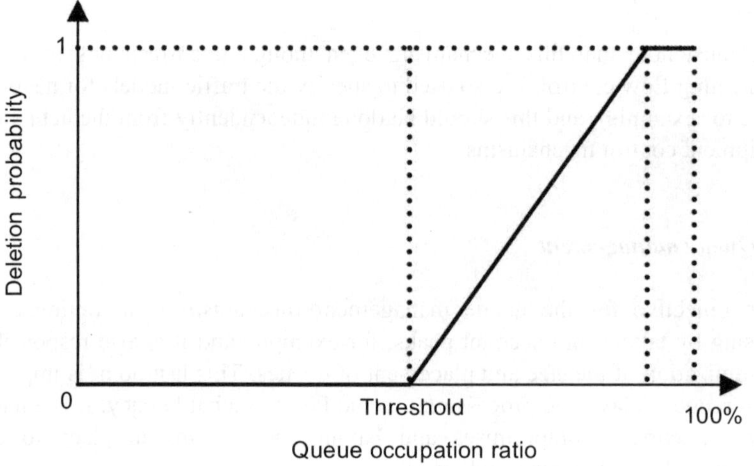

Figure 2.8. *RED mechanism*

In order to avoid congestion, we will start by deleting packets randomly before queue saturation. As we can see in the figure above, we first define a queue usage threshold. Once this threshold has been reached, each new packet arriving has an increased chance of being deleted, reaching up to 100% just before full queue saturation. The advantage to this is that not all transmitters are affected at the same time and we reduce the total number of deleted packets. Evidently, traffic must be TCP and the source must be affected by the deletions in order for this to be an efficient solution, but this technique is statistically valid.

We can name a few other techniques extended from RED:

– WRED (*Weighted RED*): multiple thresholds within the same queue, that can receive different types of traffic with different priorities, are defined. The lowest threshold is associated with the packets with the lowest priority;

– RIO (*RED with In/Out*): is a variation of WRED which defines two thresholds, the lowest being associated with the packets previously marked out of profile;

– ARED (*Adaptive*), FRED (*Flow*), etc.

Each technique addresses a particular problem and more than one can be used on the same equipment, depending on the context and the global QoS model.

In conclusion, we can point out that these mechanisms all work under the presumption that the TCP congestion control is correctly programmed, i.e. that the transmitter *actually* reduces its throughput window when packet loss is detected. There are implementations that work quite differently. This matter is discussed in RFC2309 (*Recommendations on Queue Management and Congestion Avoidance in the Internet*).

2.3.6. *Sequencing*

The last mechanism to be discussed here, sequencing, transmits in the correct order the outgoing packets through the equipment.

In the case of a unique FIFO queue, as seen previously, control is managed completely upstream. Packets are transmitted in the order in which they arrive. On the other hand, when many queues are associated with a single interface, it is important to have rules to sequence the packets and empty the queues. We will present two mechanisms that will show the issue. Many others exist and are either variations or combinations of these. You could actually find different mechanisms within the same equipment.

2.3.6.1. *Priority queuing*

In this case, there are multiple queues that correspond to different priorities. The object is to manage the transmission of the packets respecting these priorities precisely.

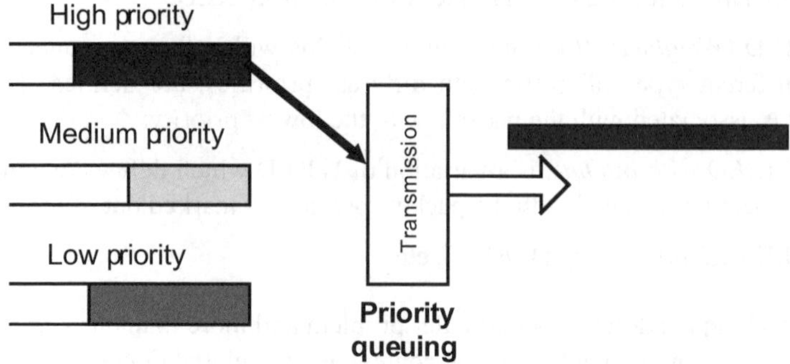

Figure 2.9. *Priority queuing*

The transmission will start with all packets in the high priority queue. Once that queue is empty, it goes to medium priority queue. It is easy to understand the problem that this mechanism brings: the priority queue can monopolize bandwidth. This mechanism makes sense if there is only one category of traffic that must take priority over everything else, for example, critical control flows.

2.3.6.2. *Weighted fair queuing*

The goal of weighted fair queuing is to share network usage between multiple applications, each having its own minimum bandwidth specifications. Each queue has a weight factor, according to its desired bandwidth.

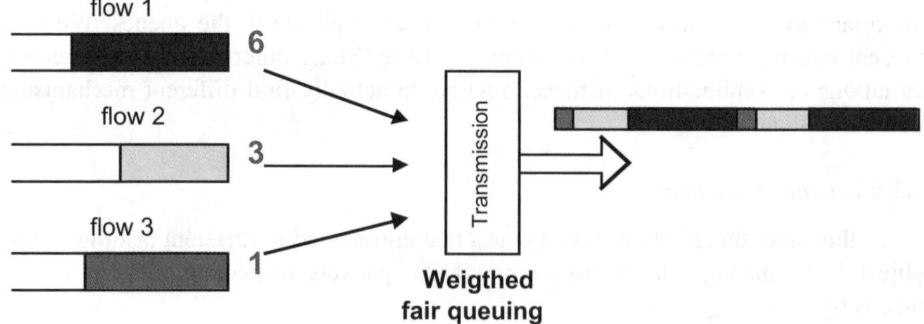

Figure 2.10. *WFQ mechanism*

In simple terms, the example in Figure 2.10 shows that when all queues are used, the sequencer transmits 6 packets from flow 1, then 3 packets from flow 2, 1 packet from the last flow and then starts again. If a queue does not have as many packets as the weight indicates, we move on to the next queue, which enables free bandwidth to be used.

In reality, this mechanism is less simple than it seems. Actually, the used bandwidth does not depend strictly on the number of transmitted packets, but also on their size. It is therefore important to consider this during weighing calculations (we will not go into further detail at this point). The theory remains the same, however, and helps to ensure a minimum bandwidth at every queue and to allocate the unused bandwidth fairly.

2.4. Overview

An overview of a QoS router is presented in Figure 2.11. On the inbound side, note the classification mechanisms, which will lead the identified flows toward the control tools. Once over this filter, and eventually marked, the packets will be associated with an outbound queue, according to the interface that will be used, as well as to the QoS associated to the flow. The queue control mechanisms will then be responsible for managing congestion, while the sequencers will be responsible for transmitting the packets on the router's outbound interfaces.

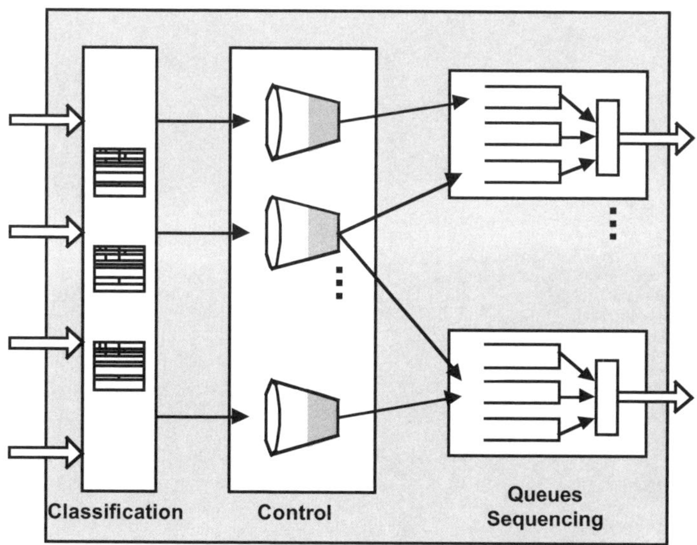

Figure 2.11. *QoS router*

These are all mechanisms that will help control and guarantee the network parameters presented earlier (availability, bandwidth, latency, jitter, loss ratio). They constitute the basic layer that must be configured and used in the framework of a global model in order to provide end-to-end QoS between the transmitter and the receiver. Architectural and integration (vertical and horizontal) problems between the different technologies need to be addressed. The next chapter describes the main models and protocols in use today.

Chapter 3

Quality of Service: Mechanisms and Protocols

3.1. QoS and IP

3.1.1. *The stack of IP protocols*

QoS in TCP/IP networks can be found on many interdependent levels:

– in the case of a local network, QoS mechanisms can be implemented at the link layer access protocols level to give priorities to some machines (the 802.1p standard over Ethernet enables the definition of 8 traffic classes);

– in the case of an extended network, the link layer can provide QoS guarantees during the connection between a user and the network (CBR classification of ATM provides a guaranteed fixed bandwidth during the time of connection);

– at the routers and IP protocol levels, searching for a specific process on certain datagrams or some identified flows (all IP packets stemming from a video will have priority in the router queue);

– the QoS approach can be based on adaptive applications: these will modify their algorithms according to the network's behavior (for example, sound coding will adapt to the network's throughput).

Chapter written by Stéphane LOHIER.

Figure 3.1 illustrates these different QoS levels by locating IP and QoS models or protocols associated at the core of the device.

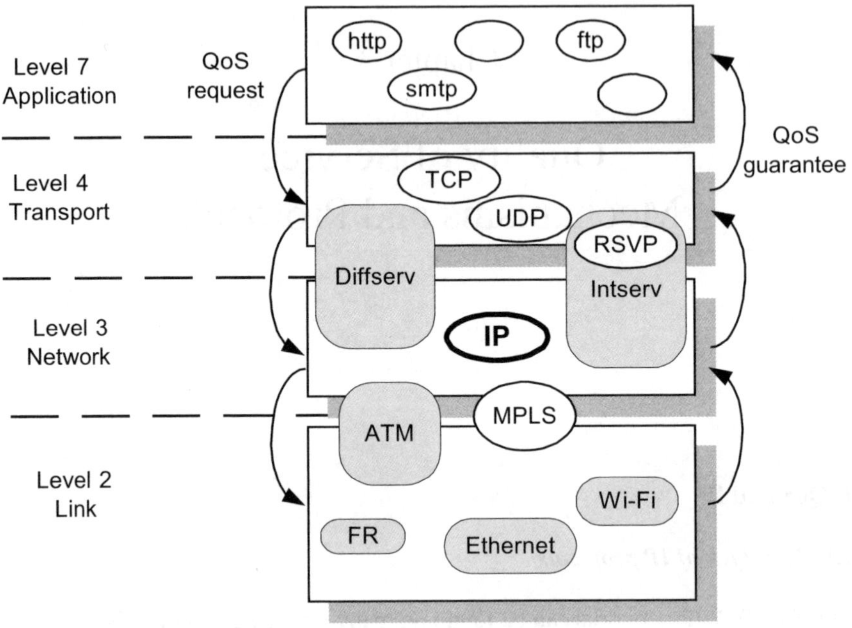

Figure 3.1. *IP and the different levels of QoS*

At the IP level, identification, class and processing of packets are executed from a reserved field in the header to code information linked to QoS.

3.1.2. *The IPv4 TOS field*

The IP packet, or IP datagram, is organized in 32 bit fields. The Type of Service (TOS) field was planned during IP protocol design to indicate with a specific code the QoS linked to a packet. Figure 3.2 positions this field in the IPv4 (RFC791) header.

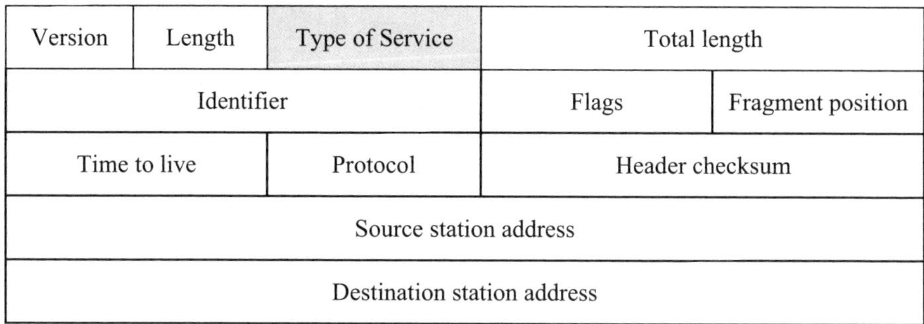

Figure 3.2. *Structure of the IPv4 header*

Detail of the TOS field over 8 bits is shown in Figure 3.3:

– the 3 bits PRECEDENCE show the priority in the datagram. This goes from 0 (PPP = 000) for a normal routing or no priority, to 7 (PPP = 111) for information with maximum priority;

– the following bits show which type of quality is required: D (*Delay*) for a short delay, T (*Throughput*) for high throughput, R (*Reliability*) for reliable routing; C (*Cost*) for a priority linked to cost (number of skips, etc.);

– the last 0 bit is not used.

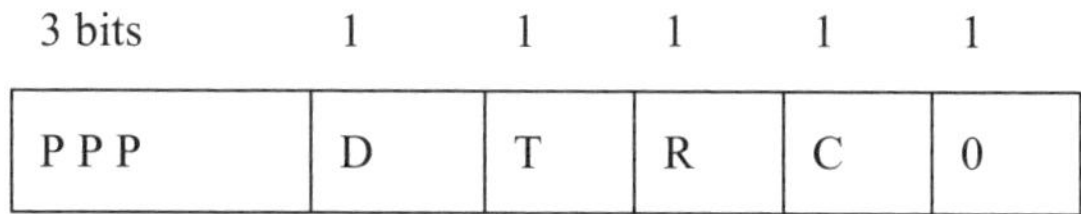

Figure 3.3. *Detail of the IPv4 TOS*

The IPv4 TOS field is rarely used in this form. Another definition proposed in the DiffServ model is explained in section 3.3.

3.1.3. *QoS on IPv6*

3.1.3.1. *Traffic class field*

Within the IPv6 header, the Traffic Class (TC) field, albeit in a different location, takes the place of the IPv4 TOS field. Figure 3.4 shows the structure of this header (RFC2460).

31 27 15 7 0

<table>
<tr><td>Version</td><td colspan="2">Traffic class</td><td colspan="2">Flow identifier</td></tr>
<tr><td colspan="3">Data length</td><td>Next header</td><td>Number of hops</td></tr>
<tr><td colspan="5">Source station address</td></tr>
<tr><td colspan="5">Destination station address</td></tr>
</table>

Figure 3.4. *Structure of IPv6 header*

As in IPv4, the TC field was designed to be used by transmitter nodes and routers to identify and distinguish the different classes or priorities of IPv6 packets. This control is managed by the DiffServ (RFC 2474) protocol, which is designed to exploit this field.

The specialized flow transmitter (multimedia, real-time, etc.) will specify a service class through the TC field. The routers, equipped with algorithms (packet classifier), interpret this field and execute a differentiated process (adaptation, queue, etc.) with the help of a packet scheduler. It is important to note that this field is excluded from checksum and can evolve.

The traffic class field is composed of two parts:

– DSCP (DiffServ Code Point) over 6 bits which contains the values of the different behaviors;

– CU (Currently Unused) over 2 bits, not used currently but would be normally used by routers to indicate a congestion risk.

Today, only two types of behaviors are standardized by the IETF (Internet Engineering Task Force):

– *Assured Forwarding* (AF): this defines 4 traffic classes and three priorities (see section 3.3 on DiffServ). Classes are chosen by the user and remain the same until they reach the recipient. Priority may be modified by routers.

– *Explicit Forwarding* (EF): its behavior is equivalent to a constant throughput leased link.

3.1.3.2. *Flow label field*

The flow label field is also linked to QoS. It can be used by one source to name sequence packets (unicast or multicast) for which a special process is requested from the feedthrough routers (choice of route, real-time processing of information, video sequences, etc.). This chosen flow label value is the same for all packets of the same application going toward the same recipient: the process is then simplified to determine what the packet belongs to. A flow is then defined: it is identified as a combination of the source address and a non-null flow ID.

3.1.3.3. *Other IPv6 contributions for QoS*

Other IPv6 functions meet QoS criteria, besides TC and flow label specific fields:

– *Simplification of the header format*: some IPv4 header fields have been deleted or have become optional. The header is now 8 fields instead of 15. This reduces packet control costs in traditional situations and limits bandwidth needs for this header.

– *Fragmentation*: IPv6 does not manage fragmentation. IPv6 requires that each internetwork link have a Maximum Transfer Unit (MTU) higher than or equal to 1,280 bytes. For each link that does not have the necessary capacity, fragmentation and re-assembly services must be supplied by the layer that is below IPv6.

– *Packet maximum time to live*: IPv6 nodes are not required to impose a maximum time to live to the packet. There is therefore a reduction of information loss because of the absence of rejected packets at the end of the span.

– *ICMPv6*: the IP control protocol has been reviewed. For IPv4, ICMP is used for error detection, tests and automatic equipment configuration. These functions are better defined by IPv6; furthermore ICMPv6 integrates multicast groups control functions and those of the ARP protocol.

3.1.4. *Processing in routers*

A QoS router integrates a specific processing logic by using different packet processing algorithms associated with output queues (Figure 3.5).

The router's first step is packet classification. It directly affects IP and can be executed in different ways:

– from the TOS field of the IPv4 header;

– from the redefined DSCP field within the IPv4 header or defined by default in the IPv6 header in a DiffServ context;

– from a multifield definition integrating, for example:

- source and destination IP addresses,

- the TOS or DCSP field,

- TCP or UDP source or destination ports, which enable a more precise classification according to the application that will be used but that makes the analysis of the headers more complex.

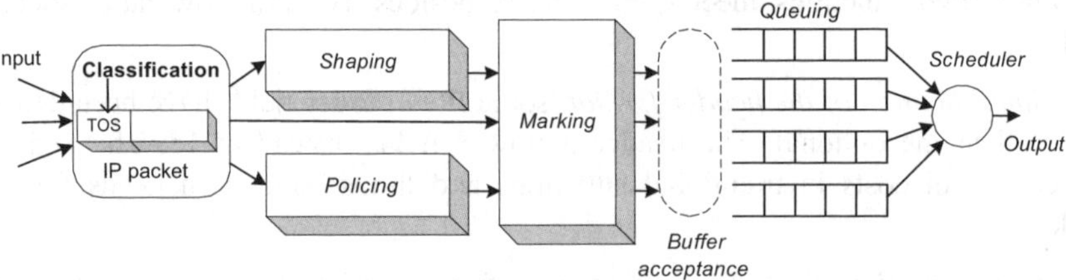

Figure 3.5. *IP processing in a router*

3.2. IntServ (RSVP) model

3.2.1. *Principle*

IntServ (Integrated Services) is an architecture model defined by IETF (RFC 1633) which proposes resource reservation in intermediary nodes (routers) before using them. Contrary to the DiffServ model, as explained in section 3.3, each application is free to request a specific QoS according to its needs. Due to the RSVP (Resource ReSerVation Protocol) associated protocol, intermediary routers check whether they can or cannot grant this QoS and accept or decline the application's reservation request accordingly. If the reservation is accepted, the application has then been granted data transfer guarantees, according to what has been negotiated

(Figure 3.6). Due to the classic Internet infrastructure, the IntServ model then adds two constraints:

– the data flow identification of an application needing QoS;

– the control of the additional information in the routers to process this QoS data flow.

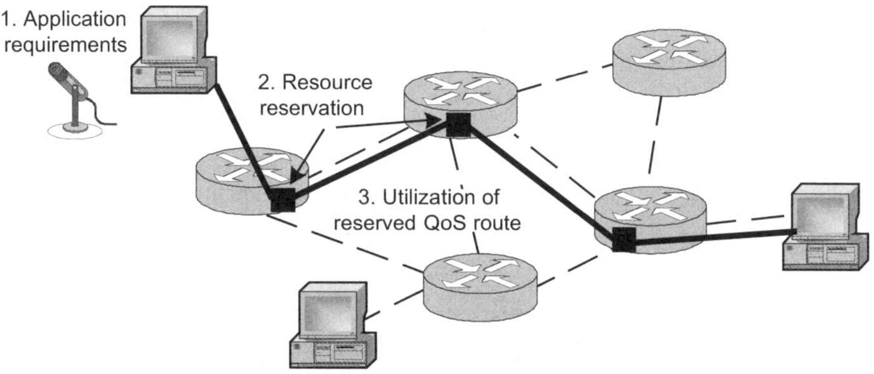

Figure 3.6. *The integrated services model*

The protocols used by IntServ are set at levels 3 and 4 of the OSI model. The RSVP messages put in place to establish and maintain a reserved route are directly transported in IP datagrams (the IP header protocol field is then positioned at 46) or through the UDP protocol if direct mode is not supported (ports 1698 and 1699 are then used). RSVP is not a routing protocol (it is meant to work with unicast or multicast routing protocols such as RIP, OSPF, RPM, etc.) and it is assimilated, in the OSI model, to a transport protocol (Figure 3.7).

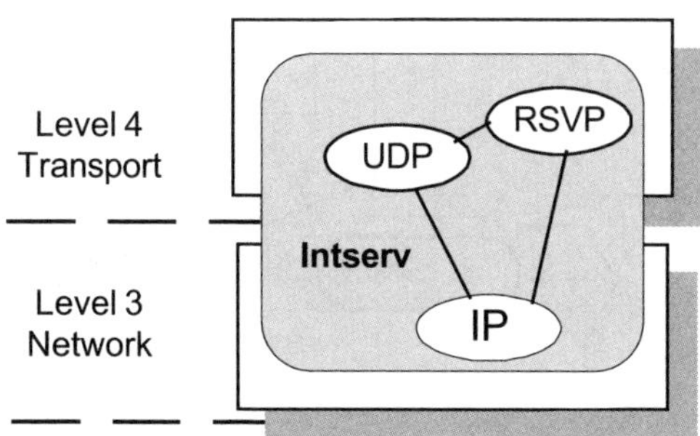

Figure 3.7. *IntServ protocols*

3.2.2. *IntServ services*

IntServ defines two types of services:

– Controlled Load (CL) service which is equivalent to the best effort service on a non-full network and controls available throughput;

– Guaranteed Service (GS) which corresponds to a dedicated virtual circuit and offers bandwidth and end-to-end delay guarantees.

3.2.3. *How an IntServ router works*

An IntServ router (Figure 3.8) must integrate supplementary control mechanisms enabling the reservation requested for QoS. Four distinct functions can be examined:

– the classifier, whose role is to classify each incoming packet according to its ownership flow (best effort packet, packet with QoS request, etc.);

– the scheduler, which controls output packet transmission by using multiple queues. Each queue corresponds to a service classification (classifier role) controlled by different algorithms (CBQ, WFQ, etc.);

– the admission control, which must decide according to the requested service classification if a new flow can or cannot be accepted based on already existing flows;

– the process linked to the RSVP reservation protocol, that drives the set of functions and allows to create and to update the relative state of a reservation within a router that is located on a route used by the QoS flow.

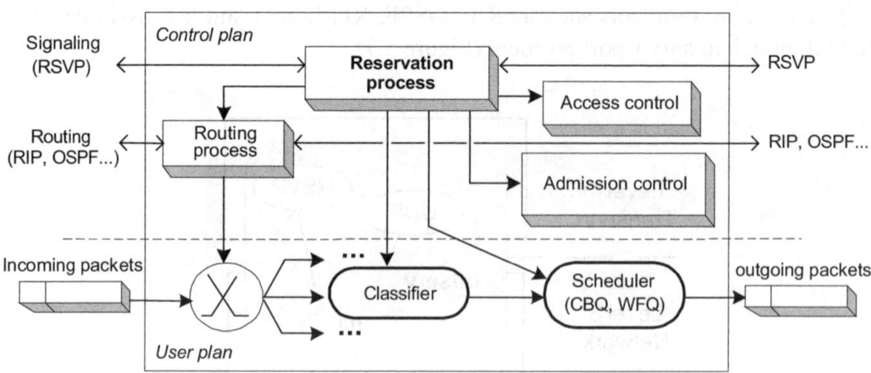

Figure 3.8. *IntServ router architecture*

3.2.4. *The RSVP protocol*

3.2.4.1. *Principle*

RSVP (RFC 2205) is primarily a signaling protocol that makes it possible to reserve bandwidth dynamically and to guarantee a delay for unicast and multicast applications. It is based on the QoS requested by the recipient and not the transmitter, which helps to prevent transmitting applications from monopolizing resources uselessly and thus jeopardizing the global performance of the network.

Routers located on the data flow route meet the RSVP requests, establish and maintain connections (RSVP messages transparently pass non-RSVP routers). Contrary to the reservation of a virtual circuit type static route, routers reserve resources dynamically by memorizing state information (soft state). When a route is no longer utilized, the resources are freed up. Likewise, if the route is modified, the state tables must be kept up to date, which causes periodic exchanges between routers.

Reservation within RSVP is executed in two steps (Figure 3.9):

– information sources periodically generate messages of Path type QoS route searches which spread according to a unicast routing protocol (RIP, OSPF, etc.) or according to a multicast distribution tree structure throughout the routers;

– unicast or multicast addressed receivers are informed of the sources requirements and respond by Resv reservation requests that execute, in the routers, requested reservations and go back up to the selected sources following the opposite route.

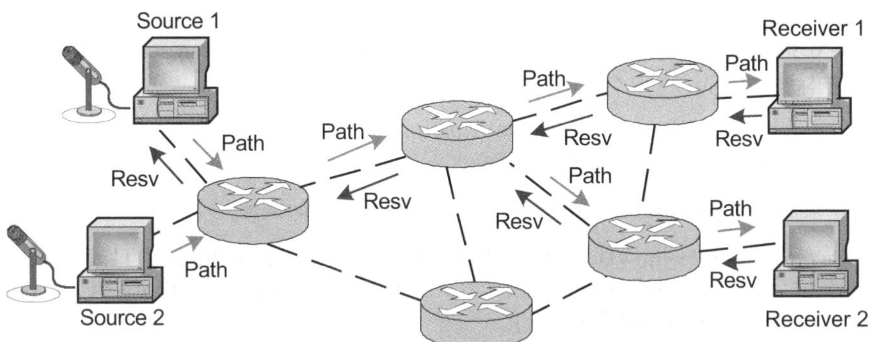

Figure 3.9. *RSVP reservation process*

The logical connection put in place by RSVP is called a *session.* It is characterized by a flow of data with a specific destination and a transport protocol. The session is therefore defined by the three elements described below:

– DestAdress: unicast IP address for a unique recipient or a group of IP addresses for multicast;

– Protocol Id: transport protocol;

– DestPort: UDP/TCP port.

3.2.4.2. *RSVP messages*

An RSVP message is made up of a 64 bit header followed by objects corresponding to the different reservation parameters (Figure 3.10).

In the header part, the explanation of the fields is:

– *Version* specifies the RSVP protocol version (presently 1);

– *Flags* is a field not used at this time;

– *Type* specifies the type of RSVP message:

- 1 for a *Path* search message,

- 2 for a reservation *Resv* message,

- 3 identifies a *PathErr* message for an error in response to a Path message,

- 4 identifies a *ResvErr* message for an error in response to a Resv message,

- 5 for a *PathTear* message that tells routers to cancel the states concerning the route,

- 6 for a *ResvTear* message that tells routers to cancel the reservation states (end of session),

- 7 for a *ResvConf* optional confirmation message sent to the receiver by the last router that received the Resv message;

– *Checksum* verifies the integrity of the full RSVP message;

– *Send TTL* gives the value of the TTL RSVP to compare with the TTL of the IP packet in order to find out if there are non-RSVP routers (TTL RSVP is not decremented by a non-RSVP router);

– *RSVP Length* gives the total length of the message in bytes (header and objects).

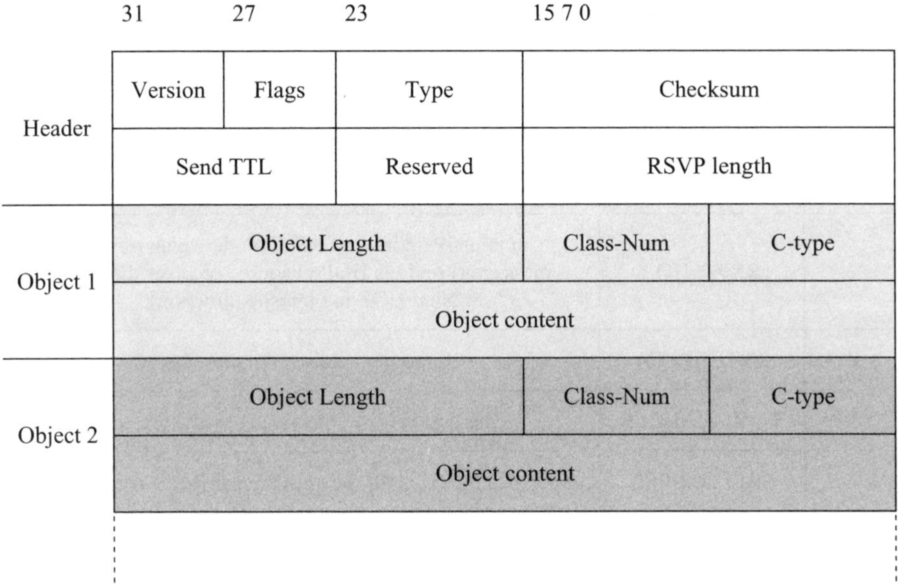

Figure 3.10. *Structure of RSVP messages*

Objects also share a header:

– *Object Length* represents the length in bytes of the object;

– *Class-Num* identifies the nature of the object (see Table 3.1);

– *C-type* groups objects with common properties (C-type = 1 for objects defined in IPv4 and C-type = 2 for objects defined in IPv6, for example).

3.2.4.3. *PATH and RESV messages*

According to the previous RSVP message description, a Path message contains, among other things, objects defined in Figure 3.11. The three objects that make up the sender descriptor describe the entire characteristics of the source (IP addresses, quantitative specifications of the requested data flow, etc.). Only one field, AD_SPEC, is modified by the feedthrough routers in order to take into account the progressive characteristics of the network at the router level, or at the links level.

Class-Num	Object name	Description
0	NULL	Content ignored by receiver.
1	SESSION	Required in all RSVP messages: contains destination IP address, destination port and transport protocol.
3	RSVP_HOP	Contains IP address of RSVP node which transmits message (Previous HOP in source-receiver direction, Next HOP in opposite direction).
4	INTEGRITY	Encrypted authentification data.
5	TIME_VALUES	Message refresh interval set by the creator.
6	ERROR_SPEC	Specifies the error for the PathErr and ResvErr messages.
7	SCOPE	Lists of hosts affected by a Resv reservation message.
8	STYLE	Defines reservation style in a Resv message.
9	FLOW_SPEC	Defines a QoS requested in a Resv message.
10	FILTER_SPEC	Defines a data packet subset within session receiving the QoS specified in FLOWSPEC.
11	SENDER_TEMPLATE	Characterizes a source in a Path message (IP address and other information).
12	SENDER_TSPEC	Defines a source's data flow characteristics within a Path message.
13	AD_SPEC	Transports in a Path message information on the state of the network used by receivers.
14	POLICY_DATA	Transports information on reservation rules defined from an administrative viewpoint.
15	RESV_CONFIRM	Contains the IP address of receiver requesting confirmation in a Resv or ResvConf message.

Table 3.1. *Objects within RSVP messages*

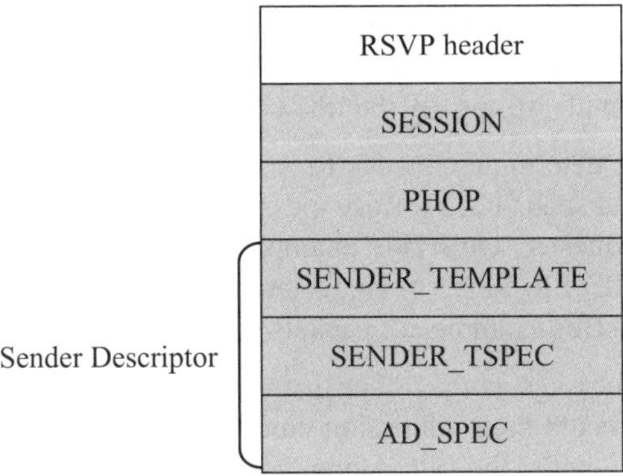

Figure 3.11. *Structure of a Path message*

The structure of a Resv message is described in Figure 3.12. The flow descriptor is made up of FLOW_SPEC, which represents quantitatively the resources requested by the receiver and of FILTER_SPEC, which identifies the packets which will receive the QoS specified in FLOW_SPEC.

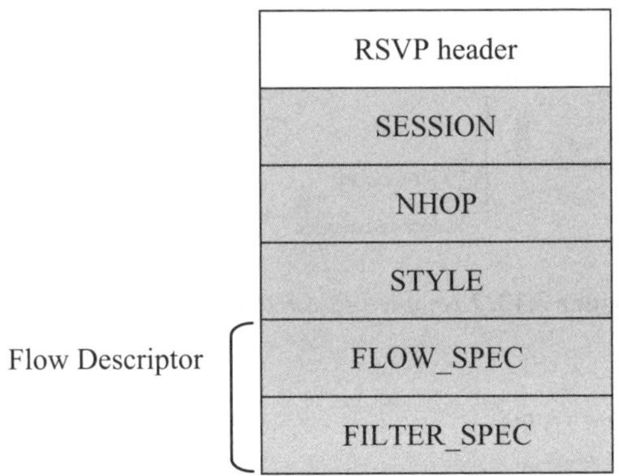

Figure 3.12. *Structure of a Resv message*

Figure 3.13 shows an example of an exchange of Path and Resv messages:

1) the source prepares a Path message in which it will specify its unique characteristics in a TEMPLATE object as well as the desired traffic in a TSPEC descriptor (throughput, variability and size of packet);

2) the Path message is dispatched to destination by applying the unicast (RIP or OSPF) or multicast routing protocol. Routers record whilst passing the path state

that will enable the return of the Resv message. At each passing in a router, the AD_SPEC object can be modified to reflect available resources which can be attributed (for example, to specify that the GS service is the only one available);

3) the receiver determines, thanks to TSPEC and AD_SPEC, which parameters to use in return and sends back a Resv message including the FLOW_SPEC objects to specify the requested QoS (for example, a GS service with a bandwidth of 512 Kbit/s) and FILTER_SPEC to characterize the packets for which the reservation will be established (for example, all packets with a port number equal to 3105);

4) the Resv message comes back using the same route as Path. The routers receiving Resv execute their admission control modules to analyze the request and possibly proceed with the allocation of resources (the packet classifier is programmed according to the FILTER_SPEC parameter and the bandwidth defined in FLOW_SPEC is reserved in the appropriate link);

5) the last router that accepts the reservation sends a ResvConf confirmation message to the receiver.

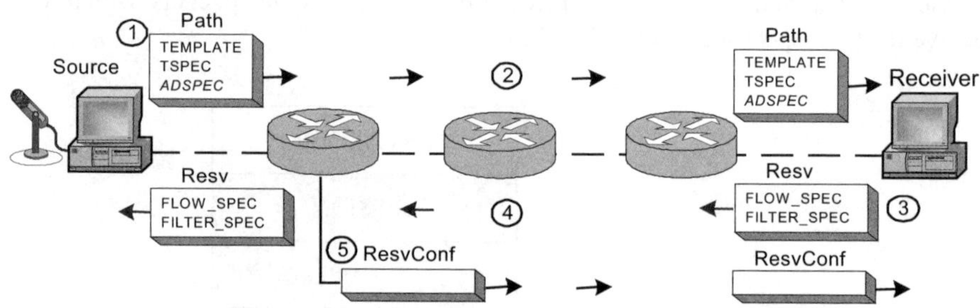

Figure 3.13. *Example of Path and Resv message exchanges*

3.2.4.4. *Reservation styles*

The reservation style of a Resv message specified in the STYLE field enables us to clarify two reservation characteristics in the routers:

– the source selection is done either by a specific list or by default according to the session's common parameters (destination addresses, transport protocol and destination port);

– the allocation of shared resources for all packets of selected sources or distinct for each source.

– Table 3.2 summarizes the different reservation styles according to a combination of these two characteristics. For the FF style, all packets from each source defined by a specific list will get their own reservation;

– SE style implies that all packets from each source are defined by a specific list and will use a shared reservation;

– for the WF style, all the packets share the same reservation whatever their source.

Source selection	**Resource allocation**	
	Distinct for each source	Shared by all sources
Explicit	FF (Fixed Filter) style	SE (Shared Explicit) style
Global (*Wildcard*)	Not defined	WF (WildCard Filter) style

Table 3.2. *RSVP reservation styles*

Figure 3.14 illustrates a request for a SE style reservation on a router with 4 interfaces. This session has three sources (S1, S2 and S3) and three receivers (R1, R2 and R3).

For each interface, a list of explicit sources and the requested shared resources is produced. Thus, the reservation request coming from R3 is made for the two sources S1 and S3 that will share the 3B bandwidth. The router reserves by keeping the maximum on the requested resources for the specified sources. The list is then limited to the sources that are downstream from the router.

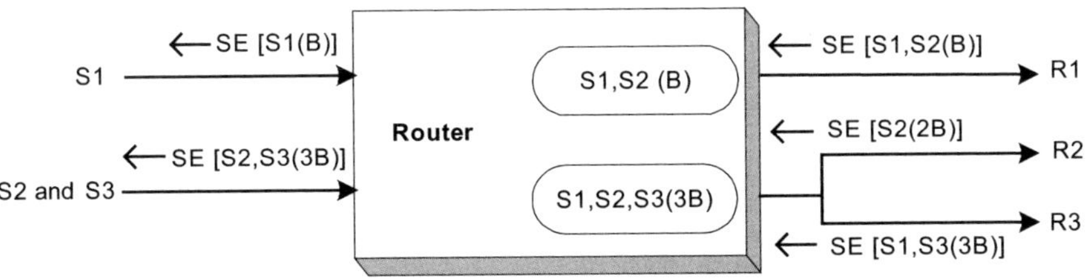

Figure 3.14. *Example of an SE reservation in a router*

3.2.5. *The disadvantages of IntServ*

RSVP requires that all state information for each node of the route linking the source to the receiver is maintained. When the number of sessions or participants of a session increases (scalability), the number of states in the routers and the refreshes between routers become sizeable and jeopardize the validity of the model in high traffic networks.

Furthermore, all the routers, including those at the core of the high throughput network, must inspect multiple fields within each packet in order to determine its associated reservation. After classification, each packet is placed in the queue corresponding to its reservation. The classification and management of queues for each individual flow make the IntServ model difficult to use in high throughput networks.

Besides, even if RSVP is expected to work with traditional routers that do not guarantee resource reservation, the process will be less efficient as the number of non-RSVP routers will be more important.

Finally, since RSVP has not been designed to define a control policy but only to manage the mechanisms ensuring this policy, a control of information exchange protocol (such as COPS) between the nodes linked with the policy servers (PDP – Policy Decision Point) must be added.

3.3. The DiffServ model

3.3.1. *Principle*

Contrary to IntServ, the DiffServ (Differentiated Services) model as defined by IETF (RFC 2475) does not propose reservations in the intermediary nodes. The basic principle consists of introducing multiple service classifications, each offering a different QoS. Depending on the application's needs, each traffic flow will then be attributed with an appropriate service classification.

This traffic classification is executed at the edge of the network, directly at the source or on an edge router, according to a preconfigured set of criteria (IP addresses or TCP/UDP ports). Each packet is marked with a code (DSCP – DiffServ Code Point) which indicates its assigned traffic classification. The routers at the core of the network (core router) use this code, transported in an IP datagram field, to determine the QoS required by the packet and the associated behavior (PHB – Per Hop Behavior), as illustrated in Figure 3.15. All the packets with the same code get the same treatment.

The packets classification criteria must reflect the source application's real requirements and therefore the information that they transmit in terms of bandwidth, sensitivity to loss of packets, sensitivity to delays and to delay variations (jitter). For example, VoIP, which by itself justifies the introduction of QoS, is sensitive to delays as well as to delay variations, much more so than to the loss of packets.

For a given service classification, a SLA (Services Level Agreement) is defined. This agreement specifies a group of rules, generally detailed according to the context, for the traffic conditioning:

– average availability rate, average loss rate;

– delay limit, average delay, average jitter;

– microflow types within each classification;

– DSCP marking value;

– allocated bandwidth, accepted data peaks;

– policing selection in case of agreement overflow:

  - packet transmission,

  - packet rejection,

  - lowering of priority level (change of class),

  - flow shaping (spreading of the exceeding traffic over time),

– buffer size in queues.

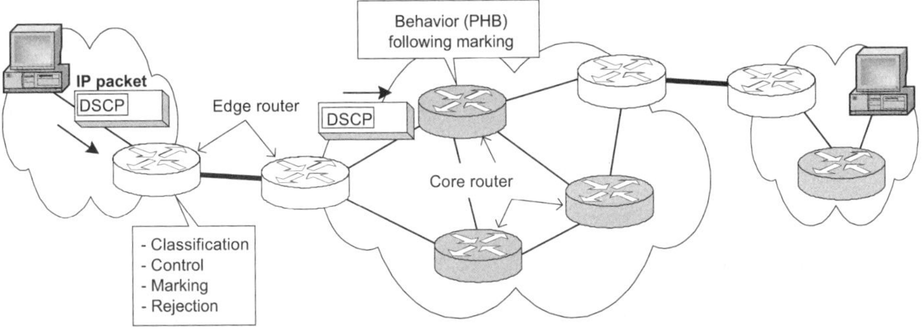

Figure 3.15. *"Differentiated services" model*

3.3.2. *Architecture*

3.3.2.1. *DiffServ domains and regions*

A DiffServ domain corresponds to a zone of contiguous nodes according to the DiffServ model and containing a common policy of control and of PHB. This policy, which defines QoS, has an SLA with the source of data flows. We generally associate a DiffServ domain to a service operator or an intranet. In order to communicate with the outside and match the PHB, these DiffServ domains execute the appropriate traffic conditionings (TCA – Traffic Conditioning Agreement) on their boundary nodes (Figure 3.16).

A DiffServ region contains a group of contiguous DiffServ domains. Each domain applies predefined SLA service contracts as well as a coherent control policy. The operator must guarantee, however, that the QoS will be ensured end-to-end on the whole DiffServ region by maintaining a constant matching between domains.

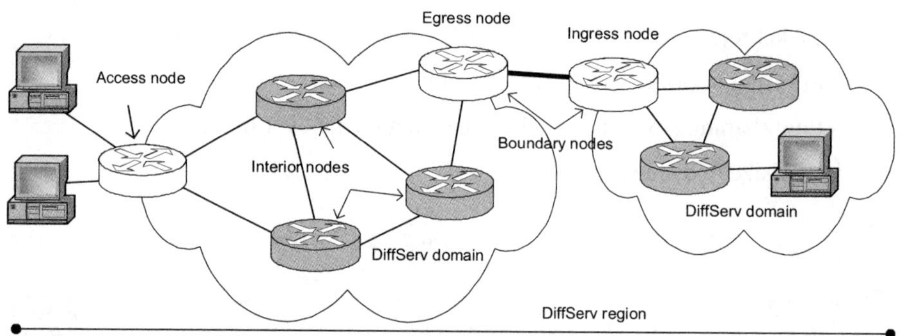

Figure 3.16. *DiffServ elements*

3.3.2.2. *DiffServ nodes*

Processes executed on the edge routers and more often on boundary nodes (these can also be servers or workstations on which the applications emitting the flow are executed) are generally the most complex and correspond to a predefined SLA. They will be able to execute a reconditioning of packets if the rules (TCA) are not the same in the destination domain. An SLA from organization X may for example classify a Gold (AF 3 service classification) flow from client Y entering as a Silver (AF 2 classification) flow within its domain. In this case, the DiffServ domain's edge router of client Y must then execute conditioning operations (TCA) for the outgoing traffic.

Depending on their situation, the boundary nodes are classified into two categories:

– the Ingress Nodes of the domain that classify traffic and verify compliance of classified traffic;

– the Egress Nodes of the domain that must aggregate the different classifications and verify the compliance with contracts negotiated downstream.

Depending on their location, different functions can be put in place in the boundary nodes (Figure 3.17):

– the classifier which detects flow service classes. It is usually based on the association of one or more characteristics of the network and/or transport layer (source or destination IP address, source or destination port, TCP or UDP protocol, etc.);

– the meter which verifies that the inbound flow classifications do not exceed the SLA defined in the router;

– the marker which works on the DSCP field. This module may, for example, decide that in the case where the contract is exceeded, exceeding flows are marked with a lower priority;

– the shaper which will slow down traffic (lower priority queue) may be activated when the flow of a classification exceed the predefined SLA;

– the dropper which comes into place to guarantee a fixed throughput for each service class.

In the case of shaping, the queues having a finite size, the packets can be dropped when profiles are exceeded by too much.

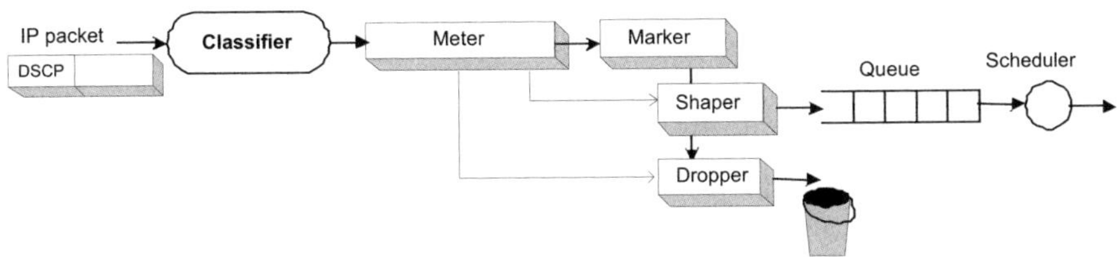

Figure 3.17. *DiffServ edge router functions*

Contrary to the IntServ model, where the major problem is the complexity of operation within the intermediary nodes, DiffServ attempts to decrease processes in the interior nodes (core router). The packet's DSCP field is analyzed and an appropriate predefined behavior (PHB) is executed.

3.3.3. *Service classes*

3.3.3.1. *DSCP codes*

The DiffServ model was designed for use with the IP network layer (see sections 3.1.2 and 3.1.3). It uses the TOS field initially defined in the IPv4 protocol but rarely used until now and is an integral part of the IPv6 protocol for which it uses the Traffic Class (TC) field.

The TOS or TC fields have been renamed as DSCP (Differentiated Service Code Point) when transferred to the DiffServ model. The explanation of each DSCP byte is detailed in Figure 3.18.

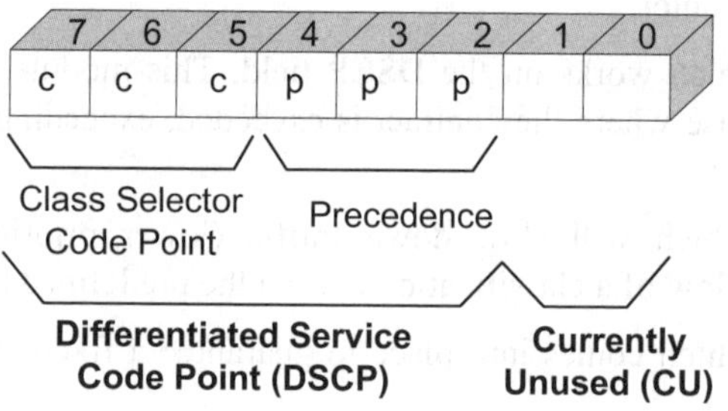

Figure 3.18. *DSCP field in TOS IPv4 or TC IPv6 byte*

– *Class Selector*: the ccc (when c can be 0 or 1) class selector helps to define the major service classes. These will be associated with PHB which will enable differentiated flow processes in intermediary routers. The higher the value of the ccc selector, the higher the priority given to its corresponding flow.

– *Precedence*: this field extends the class selectors with the help of three additional ppp bits that make it possible to define priorities. We then obtain an additional granularity (8 possible priorities per class selector);

– *Currently Unused*: this field is not used at this time and is therefore ignored by intermediary routers. Its goal is to facilitate future protocol expansions.

At this moment, three PHB behaviors are defined for DiffServ:

– EF which offers accelerated processing with bandwidth, delay, loss ratio and jitter guarantees;

– AF which guarantees high probability of packet transmitting with more options (4 traffic classes and three priority levels are defined);

– *Best Effort* (BE) which corresponds to the default service, without quality, offered over the Internet.

3.3.3.2. *EF service*

This service, which can be compared to a virtual leased line, is also named Premium because the associated traffic is serviced with high priority through routers with these characteristics:

– priority processing in queues;

– traffic parameters (bandwidth, delay, loss ratio and jitter) which are always compliant with SLA;

– outgoing traffic shaping when necessary to ensure an inter-domain QoS.

To make sure the bandwidth is adequate for the other flows, the associated traffic is usually limited to 10% of the total traffic.

This type of behavior is widely used for the transmission of real-time data such as voice or videoconferencing.

The DSCP associated with the EF service is equal to 101 110, which presently corresponds to the highest priority.

3.3.3.3. *AF service*

Where all the traffic on the Internet does not need guarantees as high as those supplied by the EF service, a second AF service is offered in order to ensure a minimum variable quality according to the applications and favoring the associated data flows compared to best effort, especially in the case of congestion.

The AF service offers multiple levels of packet transmission guarantee (see Table 3.3).

It is made up of a group of 4 service classes (AF4 to AF1), each having 3 drop precedences.

For example, when congestion occurs in service class 4, packets that have a high drop precedence value, i.e. a DSCP value of 100 110, will be dropped first.

The AF service can be used to execute the Olympic service which has three classes: bronze, silver and gold, corresponding to AF1, AF2 and AF3.

Similarly, for each Olympic class, a priority level (1, 2 or 3) can be affected.

EF – Expedite Forwarding (premium)			
101 110			
AF – Assured Forwarding			
Drop Precedence	Low ↓	Medium ↓	High ↓
Class 4	AF43 100 010	AF42 100 100	AF41 100 110
Class 3 (gold)	AF33 011 010	AF32 011 100	AF31 011 110
Class 2 (silver)	AF23 010 010	AF22 010 100	AF21 010 110
Class 1 (bronze)	AF13 011 010	AF12 011 100	AF11 011 110
BE – Best Effort			
000 000			

Table 3.3. *Point code for the different PHB*

3.3.4. *DiffServ advantages and disadvantages*

One of the major advantages is the capacity to limit processing times of intermediary routers, which helps network operators who had a hard time justifying the complexity brought about by IntServ with the reservations to now implement it on their whole infrastructure.

The PHB (behaviors) normalization constitutes a second major advantage of DiffServ by simplifying the interconnection between the different DiffServ domains.

One of the disadvantages is the obligation to establish, ahead of time, a contract (SLA) within all the equipments in the domain.

This constraint implies a thorough knowledge of the applications that will go through the network and a centralized and distributed policy from specific servers (PDP – Policy Decision Point).

Behavior based on the aggregation of flows also implies a granularity loss of the applications passing through the network, which in certain cases may be inappropriate.

3.4. MPLS architecture

3.4.1. *Principle*

MPLS (Multiprotocol Label Switching) is an architecture standardized by IETF that makes it possible to integrate and homogenize the different routing and switching protocols in place at different levels in the standard networks (Ethernet, IP, ATM, Frame Relay, etc.).

The major goal is to improve delays, and therefore QoS, in the nodes with multilevel rapid switching based on the identification of labels carried by frames or packets.

MPLS has the following characteristics:

– independent from the protocols at layers 2 and 3;

– supports layer 2 in IP, ATM, and Frame Relay networks;

– interaction with existing reservation and routing protocols (RSVP, OSPF);

– possibility of associating specific traffic profiles (FEC: Forward Equivalence Class) to labels.

Within MPLS, data transmission is done over label or LSP (Label Switched Path) switching routes.

LSPs correspond to a label sequence at each node in the route from source to destination.

Labels are specific identifiers lower layers (MAC Ethernet addresses, VPI/VCI fields in ATM cells, etc.) protocol and are distributed according to LDP (Label Distribution Protocol).

Each data packet encapsulates and transports the labels during their routing.

As long as fixed-length labels are inserted at the front of the frame or cell, high throughput switching becomes possible.

The nodes read the labels and switch the frames or the cells according to the value of these labels and of the switching tables previously established in the LSP (see Figure 3.19).

These nodes, depending on their location on the network, are either:

– LSR (Label Switch Router) for a router or switch type equipment located at the core of an MPLS network and is limited to reading labels and switching (IP addresses are not read by LSR); or

– LER (Label Edge Routers) for a router/switch at the border of the access network or of the MPLS network that can support multiple ports connected to different networks (ATM, Frame Relay or Ethernet). The LERs play an important role in the assignment or deletion of labels for incoming or outgoing packets.

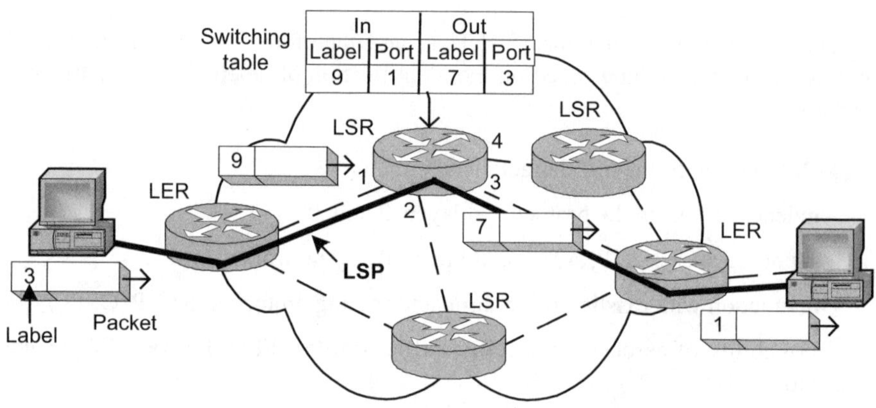

Figure 3.19. *MPLS nodes and route*

3.4.2. *MPLS label and classes*

A label, in its simplest form, identifies the route that the packet must follow. The label is encapsulated and transported in the packet's header. Once a packet is labeled, the rest of its route is based on label switching. The router that receives the packet, analyzes its label and searches for the corresponding entry in the switching table in order to determine the output interface and the new label allocated to the packet. The label values have a local impact and can be linked to a specific architecture in order to be able to determine a virtual route (DLCI type for frame relay or VCI/VPI for ATM). The generic format of a label is illustrated in Figure 3.20. It is located between layers 2 and 3 or directly in the header of layer 2 (MAC addresses for Ethernet, VCI/VPI in ATM, etc.).

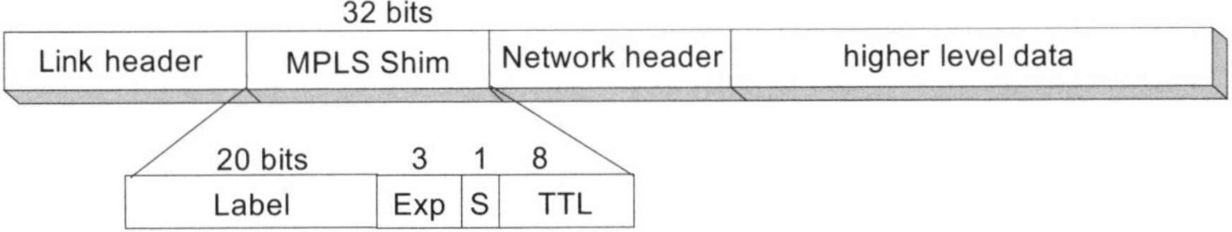

Figure 3.20. *MPLS labels' basic format*

Next to the label value, different fields are present to add functionality:

– the experimental field is not standardized and may be used to control QoS, for example, by associating a DiffServ type PHB;

– the Stack byte assumes a value of 1 when the label is located at the top of the pile in a network interconnection with multiple label levels (VPI/VCI hierarchy of an AT network, for example);

– as with IP, the TTL (Time To Live) field helps to prevent looping.

An FEC (Forwarding Equivalence Class) corresponds to a group of packets that have the same requirements, in terms of address prefixes or QoS.

Contrary to the other models, within MPLS a packet is only assigned once to an FEC, when entering the network.

Each LSR builds itself an LIB (Label Information Base) table in order to figure out how a packet must be transmitted.

The labels are therefore associated with an FEC according to a logic or a policy based on different criteria: QoS, same source or destination address prefix, packets from the same application, affiliation to a VPN (Virtual Private Network).

3.4.3. *MPLS routes*

In the example illustrated by Figure 3.21, the FEC corresponds to the destination addresses prefix 18.1. When entering the MPLS network, the LER edge router includes the label corresponding to this FEC for the incoming packet according to its LIB switching table.

The LSR central routers then exchange the label again according to the FEC and switch the packet.

The outgoing LER edge router takes away the label and ensures the routing of the packet to its destination.

The traced LSP route corresponds, in this example, to the rest of the labels 9-7-1.

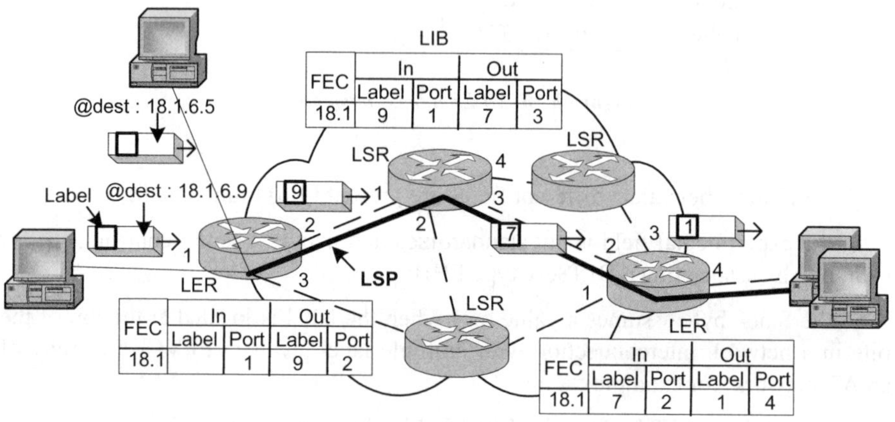

Figure 3.21. *MPLS switching example*

The LER edge routers must go back up to the network level to analyze the IP addresses and to position the labels accordingly. The LSR central routers, when the tables have been previously positioned (LPD protocol or routing protocol), play a simple label switching role (LSRs can be simple ATM switches). The routing process therefore requires two protocol levels:

– a routing protocol such as OSPF or BGP responsible for route distribution and for routing tables set-up;

– a label distribution protocol such as LDP that is responsible for the setting up of switching tables from routing tables and FECs.

3.5. QoS at level 2

3.5.1. *QoS with ATM*

The ATM technology is originally designed for QoS support. Within its model, the ATM Adaptation Layer (AAL) located over the ATM layer (ensuring multiplexing and cell switching) is responsible for supplying a QoS to applications.

Five adaptation classifications have been defined for the different flows:

– AAL1 is designed for the support of voice applications or circuit emulation requiring constant throughput flows;

– AAL2 is designed for voice or video variable throughput flows;

– AALS 3/4 for data transmission on connected or disconnected mode, very rarely used in practice;

– AAL5 class 3/4 simplified version is widely used for secure data transport.

Table 3.4 shows the correspondence between AAL classifications and the services used for different connections or applications.

Service	QoS	Applications	AAL class
CBR (Constant Bit Rate)	Guaranteed constant throughput	Non-compressed audio/video	AAL1
VBR-RT (Variable Bit Rate-Real-time)	Variable throughput, real-time traffic	Compressed audio/video	AAL2
VBR-NRT (Variable Bit Rate-Non-Real-time)	Variable throughput, batch traffic	Transactional	AAL5
ABR (Available Bit Rate)	Available throughput	File transfer, email, LAN interconnection	AAL5
GFR (Guaranteed Frame Rate)	Guaranteed frame throughput	Applications requiring minimum guaranteed throughput	AAL5
UBR (Unspecified Bit Rate)	Unspecified throughput	Best effort Internet applications	AAL5

Table 3.4. *AAL services and classes*

Moreover, when the ATM technology is used on a large scale to transport IP packets in a heterogenous network like the Internet, its QoS functionality is rarely used:

– native ATM applications able to use the QoS parameters are scarce;

– if ATM has not been deployed end-to-end, its QoS functionality might be inefficient: queuing introduced by non-ATM routers has an influence on the delay and jitter calculation;

– ATM and TCP have different behaviors when encountering congestion: ATM deletes cells and notifies the terminal system of the packet loss, TCP reduces its transmission window and modifies its throughput, whereas the ATM congestion has already been processed. This fact can be avoided, however, if the ATM network is correctly sized to avoid large scale congestion problems.

3.5.2. *QoS with Ethernet*

Some QoS functions are planned in local networks such as Ethernet, in particular within the two complementary 802.1q and 802.1p norms.

The 802.1q IEEE norm is today the *de facto* standard for the identification of Ethernet frames within a Virtual Local Area Network (VLAN). The general principle is the addition, within each Ethernet frame destined to be transmitted from one switch to another, of some additional headers containing particularly the virtual network identifier to which it belongs (VID: *VLAN Identifier*) and a field to establish priorities over the frames. The QoS mechanisms are not exactly defined for this standard but the possibility, through interconnected Ethernet switches, to group within a VLAN stations with common characteristics, enables the reduction of Broadcast domains, which are very bandwidth intensive on Ethernet.

The 802.1q fields are inserted in the Ethernet 802.3 frame as shown in Figure 3.22:

– the TPID (Tag Protocol ID) field indicates that the frame is signed according to the 802.1q protocol;

– the TCI (Tag Control Information) part contains:

- the priority field on 3 bits as defined by the 802.1p standard,

- the CFI (Canonical Format Indicator) field specifies that the MAC address format is standard,

- the VLAN ID on 12 bits identifies the VLAN to which the frame belongs.

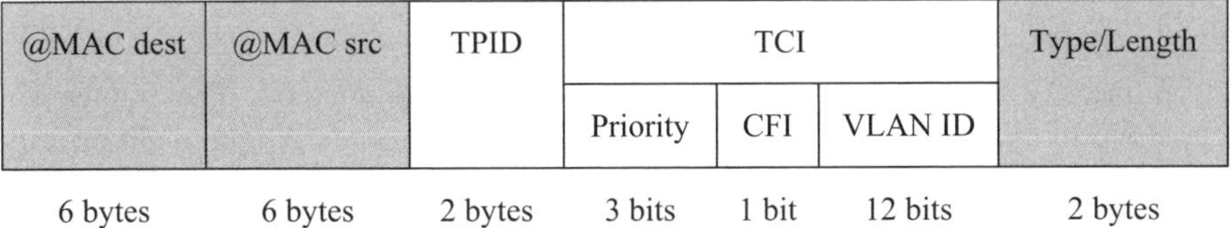

Figure 3.22. *801.1q frame format*

Priority or QoS at MAC level as defined by the 802.1p standard corresponds to different traffic classifications. The number of available traffic classifications is linked directly to the Ethernet switch capacity and its number of queues.

For a switch that has 7 queues, the 7 traffic classes that could be coded by the priority field will be available (the 0 value is reserved for a best effort traffic). The network control traffic will correspond, for example, to the highest priority class; voice and video respectively to classes 6 and 5 and the controlled load traffic to class 4.

If the switch does not have enough capacity, priorities will be grouped. For 2 queues, for example, priorities lower than or equal to 3 will correspond to traffic class 0 and priorities higher than 3 to traffic class 1.

3.5.3. *QoS with wireless networks*

The 802.11 norm used in Wi-Fi networks defines, at MAC sub-layer level, two exchange coordination functions corresponding to two different access methods:

– DCF (Distributed Coordination Function) is not based on centralized management and enables the control of asynchronous data transport with equal chances for support for all stations (best effort type).

– PCF (Point Coordination Function) is based on polling (query each terminal one after the other) from the access point (AP). This mode, rarely implemented and designed for real-time applications, will enable some form of QoS as long as the AP can control priorities by station.

The 802.11e standard is specifically designed for QoS in Wi-Fi networks and adds two new access methods: EDCF (Extended DCF) and HCF (Hybrid Co-ordination Function).

3.5.3.1. *EDCF access*

A priority control mechanism is added to the DCF method. The frames are classified in distinct queues within the originating station according to eight priority levels that correspond to eight traffic levels or Traffic Categories (TC). The lowest level will be associated with best effort traffic, whereas the highest level may correspond to video flows, for example.

For each TC, three parameters controlling support access priority are defined:

– AIFS (Arbitration InterFrame Spacing) replaces DIFS (DCF InterFrame Spacing) with an equal or higher time and controls the wait time between two frames according to the priority level (the highest priority frame will reach the minimum standard time DIFS);

– the CW (Contention Window) duration is always set according to the backoff algorithm (wait time before transmission, when the support becomes available depending on a random value and the number of attempts) while taking into account TC: a station with higher priority will have a shorter CW;

– TxOP (Transmission Opportunities) which is a fixed value timer with priority level and enables a differentiated delay in transmission when many backoff timers expire at the same time (when two classes are ready to transmit, the one with higher priority will have a lower TxOP and will transmit first).

Figure 3.23 shows an example of priority EDCF access when 3 stations want to transmit different priority frames following a previous transmission. The AIFS inter-frame wait time and the duration of the CW will be lower for the higher priority frame.

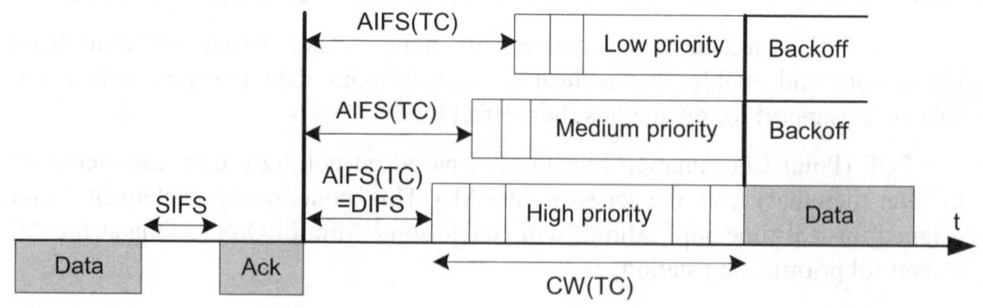

Figure 3.23. *EDCF priority access example*

3.5.3.2. *HCF access*

HCF is a hybrid access method that can be used during Contention Periods (CP) and during Contention Free Periods (CFP) and combining these two access methods with or without access point control PCF and EDCF.

In PCF mode, during CFP or CP periods, the access point controls use support with the possibility for the stations to generate successive (burst) or periodic frames, as illustrated in Figure 3.24.

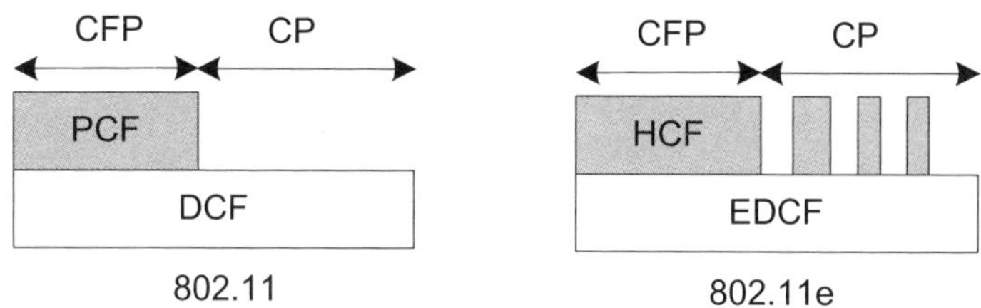

Figure 3.24. *802.11e HCF hybrid access*

Chapter 4

DiffServ: Differentiated Quality of Service

4.1. Introduction

The ever growing evolution of users of the Internet's worldwide network and the emergence of new applications needing different guarantees in terms of bandwidth, transfer delay and loss ratio have brought into focus a major gap in IP protocol: no QoS support. Currently, the type of service offered on the Internet is best effort, where all the resources are shared between all the users. Real-time applications and traditional applications are treated the same way, which is not really acceptable because a client that uses file transfers (FTP) does not care as much for how long his information will take to transfer as a client that uses videoconferencing or Voice over IP.

In order to solve the QoS problems, the 1990s have seen the emergence of adaptive applications, whose goal was to monitor the network (throughput, delays and error ratio measuring) and to modify the application algorithm's development based on observed properties, for example, modifying voice coding according to available throughput. In order for this to happen, the application must enable a return of information on the network's behavior, but most communication architectures do not supply this data and IP is inflexible. Protocols have been installed in order to put measures on the network (SNMP, RMON, RMON2) to get them back up to the network administrator. The main problem with this approach is that the application must adapt to the network. On the other hand, the network itself will not adapt to the applications and therefore the control of network resources is not useful.

Chapter written by Idir FODIL.

Later, the IETF became interested in controlling the network's resources within the IntServ group, which resulted in the RSVP protocol (Reservation Protocol). It is a signaling protocol where each application notifies the network of its requirements and, if possible, the network will guarantee the availability of the required resources for the application by reserving them for the route taken by packets of that application.

The complexity generated by RSVP weighs heavily on the performance of the routers and thus makes this approach difficult in extended networks. We cannot expect all networks to memorize all the information for all the flows in all the routes.

From these findings, the DiffServ approach of differentiated services has been developed, which consists of offering different subscription levels based on users. In short, the concept is to group all packets by traffic type instead of by application type. In this chapter we will examine the DiffServ approach.

4.2. Principles of DiffServ

The basic principle of this approach, developed within the IETF's DIFFSERV workgroup, is to provide a QoS to flow clusters rather than to individual data flows. In order to do this, the network edge routers execute classification, control (compliance with subscribed traffic) and IP packet marking (in the header) operations according to the different criteria (originating address, application type, subscription contract, transport protocol, etc.). Core routers process packets based on classification coded in the header according to an appropriate behavior, called PHB (Per Hop Behavior).

For IP packet marking, we use the IPv4 TOS (Type Of Service) field or the IPv6 Class of Service field. These fields are called DS (Differentiated Services). This is shown in Figure 4.1.

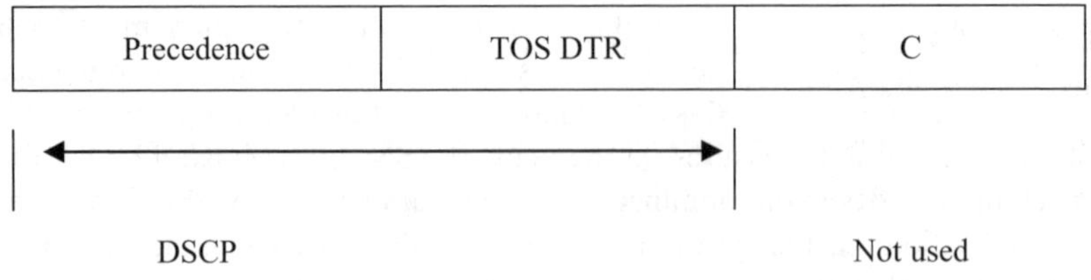

3 bits "precedence" (priority level), 3 bits of TOS field,
DTR (Delay, Throughput, Reliability), 2 bits C "Currently unused"

Figure 4.1. *The field allowing DiffServ marking*

In order for a user to have access to differentiated services from his service provider (ISP), he must have an SLA (Service Level Agreement), which is a contract specifying the supported service classifications and the volume of allocated traffic for each class.

4.3. Structure

A DiffServ domain is an administrative zone with a common group of procurement policies and of PHB definitions (routers possessing service and PHB definitions). A DiffServ region is a continuous group of DiffServ domains. In order to enable the services between different DiffServ domains, it is necessary to establish SLAs between adjacent domains.

From one domain to the other, the DS field can be re-marked and determined by the SLAs between the two domains.

4.3.1. *PHB (Per Hop Behavior)*

Three PHBs have been defined by DiffServ:

– *BE* (Best Effort): no guarantee, the network does what it can;

– *AF* (Assured Forwarding): more flexible and easier guarantees by the network. AF packets have more priority;

– *EF* (Expedited Forwarding): packets get processed more rapidly in the network (bandwidth, loss ratio, jitter and low transmission delay guarantee).

PHBs are put in place by the manufacturers in the routers with the help of queuing control mechanisms such as CBQ (Class-Based Queuing), WFQ (Weighted Fair Queuing), etc. according to the IP packet's DS field value.

The DS field can contain 64 values:

– 32 *code points* recommended. These values are assigned by the IANA (Internet Assigned Numbering Authority);

– 16 *code points* reserved for a local utilization (non-standard or experimental);

– 16 *code points* designed for a local utilization (reserved for extensions).

4.3.2. *EF Service*

Also called Premium service, the associated traffic is processed in priority in all routers. The network then guarantees the peak throughput required by the source.

The input routers must verify that the data entered by users with an EF service do not exceed the traffic contract to which they are subscribed in order to process AF and BE traffic under good conditions.

4.3.3. *AF Service*

Also called Olympic service, it prioritizes these flows based on best effort, which guarantees routing during congestion instead of the losses best effort users might encounter. In this class, conflict situations may appear between concurrent flows. In order to achieve this, four AF sub-classifications have been defined, and for each sub-classification, three loss levels have been defined, thus enabling routers to determine which packets to delete in the case of overload.

The DiffServ RFC also sets a precedence sequence on three levels, for each of the four classifications. The use of sub-classes is not fixed, on the other hand the relative importance of levels of precedence, in terms of congestion-based rejects, is referred to as AFX1<AFX2<AFX3. In reality, a RED (Random Early Detection) type congestion control algorithm is put in motion. Since the levels of precedence between the different classifications have been described in the previous chapter, we will not go over them here. Instead, we will concentrate on the implementation of DiffServ in the routers.

4.4. DiffServ in edge routers

The edge router is at the heart of the DiffServ approach, as it must control all the complex operations that will be applied to IP packets.

The edge router is responsible for:

– negotiating the traffic contract with the operator;

– verifying that the traffic submitted by the user is compliant with the subscribed contract;

– transmitting the users' IP traffic.

The mechanisms that the edge router must support in order to enable a differentiation of services can be grouped in two categories:

– a data part;

– a control part.

4.4.1. *Data part*

This service will arrange the classification and distribution of user packets to their service class, as well as data volume control that each class can manage.

In order to achieve this, it must execute five mechanisms, which are:

– classification. When a packet is received, the classifier determines to which flow or class this packet belongs to;

– conditioning. This mechanism is done with four components:

- the Meter, which verifies that the traffic is compliant with subscribed traffic,

- the Marker, which will assign a DSCP to the packet (that can be different from the one received),

- the Shaper, which will level the traffic so that it does not exceed the subscribed contract,

- the Dropper, which eliminates traffic exceeding the agreed throughput,

– queue management;

– scheduling;

– IP traffic routing.

4.4.1.1. *Classification*

There are two methods of classification:

– BA (Behavior Aggregate), which is based on the DSCP field;

– MF (MultiField), which is based on one or more IP header fields.

4.4.1.2. *Traffic conditioning*

The packet passes through a traffic conditioner, which executes multiple processes on the packet:

– the Marker marks certain fields of the IP packet (DS) in order to label it;

– the Meter measures the traffic's input/output properties according to the subscribed profile. In other words, it verifies the traffic's compliance with the subscribed contract. If packets are outside of the profile (exceeding), then it can decide to either delete them, or mark them again, or put them in a queue. The Meter is based on one or more token buckets;

– the Shaper delays the packets in order to make them compliant with the subscribed traffic profile.

4.4.1.3. *Queue management*

Since the capacity of the links can reach Gb/s, the volume of memory (queues) needed to store packets during congestion periods must be large enough not to exceed the routers' memory capacity. So, for a more efficient loss control as well as congestion control, it is necessary to implement mechanisms at queue level in order to avoid saturation.

Traditionally, packets are deleted in the order of entry or according to a predetermined algorithm, as soon as the queue becomes full. There are different queue management algorithms; the most popular are:

– *RED (Random Early Detection)*. With this algorithm, the number of packets to delete increases as the size of the queue reaches a determined threshold;

– *RIO (RED In Out)*. With this algorithm, the packets are marked by "IN" or "OUT" labels based on the service profile and during congestion the "OUT" packets are deleted.

4.4.1.4. *Scheduling*

The job of the Scheduler is to referee between packets in the different queues that are ready to be transmitted on an interface. The choice of the algorithm is based on:

– the algorithm must enable the network to guarantee a throughput by flow or a simple priority structure;

– the complexity of the algorithm, which is calculated by the number of operations per packet;

– the guarantee of service, meaning the number of communications that the network can handle within a designated classification;

– the flexibility of the algorithm when dealing with exceeding traffic.

There are different scheduling algorithms and the most commonly known are:

– *FIFO (First In First Out)*. All the packets are put in a queue and the service is FIFO. It is the simplest method to implement. However, it offers no delay or flow guarantee and, more importantly, it does not offer differentiation of services (no difference between classifications);

– *FPS (Fixed Priority Scheduling)*. With this method, we associate a queue with each class and the low priority packets are only processed when the high priority packets have been transmitted. This method only ensures that one class will be better served than the other, but offers no guarantee in terms of delays;

– *Weighted Fair Queuing (WFQ)*. The concept is to serve applications one after the other based on weight;

– *PGPS (Packetized General Process Sharing)*. This algorithm is a variation of the Weighted Round Robin Scheduling, serving low flow traffic and sharing evenly the rest of the bandwidth between large consumers. There are multiple variations of GPS (General Process Sharing);

– *Self Clocked Fair Queuing (SCFQ)*;

– *Start time Fair Queuing (SFQ)*;

– *Earliest Deadline First (EDF)*. It is a dynamic form of priority in scheduling. A deadline is assigned to each packet: the sum of arrival time plus the delay associated with the flow to which the packet belongs to. The Scheduler chooses the packet with the shortest deadline for transmission over the link;

– *Class-Based Queuing (CBQ)*. For each queue (class) a size in bytes is attributed, which the Scheduler must transmit before moving on to the next queue. Whatever bandwidth is left is evenly distributed between the other classes.

The architecture of a DiffServ edge router is illustrated in Figure 4.2.

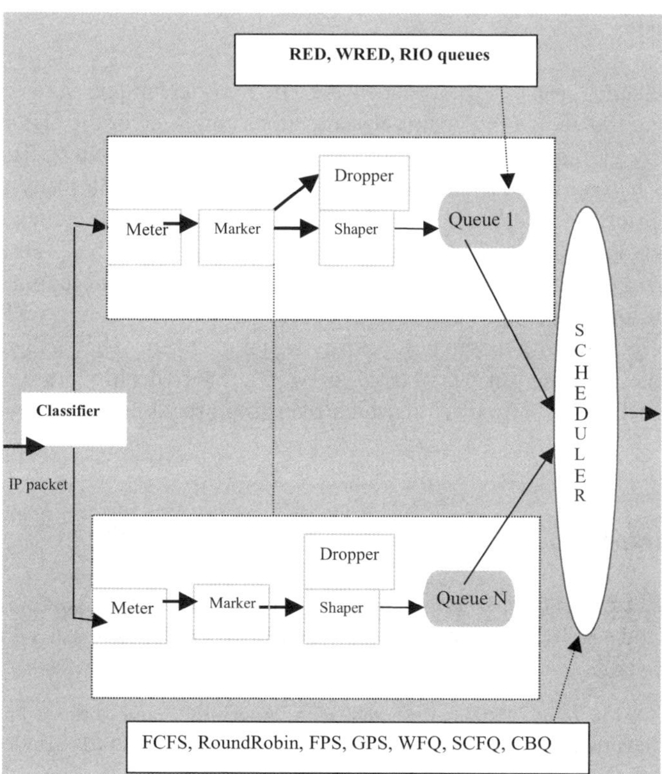

Figure 4.2. *A DiffServ router*

The control part is made up of two major parts:

– *Admission control.* There are three approaches for network resource allocation:

- deterministic. With this solution, the resources are allocated once and for all,

- statistic. The resources are allocated statistically and the allocation calculations are done by theoretical distribution,

- based on measures. The resources are allocated according to actual availability. Measures must be taken to determine the state of the network and make the allocations.

– *Policy control.* With this approach, we specify network access (resources and services) regulation according to administrative criteria. Control is done at user level and the applications or the hosts can access this service or that resource. These are based on policy-based networks, which is a technique that will be presented in the next chapters.

4.5. Conclusion

In this chapter, we have introduced the DiffServ technique, which consists of differentiating traffic in a few major classifications and to favor the classes that have constraints by giving them more or less priority in the edge routers. This DiffServ technique is based on statistical calculations: if there are not too many clients with the highest priority (*expedited forwarding*), it is obvious that these clients will see a smooth flowing network and the QoS will be easily reached. However, this technique remains statistical, i.e. that the probabilities of exceeding a network transmission time depend on the number of transmitting clients and will not ensure that a jam of clients with the highest priority will not happen. That is the reason why the operators working on very large networks have decided to use different techniques, MPLS in particular, in which overflow probability calculations can be executed.

4.6. Bibliography

[BLA 98] BLAKE S., BLACK D., CARLSON M., DAVIES E., WANG Z., WEISS W., "Internet RFC 2475: An Architecture for Differentiated Services" *RFC 2475,* December 1998.

[EGA 01] EGAN R. *et al.*, "Thales Research, an Architectural Framework for Providing QoS in IP Differentiated Services Networks", *Panos Trimintzios,* IM 2001, Seattle, May 14-18 2001.

[GUE 98] GUÉRIN R., PERIS V., "QoS in Packet Networks: Basic Mechanisms and Directions", *Computer Networks and ISDN Systems*, 1998.

[HEI 99] HEINANEN J., BAKER F., WEISS W., WROCLAWSKI J., "Internet RFC 2597: Assured Forwarding PHB Group", June 1999.

[JAC 99] JACOBSON V., NICHOLS K., PODURI K., "Internet RFC 2598: An Expedited Forwarding PHB", June 1999.

[NIC 98] NICHOLS K., BLAKE S., BAKER F., BLACK D., "Internet RFC 2474: Definition of the Differentiated Services Field (DS Field) in the IPv4 and IPv6 Headers", December 1998.

[RAJ 97] RAJ J., "A Survey of Scheduling Methods", Nokia Research Center, Boston, MA, September 17, 1997

[XIP 99] XIAO X., NI L. M., "Internet QoS: A Big Picture", *research report*, Department of Computer Science, Michigan State University, December 1999.

[ZHA 00] ZHAO W., OLSHEFSKI D., SCHULZRINNE H., "Internet QoS: an Overview", *Research Report*, Columbia University, June 2000.

Chapter 5

Quality of Service in Wi-Fi

5.1. Introduction

After the commercial flop of the third generation of mobile systems, IEEE 802.11 technology is sweeping the market thanks mostly to the simplicity of its deployment and especially to the wide use of radio band. Originally, this technology helped companies in their needs to complete or replace their wired local-area networks, but network operators have quickly felt the need to make its utilization public. The objective is for it to be used alone or to integrate it with the third generation of mobile systems and thus create the fourth generation of mobile systems.

The advantage of IEEE 802.11 networks is the large bandwidth, which is becoming more and more comparable with wired local-area networks. However, these networks have two limitations. On the one hand, this type of network has no signaling or control plan to ensure the level of service required by a user. On the other hand, the advantage of mobile cellular systems is that they do have a reliable and efficient signaling and control system which is able to provide and maintain a level of service required by a user. However, their problem is the low throughput that they can offer their users.

While we wait for a reliable and efficient integration of these technologies, improvements in wireless medium and differentiation of service in IEEE 802.11 networks are needed. IEEE 802.11 technology will be able to be used commercially only if it develops the necessary mechanisms to ensure to users QoS, security and

---

Chapter written by Yacine GHAMRI-DOUDANE, Anelise MUNARETTO and Hakima CHAOUCHI.

mobility handling. Multiple workgroups have been formed within the IEEE standardization organization in order to work on these issues. The a and g groups work on throughput increase in the initial standards which are IEEE 802.11 and IEEE 802.11b. The i group works on data security issues. The f group works on mobility issues. The h group is working on configuration problems and, finally, the IEEE 802.11e group is working on QoS issues within IEEE 802.11. In this chapter, we will go over the efforts of this last group, but before we do that, let us briefly explain the basic operation of IEEE 802.11b, as well as the major problems associated with providing QoS in the 802.11 environments.

5.2. Packets transmission with CSMA/CA access method

Bandwidth in the wireless medium is a scarce resource compared to the wired medium. MAC level protocols used in wireless networks are distributed protocols developed in order to avoid collisions and to provide network nodes with a fair access to the shared wireless medium. In the 802.11 [IEE 97] protocol, the fundamental mechanism for medium access is DCF (distributed coordination function). This approach implements a random access scheme, based on the CSMA/CA access protocol (carrier sense multiple access with collision avoidance). The retransmission of colliding packets is controlled by rules proposed within the Backoff algorithm.

The efficiency of MAC level protocols can be measured by the use of two parameters: collision probability and the fairness factor of the channel allocation to competing nodes. Protocols used for WLAN, as with any random protocol with multiple access, use the BEB (*Binary Exponential Backoff*) algorithm to treat the collision problem. BEB is a very efficient mechanism for minimizing the probability of collision which in this case is of approximately 1%.

Typically, all the variations of the CSMA/CA protocol strive to achieve fair sharing [BHA 94]. A medium access protocol is considered to be not fair if it cannot provide access to medium for individual nodes without an explicit differentiation. Consequently, when multiple nodes in a network are in competition for medium access, the probability for a node to gain access should be the same as for all the nodes in the network.

When a data packet is transmitted from the application layer, headers must be added at each layer, therefore increasing the total packet size and thus the transfer time. Figure 5.1 displays the structure of a packet transmitted via the 802.11 interface and its acronyms.

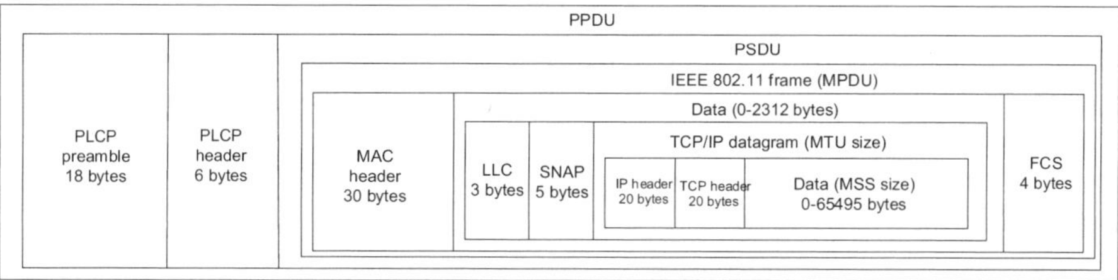

Figure 5.1. *Structure of a packet transmitted via the 802.11b interface*

FCS	Frame verification sequence
MSS	Maximum size of segment
MTU	Maximum transmission unit
SNAP	Sub-network access protocol
LLC	Link control
MPDU	MAC protocol data unit
PLCP	Physical layer convergence protocol
PSDU	PLCP service data unit (SDU)
PPDU	PLCP protocol data unit (PLCP + MPDU)

Table 5.1. *Acronyms used in Figure 5.1*

A data packet is encapsulated with a 30 byte MAC header plus 4 FCS bytes. This is the MPDU part or the PSDU part of the PPDU. The PLCP is a factor affecting bandwidth on the channel as each MPDU is transmitted. 24 bytes (192 bits) belong to the PLCP preamble; furthermore the header is independently attached to the size of the packet. Finally, the 6 bytes (48 bits) PLCP header is a preamble for PLCP of 18 bytes (144 bits) known as the long preamble. The preamble is used to signal a new data packet to the receiver. A disadvantage with PLCP is that the PLCP preamble and the header are always transmitted at a rate of 1 Mbps, independently from the data rate. This means that the time of transfer for the PLCP is constant at 192 μs with a long preamble.

The 802.11b [IEE 99b] standard offers the possibility of reducing the size of the PLCP preamble to 9 bytes (72 bits), called a short preamble. As such, it makes it possible to significantly increase the performance for higher throughputs. The version of the protocol with the short preamble enables better performances, by transmitting the short preamble at 1 Mbps and the header at 2 Mbps, which reduces transmission time to 96 µs.

A group of strict rules controls the way a transmission should behave within CSMA/CA. First, the sender must wait for a DIFS (Distributed Coordination Function InterFrame Space) period of 50 µs before the medium can be considered available. Only after this period can a data frame or a request to send be transmitted. The receivers respond with an acceptance (ACK) or a clear to send. The receiver must wait for a SIFS (short interframe space) period of 10 µs before responding.

5.2.1. *Performance degradation*

Wireless networks provide comparable performance with cabled networks. However, radio links introduce additional problems. This is due to the quality of the signal on the radio medium. These problems linked to the radio interface and therefore not pertinent to wired networks are: transmission errors, radio signal interference due to simultaneous transmissions, signal strength that diminishes greatly due to the distance, loss of carrier and the impossibility to detect collisions via transmission monitoring techniques. Furthermore, effects caused by the reduction of the signal, the interferences from other users and the objects are also deteriorating the performance of the channel [JAK 94]. Measures presented in [DUC 92] show that packet error ratio strongly depends on the distance between the sender and the receiver. They do not, however, follow the same distribution by increasing the distance. Hence, transmission on radio links is not reliable enough. Multiple research studies [CHA 04, CRO 97, WEI 97, BIA 00, TAY 01] for the evaluation of the IEEE 802.11 DCF performance show that the network's performance is very sensitive to the number of stations trying to access the channel.

5.2.2. *Support for speed changes*

IEEE 802.11 [IEE 97] networks are becoming the most widely used wireless access networks because they can offer a comparable performance to wired networks. Nevertheless, these networks have problems such as errors due to the radio resource, signal interference, mobility of the user and fair sharing method of the radio resource which is the CSMA/CA access mechanism. In order to adapt the transmission throughput with the quality of the link, a theoretical throughput variation function has been developed. This functionality progressively decreases theoretical throughput from 11 Mbps to 5.5 Mbps, 2 Mbps or 1 Mbps when a node detects unsuccessful frame transmissions. The multiple data transfer rate capacity represents a dynamic change of the throughput with the objective of improving the performance for this station. However, in order to ensure coexistence and interoperability between stations capable of implementing the theoretical throughput variation functionality, a set of rules that must be followed by all stations is defined within the standard.

The probability of access among others is a second functionality implemented within the CSMA/CA access method. The access method for CSMA/CA support guarantees that in the long term all stations have the same access probability. When a station captures the medium, if it has a low throughput (i.e. throughput lower than 11 Mbps) therefore monopolizing the medium for a long period of time, it then penalizes the other stations by requiring a higher throughput. In this case a degradation of the performance of the cell throughput is identified.

This performance degradation caused by the influence of a low throughput station has been studied in [HEU 03]. The authors show that the useful throughput is much lower than the theoretical throughput. They also analyze the effects of a low theoretical throughput station over the other stations that share the radio medium.

Figure 5.2 shows the degradation within the throughput value of the 802.11b cell when a single station decreases its theoretical throughput.

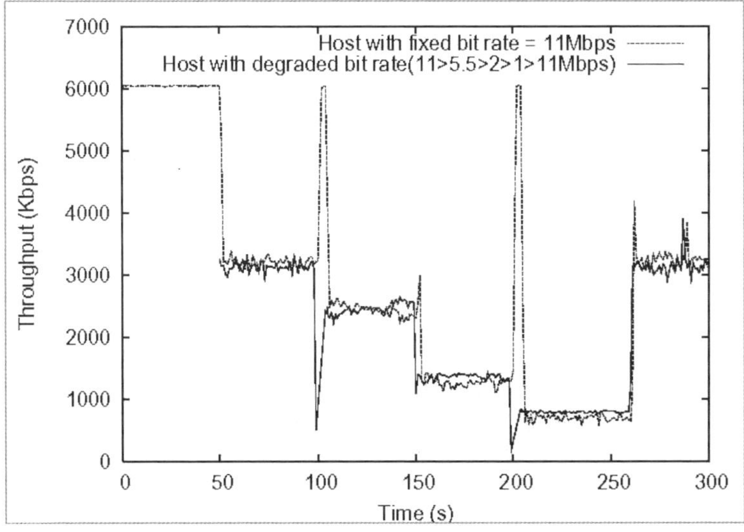

Figure 5.2. *Degradation of bandwidth in an 802.11b cell*

5.3. MAC level QoS in IEEE 802.11

In wireless networks, offering guarantees in terms of delay or jitter can only happen if the medium access method enables it. The IEEE 802.11 [IEE 97] standard respects this rule. In this section, we will describe the different access methods which make it possible to offer QoS support in the MAC layer of IEEE 802.11. These access methods can be divided into two categories:

– centralized approaches: approaches based on the invitation of stations to transmit. These invitations are transmitted by a "Coordinator" node;

– distributed approaches: or service differentiation approaches. This service differentiation is obtained by an adequate spacing between transmissions. This spacing represents the priority for medium access.

5.3.1. *History*

In its initial standard [IEE 97], also known as the basic standard, the IEEE's 802.11 committee proposed two medium access methods. The first is the Distributed Coordination Function (DCF). This method is similar to the Ethernet access method. It has been designed to support the transport of asynchronous data and enables all users who want to transmit data to have an equal chance to access the wireless medium. So DCF offers a fair access to medium (i.e. one station cannot have priority over another). The second access method is the Point Coordination Function (PCF). With this one, the different data transmissions between network stations are managed by a central coordination point. This access method is generally located in the AP. It has been designed to enable the transmission of delay sensitive data (i.e. to guarantee QoS to stations that require it).

In order to obtain a QoS and improve DCF and PCF access methods, the *e* group of IEEE 802.11 [IEE 02] is defining new priority mechanisms for the IEEE 802.11 standard. This group introduces two new access methods: the Enhanced Distribution Coordination Function (EDCF) and the Hybrid Coordination Function (HCF). A station using 802.11e will then execute these two access protocols instead of the DCF and PCF methods.

In the next section, we will examine these different access methods by separating between the distribution techniques description from the centralized techniques. Apart from these access methods proposed within the frame of standardization of wireless local networks, many other improvements exist. However, in this section we only describe studies linked to standardization.

5.3.2. *Distributed approaches*

In the following, we describe both distributed access methods proposed or discussed within IEEE 802.11: DCF and EDCF.

5.3.2.1. *DCF*

The DCF protocol is the access method which enables the transfer of asynchronous data in best-effort mode. The DCF method is not a MAC layer offering QoS. However, this access method is described here because it is at the core of the other distributed access methods which offer a certain degree of separation between flows and frames (QoS).

5.3.2.1.1. Description of DCF

The DCF is based on Carrier Sense Multiple Access with Collision Avoidance (CSMA/CA). The general principle of CSMA/CA is to listen to the medium before transmitting and then to attempt to obtain access: if the link is unoccupied when a node wants to transmit data, the node transmits its frames. If, on the other hand, the link is occupied, the node waits for the end of the current transmission to earn the right to access the medium.

With DCF, each station watches for activity from the other stations within the same BSS (cell). When a station transmits a frame, the other stations perceive that a frame has been sent in the network. Then, in order to avoid that all the other stations continue to listen to the medium until the end of the transmission, they will update a timer called Network Allocation Vector (NAV) that makes it possible to delay all expected transmissions. The NAV is calculated with regard to the information located in the shelf life field within the transmitted frame. The other stations will not have the capacity to transmit data until the end of the NAV.

The access to the wireless medium is controlled by the use of a space between the transmission of two consecutive frames within the same BSS. These spaces, called InterFrame Space (IFS), correspond in fact to periods of inactivity on the transmission medium. The standard defines two IFS types for the DCF mode:

– Short IFS (SIFS): this is the smallest of the IFSs. It is used to separate transmissions within the same dialog (between fragments of frames, between a data frame and its ACK, etc.). This enables the station to have priority over all the other stations for accessing the medium;

– DCF IFS (DIFS): this space, larger than SIFS, is used when a station wants to start a new transmission. It must wait a DIFS time before transmitting its data on the medium.

It is important to note that the fact that SIFS is smaller than DIFS makes it possible for the access to medium to give priority to the frames of a current transmission instead of a new transmission.

So, before each data transmission, each station must sense the medium and make sure that it is available. If the medium is available during a time interval equal to DIFS, the station can then transmit. If not, the transmission is deferred until the medium becomes available. This can bring the risk of contention for access to the medium, i.e. two stations or more wait for the end of a transmission to take their turn. In order to avoid as much as possible a risk of collision caused by two almost simultaneous transmissions, a procedure, called Backoff procedure, is launched once the medium becomes available for a length equal to DIFS. More specifically, the station calculates a random timer called Backoff timer, which is called $T_{Backoff}$ in [5.1]. This is between 0 and a maximum value, called Contention Window (*CW*), and is calculated as follows:

$$T_{Backoff} = Rand(0, CW) \times SlotTime \qquad [5.1]$$

where *SlotTime* is an interval of time used to define the IFS intervals as well as the timers for the different stations. Its value is dependent on the physical layer used.

If the timer has not reached the 0 value and the medium becomes busy once again, then the station freezes the timer. This timer will be unfrozen when the medium becomes available again for a time equal to DIFS. In this way, the stations that have randomly chosen high *Backoff* time have priority for access to the medium when it becomes available once more. When the timer reaches the 0 value and the medium is still available, the station transmits its frame. If two or more stations reach the 0 value at the same time, a collision happens and each station must generate a new random timer between 0 and 2**CW*. The *CW* value moves between a minimum (*CWmin*) and a maximum (*CWmax*). A maximum of tries is also defined by 802.11; if this number is reached before the correct transmission, then the current transmission is simply abandoned. *CWmin* and *CWmax* values have also been the object of a standardization by the 802.11 group; they depend on the physical layer used.

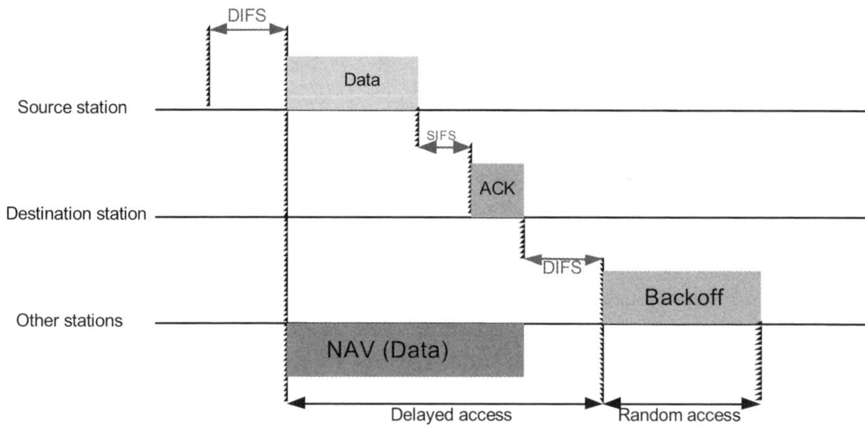

Figure 5.3. *Data transmission with basic DCF access method*

If the transmitted data has been correctly received, the destination station waits for a time equal to SIFS and then transmits an acknowledgement (ACK) to confirm correct data reception. If no ACK is received by the transmitting station after a predefined time, it is then assumed that a collision has taken place and a data retransmission attempt, by using the Backoff algorithm, is executed.

It is important to note that after each successful transmission, the Backoff process is executed again by the stations in competition. This is done even when no frame is waiting for transmission. Since this procedure is executed after each transmission, it is called *post-Backoff*. A new procedure for the transmission of a new frame can be initiated only after the expiration of that time.

During the transmission time between the source station and the destination station, the other BSS stations update their NAV. The value calculation of the NAV includes the transmission time of the data frame and the InterFrame Spaces as well as the ACK (see Figure 5.3).

5.3.2.1.2. Discussion

Due to Backoff's algorithm, stations have the same probability of accessing the wireless medium. However, as it is defined, DCF does not guarantee a minimum delay and therefore cannot be used with real-time applications.

5.3.2.2. *EDCF*

The EDCF access method is, as its name indicates[1], an improvement of the basic DCF access method. This improvement is proposed in order to enable a differentiation of services between flows, at the wireless interface level.

5.3.2.2.1. Description of EDCF

QoS support within EDCF is possible due to the introduction of traffic categories (*TC*). In fact, the frames are delivered using different Backoff instances within the same station, each Backoff instance configured with parameters specific to a particular traffic category. In the contention period (CP; see section 5.3.3.1.1), each TC within the station has to be able to transmit its data and to start, if needed, a Backoff procedure independently from the other TC in the same station. The Backoff procedure in EDCF is initiated when it is detected that the medium is available for a period equal to an IFS time called arbitration IFS (*AIFS*). The AIFS can take any value, higher than or equal to DIFS, therefore participating to the TC characterization to which it belongs. After a wait time equal to AIFS, a Backoff timer is initialized with a value within the [1, *CW* + 1] interval. The minimum (*CWmin*[*TC*]) and maximum (*CWmax*[*TC*]) sizes of the Backoff window are also parameters that characterize a traffic category. The counting down procedure of the Backoff timer is the same as DCF with one exception: when the Backoff timer is initiated after having been interrupted, since the medium had become occupied again, it is decremented by 1 at the end of the AIFS time and not at the end of the first time slot following DIFS (or AIFS), as with DCF. This makes it possible to accentuate the priority of TCs that had to block their Backoff timers. After each collision a new contention window CW is calculated according to a persistence factor PF [TC]. This persistence factor is the third parameter characterizing a traffic category. In fact, contrary to the basic standard where the contention window size is simply doubled (equivalent to $PF = 2$), 802.11e uses the PF[2] parameter to increment the contention window size for each TC, as follows:

$$newCW[TC] >= ((oldCW[TC]+1)\times PF)-1 \qquad [5.2]$$

As shown in Figure 5.4, each station can implement up to eight queues, each corresponding to a specific traffic category [TC]. If the timers of one or more TC within the same station reach the 0 value at the same time, the scheduler within the station would give priority to the traffic category with the highest priority in the

1 EDCF: Enhanced DCF.

2 Note that this possible choice of the persistence factor has been removed from the latest draft but we keep it here for completeness.

station, thus avoiding a virtual collision. Other collisions can, however, happen between stations.

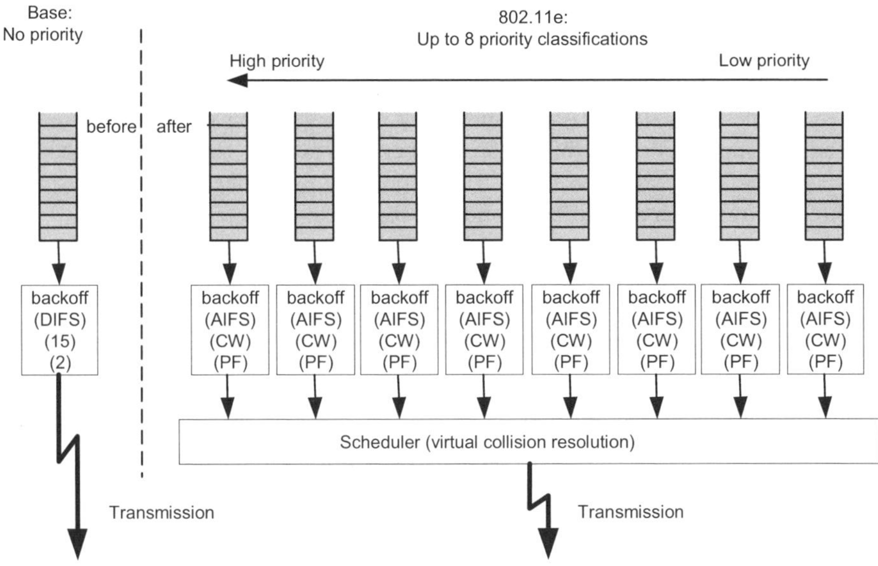

Figure 5.4. *Virtual Backoff with 8 traffic categories [TC]: (a) on the left, DCF, equivalent to EDCF with AIFS = DIFS, CWmin = 15, CWmax = 1,023 and PF = 2; (b) on the right, EDCF*

In summary, each TC is characterized by three different parameters that specify its relative priority degree:

– arbitrary interframe space time, *AIFS[TC]*, higher than or equal to DIFS;

– progress intervals of the particular Backoff window (*CWmin[TC]* and *CWmax[TC]*);

– a persistence factor, *PF[TC]*, for the progress of the Backoff window.

Another crucial parameter in 802.11e is the *TXOP*[3]. A *TXOP* is defined as the time interval during which a station has the right to transmit. It is characterized by an instant of transmission start and a maximum transmission time and it is obtained after contention (*EDCF-TXOP*) or with the help of the hybrid coordinator HC (*polled-TXOP*)[4]. In the case of the EDCF access method, the size of *EDCF-TXOP* is

3 Transmission Opportunity.

4 See HCF access method (section 5.3.3.2.1).

limited within a cell for each TC; this maximum values are transmitted inside the frame tags.

5.3.2.2.2. Service differentiation at the MAC level

In order to introduce a service differentiation in wireless local-area networks, a certain degree of separation between different types or service classes must be offered by the MAC layer. As illustrated by the description of the EDCF access method, the introduction of this separation between service classes corresponds to the assignment of priorities to flows. This is done by using one or a combination of the three following service differentiation mechanisms:

– Backoff incrementation function;

– differentiation by the Backoff windows (CWmin);

– differentiation by DIFS (called AIFS in EDCF).

a) Backoff incrementation function

As illustrated in the previous section, each time a transmission fails due to collision, the contention window size is doubled.

$$newCW = (oldCW) \times 2 \quad [5.3]$$

By considering that the only configurable term in equation [5.3] is the multiplier 2, one possibility for introducing a certain priority is to replace, for each service class j, this multiplier by a priority (P_j) assigned to each service class. In other words, for each transmission attempt, instead of multiplying the Backoff window size by 2 it will be multiplied by P_j. Evidently, the higher the factor $P_{j,}$, the larger the Backoff window and therefore the smaller the probability that a frame belonging to this class will be transmitted quickly, and consequently the throughput will be slower.

According to [AAD 01, AAD 03], this mechanism seems to yield good results only when we consider UDP traffic and when the number of terminals in contention is high. On the other hand, when the number of terminals in contention is low, the CW value is usually at its minimum ($CW = CWmin$) and it is rarely incremented. Therefore, no service differentiation is noticed.

b) Differentiation by Backoff windows (CWmin)

This second method was proposed due to the limitations of the previous method [AAD 03, VER 01]. Indeed, when the number of terminals in contention is low and when the *CW* value is mostly equal to *CWmin*, the idea is then to make an adequate choice of this initial value (*CWmin*) for each service class in order to obtain a service differentiation between these classes.

This second mechanism partially resolves the problems of the previous method. In fact, a clearer service differentiation is obtained for UDP flows when the number of terminals is small. However, this mechanism does not always make it possible to obtain a strict and clear service differentiation for TCP flows [AAD 03].

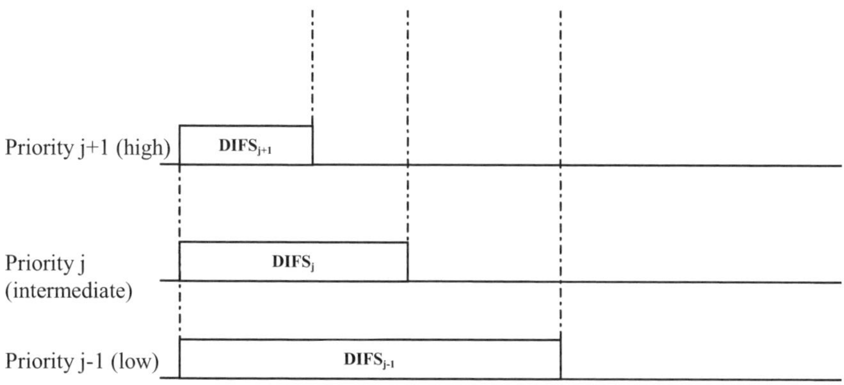

Figure 5.5. *Differentiation by DIFS spaces*

c) Differentiation by DIFS

As we have seen earlier in this chapter, in DCF, priority is given to an ACK transmission compared to data frame or RTS frame transmission. This is done due to an adequate choice of InterFrame Space (IFS). In fact, the transmission of an ACK is only delayed for a time equal to SIFS when the medium becomes available again, whereas the transmission of a data or RTS frame must wait longer (time equal to DIFS). Consequently, if a station that wants to transmit is in contention with a station that must send an ACK, the ACK will access the medium in priority. This same idea has been used in order to introduce a service differentiation for data transmission. This can be made possible by using different $DIFS_j$ (or $AIFS_j$) values, each corresponding to a service class j. In order for the priority to be given to class j instead of class j-1, $DIFS_j$ must be lower than $DIFS_{j-1}$ (see Figure 5.5).

This method, more deterministic, has yielded better results than the previous methods because it enables stricter priorities between UDP flows or between TCP flows [AAD 01, AAD 03]. We have noted, however, that this differentiation was still limited within TCP flows, especially in comparison with the one obtained within UDP flows. The curves in Figure 5.6 clearly show this limitation. These curves, obtained by simulation, consider three flows from three different stations. Each one of these flows overloads the radio link, using UDP traffic (Figure 5.6(a)) for one and using TCP traffic for the other (Figure 5.6(b)). Different priority values ($DIFS_j$) are allocated to these three flows.

5.3.2.2.3. Discussion

Service differentiation techniques presented here and, especially the differentiation by DIFS, enable the introduction of an efficient priority between service classes for UDP traffic. However, differentiation of services for TCP traffic is either inefficient, when talking about differentiation mechanisms based on the Backoff algorithm, or insufficient, in the case of differentiation by DIFS.

Variations of these algorithms [NIA 02] have also been proposed in other works, but the authors reach the same conclusions with regards to the service differentiation within TCP traffic.

5.3.3. *Centralized approaches*

In the following section we explain the two centralized access methods proposed or discussed within IEEE 802.11: PCF and HCF.

5.3.3.1. *PCF*

For the support of services with time constraints, the basic IEEE 802.11 standard defines a centralized access method, known as PCF.

5.3.3.1.1. Description of PCF

Coordination is done by a station called Point Coordinator (*PC*). PCF has by definition priority over DCF. Indeed, the time space (*IFS*) before each transmission from a station using PCF is lower than DIFS but remains higher than SIFS; this space is called PCF Interframe Space (*PIFS*).

In addition to the division in basic time units (*SlotTime*), the time is equally separated in what we call superframes. When PCF is used, the superframe is made up of two periods: Contention Free Period (CFP) and Contention Period (CP). These two periods continuously alternate (see Figure 5.7). During the CFP period, the PCF access method is the one used and the DCF method will be used during the CP period. A superframe may not contain a CFP period, yet it is mandatory that it includes a CP period of a length which enables the transmission of at least one MSDU with DCF.

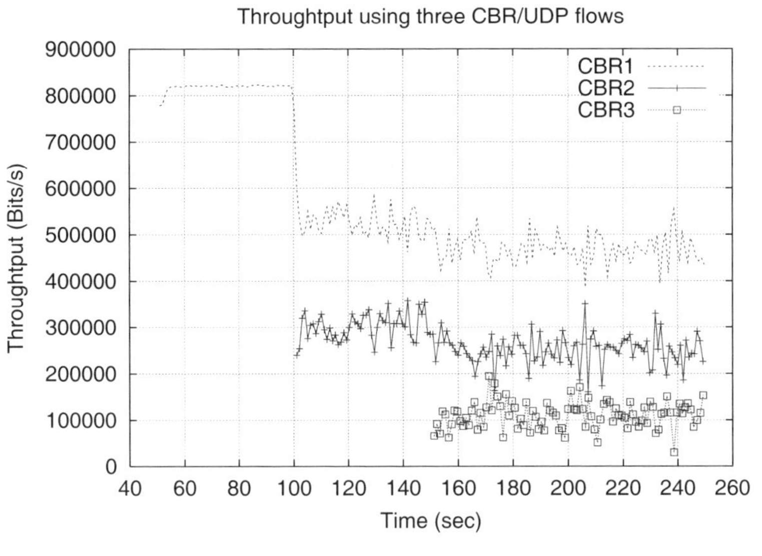

a)

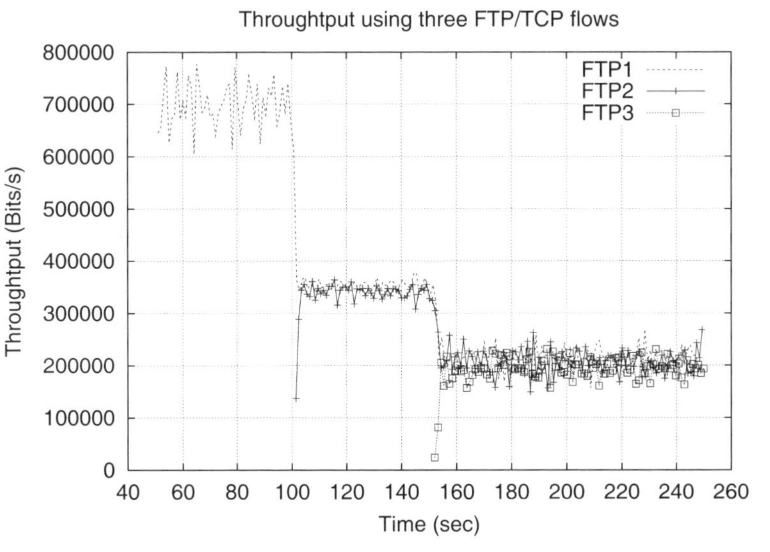

b)

Figure 5.6. *Throughputs obtained by using a differentiation by DIFS when the medium is overloaded by 3 stations with different priorities ($DIFS_1$=50µs, $DIFS_2$=130µs, $DIFS_3$=210µs) – (a) UDP traffic and (b) TCP traffic*

The superframe starts with the transmission of a beacon frame; this transmission is sent even if PCF is not activated. It is the PC, generally located in the AP, which generates, at regular intervals, the beacon frames. In addition to the announcement of the start of the superframe, the beacon frame also enables the synchronization between station clocks enabling them to know when the CFP period ends, i.e. the time (TBTT[5]) after which a control frame *CF-End* will be sent (Figure 5.7).

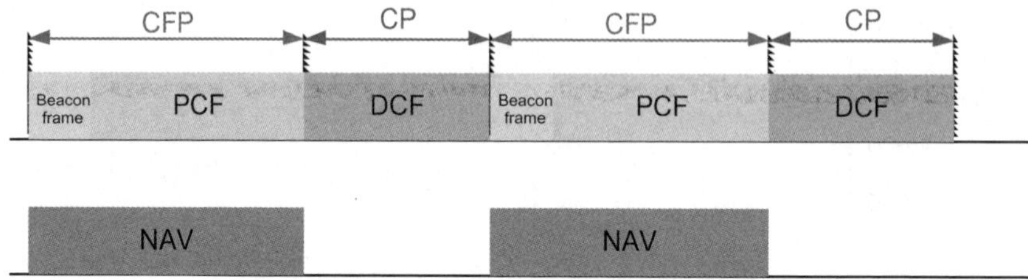

Figure 5.7. *Alternation between periods, CFP and CP*

During CFP there is no contention between stations. Indeed, the stations are invited to transmit through the PC. Thus, the PC requests particular stations, one after the other, to transmit one of its queued frames. Since the CP may also have frames to transmit, it uses, in addition to invitation frames, data frames that it piggybacks to invite the other stations to transmit. A station receiving an invitation to transmit responds by sending a data frame; in the case where the station does not respond to the invitation after a time equal to PIFS, the PC invites another station to transmit. In this way, no inactivity period higher than PIFS is authorized during CFP. The PC continues to transmit invitations until the TBTT time, allocated at CFP level, expires. A control frame (*CF-End*) is then transmitted to announce the end of the CFP period and the beginning of the CP period. Figure 5.8 shows an example of operations within CFP.

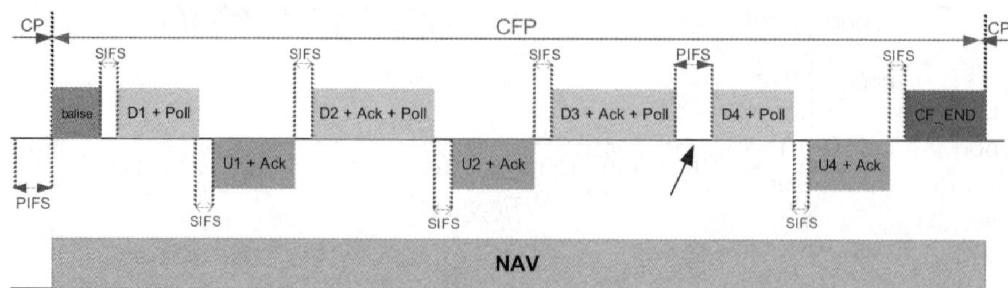

Figure 5.8. *Transmissions of data in CFP period*

5 Target Beacon Transition Time.

5.3.3.1.2. Discussion

The PCF access method has neither been integrated in commercial products nor implemented at this time. The current cards only use the DCF method. In fact, the PCF access method has many disadvantages, such as:

– the impossibility to recognize delays between control frames. When the TBTT time is reached, the coordinator point is poised to send a control frame (CF-End). However, this can be executed only if the medium is available for a period longer than or equal to PIFS. So, if the medium is not available close to the expiration period of the TBTT timer, transmission of the CF-End frame is delayed. This means that within each CFP unexpected delays can be introduced;

– the impossibility to recognize transmission duration of stations asked to transmit. A station invited to transmit is authorized to transmit only one frame, which can be fragmented. This, added to the fact that the station can use different coding and modulation mechanisms, makes the duration of the fragmented frame transmission unpredictable and uncontrollable by the coordinator point;

– a hidden station that misses the beacon frame will have no information concerning the TBTT or the beginning of the CFP period. It might then continue to transmit using the DCF access method that is causing collisions.

These three problems can clearly deteriorate the QoS offered. Due to these three disadvantages, no strong guarantees can be offered for the transmission of a frame within the CFP.

5.3.3.2. *HCF*

To make up for the disadvantages of the PCF access method, a more deterministic access method, in terms of transmission time given to stations, is currently being standardized by IEEE 802.11. This technique is called HCF which stands for *Hybrid Coordination Function.*

5.3.3.2.1. Description of HCF

The HCF method is an extension of the access rules of the EDCF method. In fact, the hybrid coordinator, HC, may, at any time, self-allocate a Transmission Opportunity (TXOP) to start an MSDU transmission. This can obviously be done only after making sure that the medium is free of any transmission during a time higher than or equal to PIFS, which is of course lower than DIFS, therefore giving it priority over EDCF.

During a CP period, each TXOP starts either when the medium has been declared free according to the EDCF rules (i.e. AIFS + Backoff timer) or when the station receives an invitation (*QoS CF-Poll*) from the HC to transmit. Thus, the HC may allocate a TXOP to any station during the contention period by using its priority access method to transmit the control frame *QoS CF-Poll*. During the CFP period, the moment of the transmission start and the maximum transmission length of each TXOP are also specified by HC still using *QoS CF-Poll*. The CFP period then ends after the expiration of time announced by the beacon frame (TBTT) or by sending the control frame *CF-End* by the hybrid coordinator.

For the hybrid coordinator to invite the stations to transmit their frames, it needs information concerning the stations' requirements. Furthermore, this information must be updated frequently. An additional random access protocol enabling a quick resolution of collisions has therefore been defined by IEEE 802.11e to enable the HC to acquire this information. This protocol, called Controlled Contention (CC), tells the HC which station requires an invitation to transmit, when and for how long. This is done when the stations send resource requests (RR), using the Controlled Contention. The Controlled Contention happens only during specific intervals of time, called Controlled Contention Intervals or CCI. During these intervals, RR messages are the only ones authorized. Consequently, no contention is possible between these control frames and other types of frames. One last note, a CCI can be initiated by the HC at any time (CP or CFP period) by the transmission of a CC frame.

5.3.3.2.2. Discussion

The introduction of the TXOP time, as well as its associated resource management mechanisms, is a major improvement compared to the PCF access method. Indeed, this helps the HC to manage with more ease, and more efficiency, the allocation of resources.

Previously, the resource allocation techniques had to operate without knowing beforehand how long the transmissions would last. Consequently, it was harder to efficiently share the CFP period between the different stations.

In order for the HCF to function as efficiently as possible, CCI frequency must be high. This could, however, generate a significant signaling overload.

5.4. Summary and conclusion

As explained in the previous sub-sections, the introduction of QoS in 802.11 networks can be divided into two categories: 1) medium access techniques that rely on invitation, PCF and HCF, and 2) techniques based on the introduction of a service differentiation within the distributed medium access technique function, DCF.

The introduction of service differentiation techniques in the DCF function makes up for the complexity in management. However, service differentiation in the case of TCP traffic is either inefficient, in the case of differentiation mechanisms based on the Backoff algorithm, or insufficient, for differentiation by DIFS. Other variations [NIA 02] have also been proposed in other works, but the authors reach the same conclusions with regard to service differentiation within TCP traffic.

With regard to the centralized techniques, on the one hand, the PCF centralized access method does not really make it possible, in certain cases, to offer a guarantee on packet transmission. On the other hand, the HCF access method can generate a signaling overload and a considerable complexity in management. In fact, the choice for optimal size of the superframe [LIN 03] on the one hand, and the allocation of rights to transmit to the different requesting stations on the other hand, necessitate considerable signaling (during CCI periods where frequency has not been determined yet) as well as the development of complex techniques for bandwidth dispatching between stations.

5.5. Bibliography

[AAD 01] AAD I., CASTELLUCCIA C., "Differentiation mechanisms for IEEE 802.11", *IEEE Infocom'01*, Anchorage, Alaska, April 2001.

[AAD 03] AAD I., CASTELLUCCIA C., "Priorities in WLANs", *Computer Networks*, vol. 41, no. 4, p. 505-526, March 2003.

[BHA 94] BHARGAVAN V., DEMERS A., SHENKER S., ZHANG L., "MACAW: A Media-Access Protocol for Packet Radio", in *Proceedings of ACM SIGCOMM*, 1994.

[BIA 00] BIANCHI G., "Performance Analysis of the IEEE 802.11 Distributed Coordination Function", *IEEE Journal of Selected Areas in Telecommunications*, Wireless Series, vol. 18, March 2000.

[CHA 04] CHAOUCHI H., MUNARETTO A., "Adaptive QoS Management for IEEE 802.11 future Wireless ISPs", *ACM Wireless Networks Journal*, Kluwer, vol. 10, no. 4, July 2004.

[CRO 97] CROW B. P., WIDJAJA I., KIM J. G., SAKAI P. T., "IEEE 802.11 Wireless Local Area Networks", *IEEE Communications Magazine*, September 1997.

[DUC 92] DUCHAMP D., REYNOLDS N. F., "Measured Performance of a Wireless LAN", in *Proc. of 17th Conf. on Local Computer Networks*, Minneapolis, 1992.

[HEU 03] HEUSSE M., ROUSSEAU F., BERGER-SABBATEL G., DUDA A., "Performance Anomaly of 802.11b", in *Proceedings of IEEE INFOCOM 2003*, San Francisco, United States, 2003.

[IEE 97] "LAN MAN Standards of the IEEE Computer Society", *Wireless LAN media access control (MAC) and physical layer (PHY) specification*, IEEE Standard 802.11, 1997.

[IEE 99a] "High-Speed Physical Layer in the 5GHz Band", supplement to 802.11-1999, IEEE Standard 802.11A, 1999.

[IEE 99b] "Higher-Speed Physical Layer Extension in the 2.4 GHz Band", supplement to 802.11-1999, IEEE Standard 802.11B, 1999.

[IEE 02] IEEE 802.11 Task Group E, "Wireless MAC and Physical Layer Specifications: MAC Enhancements for Quality of Service: part 11", draft supplement to IEEE Standard 802.1, 2002.

[JAK 94] JAKES W. C., "Microwave Mobile Communications", *Wiley-IEEE Press*, May 1994.

[LIN 03] LINDGREN A., ALMQUIST A., SCHELÉN O., "Quality of Service Schemes for IEEE 802.11 Wireless LANs – An Evaluation", *Journal of Special Topics in Mobile Networking and Applications (MONET)*, special issue on Performance Evaluation of QoS Architectures in Mobile Networks, vol. 8, no. 3, June 2003.

[NIA 02] NIANG I., ZOUARI B., AFIFI H., SERET D., "Amélioration de schémas de QoS dans les réseaux sans fil 802.11", *French Colloquium on Protocol Engineering, CFIP'02,* Montreal, Canada, May 2002.

[TAY 01] TAY Y. C., CHUA K. C., "A Capacity Analysis for the IEEE 802.11 MAC Protocol", *ACM Baltzer Wireless Networks*, vol. 7, no. 2, March 2001.

[VER 01] VERES A., CAMPBELL A. T, BARRY M., SUN L.-H., "Supporting Service Differentiation in Wireless Packet Networks Using Distributed Control", *IEEE Journal of Selected Areas in Communications (JSAC)*, vol. 19, no. 10, p. 2094-2104, October 2001.

[WEI 97] WEINMILLER J., SCHLAGER M., FESTAG A., WOLISZ A., "Performance Study of Access Control in Wireless LANs IEEE 802.11 DFWMAC and ETSI RES 10 HIPERLAN", *Mobile Networks and Applications*, vol. 2, 1997.

Chapter 6

Quality of Service: Policy-based Management

6.1. Introduction to policy-based management in IP networks

For the last 10 years, the Internet has grown rapidly and the traditional approach based on the management of individual equipment has become insufficient and difficult. Due to the complexity and the increase in the size of IP networks in their transition toward multiservice, it is necessary to have an infrastructure which enable the automatic control of the networks' behavior. Policy-based management is seen as an effective solution for company networks and network suppliers. The purpose of this chapter is to introduce policy-based management. The two protocols that can be used for the distribution of policy rules are respectively COPS (Common Open Policy Service) and SNMP (Simple Network Management Protocol) and are presented in detail.

The difference between policy-based management and traditional network management is important: policy-based management enables the network administrator to define policy rules to manage the behavior of a complete network instead of individually manipulating each piece of network equipment [KOS 01]. The policy-based management system is responsible for distributing these rules to the appropriate network components as well as at appropriate moments. Networks controlled by this approach are called policy-based networks.

For a comprehensive view of policy-based management, a simple example is presented in Figure 6.1.

Chapter written by Thi Mai Trang NGUYEN.

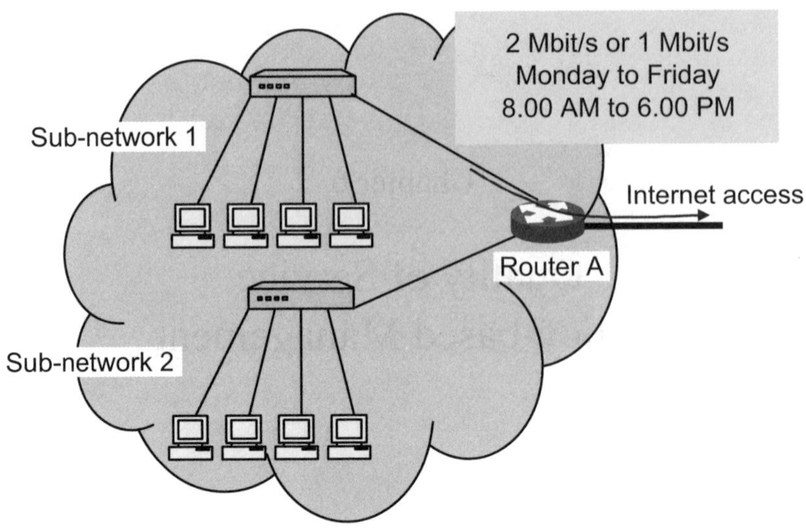

Figure 6.1. *Example of policy-based management*

Figure 6.1 shows a corporate network with two sub-networks and an Internet access router. The administrator can define policies such as "Between 8.00 AM and 6.00 PM Monday to Friday, total sub-network 1 traffic toward Internet is guaranteed at 2 Mb/s with the condition that there is no congestion, and at 1 Mb/s if there is congestion for Internet access". The policy-based management system is responsible for communicating these policies to the access routers. In this case, the management system will distribute these policies to Router A. These policies can take the form of conditions and actions as follows:

– if <after 8.00 AM> and <before 6.00 PM> and <no congestion> then <sub-network 1 traffic toward Internet ≥ 2 Mb/s>;

– if <after 8.00 AM> and <before 6.00 PM> and <congestion> then <sub-network 1 traffic toward Internet ≥ 1 Mb/s>.

The management system is aware of the access link state (via a monitoring system) and knows if there is congestion or not. In accordance with the time and date, the policy-based management system determines the allocation of resources in router A. Between 8.00 AM and 6.00 PM, from Monday to Friday, the management system configures a classifier in router A to filter traffic coming from sub-network 1 and going to the Internet. The flow's packets are transmitted to a queue that will be served at a minimum throughput of 2 Mb/s if there is no congestion on the access link to the Internet. If there is congestion on the access link, the management system modifies the parameters of router A's scheduler in order for the queue, where the packets from sub-network 1 and going to the Internet are memorized, to be served at a minimum throughput of 1 Mb/s. Before 8.00 AM or after 6.00 PM and during the weekend, the network processes all traffic in the same way and does not guarantee any minimum throughput for any traffic.

6.2. Architecture and protocols for policy-based management

The IETF (Internet Engineering Task Force) and the DMTF have proposed an architecture for policy-based networks [VER 01]. This architecture is presented in Figure 6.2.

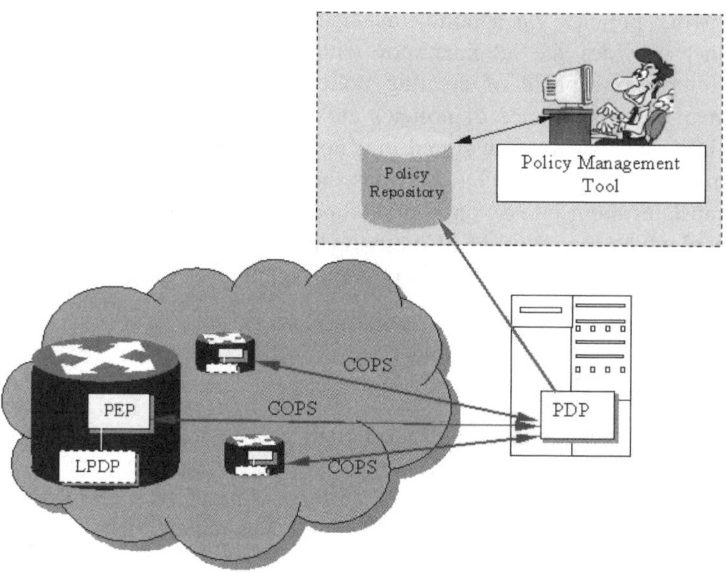

Figure 6.2. *Architecture of policy-based networks*

The components of this architecture and the associated functionalities are:

– policy management tool, which is an interface enabling the administrator to create network policies;

– policy repository, which stores policies in the form of standardized information models [MOO 01]. The policy repository can be an LDAP (Lightweight Directory Access Protocol) repository or a relational database depending on the implementation;

– the policy enforcing component (PEP – Policy Enforcement Point), which is an entity that receives the policies and must apply them in its packet processing. A PEP can be an RSVP router, a DiffServ router, a firewall, etc.;

– the decision making component (PDP – Policy Decision Point), which is responsible for choosing the policy rules in the policy repository and for distributing them to the PEP;

– the local PDP (LPDP – Local PDP), which is an optional component located in a PEP to share decision making with the remote PDP and maintain the PEP functioning when the remote PDP is absent.

The typical function of policy-based networks is as follows. The administrator defines network policies via a policy management tool. This tool is a piece of software that provides the administrator with components (for example, via a graphical interface) capable of creating policies. It is also responsible for the verification of the compliance of policies and the detection of conflicts between policies. These policies are then stored in a Policy Repository and are ready to be executed in the network. The PDP accesses the policy repository and chooses the policies, translates them into an appropriate code and distributes them to the PEPs via a protocol which can be COPS (Common Open Policy Service), SNMP (Simple Network Management Protocol), CLI (Common Line Interface), etc.

6.3. The COPS protocol

The IETF (Internet Engineering Task Force) has defined the COPS (Common Open Policy Service) protocol [DUR 00] as the policy communications protocol between the PEP and the PDP. The COPS protocol is based on TCP (Transmission Control Protocol) and uses the client/server model. The policy client corresponds to the PEP. The policy server corresponds to the PDP.

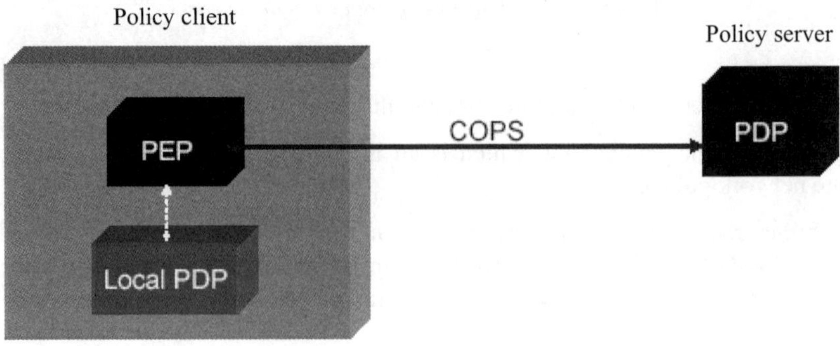

Figure 6.3. *COPS protocol*

The COPS protocol has defined 10 messages for all protocol operations: OPN, CAT, CC, REQ, DEC, RPT, DRQ, KA, SSQ, SSC. These messages are presented in Figure 6.4.

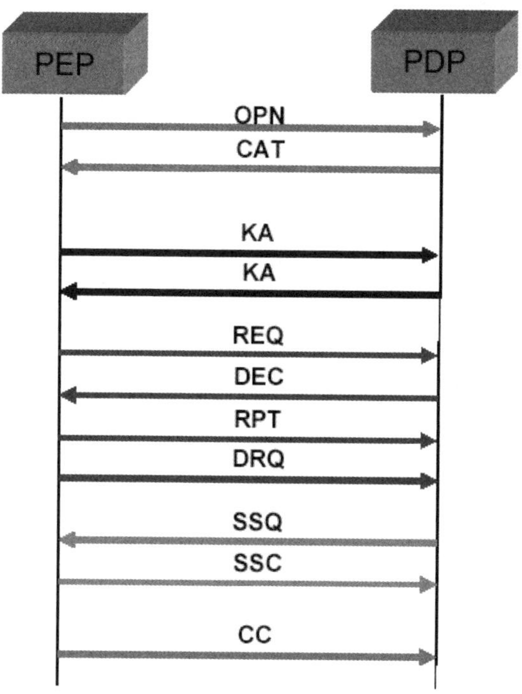

Figure 6.4. *COPS messages*

Three messages, OPN (Client-Open), CAT (Client-Accept) and CC (Client-Close), execute the opening and closing of the COPS connection. In order to establish a COPS connection, the PEP sends an OPN message indicating its identity for authentification. The PDP sends a CAT message if it accepts the PEP. If not, it sends a CC message to reject the PEP. The CC message can also be sent by the PEP or the PDP to close a COPS active connection.

After the connection has been established, the PEP and the PDP can exchange policies via REQ (Request), DEC (Decision), RPT (Report) and DRQ (Delete Request State) messages. In order to obtain policies to apply, the PEP sends an REQ message to solicit policy delivery from PDP. The PDP responds to the REQ message with a DEC message indicating the policies to apply. After enforcing these policies, the PEP sends an RPT message to report the result of these policy enforcements.

COPS is a stateful protocol. Requests sent by the PEP are controlled by states maintained in the PDP until they are explicitly removed by the PEP. In order to delete a state, the PEP sends a DRQ message indicating the state to remove.

The COPS protocol also supports mechanisms to guarantee the reliability of the protocol and the robustness of the system. When there is no message to send through the connection, the KA (Keep Alive) message is sent by the PEP and echoed by the PDP in order to maintain an open connection. If an interruption is detected, (for example, a KA message loss), the PEP can terminate the connection with the current PDP and connect to a secondary PDP. During this transition, the local PDP serves as cache in order to maintain normal system function. The PEP periodically tries to reconnect to the primary PDP. After an interruption, the PDP can send an SSQ (Synchronization Request) message to the PEP to request that the PEP synchronizes all current states. The PEP resends the requests to keep states active and ends with the SSC (Synchronization Complete) message.

Each use of the COPS protocol for a specific policy type is defined in a COPS extension, again called COPS client-type. For example, the use of the COPS protocol controlling the allocation of resources in an RSVP/IntServ network is defined by COPS-RSVP client-type; the use of the COPS protocol that supplies policies in the DiffServ routers is defined by COPS-PR DiffServ client-type, etc. Figure 6.5 shows COPS client-types proposed in other works.

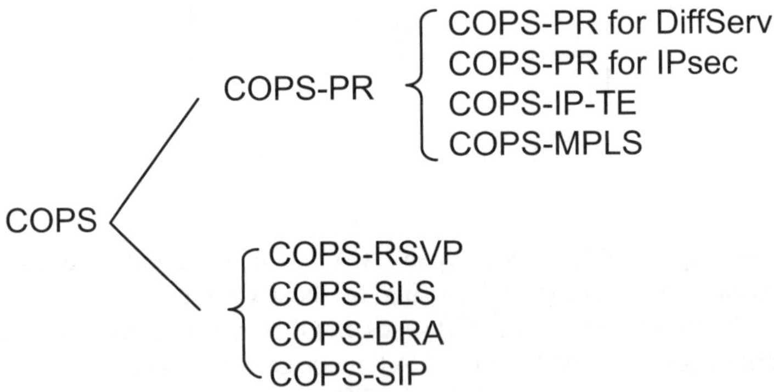

Figure 6.5. *COPS client-types*

The IETF has standardized COPS-RSVP for resource allocation control in the IntServ/RSVP network and COPS-PR for the supply of policies in the equipment configuration. COPS-RSVP and COPS-PR are presented in detail in sections 6.4 and

6.5. Other uses for the COPS protocol were proposed by researchers. These client-types are briefly presented here.

– COPS-PR for IPSec [LI 03] uses COPS-PR to manage IPSec security associations;

– COPS-IP-TE [JAC 01] uses COPS-PR to configure traffic engineering policies;

– COPS-SLS [NGU 03] uses both models Outsourcing and Provisioning of the COPS protocol for policy negotiations in the form of an SLS (Service Level Specification);

– COPS-DRA [SAL 02] uses the COPS protocol to request allocation of resources in the DiffServ network;

– COPS-SIP [GRO 00] uses the COPS protocol for AAA operations in services based on SIP;

– COPS-MPLS [REI 00] uses COPS-PR to manage the MPLS network.

6.4. COPS-RSVP

COPS-RSVP [RFC2749] is a COPS client-type that controls the allocation of resources in the Intserv/RSVP network.

The architecture and the operation of the COPS-RSVP are presented in Figure 6.6.

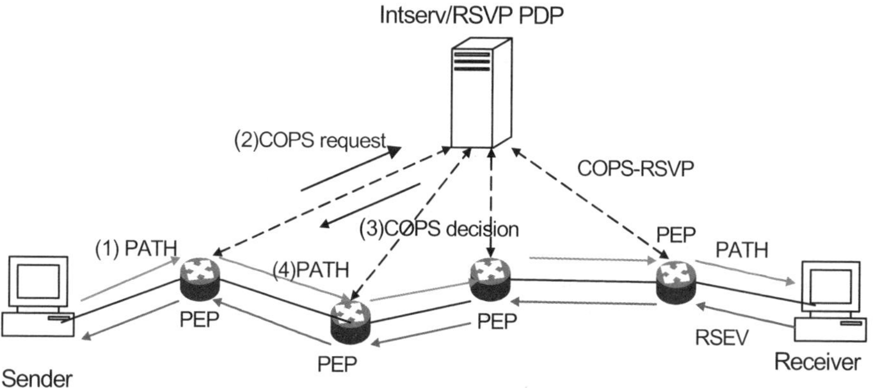

Figure 6.6. *COPS-RSVP*

Policy-based management is executed by the Path, Resv, PathErr and ResvErr messages of RSVP [BRA 97]. When one of these RSVP messages is received, the PEP sends a COPS request to the PDP, encapsulating the received RSVP objects. The PDP consults the policy repository and communicates to the PEP its policy-based management decision. Since the PEP has to send a COPS request to the policy server each time it receives an RSVP message, the COPS-RSVP operation follows the Outsourcing model.

In the Outsourcing model, the PEP sends a policy request to execute each time it must process an event. The PDP sends its decision and the PEP applies this decision in the processing of packets. This model is illustrated in Figure 6.7.

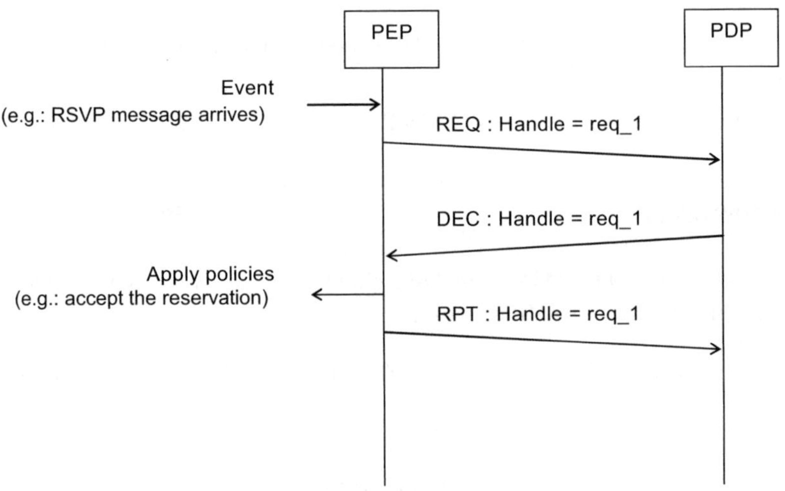

Figure 6.7. *Outsourcing model*

In RSVP, a new request for resource reservation must be approved by two controls: a resource control and a policy control [BRA 97]. Resource control will verify if the router has sufficient available resources in terms of memory and bandwidth to satisfy the request. Policy control will determine if a user can request that reservation. Information for policy control concerns information identifying the user or user class, rights and limits in terms of resources. This information is transported in the POLICY_DATA [HER 00b] object in the RSVP message. Based on this information, the PDP can accept or refuse a request for resource reservation, or modify the priority level of the RSVP session.

6.5. COPS-PR

COPS-PR (COPS usage for policy provisioning) [CHA 01] is defined for supplying policies to the network's equipment and for installing policies in its configuration. The architecture and operation of COPS-PR are illustrated in Figure 6.8.

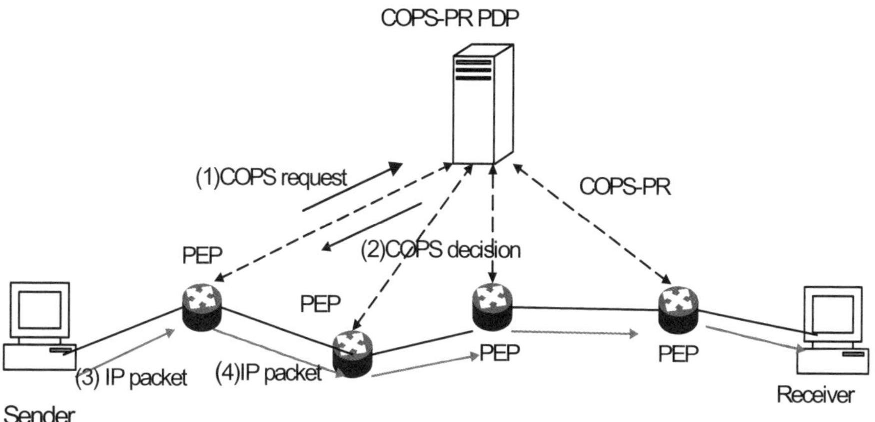

Figure 6.8. *COPS-PR*

In COPS-PR, the PEP does not send a policy request for each IP packet's arrival. Instead, policies are supplied and installed in the equipment at the start. When the PEP processes DiffServ packets, it applies the policies in the processing of IP packets without sending a request to the PDP. This model of operation is called the provisioning model. The provisioning model is also called the configuration model as the policies are installed in the configuration of the equipment. For example, policies are used to configure classifiers, meters, queues, or schedulers. In order to have policies to install, the PEP sends a request to the PDP during the start-up. The PDP supplies all the necessary policies in order for the PEP to install them in its configuration. This model is illustrated in Figure 6.9.

COPS-PR is designed to transport any type of policy (QoS policies, security policies, etc.). PEP can be a DiffServ router, a firewall, a VPN gateway, etc. Due to the variety of policy types, COPS-PR uses a data structure called PIB (Policy Information Base) to transport policies. Each PIB is a group of classifications and their attributes and represents information of a specific type of policy. For example, the DiffServ PIB defines policy information for controlling the DiffServ network and the IPSec PIB defines policy information for managing IPSec tunnels. Figure

6.10 shows a DiffServ PIB class which defines a Meter [CHA 03]. A defined class within a PIB is called PRC (Provisioning Class) [CHA 01].

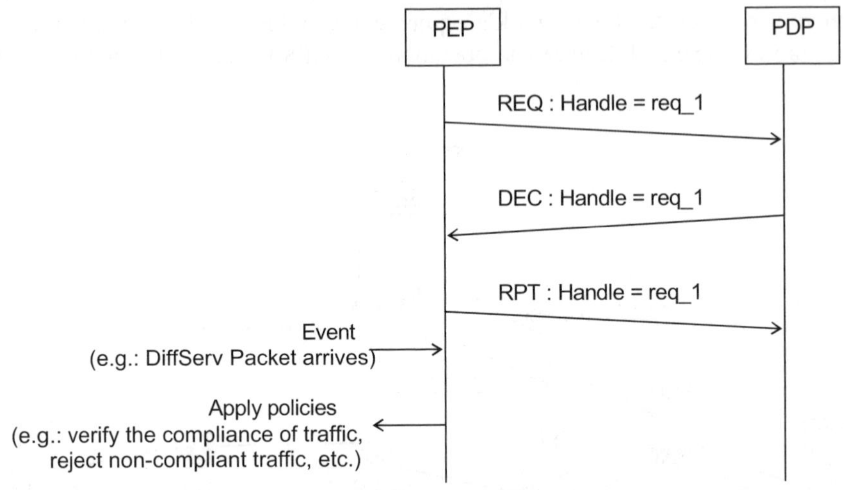

Figure 6.9. *Provisioning model*

```
dsMeterEntry OBJECT-TYPE
    SYNTAX        DsMeterEntry
    STATUS        current
    DESCRIPTION
       "An entry in the meter table describes a single
       conformance level of a meter."
    PIB-INDEX { dsMeterPrid }
    UNIQUENESS { dsMeterSucceedNext,
                 dsMeterFailNext,
                 dsMeterSpecific }
    ::= { 1.2.1.1.6.2.1 }

DsMeterEntry ::= SEQUENCE  {
    dsMeterPrid                 InstanceId,
    dsMeterSucceedNext          Prid,
    dsMeterFailNext             Prid,
    dsMeterSpecific             Prid
}
```

Figure 6.10. *PRC defining a Meter in a DiffServ router*

As shown in Figure 6.10, the name for this PRC is dsMeterEntry. This class has four attributes: dsMeterPrid, dsMeterSucceedNext, dsMeterFailNext and dsMeterSpecific. The dsMeterPrid attribute is simply an identifier to identify an instance of this class. The dsMeterSucceedNext attribute is a pointer pointing to another instance (a Marker) that processes traffic compliant with the traffic contract. The dsMeterFailNext attribute is a pointer pointing to another instance (a Dropper) which processes traffic non-compliant with the traffic contract. The dsMeterSpecific attribute is a pointer pointing to an instance defining the traffic profile. The simple Token Bucket type Meter with two parameters determines throughput and burst size. Each class has an OID value only for identifying the class. For example, dsMeterEntry class has an OID of 1.2.1.1.6.2.1.

Figure 6.11 shows an example of the use of the Meter class. In this example, a simple Token Bucket type Meter is used to control the compliance of traffic with a 64 Kb/s throughput and a maximum burst size of 1,500 bytes. Packets compliant with this profile are marked as "AF11", a value of the DSCP. Non-compliant packets are marked as Best Effort.

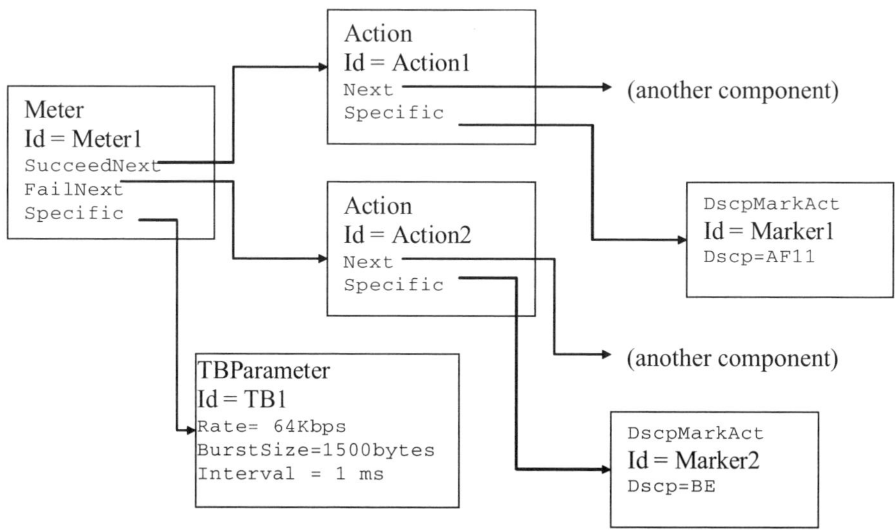

Figure 6.11. *Example of a part of the DiffServ router's configuration*

An instance of a PRC is a PRI (Provisioning Instance). A PIB can be described in a conceptual manner as a tree in which the branches represent the PRCs and the leaves are PRIs. This concept is presented in Figure 6.12.

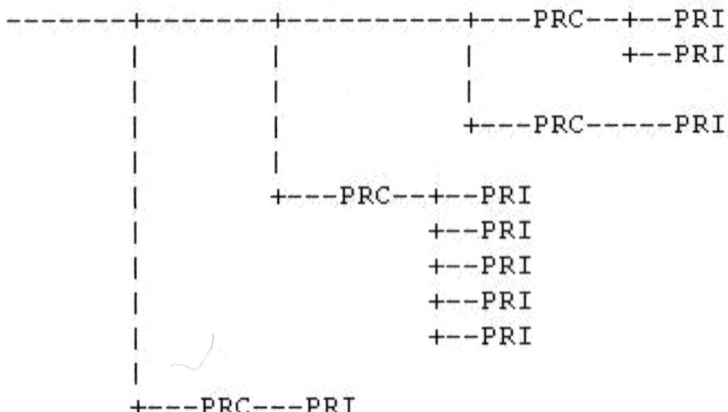

Figure 6.12. *The PIB tree*

If we look at the configuration example presented in Figure 6.11, we can see a PIB tree, as shown in Figure 6.13.

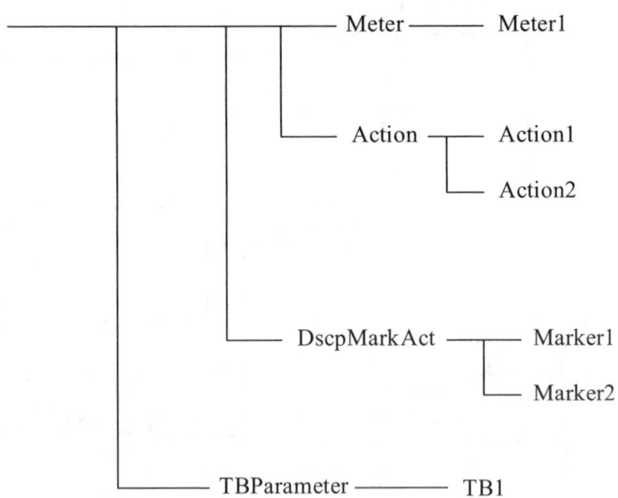

Figure 6.13. *Example of a PIB tree*

COPS-PR has defined objects to transport PIB information [CHA 01]. For example, a PRID can be transported by using two COPS-PR objects: the PRID object and the EPD object, as shown in Figure 6.14.

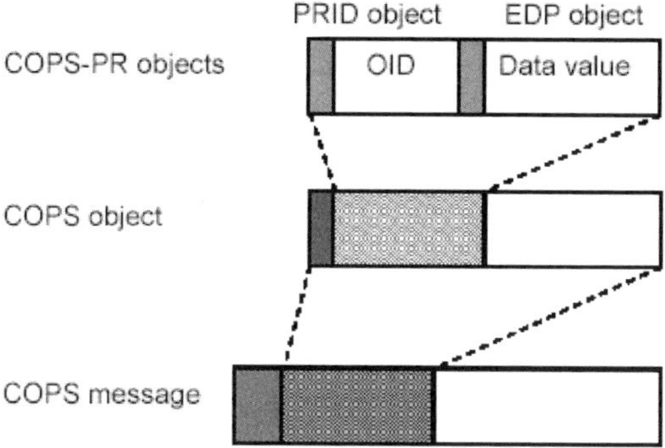

Figure 6.14. *Transport of PIB information in COPS-PR*

In order to transport a PRID, the OID of the class is transported by a PRID object and the attribute values are transported by an EPD object. These objects are encapsulated in a COPS object (Named ClientSI) which is then encapsulated in a COPS message and sent over the COPS connection. PRI information is coded by using, for example, BER (Binary Encoding Rule) or XML (Extensible Markup Language).

6.6. SNMP

The SNMP (Simple Network Management Protocol) may also be used for policy communication between PEP and PDP. The SNMP is associated with an architecture containing four basic components [PER 97] which are shown in Figure 6.15.

In this figure, the following entities are presented and are defined as:

– *agent*, which is as SNMP entity integrated in a network equipment;

– *manager*, which is as SNMP entity with a management application;

– *management information*, which defines information for the management of network equipment;

– SNMP, which is the protocol that enables the management application to contact the agent in order to access management information in network equipment.

In policy-based management SNMP is used for the policy configuration within network equipment [BOR 00]. In other words, the SNMP is an alternative to the COPS-PR protocol. In comparison with the policy-based management architecture, the *SNMP manager* corresponds to the PDP and the *SNMP agent* corresponds to the PEP. Policy information is defined and transported in the form of an MIB (Management Information Base) [BAK 02].

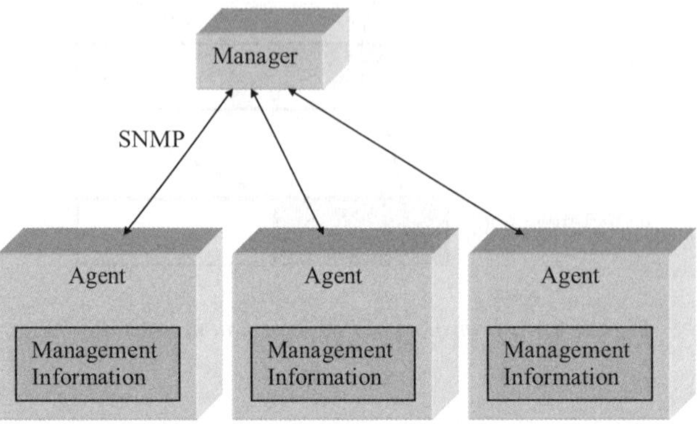

Figure 6.15. *SNMP model*

6.7. Conclusion

The objective of this chapter was to present the policy-based management concept in IP networks. The policy communications architecture and protocols have been presented in detail. In policy-based management architecture, PDP is a central entity managing the behavior of the whole network. Policies are defined by the network administrator, translated in an acceptable form by the routers and communicated to PEPs which apply them in the processing of IP traffic. The COPS protocol is a standard and flexible protocol in the communication between PEP and PDP. The COPS protocol supports two policy-based management models: outsourcing and provisioning. COPS-RSVP is used to control the allocation of resources of an Intserv/RSVP network. COPS-PR is defined to configure policies in the network equipment. The SNMP is an alternative to COPS-PR, for the configuration of policies in equipment. Policy-based management has become extremely important in IP networks because it enables for the control of a complete network as an entity instead of controlling each component individually. This approach is necessary in order to control large IP networks, introducing QoS and thus new services.

6.8. Bibliography

[BAK 02] BAKER F., CHAN K., SMITH A., "Management Information Base for the Differentiated Service Architecture", *RFC 3289*, May 2002.

[BOR 00] BOROS S., "Policy-Based Management with SNMP", *6th EUNICE Open European Summer School*, Enschede, Netherlands, September 2000.

[BRA 97] BRADEN R., ZHANG L., BERSON S., HERZOG S., JAMIN S., "Resource ReSerVation Protocol (RSVP) – Version 1 Functional Specification", *RFC 2205*, September 1997.

[BRU 01] BRUNNER M., QUITTEK J., "MPLS Management Using Policies", *Proceedings of IEEE/IFIP International Symposium on Network Management*, p. 515-528, May 2001.

[CHA 01] CHAN K., SELIGSON J., DURHAM D., GAI S., MCCLOGHRIE K., HERZOG S., REICHMEYER F., YAVATKAR R., SMITH A., "COPS Usage for Policy Provisioning (COPS-PR)", *RFC 3084*, March 2001.

[CHA 03] CHAN K., SAHITA R., HAHN S., MCCLOGHRIE K., "Differentiated Services Quality of Service Policy Information Base", *RFC 3317*, March 2003.

[DUR 00] DURHAM D., BOYLE J., COHEN R., HERZOG S., RAJAN R., SASTRY A., "The COPS (Common Open Policy Service) Protocol", *RFC 2748*, January 2000.

[GRO 00] GROSS G., RAWLINS D., SINNREICH H., THOMAS S., "COPS Usage for SIP", draft-gross-cops-sip-00.txt, Internet Draft, November 2000.

[HER 00a] HERZOG S., BOYLE J., COHEN J., DURHAM D., RAJAN R., SASTRY A., "COPS Usage for RSVP", *RFC 2749*, January 2000.

[HER 00b] HERZOG S., "RSVP Extensions for Policy Control", *RFC2750*, January 2000.

[JAC 01] JACQUENET C., "A COPS Client-type for IP Traffic Engineering", draft-jacquenet-ip-te-cops-02.txt, Internet Draft, June 2001.

[KOS 01] KOSIUR D., *Understanding Policy-Based Networking*, John Wiley & Sons, 2001.

[LI 03] LI M., "Policy-Based IPsec Management", *IEEE Network* 17, 6, November/December 2003.

[MOO 01] MOORE B., ELLESSON E., STRASSNER J., WESTERINEN A., "Policy Core Information Model – Version 1 Specification", *RFC 3060*, February 2001.

[NGU 03] NGUYEN T. M. T., BOUKHATEM N., PUJOLLE G., "COPS-SLS usage for dynamic policy-based QoS Management over heterogeneous IP networks", *IEEE Network* 17, 3, p. 44-50, May/June 2003.

[PER 97] PERKIN D., MCGINNIS E., *Understanding SNMP MIBs*, Prentice Hall, 1997.

[REI 00] REICHMEYER F., WRIGHT S., GIBSON M., "COPS Usage for MPLS/Traffic Engineering", draft-franr-mpls-cops-00.txt, Internet Draft, July 2000.

[SAL 02] SALSANO S., VELTRI L., "QoS Control by Means of COPS to Support SIP-Based Applications", *IEEE Network*, March/April 2002.

[VER 01] VERMA D. C., *Policy-Based Networking: Architecture and Algorithms*, New Riders, 2001.

Chapter 7

Inter-domain Quality of Service

7.1. Introduction

Today, a number of Internet service providers (ISP) offer an IP infrastructure with Quality of Service (QoS) enabling new types of services such as voice over IP, video, virtual private networks, etc. However, these services are only available within the ISP domain. In fact, deployment of services over a group of domains belonging to different providers is very complex. Indeed, configuration tasks become very difficult due to the heterogenity of the operators' network technologies and their strategy differences. This complexity is heightened by the fact that each provider has its own definition of service concepts (premium, gold, silver, bronze, etc.). Today, these tasks are mainly accomplished by human interaction: phone calls and fax between different organizations and manual system configuration. This approach introduces high risks of configuration mistakes, as well as long delays for procurement of services.

Recent studies in Internet traffic have shown that over-provisioning is not a good approach to ensure quality of communications with a QoS. The current approach consists of differentiating between client traffic based on Service Level Agreements (SLA). An SLA is an agreement (or part of an agreement) between a service provider and a client. It is designed to create a common understanding on what the client has requested and what the provider has accepted to supply and at what cost [SLA 01].

Chapter written by Mauro FONSECA.

Providers understand, on the one hand, that the SLA guarantee is a differentiation factor in a very competitive telecommunications market, as well as in their competition with other ISPs, in order to obtain a larger market share. On the other hand, ISPs understand the importance of a strategic cooperation with other ISPs, be it to provide a global network with QoS or to guarantee end-to-end SLAs.

In fact, it is impossible for an ISP to deploy a worldwide network, or to control end-to-end communications over a number of administrative domains. In this approach, the only way to offer end-to-end guarantees for client communications is to put agreements (SLA) in place between clients and the supplier or ISP. However, due to the versatility of clients and ISP strategies, it is important to execute these agreements (SLA) automatically. Today, it is possible to envisage such a solution, with the maturity of underlying technologies which facilitate the execution of such a service.

On the one hand, the management of networks with policies (PBN – Policy-Based Networking) enables the automation of the network configuration based on high level policies indicating the provider's commercial strategy. On the other hand, the agent technology has shown, in many sectors, its potential in resolving complex distributed problems.

The goal of this chapter then is to examine the use of inter-domain network management based on policies, associated with the mobile agent technology in order to enable inter-ISP cooperation and to provide end-to-end SLAs to clients.

Let us suppose that a group of ISPs want to cooperate in offering a virtual network ensuring a global QoS. Clients can then request SLAs between any access point within that global network in order to deploy different types of services: pipe, channel, VPN, etc. This approach implies that the ISPs have already installed a management system for the configuration and monitoring of the network equipment, to provide different service classifications (we are talking about DiffServ classes here). The goal is to improve this system by adding a layer of mobile agents for the negotiation of SLAs. The complex aspects of this negotiation are the many service possibilities, as well as the semantics of the QoS parameters used in each ISP domain [MAR 00]. For example, a premium service in the domain of one ISP can be different from the premium service of another ISP.

7.2. Goal

The goal of this chapter is to present the negotiation and configuration procedures in order to enable the introduction of inter-domain services which guarantee end-to-end QoS. The starting point for the global process is a client wanting to use a service through different administrative domains (a global network), as described in Figure 7.1. Each domain belongs to a specific service provider with its own objectives and its own strategy. However, this ISP must cooperate with other ISPs in order to offer end-to-end services guaranteeing QoS.

Figure 7.1. *An end-to-end service request*

The ISP probably uses heterogenous network technologies. Nevertheless, we presume that each domain of each organization has been able to deploy a policy-based network management system, taking over the role of bandwidth broker (BB) for its domain. This negotiator thus supplies a uniform interface to configure the network. Here, we are concerned with the automation component regarding inter-domain interactions between the domains of the user and the supplier, and not with the inter-domain network's management.

Based on these assumptions, the goal is to improve the BB's function with abilities which enable the interaction with other BBs, in order to negotiate inter-domain service. The interaction is based on a group of predefined agreements, established between the different ISP. These agreements must be grouped within an SLA called Service Intersuppliers SLA (I2I-SLA – ISP to ISP SLA). These agreements mark the formal negotiated limits between the provider ISP and the client ISP for a specific service. The I2I-SLA is designed to create a common agreement between two ISPs about services, priorities, responsibilities, etc. In the same way, the user can set an SLA with a specific ISP for the supply of an end-to-end service. The characteristics of the agreement are grouped in an SLA between the client and the ISP (C2I-SLA – Customer to ISP SLA).

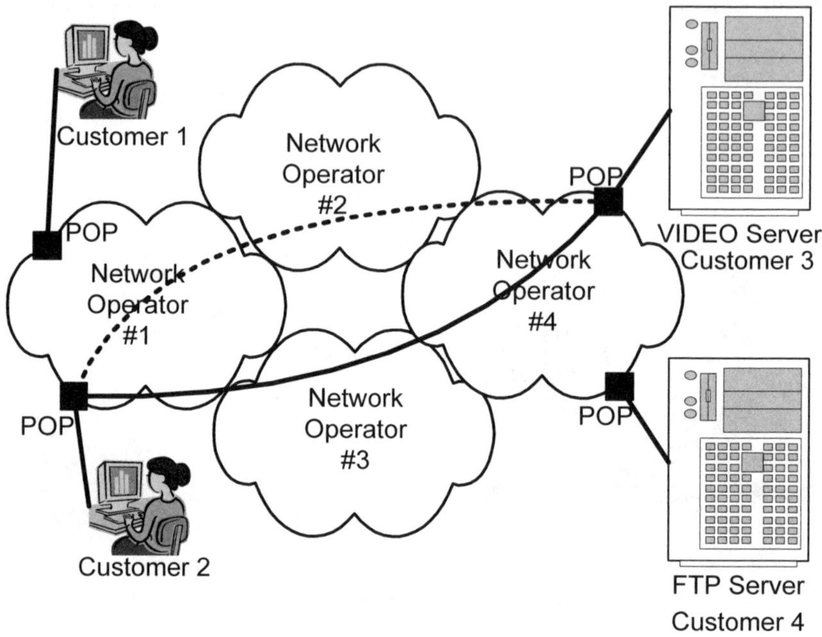

Figure 7.2. *End-to-end services through multiple domains*

When the requested service goes through the domains of a certain number of ISPs, as shown in Figure 7.2, a negotiation process must be sent between the ISPs involved, in order to put the end-to-end service in place and thus ensure the best deal for the client's request. For example, if the client requests a service through multiple domains, the starting ISP must launch a search process in order to find different routes from the client to the destination's access point. If multiple routes are possible, then the process must identify the best solution according to the client's preferences in terms of QoS, cost or any other characteristic defined in the requested SLA. Therefore, improving the architecture of PBM is necessary in order to take into account the multipart process of inter-domain control as described in Figure 7.2.

The I2I-SLAs and the C2I-SLAs are the same conceptually but are different in terms of execution. They do not address the same level of granularity of service and of time proportion. A C2I-SLA has a fine granularity for the definition of service, whereas an I2I-SLA has a wider granularity. This means that the negotiation process between ISPs is usually less frequent and is based on large bandwidth quantities for each negotiated service because usually the high volumes negotiated concern long-term usage. Besides, the negotiation between a client and his ISP is generally more frequent and concerns a small bandwidth volume for each requested service. The ISPs can then reach higher profitability with this mechanism.

7.3. Motivations for the use of mobile agents to offer inter-domain QoS

The goal for automation between management systems based on policies (PBM) of ISPs is to conceal the complexity of the negotiation process for end-to-end services. Each ISP can in fact have different agreements (I2I-SLA) with other ISP in order to provide connectivity to a given destination, as shown in Figure 7.2, thus making it possible to easily have a competitive market.

The PBM approach facilitates the representation of ISP strategies with the use of policies. However, this approach does not facilitate the automation of the negotiation process itself. For this reason, we are proposing the use of mobile agents as a flexible approach introducing PBM over inter-domain IP networks. The use of mobile agents is motivated by the fact that the negotiation between a client and a ISP or between two ISPs can be very complex. It will be made considerably easier if a "delegation" is executed in order to minimize the large number of interactions caused by the use of a traditional client/server protocol. In this way, mobile agents constitute an interesting approach for helping to automate the negotiation process, while avoiding a complex scheme of interactions linked to a traditional client/server protocol. Consequently, the policy-based management philosophy associated with the mobile agent approach enables the local execution of negotiation and decision, since the mobile agents are capable of transporting information such as behavior policies.

7.3.1. *Control of inter-domain QoS parameters*

In order to have a good understanding of the problem, the topology presented in Figure 7.2 is considered as the reference example for this chapter. This figure presents four service providers wanting to cooperate in order to provide end-to-end services (a global network). We have complemented the example with network level information. Let us presume that the service provider has put in place quality services, distributed in 3 service classes:

– Gold: traffic in this category is limited to 20% of the available throughput, delay of 10 ms, 1 ms of jitter, 10^{-12} probability for loss of packets.

– Silver: traffic in this category is limited to 40% of the available throughput, delay of 20 ms, 5 ms of jitter, 10^{-6} probability for loss of packets.

– Bronze: traffic in this category is limited to 40% of the available throughput, with no delay guarantee, no jitter guarantee, no guarantee for loss of packets.

The three services are defined by using the following QoS parameters: delay, jitter, packet loss and throughput. These parameters affect the client's traffic and are an integral part of the negotiated SLA. For simplicity purposes, we are currently not taking into account other SLA parameters, which are described in [SMI 01] or [VEN 01].

If the client requests a service between his site and a remote site, the perceived delay by the client's applications constitute end-to-end delay. This is introduced throughout the nodes on the route by the queues, processing or congestion in each intermediary node. Since we must process the networks from multiple domains [WAN 96], delay is then considered additive, i.e.:

$$D_{end-to-end} = \sum_{i=1}^{n} D_i \qquad [7.1]$$

Jitter corresponds to the alteration of inter-packets arrival times compared to the inter-packet times of initial transmission (delay variation). Jitter is particularly harmful to multimedia traffic. In the case where a connection travels through multiple domains (inter-domain), jitter will accumulate on the root mean square (RMS) basis, i.e.:

$$G_{tot} = \sqrt{G1 + G2 + \ldots + Gn} \qquad [7.2]$$

Di represents the one-way delay of domain n and Gn represents jitter or the standard deviation of the delay variation in domain n.

The loss of packet refers to the failure to receive a transmitted packet. It is usually due to the drop of packets at certain points along the network's route due to a congestion problem. Then, by traveling through multiple domains, the probability of loss accumulates on a probability base, i.e.:

$$PP_{tot} = 1 - \left[(1 - PP_1)(1 - PP_2)\ldots(1 - PPn)\right] \qquad [7.3]$$

The way the ISP shares the bandwidth and the way it will distribute it between the different classifications is independent from clients' requirements. However, the ISP must evaluate its choices at medium term according to the network's usage and the QoS failure within the network (verified by the monitoring loop). The domain administration is responsible for guaranteeing that the necessary resources will be supplied and/or reserved to support SLAs offered by the domain [BLA 98].

Now let us presume that the goal of ISP1 is to maintain this service within the following limits:

Gold:

DSCP: EF

Delay: Max = 10 ms

Jitter: Max = 1 ms

Packet loss: Max 10^{-12}

Silver:

DSCP: AF

Delay: Max = 20 ms, Probability = 10^{-3}

Jitter: Max = 5 ms, Probability = 10^{-3}

Packet loss: Max 10^{-6}, Probability = 10^{-3}

Bronze:

DSCP: BE

If these services have to be supplied end-to-end, the service provider is not in a position to ensure that all feedthrough networks throughout the route to destination will guarantee these services. Thus, there is a necessity to collaborate with remote service providers in order to identify services available locally and the associated QoS, and to define corresponding rules. For this collaboration to happen, we will move to a negotiation based on SLAs.

7.4. Negotiation of inter-domain QoS

As presented previously, this solution introduces two types of SLA: the SLA between a client and an ISP (C2I-SLA) and the SLA between two ISPs (I2I-SLA). The example in Figure 7.2 presumes that the ISP1 controls two I2I-SLAs with ISP2 and ISP3, in order for the services of ISP4 to be negotiated by ISP2, ISP3 or both. This simple mechanism ensures that the services going through domains of neighboring ISPs are transparent. It is therefore easier to control these services (Figure 7.3).

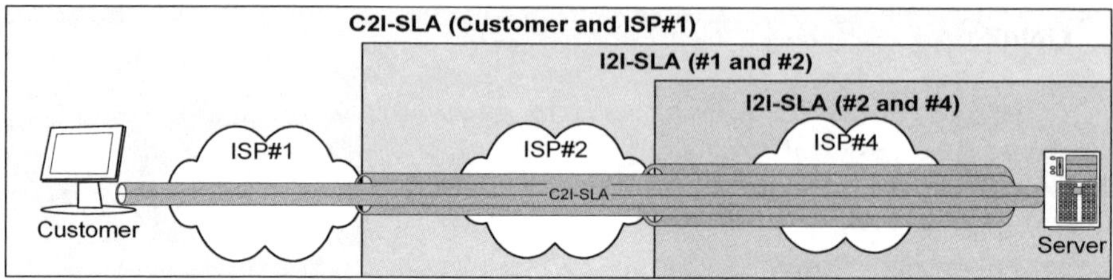

Figure 7.3. *C2I-SLA and I2I-SLA*

7.4.1. *Inter-domain SLA*

An important aspect in the contractual connection between a client and his provider is the SLA [SLA 01]. It represents a formal agreement negotiated between the two entities. It is the reference document which makes it possible to verify if the service supplied by the provider is compliant with the client's request. The SLA is the contract which exists between the service provider and the client. It is designed for the creation of a joint agreement in terms of services, priorities, responsibilities, etc. The SLA is defined by an SLS and SLOs [WES 01].

The SLA must explain the agreements between the client and the ISP. It mainly contains the following information:

– who the client is;

– which service the client has subscribed to;

– when the client is ready to use the service;

– from where the client will use the service;

– how to monitor the service supplied in order to verify QoS, for purposes of invoicing as well as refunding in case of failure.

We have used the approach presented in [CEL 00] which is summarized in the following five levels of agreement:

– *Basic* agreements, which contain descriptive information about the provider, the client and the required service; they also contain information the required period of usage and the validity of this period (permanent or at specific dates/times). This information may contain names as well as signatures for non-repudiation, bank identifier, etc. Furthermore, Basic agreements contain the applicability period of the SLA.

– *Topology* agreements, which describe the number and nature of terminal points that a service may use, and the report of generation and consumption of traffic between these points. The topology agreements are made up of sub-unit service access points (SAP) and of the graphic sub-units.

– The sub-unit *SAP* is a list of terminal points used in the construction of the topology. The reason for maintaining this list separately is that the terminal points descriptions can be complex, using IP addresses or attributes specific to ISP. There is no real need to repeat these descriptions within each topology agreement or in the QoS agreements. The use of SAP allows us to assign, within the document, serial numbers specific to each service access point, reused whenever necessary.

– The sub-unit *Graphic* is used to describe a topology, terminal points can be a source, a destination or both. A source terminal point produces, but does not consume traffic, whereas a destination consumes traffic without production. The sub-unit graphic contains a list of sources and destinations, with special semantics which describes their correlations ex: 1-1; 1-M; M-1; 1-any; any-1, etc.

The limits of the pipe, the funnel or of the tube can be specific IP addresses, or the keyword "ANY", which indicates that the traffic passing through any server is part of the topology. Please note that the funnel topology is more common than pipe topology, and that tube topology is more common than the first two. The reason for specifying them is to provide the correct model.

7.5. An architecture for inter-domain negotiation

After defining the different models necessary to maintain information on resources and the SLA, we will now present an agent-based architecture in order to improve the negotiation process between the different domains.

7.5.1. *Mobile agent advantages*

Mobile agent technologies offer many advantages. The agents can help users, service and network providers in accomplishing negotiation tasks. These tasks include the choice of service, QoS specifications, evaluation of cost of negotiations, etc. The agents are particularly appropriate for the tasks where rapid decision making is essential. The mobile agents satisfy the two most important performance aspects in an SLA: availability and response time [VER 99]. The following attributes show the agents' capabilities in the negotiation execution tasks [CHI 00]:

– *Autonomy*: agents can be reactive or proactive. They can execute tasks autonomously following predefined rules or constraints. Their intelligence level will depend on the indicated roles or tasks.

– *Communication*: with this capability, negotiations can be established between two agents. The agent communication language (ACL) within FIPA has largely been adopted.

– *Cooperation*: agents can cooperate in the execution of a task. Consequently, the agents representing users, service providers and network providers can now cooperate for the installation of end-to-end service throughout multiple domains.

– *Mobility*: agents based on Java can migrate throughout networks and heterogenous platforms. This attribute differentiates the mobile agents from other static agent forms. Mobile agents can migrate their execution and calculation processes to remote servers. This, compared to conventional client/server systems, protects local and shared resources, such as bandwidth and central unit usage. In this way, the SLA's process of negotiation can migrate to the service provider or the network provider's domain.

The idea is to enable the agents enough autonomy to negotiate the best service in the name of clients or service providers, according to their predefined strategy. The management system of the SLA's ISP is also based on a group of agents. However, these agents are static and only execute local operations.

The agent approach makes it possible to break down the management system in components, which can easily be combined with the PBM components.

The PBM system enables a control and efficient configuration of the ISP's network equipment in order to represent higher level agreements, as shown in Figure 7.4.

The architecture presented in Figure 7.4 helps in the execution of the negotiation process between a client and his ISP and between two different ISPs. It is based on a group of agents that move from a policy-based management system to another, in order to execute the negotiation process in the name of the activation party (client or ISP).

The SLA is at the core of the architecture, since it is the subject of the negotiation. Each PBM system is also based on a mobile agent architecture whose goal is to enable the initial PBM system in the transformation of its role as bandwidth negotiator in the domain into a role of inter ISP (inter-domains) SLA negotiator, as described in Figure 7.4.

The architecture is based on a group of mobile agents and static agents. Static agents are used within a domain to interact with the local PBM system and to improve it with new functionality. Mobile agents are used for the interaction between the different domains (between a client domain and an ISP domain or between two ISP domains).

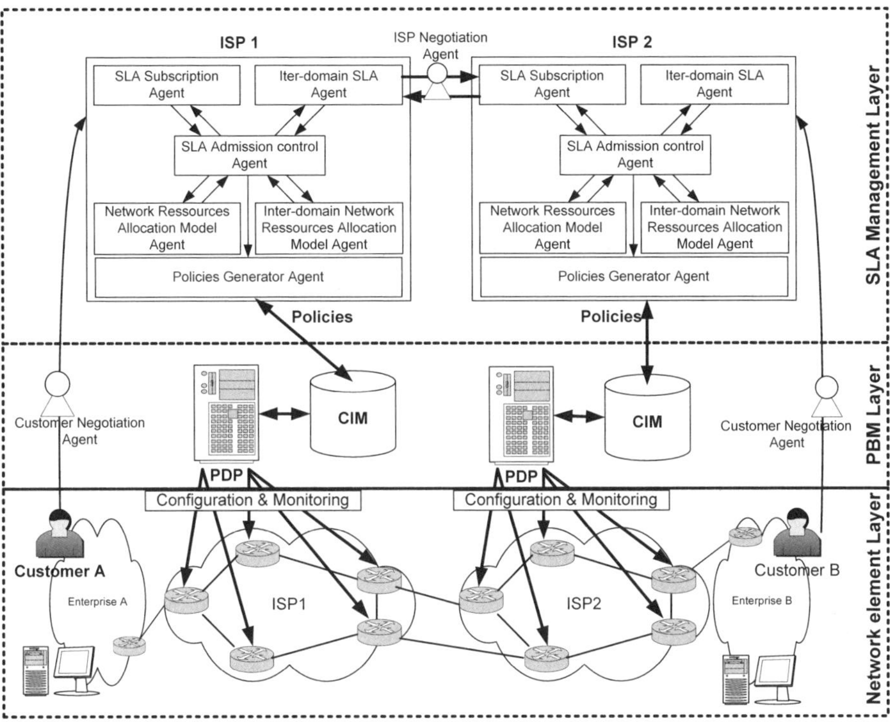

Figure 7.4. *The architecture of the dynamic management system of SLAs*

The group of agents is distributed as follows:

– *Customer Negotiation (CN) agent*: it is an agent instantiated by the client. It is responsible for the dialog with the client's ISP, in the name of the client. It transports the required C2I-SLA as well as the authorization and the necessary policies for client negotiation.

– *SLA subscription (SSU) agent*: it is the agent responsible for the interaction with the CN agent for the subscription of a new SLA. Consequently, it creates C2I-SLA objects stored in the Common Information Model (CIM) repository as well as objects relative to the SLA.

– *Inter-domain SLA subscription (ISSU) agent*: this agent is responsible for the entire SLA negotiation process with the pair ISPs. It uses one ISP negotiation agent to interact with remote ISPs. Consequently, it creates I2I-SLA objects and sends them to an SAC agent.

– *SLA Admission Control (SAC) agent*: this is responsible for:

- controlling whether the new SLA can be accepted or not according to the available resources and existing SLA,

- interacting with SSU and ISSU agents for the verification of terms of the new SLA according to available resources in the domain and at inter-domain level,

- obtaining available resources in the domain with the NRAM agent,

- obtaining available resources in the inter-domain with the INRAM agent,

- sending C2I-SLA objects accepted for the NRAM agent,

- sending I2I-SLA objects accepted for the INRAM agent,

- sending SLA objects accepted as well as relative to the PG agent objects.

– *Network Resource Allocation Model (NRAM) agent*: this agent maintains information on available resources in the ISP's network as well as on future resource allocations.

– *Inter-domain Network Resource Allocation model (INRAM) agent*: this agent maintains information on available resources between pair ISP networks (inter-domain). It also maintains future inter ISP resource allocations.

– *Policy Generator (PG) agent*: its main role is to translate accepted SLA in operational policies and to store them in the policy repository (CIM).

– *ISP Negotiation (IN) agent*: it is an agent created by the ISSU agent for the negotiation of the SLA with the ISP pair.

– *Policy Decision Point (PDP)*: it is the component responsible for decision-making in terms of stored policies in the policy repository (CIM). It then adopts the configuration decisions contained in the activated rules action.

– *Common Information Model Repository (CIM)*: is the database containing CIM model classes and other information classes. It ensures persistence of the PBM system. It is also manager for CIM objects.

7.5.2. *Interaction protocol between the client and the ISP*

In the case where a service requested by the client can directly be accomplished within the ISP domain, the interaction protocol between a client and his ISP is based on agent interactions limited to the ISP domain.

In order to show these interactions, we have used the approach proposed in [BAU 01]. The SLA negotiation protocol is made up of 6 services: submit, refuse, accept, propose, confirm and cancel. These interactions are explained further in Figure 7.5.

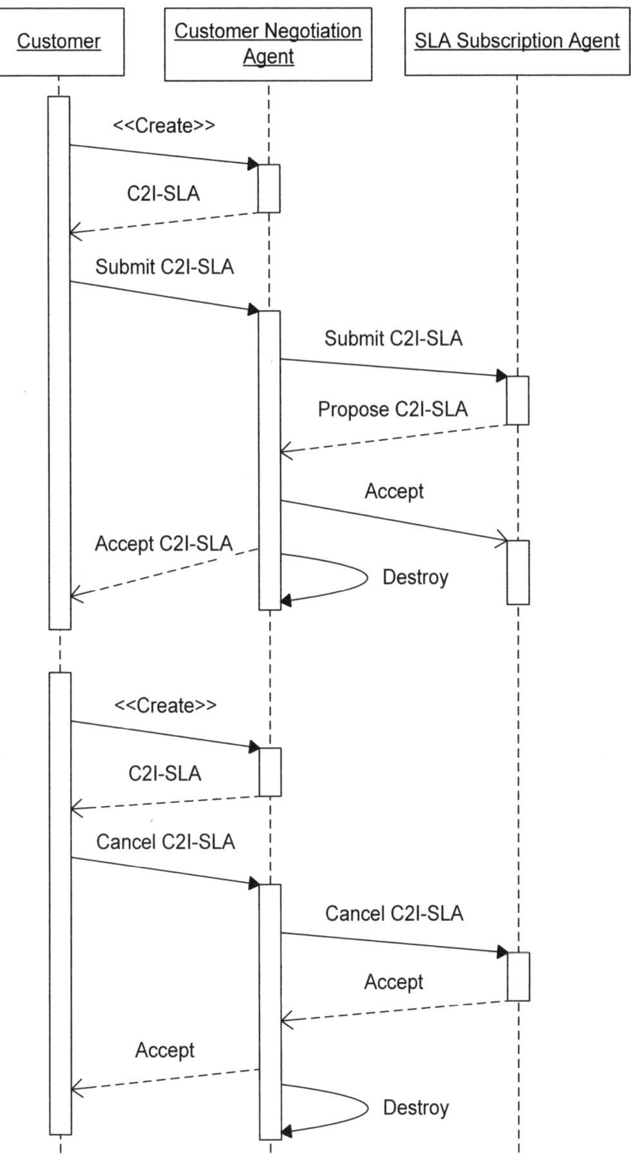

Figure 7.5. *Interaction protocol between client and ISP*

The interaction diagram presented in Figure 7.5 shows a scenario where the client gives his agent the authority to negotiate locally any proposition from the ISP which is different form the initial request. The client indicates the maximum and minimum limits for the SLA's negotiated parameters, as well as the priority between these parameters during the process of negotiation.

Simple rules can be used to represent such a policy and they are explained in section 7.5.3.1.

7.5.3. *Interaction protocol between two ISPs*

In the case where the ISP is not capable of satisfying end-to-end clauses of the requested service, i.e. when the range of service exceeds a certain number of IPS domains, the process of negotiation becomes more complex. In fact, the interaction protocol between two SLA management systems from two remote ISPs must be initiated in order to identify the different possible routes that will support the required QoS. This process is initiated in two cases: when no inter-ISP resource allocation is executed yet or when available resources for inter ISP communications are not sufficient to complete the request.

Similarly to the previous negotiation process, interactions between ISP domains require analog phases, as demonstrated in Figure 7.6. Even so, an ISP cannot negotiate the exact required values from the SLS parameters with another ISP because the negotiation of resources for inter-ISP communications is not done for a small unit of resources but for aggregates of resources. Thus, the ISP must define a negotiation strategy to be delegated to its negotiation agent. For example, the ISP can set different constraint rules regarding: maximum cost by Mb/s that it wants to pay, limits for allocated bandwidth, jitter and loss probabilities, maximum service time, etc.

Consequently, the agent must indicate to the remote ISP if the local domain will be used as a pass-through domain or as a final domain. If it is a pass-through domain, this means that the final SAP is in a different network; otherwise it is directly linked to the ISP domain. In order to record this information in a neutral way and to avoid heterogenity in the representation of inter-domain data, the requested SLA is represented by XML. We have then determined new tags in order to define the agent's strategy for the negotiation between ISP domains; this is explained in section 7.5.3.1.

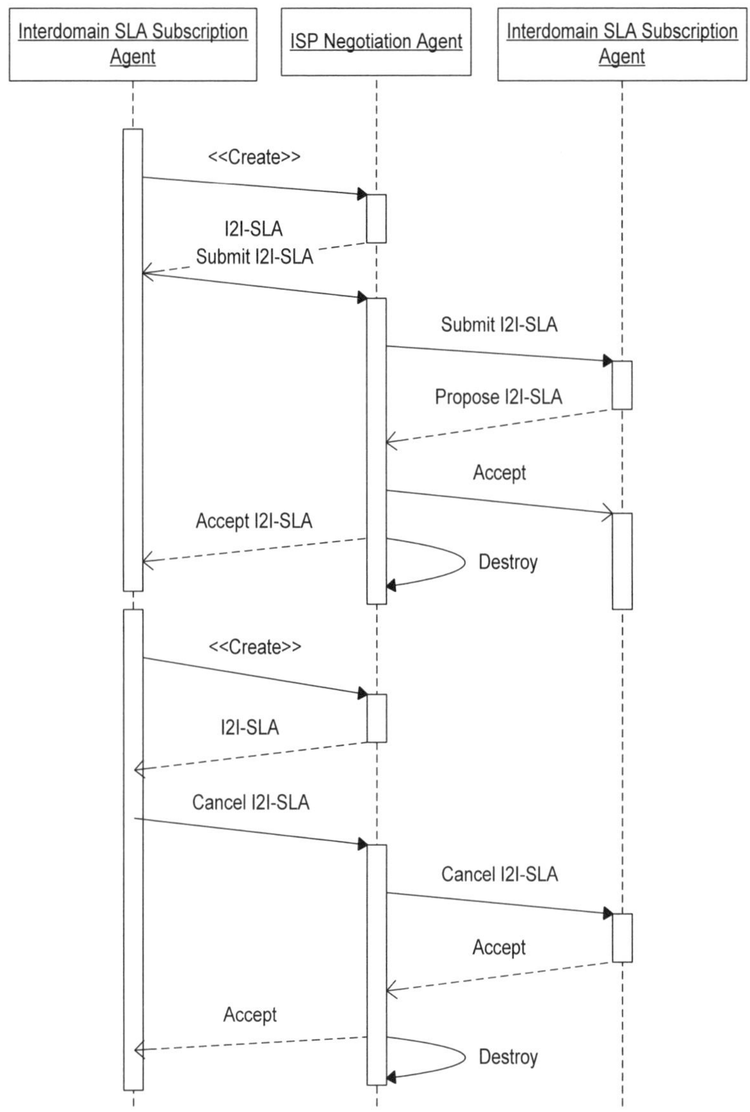

Figure 7.6. *Interaction protocol between two ISPs*

7.5.3.1. *Agent delegation policy*

Clients and ISPs can delegate SLA negotiation responsibility to their agent. However, they can limit the level of the responsibility. This is detailed in terms of authorization and refusal rules.

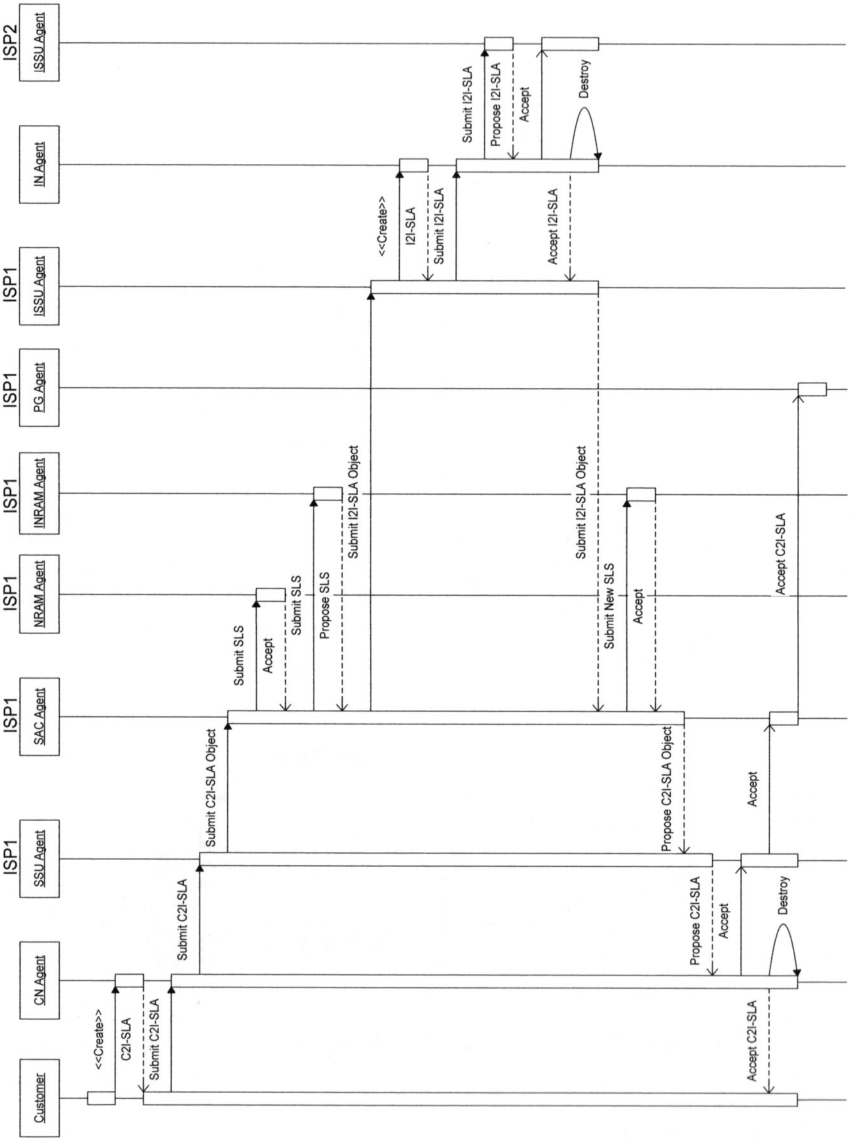

Figure 7.7. *Negotiation protocol between client, ISP and between two ISPs*

An example of such an agent delegation policy represented by XML is described here:

```
< !-- This is the SLA Negotiation in an XML file. -->
<Agent_Behavior>
<SLA_Behavior>
<IF> <Compare type = “Different”>
<SLA type = “Proposed”> <SLA type = “Requested”>
<THEN> <Negotiation start = “1”> </THEN>
< !-- agent can negotiate locally -->
<IF> <Negotiation start = “1”>
<THEN> <Priority high = “BANDWIDTH” medium = “DELAY” low = “JITTER”>
<IF> <Compare type = “Minor”>
<SLA type = “Proposed” item = “BANDWIDTH”
value = “10” unit = “Mb/s”>
<THEN> <SLA type = “Proposed” status = “Refuse”> </THEN>
<IF> <Compare type = “Minor”>
<Proposed_SLA item = “DELAY” value = “130” unit = “ms”>
<THEN> <SLA type = “Proposed” status = “Accept”> </THEN>
<IF> <Compare type = “Minor”>
<Proposed_SLA item = “JITTER” value = “5” unit = “ms”>
<THEN> <SLA type = “Proposed” status = “Accept”> </THEN>
</THEN>
</SLABehavior>
</Agent Behavior>
```

In this example, the client has allowed the agent to negotiate in his name. He has classified the bandwidth negotiation as having the highest priority, then the delay and finally the jitter. The bandwidth has a minimum value to ensure, any delay lower than 130 ms is acceptable and any jitter lower than 5 ms is also acceptable.

7.5.3.2. *SAC agent*

The SAC agent is responsible for the acceptance or the refusal of new SLA requests based on the ISP’s capacity, the ISP’s policy, the existing SLAs (i.e. current allocations and future allocations) and the available resources.

When an SLA subscription agent receives a request from a negotiation agent of a client for a new SLA, it transmits the request to the SLA admission control agent, after initial verification such as the identity of the client and the terms of the commercial policy.

So, the role of the SLA admission control agent is to transform the requested SLS contained in the SLA into offered and existing services for the ISP, and to verify if it is possible to satisfy these parameters or not according to the available resources at the time of service deployment.

If some access points are out of the ISP domain, it then must identify possible routes in order to reach destination as well as ISP domains to pass-through. Based on this list, it asks the INRAM if an agreement has already been established with all ISPs to destination. If this is not the case, SLA inter-domain subscription agents must be created in order to negotiate with the remote ISP. These agents must transport an SLA request corresponding to the type of service to put in place (for example, acceptable access points as well as maximum cost).

SLS parameters must take into account a global delay that is additive for the service, a jitter that is calculated based on RMS and a loss that is calculated on a probability base, as explained in section 7.3.1.

When all the negotiation agents have reached a successful negotiation, the SLA subscription agent can process all the results for a final decision and the transmission of the decision to the client's negotiation agent. If the client accepts the terms of the C2I-SLA, the SAC agent sends the C2I-SLA objects to NRAM and INRAM agents to confirm the subscription.

In this way, the ISP's client is also a client of all the other ISPs that have committed to maintain the I2I-SLA agreements. Together, they are responsible for the client's service and for the initiation of a fault management process for the offered service to the client, in order to identify which ISP is responsible for the SLA's violation.

7.5.4. *Generation of policy rules*

When the SLA is accepted, the SSU agent informs the PG agent of the decision. Then, the PG agent creates corresponding policy rules (as described in PCIM) in the CIM repository. These policies will be used by the PDP in order to deploy expected services within the ISP domain.

The policies are defined by a group of rules described in [STA 99]. Policy rules are created according to the accepted SLA's values. They are expressed in terms of conditions (*PolicyCondition*) and actions (*PolicyAction*). When the condition part is verified, the action part is executed. These policies are represented in CIM by using three main classes (as defined in PCIM):

IF ((VendorPolicyCondition == True) AND
(PolicyTimePeriodCondition == True))
THEN execute (VendorPolicyAction).

These classes can be extended in order to receive the different agreement terms specified in the proposed SLA's information model.

Rule 1: concerning the ISP's commitment for delivery of the service. The parameters are as follows:

– common agreements (CommonAgreements);

– validity period (ValidityTimePeriod);

– topology agreements (TopologyAgreements);

– QoS agreements (QoSAgreements);

– agreements monitoring (MonitoringAgreements);

– cost agreements (CostAgreements).

The rule will then be:

IF ((PolicyTimePeriodCondition == True) AND
(VendorPolicyCondition == True))
THEN execute(VendorPolicyAction1 AND VendorPolicyAction2 AND
VendorPolicyAction3)

or:

PolicyTimePeriodCondition = (PolicyTimePeriod in ValidityTimePeriod)
VendorPolicyCondition = (CustomerID have a SLA based in CommonAgreements)
VendorPolicyAction1 = (provide AgreedSAPs with AgreedService
how committed in TopologyAgreements and QoSAgreements)
VendorPolicyAction2 = (Start Monitoring according to MonitoringAgreements)
VendorPolicyAction3 = (Start Accounting according to CostAgreements)

VendorPolicyAction1 will be transformed by PDP into a precise action object according to basic technology. For example, in the case of DiffServ, action classes will be transformed into DiffServ configuration actions based on the QoSAgreements.

VendorPolicyAction2 can be a simple packet accounting action to the interface in case of fixed cost or of complex action in the case of multiple invoicing schema values (for example, session period, quantity of data exchange, etc.).

Rule 2: regarding the client's commitment to send traffic of a certain profile. The rule that affects the client's traffic filter will be defined as follows:

```
IF ((PolicyTimePeriodCondition == True) AND
(VendorPolicyCondition == True))
THEN execute (VendorPolicyAction3)
```

or:

```
PolicyTimePeriodCondition = (PolicyTimePeriod in ValidityTimePeriod)
VendorPolicyCondition = (Customer1InputTraffic >
Customer1AgreedInputTrafic in LoadDescriptor)
VendorPolicyAction = (Action in ExcessTraffic in LoadDescriptor)
```

To simplify matters, we are only describing the simple rules.

7.5.4.1. *NRAM*

The NRAM agent is responsible for updating resource usage in time table.

In fact, when an ISP accepts an SLA, it must activate the corresponding service in a permanent way or for a specific day/time/period.

We have then defined an agent responsible for the maintenance of this model in order to control the acceptance or refusal of new reservation requests according to available resources.

In this chapter, we present a simple approach for a resource control model using tables. We presume that the ISP has installed a series of services (presented here as Gold, Silver and Bronze).

Each service has a specific bandwidth quantity assigned initially.

When an SLA is accepted with a client or a remote ISP, the corresponding service's available resource may decrease or increase according to the SLA.

The goal of the different tables is to show a forecast model taking into account services that are not permanently activated. A more effective approach would have been to use a symbolic approach. However, this type of approach is harder to define.

In this allocation table, we presume that the minimum reservation time is one hour. Each time an SLA has been accepted to provide a service for a certain day/time/period, the corresponding cell in the table is marked.

Thus, we have specified a table with services between ISP pairs in the ISP domain. This table defines for each day of the year the resource allocation program for each agreed SLA (we presume that the ISP will allow the reservation for a whole year) (see Figure 7.8).

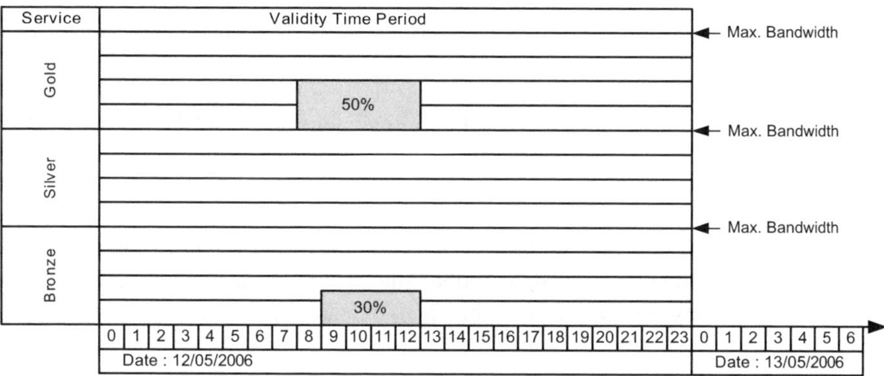

Figure 7.8. *Services timetable*

In Figure 7.8, we have presented the program for a Gold level service dated 05/12/2002 starting at 8:00 AM and ending at 12:00 PM with a reserved bandwidth of 0.5 Mb/s (50% of 1 Mb/s) between SAP1 and SAP2.

Figure 7.8 shows a reservation of 30% of the bandwidth allocated to the Bronze level service between SAP1 and SAP2. The ISP can accept any future reservation request as long as the bandwidth allocation is not completely used by the service.

The same process applies to the inter-domain resource model and its corresponding agents.

7.6. Conclusion

In this chapter, we have presented a frame that enables the deployment of end-to-end IP services with negotiated SLAs. A group of ISPs cooperate in order to provide their customers with end-to-end value added services. This is based on the presumption that the next ISPs will deploy PBM to control networks and services.

The proposition must reinforce this PBM system by incorporating it into a mobile agent system in order for them to interact automatically.

In order to provide this end-to-end SLA, which is mandatory for the deployment of worldwide services, the client ISPs negotiate each SLA with remote ISPs for their clients. This automatic interaction in executed by a group of mechanisms to avoid ambiguity with automatic negotiation.

In this context, we have addressed the problem of end-to-end service assurance in the context where services will go through a certain number of ISP networks. We have presented multiple aspects of negotiation between a client and an ISP and between two ISPs.

This multiagent architecture facilitates interactions in a flexible and dynamic way. The multiagent system is established over a management system, based on policies, which controls the behavior of an ISP network.

The client indicates the requested QoS from his ISP in terms of C2I-SLA. Among other information, the SLA defines the maximum acceptable cost that the client is ready to pay for the service. The client delegates to an agent the negotiation process with the ISP, which will decide if it will accept or refuse the request. Different interaction strategies between the client's agent and the ISP have been defined according to the level of delegation that the client has provided to his agent.

7.7. Bibliography

[BAU 01] BAUER B., MULLER J., ODELL J., "Agent UML: A Formalism for Specifying Multi-Agent Interaction", in *proceedings of the 1st Int. Workshop on Agent-Oriented Software Engineering, AOSE'00*, p. 91-104, 2001.

[BLA 98] BLAKE S., BLACK D., CARLSON M., DAVIES E., WANG Z., WEISS W., "An Architecture for Differentiated Services", *IETF RFC 2475*, December 1998.

[CEL 00] CELENTI E., RAJAN R., DUTTA S., "Service Level Specification for Inter-domain QoS Negotiation", http://www.ist-tequila.org/standards/draft-somefolks-sls-00.txt, November 2000.

[CHI 00] CHIENG D., MARSHALL A., HO I., PARR G., "A Mobile Agent Brokering Environment for the Future Open Network Marketplace", *7th International Conference On Intelligence in Services and Networks (IS&N2000)*, Springer-Verlag LNCS Series, vol. 1774, p. 3-15, February 2000.

[MAR 00] MARSHALL A., "Dynamic Network Adaptation Techniques for Improving Service Quality", *Networking 2000*, Springer Verlag, May 2000.

[SMI 01] SMIRNOV M. *et al.* "SLA Networks in Premium I", CADENUS-WP1, www.cadenus.org/deliverables/d11-final.pdf, March 2001.

[STA 99] STARDUST, *Introduction to QoS Policies*, White Paper, September 1999.

[TMF 01] "SLA Management Handbook", *TeleManagement Forum 2001* – GB 917, June 2001.

[VEN 01] VENTRE G. *et al.*, "QoS Control in SLA Networks", *CADENUS-WP*, www.cadenus.org/deliverables/d21-final.pdf, March 2001.

[VER 99] VERMA D., *Supporting Service Level Agreements on IP Networks*, Macmillan Technical Publishing, 1999.

[WAN 96] WANG Z., CROWCROFT J., "Quality of Service Routing for Supporting Multimedia Application", *IEEE JSAC*, September 1996.

[WES 01] WESTERINEN A., SCHNIZLEIN J., STRASSNER J., SCHER-LING M., QUINN B., HERZOG S., HUYNH A., CARLSON M., PERRY J., WALDBUSSER S., "Terminology for Policy-Based Management", *IETF RFC 3198*, November 2001.

Part 2

The Evolution of IP Networks

Chapter 8

An Introduction to the Evolution in the World of IP

8.1. Introduction

The world of IP has been in continuous flux since its beginnings and nothing seems capable of stabilizing it. IP protocol is everywhere, from the terminal to the heart of the most powerful networks. It is as easily installed on static and dynamic networks.

One of the most expected evolutions concerns the transition from IPv4 to IPv6. Solutions for this migration are detailed in Chapter 9. The management and control of the new generation are explained in Chapter 12. The security that characterizes the most important improvements, related to the delays in the IP environment, is examined in Chapter 13. One of the major developments concerns the appearance of fiber optic networks that transport IP packets, requiring a more precise implementation than for a simple fiber optic connection from end to end. These improvements are examined in Chapter 26. Finally, network IP addressing poses numerous problems now that billions of addresses have been assigned. This is the topic of Chapter 10. Finally, new and essential applications in the development of the IP world concern telephony and video over the Internet, which are addressed later on, but other great trends are becoming evident today and we will approach them in this chapter.

Chapter written by Guy PUJOLLE.

We will therefore take a first look at the evolution of IP networks in this chapter and we will detail certain important elements that are not considered in the other chapters.

8.2. Great evolutions

The IP world is a part of the transportation of data in packets between computers and for many years, its reason for being was electronic messaging and file transfers in a world without connection, that is to say in a network in which no path is delineated, the packets following unspecified paths as a result of the routing tables located in the nodes. These nodes that forward the packets are called routers. The router has a shunting function which allows the configuration, with regards to the final destination, of the exit port in the router. A summary schema for this solution in Figure 8.1 describes this technique using the example of a router and its routing table. The addresses indicated in the routing table are IPv4 addresses.

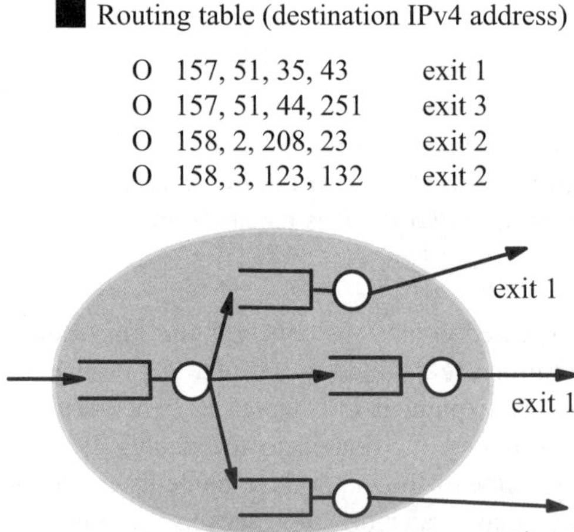

Figure 8.1. *An IP packet delivery network*

The first characteristic in this Internet technology is derived from routing properties: first, two packets coming from the same segment of the same message can follow different paths from the sender to the receiver. This characteristic certainly has advantages and disadvantages. The advantage is the great flexibility that this routing presents, which can change according to the router. We can then easily avoid congestion in the network as well as line interruptions or breakdowns in the nodes. The disadvantage is that the QoS suffers as it is difficult to predict where each packet of a same message will travel. The first main goal is to find ways of maintaining the QoS; we will come back to this.

A second great trend concerns the transition from IPv4 to IPv6. In the past 10 years, this transition has been approached and pushed back as with every attempt to go towards IPv6, IPv4 proves its ability to replace it. The number of available addresses remains the best argument for IPv6. In effect, each dynamic network will be assigned at least one IP address, particularly for new generations of dynamic IP networks, whether they are dedicated to telecommunications norms or wireless norms such as IEEE 802.20. This is also one of the reasons for which IPv6 is currently penetrating more densely populated areas in Asia. We will examine a few elements from the above in Chapter 9, and some mobility elements in this chapter.

Security makes up another important branch in the evolution of the IP world. However, these leading ideas are not entirely clear in the debate between the demand of non-authentication, which enables some liberties crucial for the user, and systematic authentication. Spam and virus problems must also affect significant security requirements. The main solution elements are in Chapter 13. A non-technical solution for these requirements could be the presence of two independent networks, one for the general public with little security and anonymous clients in the network and one for the business world, which would require valid authentication of the clients.

Another debate concerns IP network addresses. Since IP is looking to be the universal network, many applications require a connection with the receiver and therefore the need to establish this contact beforehand to enable the installation of elements necessary for a smooth communication between the entities of the application. The address can also help determine the best path and the scheduling of associated resources to improve the QoS. This vision, for the development of a strong address, is strongly debated but seems to be acquired today with the NSIS (next step in signaling) working group.

Another direction resulting from the operators of large networks concerns the automatic configuration of IP networks without the assistance of specialized personnel. In fact, every time a new client subscribes to the network, we must make room for him in the network and must therefore anticipate the configuration to apply in the nodes he will go through. Different techniques also develop very rapidly in this direction like the policy-based management, which enables a center to configure the network routers. This evolution is one course, but a more long-term vision is defined here: why would not a client have the possibility of choosing a network among those in the surrounding environment and ask for an immediate connection without subscribing? If the network has a way to auto-configure, this solution is indeed possible. It is a possible extension of policy based management techniques but within the framework of outsourcing instead of provisioning as is done today.

We will also address, in this chapter, multicast applications that function with a large number of users who must be reached in parallel to avoid inconsistencies in the system. For example, electronic network games with a large number of players are a part of these environments. We will also address filters, which are essential elements in the world of IP. In order to control the network, we must identify the streams we want to exploit; once these are understood, we can improve the QoS by accelerating certain streams and slowing down others. We can equally block streams that are from potential attackers and we can also determine the costs to be assigned to each client according to the streams and their characteristics.

Large volumes of research are underway to introduce intelligence in the networks and we will finish this chapter by examining the great possibilities currently available.

8.3. Quality of Service

To obtain a good QoS, four main directions can be considered: two using routing techniques and two using switching techniques. Within the framework of routing, since different packets from a same message do not necessarily travel the same route and since it is almost impossible to schedule resources for a client's packets, the only solution is to carry out an oversizing of the network. For this, we only need examine the streams in the network connections and then double the capacity every time a threshold, for example of 60% of the connection's capacity, is exceeded at peak traffic times.

The solution of oversizing was completely reasonable between 2000 and 2005, given the increase in capacities of the lines of communication using WDM (wavelength division multiplexing) techniques in which we multiplex a large number of wavelengths or, a bit less precisely, distinct lights in a fiber optic's core. This solution became quite onerous from 2005 because traffic was increasing rapidly while technological advancements were falling behind, incurring large costs for doubling capacity.

A second solution was put in place in the last few years with routing technology, oversizing, but only for a small portion of the traffic: the traffic from users who wish to have a good QoS and who are ready to pay more for their subscription. To implement this solution, we only need introduce classes of clients and oversize the network according to the class with highest priority. If this class is heavily limited, for example, by the expensive cost of subscription, clients in the highest priority class have the impression of being in an almost empty network and have no delays in the intermediate nodes. Of course, clients in the higher class have a greater priority in the network nodes.

If the stream is particularly important, it is possible to call on gigarouters and terarouters. Once again, this solution requires oversizing or DiffServ-type technologies, in other words, the introduction of three large classes for clients with high priority to clients who pay the most.

The last two solutions are associated with switching technology. In a switching technology, a path is laid out by an address that references an entry and exit for each node in such a way that an entry connection and an entry reference correspond to an exit connection and an exit reference. Switching tables located in the enable the establishment of paths in the network. The advantage of this solution is that we are able to schedule resources along the path determined by these references.

A first solution is determined by the MPLS (multiprotocol label switching) technique, which sets up paths thanks to an IP address. Nodes are LSP (label switch routers) that use the IP routing to effect paths and undo them. Once the paths are in effect, IP packets encapsulated in an ATM-type or Ethernet-type weave are switched. The advantage of this solution is that we can plan and design the traffic because the streams are determined in the network subscriptions and we know how the packets are traveling.

Finally, a fourth solution comes from policy management and control technologies. In this solution, a central server, the PDP (policy decision point), takes charge of the authorizations and the defining of user packet transportation paths. The path assignment is often carried out due to a bandwidth broker. This solution has the advantage of enabling a dynamic configuration. In fact, once the resources have been assigned to the client, the PDP configures the routers by sending the configuration code via the COPS (common open policy service) or the NetConf protocol.

8.4. IP mobility

The IP protocol is more often presented as a possible solution for problems arising from mobile users. Mobile IP protocol can be used in IPv4, but the lack of potential addresses complicates the management of communications with the mobile unit. IPv6 protocol is preferable, due to the large number of available addresses, which makes it possible to assign temporary addresses to stations as they move.

8.4.1. *Mobile IP*

Mobile IP functionality is as follows: a station has a basic address; an agent is assigned to it and its function is to follow the communication between the basic address and the temporary address; in a call to the mobile station, the request is forwarded to the database that contains the basic address; thanks to the agent, it is possible to complete the communication between the basic address and the temporary address and to complete the connection request to the mobile.

This solution is similar to that used in mobile networks, whether it is the European GSM version or the American IS 95 version. The terminology used in mobile IP is as follows:

– *mobile node*: terminal or router that changes its connection point from one sub-network to another sub-network;

– *home agent*: sub-network router on which the mobile node is registered;

– *foreign agent*: sub-network router visited by the mobile node.

The mobile IP environment comprises three relatively disjoined functions:

– *agent discovery*: when the mobile arrives at a sub-network, it searches an agent that is likely to deal with it;

– *registration*: when a mobile is outside its basic domain, it registers its new address (*care-of-address*) with its home agent. Based on the technique used, the registration can be made either directly with the home agent, or by proxy with the foreign agent;

– *tunneling*: when a mobile is outside of its sub-network, packets must be delivered to it via tunneling, which enables the home agent to connect to the *care-of-address*.

Figures 8.2 and 8.3 illustrate mobile IP communication diagrams for IPv4 and IPv6.

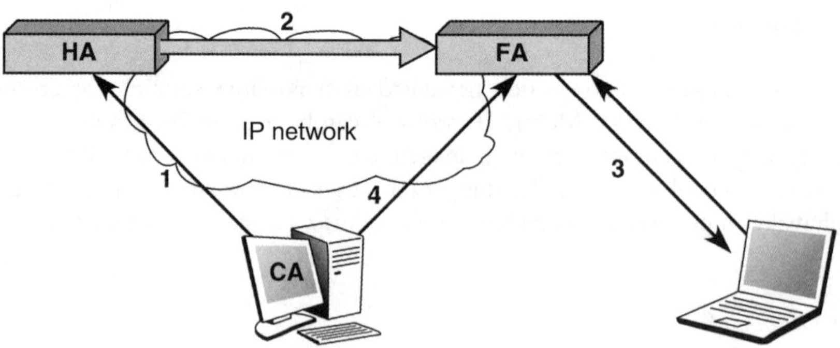

Figure 8.2. *Mobile IP for IPv4*

The management of the mobility indicates the possibility to continue the communication in the best possible conditions, even when a terminal is moved. Typically, mobility management is effected with the help of a home agent, which holds the characteristics of the user, and of a foreign agent who manages the client locally. In the mobile Internet environment, where cells can become miniscule, it is no longer conceivable that the visited agent alert the home agent at every cell change, at the risk of inordinately overloading the network. It is therefore necessary to hide the terminal relocation from the home agent. We will present succinctly four propositions leading to the mastery of these intercellular changes: Cellular IP, Hawaii, TeleMIP and EMA.

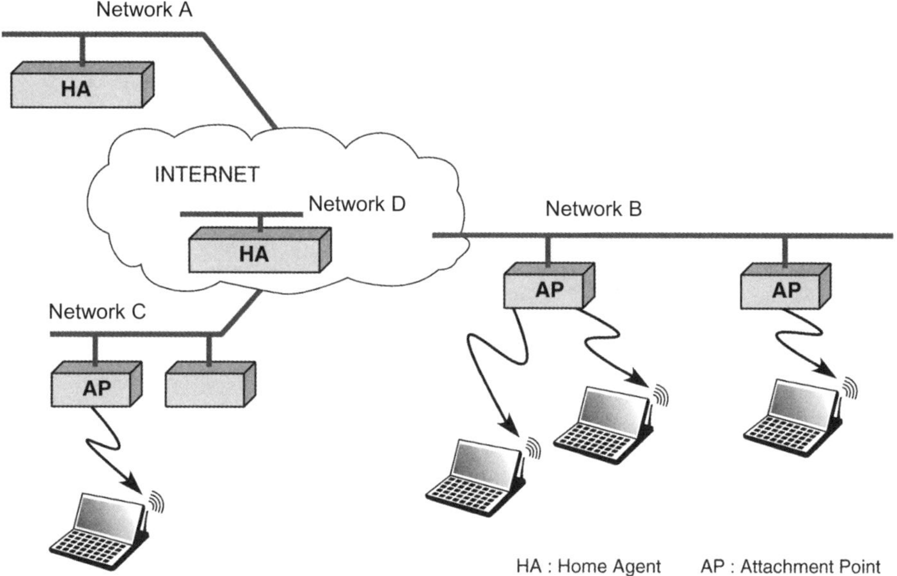

Figure 8.3. *Mobile IP for Ipv6*

To properly reframe the problem, let us recall that mobile IP enables mobility management, but by going through the home agent and that this technique only applies to slow relocations or to user relocations, and not to terminal relocations. A mobile user does not have a terminal, or at least does not move with the terminal, but reconnects on a terminal within the visited location. This is a case of micromobility. In mobile terminal applications, the terminal stays with the user and the application runs on the same terminal. This is a case of macromobility.

8.4.2. *Micromobility protocols*

Many micromobility protocols have seen the light of day in recent years. In the following sections we will present some elements of the main protocols such as Cellular IP and Hawaii.

8.4.2.1. *Cellular IP*

The objective of Cellular IP, which is an IETF workgroup, is to enable the rapid management of intercellular changes or *handovers*. Cellular IP is based on mobile IP when it concerns the management of inter-domain mobility.

Cellular IP architecture is illustrated in Figure 8.4. The interface between the external world and the Cellular IP domain is affected by the means of a gateway.

The visited network is made up of small cells in which the mobile terminal moves. Mobility is managed by the gateway both for entry and for exit. This gateway regularly transmits supervision messages called *beacons*, which enable mobile stations to map out the best route to reach the gateway. The information is memorized in routing caches enabling the stations to know at any moment the paths to the gateway. When the mobile station is inactive, when it receives a call, the gateway transmits a paging message to locate the targeted station. The station wakes up and uses routing caches to determine the path along which it will send its response.

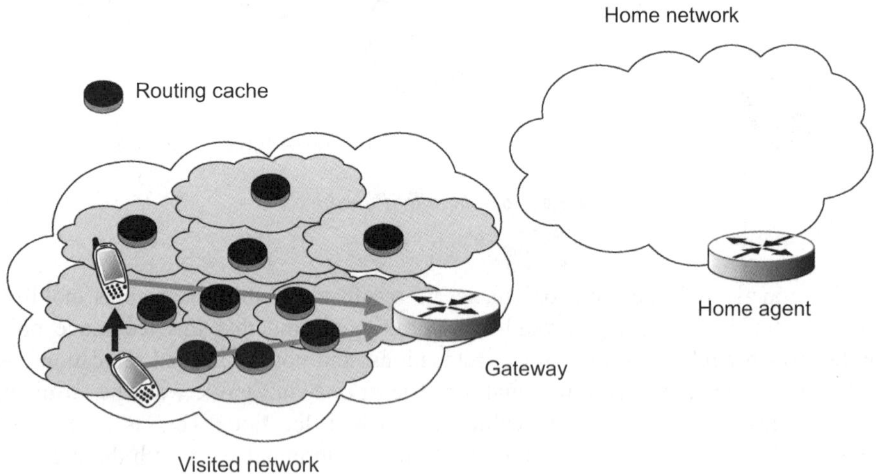

Figure 8.4. *Cellular IP operation*

8.4.2.2. *Hawaii*

The Hawaii technique is quite similar to Cellular IP, the only difference being in the identification of a terminal in an external call. This identification is still effected by a paging message but with the use of an Internet *multicast* address. The whole of the stations belonging to the domain have a multicast address making it possible to carry out a broadcast. The advantage of this system compared to Cellular IP, which automatically floods, lies in the use of selective multicast addresses corresponding, for example, to classes of clients with distinct QoS. Multicast transmission can therefore be more restricted.

Hawaii enables, moreover, the use of RSVP with the multicast address to control the QoS in the domain.

8.4.2.3. *TeleMIP (Telecommunication-enhanced Mobile IP)*

TeleMIP is a third solution that gives a new temporary IP address, called CoA (*care-of-address*) to a mobile station every time it is relocated within a gateway-managed domain.

Each cell comprises a visited agent station. The mobile station is assigned two temporary CoA addresses, one for the domain and the other for the cell. This temporary address change for every cell change enables the mobile station to reacquaint the gateway with the new cell in which the mobile terminal is located. In this way, when a new call comes in, the gateway can communicate with the targeted mobile via the visited agent. This strategy is illustrated in Figure 8.5.

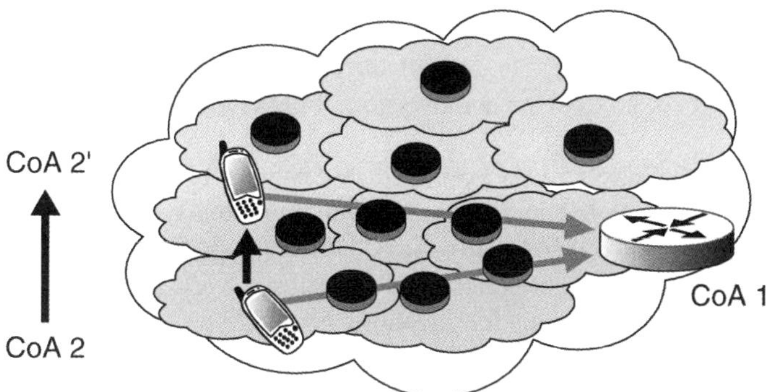

Figure 8.5. *TeleMIP Operation*

8.4.2.4. *EMA (Edge Mobility Architecture)*

Available industry-wide in 2005, EMA technology consists of identifying the geographical location of the mobile terminal via an ad hoc network. When a call comes in, the whole of the cells that rely on the gateway act as if they were a group of ad hoc stations. Many algorithms have been tested for the routing in this ad hoc network.

8.4.3. *Video and voice transmissions in mobile environments*

In the first mobile Internet networks, voice and video transmissions had to make use of very low-flow lines of communication. With the strength increase in the flow of these networks, the problems encountered are very similar to those in ground networks.

The progressive contribution of GSM, GPRS and UMTS generations concerns compression. This compression must succeed in considerably lowering the bandwidth of applications. The solutions consist in degrading quality, whether for telephony or for video transmissions. The lowest possible bandwidth to maintain an acceptable quality is 9.6 kbps for a telephone call, with a possible degradation to 4.8 kbps. In this case, the voice is altered and of a much lesser quality than the proposed traditional quality of switched telephone networks.

Video streaming, or constant streaming with temporary constraints, will not be as easily integrated due to the volume of data necessary for the dynamics of the image. Video in non-real-time is certainly simpler than real-time streaming applications that require a circuit type mode of data transfer. The real development of video on a GSM terminal is not foreseeable until the arrival of either the availability of higher capacity networks or an extreme compression of the video stream.

Video and telephony will be examined in Chapters 17 and 18 respectively. We can however point out the increase in power with this type of application even though the world of IP was not ready for it. To resolve particular constraints such as working through firewalls or giant NAT machines, various proprietary products have appeared, the most famous for telephony being Skype.

8.5. IP multicast

Internet authorities have invested a lot in multicast applications that simultaneously address numerous targets and many possibilities are currently offered in this field.

A multicast application is illustrated in Figure 8.6. PC 1 must send a multicast message to PCs 2, 3, 4, 5 and 6. For this purpose, PC 1 sends a message to the first network router to which it is connected. The router makes two copies and sends them to two routers, etc. This solution greatly reduces traffic in the case where we would have to send, from terminal 1 to the network, five copies destined to five multicast terminals.

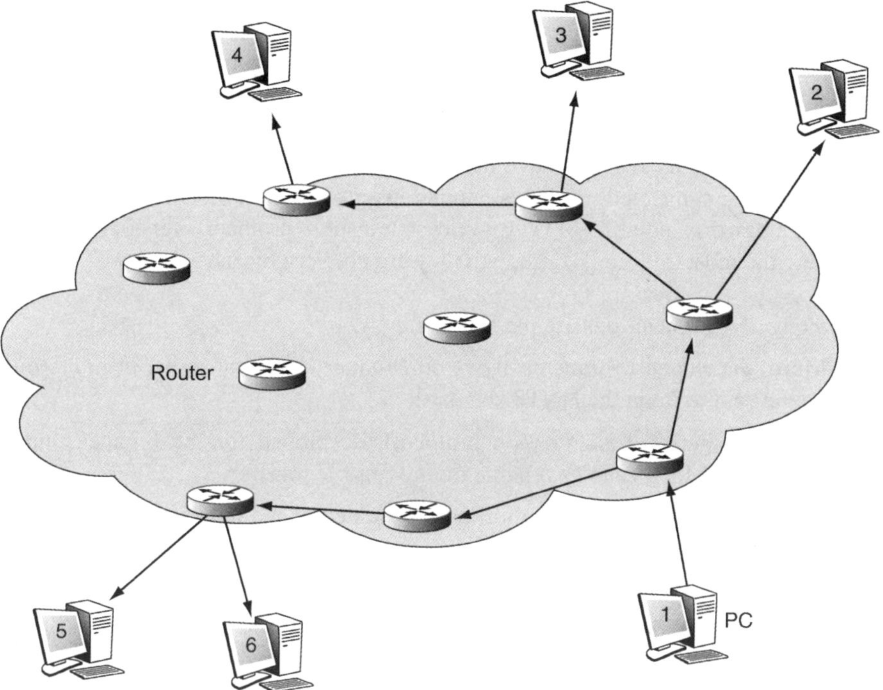

Figure 8.6. *Multicast application*

The basic network is called Mbone (multicast backbone). Mbone users can send and receive packets from multipoints applications. Mbone is particularly geared toward interactive teleconference applications.

Integrated with the Internet, Mbone is a network designed to support multipoint applications. In real-time, the service is of best effort type and in multipoint between network hosts. It can be used, among other things, for conferences, seminars and workgroups, for distributed gaming software or for group management or simply for the broadcasting of audio programs. A directory of running sessions is maintained to enable potential users to join the multipoint. The difficulty lies in the management of applications, which must keep track of the stream of participants and their geographical location.

Mbone was launched in 1992 with 20 modes and routers capable of managing multicast. Mbone router connectivity is assured by tunnels implemented between stations. IP packets exiting these tunnels can be broadcast on the associated local network. Routing between multicast routers is controlled by DVMRP (distance-vector multicast routing protocol). New protocols have appeared to improve multipoint management, in particular PIM (protocol-independent multicast), which manages routing, but also application protocols such as RTP (real-time transport protocol) for voice delivery.

Important security features have been added to Mbone to enable the recognition of users joining connected machines. Numerous multipoint protocols were proposed in this framework, notably MTP (multicast transport protocol) versions 1 and 2, which are the most traditional, although they are not very recent.

Among new propositions are the following:

– Muse, developed to transmit news on Mbone which transmits, in multipoint, information sent to it via the NNTP standard;

– MDP (multicast dissemination protocol), developed for the broadcasting of satellite images, which enables reliable multipoint file transfer;

– AFDP (adaptive file distribution protocol), developed to offer file-sharing via UDP;

– TMTP (tree-based multicast transport protocol), a proposition for reliable multipoint transmission of data found in shared applications;

– RMTP (reliable multicast transport protocol), which, similarly to the preceding protocol, deals with reliable multipoint distribution of blocks of data, but differing in the organization of group participants to minimize data compromise;

– MFTP (multicast file transfer protocol), quite similar to the preceding protocol for reliable transmission of blocks of data via broadcasting, multipoint or unicast;

– STORM (structured-oriented resilient multicast), also proposing a reliable and interactive protocol, and whose newness is in the organization of dynamic group members.

8.6. VPN IP

The new packet (level 3) now being an IP level, VPN level 3 is also called VPN IP. This VPN generation dates back to the beginning of 2000. It makes it possible to gather all the properties found in intranet and extranet networks, notably the computer system in a shared enterprise. The IP solution enables the integration of both fixed and mobile terminals. We will see, initially, the proper IP level VPN and the VPN resulting from MPLS protocols.

8.6.1. *IP level VPN*

An IP level VPN is illustrated in Figure 8.7. Businesses A, B and C are IP level VPN. Their access points are IP routers that enable IP packets to enter into and exit to other branches.

In this figure, clients of the same VPN use the IP network to go from one access point to another belonging to the VPN. Service and security are controlled by the user. Since security is an essential consideration for these networks, the first IP VPN generation used IPSec protocols to put communications in place. This protocol creates encrypted tunnels by proposing to the user a choice of encryption and authentication algorithms. VPN access points communicate with each other via these encrypted tunnels.

A new obstacle for the user is in dynamically configuring the nodes so that the service is within the SLA. One solution that is currently being developed, which we examine a little further, relies on configuration using policies. Before examining the configuration technique, we will introduce other VPN types.

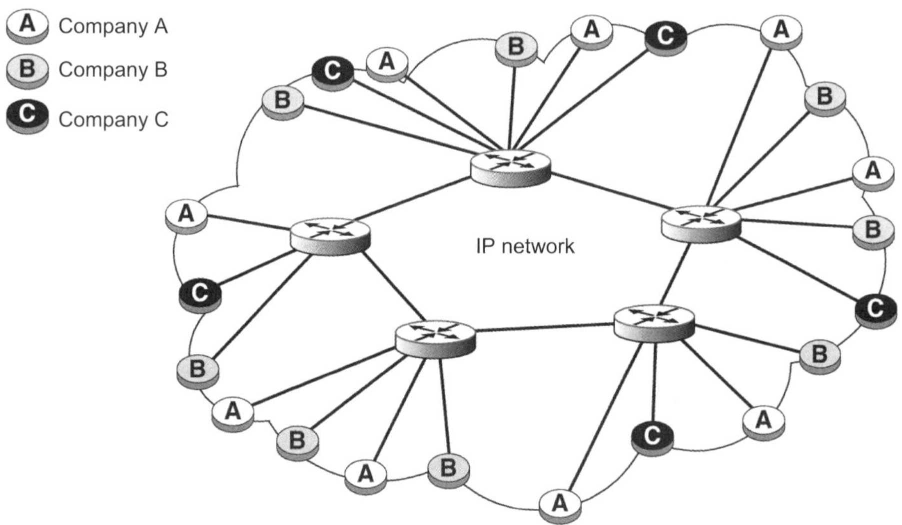

Figure 8.7. *Example of IP VPN*

8.6.2. *MPLS VPN*

A great tendency in VPN at the turn of the century consisted in using MPLS networks. MPLS flexibility authorizes the use of level 2 and level 3 functionalities. This is why we sometimes say these virtual private VPN networks are of level 2.5.

MPLS creates tunnels called LSP (label switched path) that are nothing but virtual circuits. These LSPs are much more flexible, however, than virtual circuits and offer, moreover, a good QoS. Figure 8.8 illustrates the first MPLS VPN model used: the overlay model.

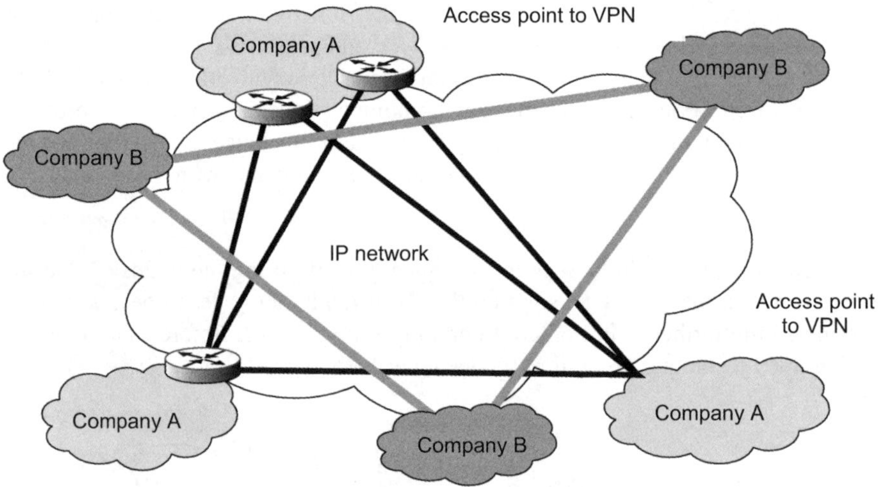

Figure 8.8. *Example of VPN MPLS overlay*

It is similar to corporate VPN in that it entrusts to an outside user a network in a shared infrastructure. In the figure, two businesses share an operator network with different LSPs. One of the businesses has an access point through which both LSPs come in from two different sites for reliability reasons: if one of the access points fails, the second continues to provide service. The fundamental restriction in this model lies in the scalability, which is quite expensive since, if 10 sites already exist and we want to add an 11th, we must add 10 new LSPs.

A second MPLS VPN solution, called peer model, enables scaling by increasing the number of interconnected sites. This model must enable the VPN operator to increase the number of VPN up to several 100,000s of sites if necessary. Another underlying function of the VPN peer model concerns the possibility for the operator to offer new custom-made services to the user and to control his network infrastructure, especially if the user does not have a strong knowledge of Internet routing.

This solution combines features from several technologies:

– routing information is partially distributed;

– routing tables are multiples.

The addresses used are of IP VPN type.

Figure 8.9 illustrates a peer VPN model. The essential difference between this and the overlay model is in the router's MPLS access. In the overlay model, the MPLS access point is in the user's network. In the peer VPN model, the access router is with the operator such that a data stream from a VPN user can move at will towards any VPN point by using the MPLS network.

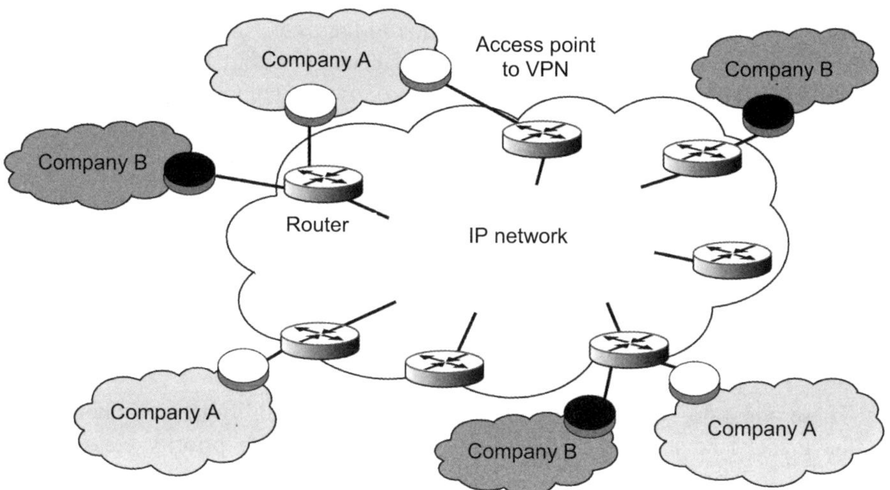

Figure 8.9. *Example of MPLS VPN Peer*

The first routing algorithm for packet routing must ensure connectivity between sites. This routing can only be defined for company A, since company B can have a different routing protocol or can work with a different routing algorithm. The most common routing algorithm in MPLS VPN uses the BGP protocol (border gateway protocol).

The following four steps define the distribution of routing elements:

– routing information comes from the client and is sent to the operator's access point. RIP (routing information protocol), OSPF (open shortest path first) and BGP protocols can be used by the VPN client;

– at the operator's point of entry, routing information is exported into the operator's BGP algorithm;

– at the point of exit, BGP information is compiled for transmission to the user;

– also at the point of exit, routing information is exported to the user in a form specific to that user.

This solution makes it possible to transport information from one VPN point to any other VPN point via an MPLS VPN BGP. Thanks to this mechanism, specific routing tables can be put into place for each VPN.

An addressing problem exists due to the fact that BGP routing used by an operator assumes that the IP address is unique to the user. However, this is not true because a client can use several VPN simultaneously within the same operator's network. One solution to this problem consists of creating unique addresses for each VPN by creating new IP VPN addresses. These addresses are created by concatenation of a fixed-length field called the route distinguisher and of a standard IP address. This field creates a unique global address for the user from the point of view of the operator. To ensure that this global address is unique, the route distinguisher contains three fields: the type, SUR 2 bytes, the number of the autonomous system, SUR 2 bytes as well, and a number determined by the VPN operator, SUR 4 bytes. For the BGP algorithm there is no difference between a single IP address and the IP VPN address.

These operators' MPLS BGP VPNs offer a great flexibility to their clients. One of the greatest advantages is that they can use LSPs with priority classes. This enables a user to work with classes of service corresponding to these priorities. These VPN provide the user with specific functionalities from the configuration to the treatment of various applications. The advantage for the business is the ability to outsource numerous services to the VPN operator.

8.7. Filtering

Filters are omnipresent in the world of IP. In fact, nothing can be controlled without having a precise knowledge of the transiting streams. It is therefore imperative to recognize these streams whatever their encapsulation in tunnels is. It is, however, obvious that if the stream is encrypted, its recognition is no longer possible. Therefore we only address unencrypted streams here.

Filters are essentially applied on port numbers. Port numbers are associated with each type of application and the IP address plus the port number form the socket number. Managing these port numbers, however, is not so simple. In fact, more and more ports are dynamic. With these ports, the sender transmits a request on the standard port, but the receiver chooses a new port available for communication. For example, the RPC (remote procedure call) application dynamically affects port numbers. Most P2P (peer-to-peer) applications or telephony signals are equally dynamic. A hacker will use a traditional port number while using a non-standard stream. In conclusion, one of the more common research topics now is finding a way to determine streams without using port numbers.

Dynamic port assignment is controlled with an astute firewall. Communication can then be traced and it is possible to find the new port number upon the return of the TCP message request transmission. Upon the arrival of the response indicating the new port, it is necessary to detect the number of the port that is replacing the standard port. An even more complex case is possible, in which the transmitter and the receiver are in agreement over a port number. In this case, the firewall cannot detect the communication unless all ports are blocked. This is the essential reason for which firewalls do not accept pre-determined communications.

This filtering and recognition of dynamic ports is not sufficient, however, as it is always possible for a hacker to upload his own data within a standard application on an open port. For example, a tunnel can be set-up on port 80 that manages the http protocol. Within the http application, a stream of packets from another application can pass through. The firewall sees an http application that, in reality, is delivering packets from another application.

A business cannot block all ports without restricting its own applications. We can certainly try to add other methods of detection, such as ownership of known IP address groups, in other words pre-defined IP addresses. Once again, borrowing a known address is easy enough. Moreover, the most damaging attacks are made in ports that are impossible to block, such as the DNS port. One of the most damaging attacks is done by a tunnel in the DNS port. Yet again, the company's network machine that manages DNS must have weaknesses so that the tunnel can be completed and the invading application can make its presence known within the company. We will see in the next section how the firewall's security can be reinforced.

To secure access to a company's network, a more powerful solution consists of filtering not only at either level 3 or 4 (IP address or port address) but also at the application level. This is called an application filter. The idea is to recognize the identity of the application within the packet stream rather than to rely on port numbers. This solution enables the identification of an application inserted into another and to recognize applications on non-conforming ports. The difficulty with this type of filter lies in updating the filters every time a new application appears. A firewall equipped with such an application filter can, however, restrict any unrecognized applications, thereby maintaining a certain level of security.

8.8. Intelligent IP networks

A great future lies in wait for networks that are intelligent enough to provide a variety of services, the most important of which is to maintain a level of quality and of security. These intelligent networks are able to predict requirements and adapt to them, or at least adapt when a problem arises. Training is to be developed to bring network equipment to a level of real intelligence.

The design of a network capable of intelligently managing itself consists of interconnected routers controlled by a special node that can be called, in this case, a GDP (goal decision point) which we will describe later on. It should be noted that we do not use the term intelligent network to designate telecommunication architectures adaptable to providing a service without any specific intelligence.

For example, clients can start by negotiating their SLA via a web interface that considers all the requirements as well as the operator's constraints. The synthesis of the requirements and the constraints is affected by a GDP. The objective of this equipment is to determine the network's goal, which can change with time. For example, this goal can be to satisfy as best as possible all the users' requirements or, on the contrary, to optimize the global flow of the network to meet the operator's requirements. Once the general goal is identified, it is broadcast within all network equipment, which can include routers, switches, LSRs and also firewalls or other hardware. Each of these network parts is called GEPs (goal enforcement points).

The goal is sent to the GEPs and more particularly to a software application agent or intelligent agent. The whole of the intelligent agents constitutes a multiagent system. This multiagent system is responsible for determining the configuration to be applied to the various nodes based on behavior. These behaviors enable the automatic configuration and dimensioning of network equipment parameters.

Two types of feedback can be carried out in such a set-up. Real-time feedback comes from local hardware such as a router and influences the local intelligent agent. The local intelligent agent can modify the value of the node's parameters. Feedback on the GDP goals is equally possible. New clients or a change in a client's SLA in a dynamic framework can influence the goal decided by the GDP.

When a communication travels through many networks, the GDPs agree on a common goal and pass this goal on to each network's multiagent. These multiagent systems implement specific configurations in each network. These configurations can be different within these sub-networks while having the same global goal.

The architecture we just described is being commercialized by a small French company: Ginkgo-Networks. It should be noted, finally, that this type of systems could equally become responsible for the automatic configurations for security goals. In the longer term, a broadcasting of the GDP in network nodes is certainly foreseeable in a return to completely shared architectures.

A second extension of this model should be possible with the arrival of a new protocol environment called SCP/SP (smart control protocol/smart protocol) whose objective would be to optimize communications within a sub-network. The GDP could very well choose a protocol environment adapted to its goal. For example, if a sub-network is an ad hoc network or a network of sensors, nothing would stop the network from choosing more energy-efficient protocols. The IP protocol would remain, in this case, the protocol for internetwork connectivity. We could even say that IP would return in this case to its original function of interconnecting heterogeneous networks.

8.9. Conclusion

Without being exhaustive, this chapter has described a number of important points regarding the strategic development of IP networks. These points concern the existence of a stronger dynamic in the network, which enables the IP networks to better adapt to the clients' constraints. In particular, these networks, in the years to come, should see the arrival of auto-control software, that is to say, the implementation of mechanisms capable of automatically configuring network equipment in real-time, in order to enable them to respond dynamically to users' requirements.

8.10. Bibliography

AFIFI H., ZEGHLACHE D., *Applications & Services in Wireless Networks*, Stylus, 2003.

AMMANN P.T., *Managing Dynamic IP Networks*, McGraw-Hill, 2000.

ARMITAGE G., *QoS in IP Networks*, Pearson Education, 2000.

BLACK D.P., *Building Switched Networks: Multilayer Switching, Qos, IP Multicast, Network Policy, and Service Level Agreements*, Addison Wesley, 1999.

BLACK U.D., *Internet Security Protocols: Protecting IP Traffic*, Prentice Hall, 2000.

BLACK U.D., *IP Routing Protocols: RIP, OSPF, BGP, PNNI and Cisco Routing Protocols*, Prentice Hall, 2000.

BLANK A., *TCP/IP Jump Start: Internet Protocol Basics*, Cybex, 2002.

BURKHARDT J. *et al.*, *Pervasive Computing*, Addison Wesley, 2002.

CHAO H.J., GUO X., *QoS Control in High-Speed Networks*, Wiley, 2001.

CHAPPELL L.A., *Guide to TCP/IP*, Tittel, 2003.

COLLINS D., *Carrier Grade Voice Over IP*, McGraw-Hill, 2000.

COMER D.E., *Internetworking with TCP/IP Vol.1: Principles, Protocols, and Architecture*, 4th edition, Prentice Hall, 2000.

COMER D.E., *TCP/IP : architecture, protocoles et applications*, 4th edition, Dunod, 2003.

CYPSER J., *Communication for Cooperating Systems: OSI, SNA, and TCP/IP*, Addison Wesley, 1991.

DAVIDSON J., PETERS L., GRACELY B., *Voice over IP Fundamentals*, Cisco Press, 2000.

DORNAN A., *The Essential Guide to Wireless Communications Applications, From Cellular Systems to WAP and M-Commerce*, Prentice Hall, 2000.

EVANS H., ASHWORTH P., *Getting Started with WAP and WML*, Cybex, 2001.

FARREL A., *The Internet and Its Protocols: A Comparative Approach*, Morgan Kaufmann Publishers, 2004.

FOROUZAN B.A., HICKS T.G., *TCP/IP Protocol Suite*, McGraw-Hill, 2003.

FURHT B., ILYAS M., *Wireless Internet Handbook: Technologies, Standards, and Applications*, Auerbach, 2003.

GOLDING P., *Next Generation Wireless Applications*, John Wiley & Sons, 2004.

GUTHERY S., CRONIN M., *Mobile Application Development with SMS and the SIM Toolkit*, McGraw-Hill, 2003.

HALL E., *Internet Application Protocols: The Definitive Guide*, O'Reilly, 2003.

HARDY W., *QOS: Measurement and Evaluation of Telecommunications: QoS*, Wiley, 2002.

HASSAN M., ATIQUZZAMAN M., *Performance of TCP/IP Over ATM Networks*, Artech House, 2000.

HASSAN M., JAIN R., *High Performance TCP/IP Networking*, Prentice Hall, 2003.

VAN DER HEIJDEN M., TAYLOR M., *Understanding WAP: Wireless Applications, Devices, and Services*, Artech House, 2000.

HELD G., *Data Over Wireless Networks: Bluetooth, WAP, and Wireless LANs*, McGraw-Hill, 2000.

HOLMA H., TOSKALA A., *UMTS: les réseaux mobiles de troisième génération*, Osman Eyrolles Multimédia, 2001.

HOUGLAND D., ZAFAR K., BROWN M., *Essential WAP for Web Professionals*, Prentice Hall, 2001.

HUITEMA C., *Le routage dans l'Internet*, Eyrolles, 1994.

HUITEMA C., *IPv6 the New Internet Protocol*, Prentice Hall, 1998.

HUNT C., *TCP/IP Network Administration*, O'Reilly, 2002.

INSAM E., *TCP/IP Embedded Internet Applications*, Newnes, 2003.

JAMOIS-DESAUTEL D., *Guide des services Wap : GSM et GPRS*, Eyrolles, 2001.

JANCA T.R., *Principles & Applications of Wireless Communications*, Thomson Learning, 2004.

KAARANEN H. *et al.*, *UMTS Networks: Architecture, Mobility and Services*, Wiley, 2001.

KALAKOTA R., ROBINSON M., *M-Business: The Race to Mobility*, McGraw-Hill, 2001.

LAGRANGE X., GODLEWSKI P., TABBANE S., *Réseaux GSM-DCS*, Hermès, 2000.

LIN Y.B., CHLAMTAC I., *Wireless and Mobile Network Architectures*, Wiley, 2000.

LONG C.S., *IP Network Design*, Osborne McGraw-Hill, 2001.

MCDYSAN D., *QoS and Traffic Management in IP and ATM Networks*, McGraw-Hill, 1999.

MAKKI K., PISSINOU N., PARK E.K., *Mobile and Wireless Internet: Protocols, Algorithms, and Systems*, Kluwer Academic Publishers, 2003.

MALHOTRA R., *IP Routing*, O'Reilly, 2002.

MINOLI D. *et al.*, *Internet Architectures*, Wiley, 1999.

MISHRA A., *QoS in Communications Networks*, Wiley, 2002.

MOY J.T., *OSPF: Anatomy of an Internet Routing Protocol*, Addison Wesley, 1998.

REYNDERS D., WRIGHT E., *Practical TCP/IP and Ethernet Networking for Industry*, Newnes, 2003.

STEWART J. W., *BGP4: Inter-Domain Routing in the Internet*, Addison Wesley, 1998.

SANTIFALLER R., *TCP/IP and NFS, Internetworking in a UNIX Environment*, Addison Wesley, 1991.

SINGHAL S. *et al.*, *The Wireless Application Protocol: Writing Applications for the Mobile Internet*, Addison Wesley, 2001.

SOLOMON J.D., *Mobile IP the Internet Unplugged*, Prentice Hall, 1998.

THOMAS S.A., *IP Switching and Routing Essentials: Understanding RIP, OSPF, BGP, MPLS, CR-LDP, and RSVP-TE*, Wiley, 2000.

TOMSU P., SCHMUTZER C., *Next Generation Optical Networks: The Convergence of IP Intelligence and Optical Technologies*, Prentice Hall, 2001.

TONG-TONG J.R., *NFS: système de fichiers distribués sous Unix*, Eyrolles, 1993.

VEGESNA S., *IP QoS*, Cisco Press, 2001.

WANG Z., *Internet QoS: Architectures and Mechanisms for QoS*, Morgan Kaufmann Publishers, 2001.

WILLIAMSON B., *Developing IP Multicast Networks: The Definitive Guide to Designing and Deploying CISCO IP Multi-cast*, Cisco Press, 2000.

Chapter 9

IPv6, the New Internet Generation

9.1. Introduction

The IPv4 protocol, which is at the core of the Internet, was conceived in the beginning of the 1970s in order to cater for the need to share data between different workstations. The democratization of the Internet and its success have made IPv4 the protocol of convergence for all network designs and for current and future applications. However, IPv4 was not conceived to meet the needs of millions and millions of users in terms of services such as mobility, multicast, quality of service and flexible and easy management.

The IETF has been researching and designing solutions, such as the introduction of NAT to address the lack of addresses, of MobileIPv4 for mobility, of DiffServ or RSVP to introduce quality of service. But all these solutions, proposed independently from each other, do not address the global needs. Therefore, IETF has decided to profit from their experience in the deploying of IP networks and in problem solving, to create a new IP protocol. This protocol, IPv6, has the goal of improving IPv4 in terms of scaling, security, ease of configuration, mobility, and lack of addresses.

In this chapter, we will first discuss the IPv6 protocol by detailing the header format, the different types of IPv6 addresses as well as their configuration. Then, we will elaborate on the mechanisms that enable the transition from the current IPv4 protocol to the IPv6 protocol.

Chapter written by Idir FODIL.

9.2. IPv6 characteristics

Large addressing space: IPv6 disposes of a 128-bit address instead of IPv4's 16 bits. Moreover, the large addressing space in IPv6 was designed to enable several hierarchies (levels) in allocating addresses from the core of the Internet to the individual sub-networks of a company. With more available addresses, address conservation techniques such as NAT no longer have a purpose.

New header format: the new IPv6 format was conceived to have minimal overheads. This is achieved by moving all the fields and unnecessary options into options put after the IPv6 header. This enables greater efficiency and better router performance. IPv6 and IPv4 headers are not compatible, so a terminal attempting to use both protocols will need an implementation that supports both protocols.

Hierarchical and efficient addressing and routing: IPv6 global addresses were conceived to create an efficient and hierarchical routing.

Automatic auto-configuration: in the interest of simplifying terminal configuration, IPv6 offers two configuration mechanisms as follows:

– DHCP server configuration (stateful);

– automatic configuration without DHCP server (stateless): terminals configure their global IPv6 address with the help of routers that broadcast the network prefix to the network.

Security: IPSec support is a requirement in the IPv6 standard. This enables a standard and interoperable network security.

Mobility: mobility was taken into consideration in the design of IPv6 and enabled the creation of MobileIPv6 mobility mechanisms, which are more efficient than MobileIPv4.

Multicast: like mobility, multicast was considered in the design of IPv6 and is mandatory for all implementations.

Protocol for interaction between nodes: NDP (neighbor discovery protocol) based on ICMPV6 replaces ARP, ICMPv4 Router discover and ICMPv4 Redirect.

Extendibility: new functionalities can easily be added to IPv6 due to the extension headers that follow the IPv6 header.

9.3. IPv6 packet header

Ver = 6	Traffic class	Flow label	
Payload length		Next header	Hop limit
Source address			
Destination address			

Figure 9.1. *Presentation of the IPv6 packet header*

Ver (4 bits): version of protocol IP = 6.

Traffic class (8 bits): class identifier, used by DiffServ.

Flow label (20 bits): used to identify traffic. They are not currently used.

Payload length (16 bits): length of the useful information of the IPv6 packet.

Next header (8 bits): identifies the type of header that immediately follows the IPv6 header. It is equivalent to the protocol field in IPv4.

Hop limit (8 bits): equivalent to the TTL field in IPv4. It is decremented by 1 for each node that transfers the packet. The packet is discarded if the hop limit is equal to zero.

Source address (128 bits): IPv6 address of the packet sender.

Destination address (128 bits): IPv6 address of the packet receiver.

9.3.1. *Extension headers*

Extension headers are options separately placed between the IPv6 header and the transport protocol header (TCP/UDP). Most extensions are only processed by the IPv6 packet receiver, thus relieving the routers of useless processing. Extension headers are not of fixed length, which enables the support of diverse functionalities. Currently defined extensions are:

– *hop by hop*: extensions requiring packet processing at each hop;

– *routing* (like *loose source* in IPv4): this routing is mainly used by MobileIPv6;

– *fragmentation*: fragmentation and reassembly;

– *authentication*: IPSec integrity and authentication;

– *encapsulation*: IPSec confidentiality;

– *destination option*: optional information examined by the receiver.

9.4. IPv6 addressing

9.4.1. *Address format*

An IPv6 address is represented by 8 hexadecimal numbers of 16 bits each, as follows:

3FFE:0000:0000:0000:0003:F8FF:FE21:67CF

In the interest of making the writing of IPv6 addresses easier, compressions can be used as follows: ":3: is equivalent to 0003":

3FFE:3:F8FF:FE21:67CF

is equivalent to:

3FFE:0000:0000:0000:0003:F8FF:FE21:67CF

An IPv6 address prefix is represented as follows: *IPv6 address/prefix length*. The length of the prefix is a decimal value specifying the number of bits that make up the prefix (example: 3FFE:1234::/64).

9.4.2. *Address types*

Unicast address: identifies a single interface.

Anycast address: identifies a group of interfaces. An IPv6 packet sent to an anycast address will be delivered to a single member of this group.

Multicast address: identifies a group of interfaces. An IPv6 packet sent to a multicast address will be delivered to all the members of the group.

Broadcast addresses no longer exist in IPv6 and their functions have been replaced by multicast addresses.

9.4.2.1. *Unicast addresses*

There are three types of IPv6 unicast addresses: 1) the global address, 2) the site address, and 3) the link address and the compatible IPv4 address (IPv6 addresses containing IPv4 addresses).

Global unicast address: address used for global communication. Its format is as follows:

001 3 bits	Global routing prefix 45 bits	Sub-network ID 16 bits	interface ID 64 bits

The global routing prefix is routable over the Internet. This kind of prefix is provided by each Internet service provider to its clients. The sub-network ID is the identifier of a link in a site, thus enabling the identification of sub-networks and having a local addressing hierarchy. The interface identifier is used to identify interfaces on a link. It is necessary for it to be unique on a link. In certain cases it can be derived from the addresses of the link layers.

Local unicast address: these are addresses that can be routed only in a sub-network or in a client network. Their goal is to be used in a site for local communication and for creating global addresses. There are two types of local addresses: link and site. The local link address is used on the same link and the local site address on the same site. The local link address has the following format: FE80::Interface ID.

1111111010 10 bits	0 50 bits	Interface ID 64 bits

The local site address has the following format: FEC0:0:0:SLA_ID: interface ID.

1111111011 10 bits	0 38 bits	Sub-network ID 16 bits	interface ID 64 bits

IPv6 addresses containing IPv4 addresses: the goal of these addresses is to enable IPv6 supporting equipment to communicate through an IPv4 infrastructure. The machines assign a unicast IPv6 address that contains their IPv4 address. These types of addresses are called IPv6 compatible IPv4 and have the following format:

0:0:0:0:0 80 bits	Compatibility bits 0000 (16 bits)	IPv4 Address 32 bits

Another type of IPv6 address containing IPv4 addresses has also been defined. It enables the representation of an IPv4 address as an IPv6 address. These address types are called IPv4 mapped IPv6 and have the following format:

0:0:0:0:0 80 bits	Compatibility bits FFFF (16 bits)	IPv4 Address 32 bits

9.4.2.2. *Anycast addresses*

An anycast IPv6 address is assigned to more than one interface (generally to several nodes). A packet sent to an anycast address is routed to the nearest interface that has this anycast address, according to the routing protocol metrics.

Anycast addresses can be used for different purposes. Currently they are used to represent the group of routers belonging to a particular sub-network. The routers belonging to an Internet service provider all have an anycast address derived from the provider's prefix. These anycast addresses can be used to locate the nearest distributed network resource (for example, DNS). These anycast addresses are allocated outside of the unicast addressing space and use any format defined for unicast addresses.

9.4.2.3. *Multicast addresses*

A multicast IPv6 address enables the identification of a group of interfaces. An interface can belong to an unspecified number of multicast groups. The format of multicast addresses is as follows:

11111111 8 bits	Flags 4 bits	Scope 4 bits	ID group 112 bits

The multicast address scope represents the part of the network in which the multicast is valid. It includes the node (0x1), the link (0x2), the site (0x5), the organization (0x8) and global (0xE). A number of multicast addresses have been pre-defined as follows:

– reserved multicast address: this is reserved and never assigned to a multicast group. These addresses have the following format: FF0x:0:0:0:0:0:0:0;

– node addresses: they identify all the IPv6 nodes in the given scope. These addresses have the following format: FF0y:0:0:0:0:0:0:1, where y = 1 (local node) or 2 (local link);

– router addresses: they identify all the IPv6 routers in the given scope. These addresses have the following format: FF0y:0:0:0:0:0:0:2, where y = 1 (local node) or 2 (local link).

9.4.3. *Address configuration*

There are several types of IPv6 address configuration:

– *manual*: addresses are written in a local configuration folder for each terminal. This method does not require the dedicated configuration protocol implementation but is not adapted for reconfiguration;

– *automatic (stateless)*: an IPv6 address is composed of a prefix and of an interface identifier. Prefixes are generally broadcast by routers on each link, whereas the interface identifier is locally constructed by the machine based on the MAC address or on a random number. From there, each IPv6 node can construct its own IPv6 address;

– *by server (stateful)*: in this method, the nodes obtain configuration information from a server. The servers maintain a database containing machines and their assigned IPv6 addresses. The protocol used for this model is DHCPv6, which is a client/server protocol that enables managing equipment configurations. DHCPv6 is not an extension of DHCPv4 but is a new protocol.

9.5. Transition from IPv4 Internet to IPv6 Internet

The Internet is currently built on IPv4 protocol and the new Internet generation will be built around the new IPv6 protocol. Nonetheless, IPv6 will not replace IPv4 form one day to the next, but will co-exist during several years with IPv4. The IETF has specified several transition mechanisms for IPv6 nodes and routers, which will enable them to integrate into existing IPv4 networks and will facilitate the transition to the new IPv6 Internet.

Transition mechanisms can be classified in three categories:

– double pile: based on the use of a double pile IP supporting IPv4 and IPv6 protocols in routers and terminals;

– encapsulation: based on the encapsulation of IPv6 packets into IPv4 packets;

– translation: based on translating IPv6 packets into IPv4 packets and vice versa.

9.5.1. *Double pile*

The goal of double pile is to enable machines to preserve their traditional access to IPv4 networks and to have access to IPv6 networks and services.

Currently, the majority of routers in company networks are multi-protocol in nature. Many workstations also support a combination of IPv4, IPX, AppleTalk, Net Bios, SNA, DECnet and other protocols. Adding a new protocol (IPv6) on a terminal or a router is a known and controlled issue. By executing a double IPv4/IPv6 a terminal will have access to IPv4 and IPv6 resources. The routers executing the two protocols can transfer traffic to IPv4 and IPv6 nodes.

Double pile enables the support of IPv4 and IPv6 applications and services, however, it suffers a large inconvenience. Each machine on the network must have a public IPv4 address, which means the problem of using up IPv4 addresses will not be solved.

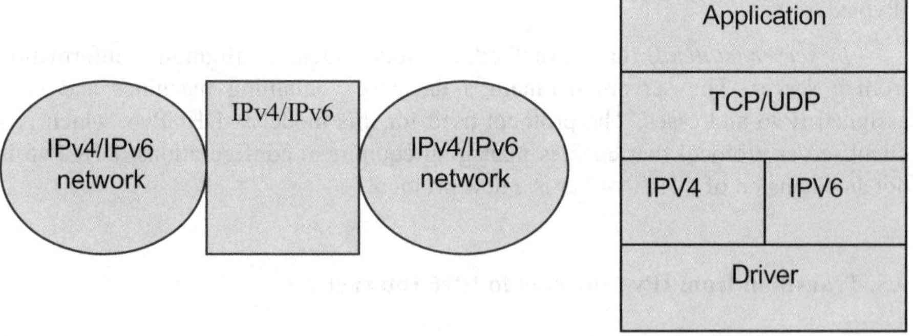

Figure 9.2. *Double pile*

9.5.2. *Encapsulation*

The interconnection of the first IPv6 networks appearing during the first deployments of different parts of the Internet necessitated the use of the IPv4 Internet infrastructure. Encapsulation techniques made it possible to connect IPv6 networks without any native IPv6 connection.

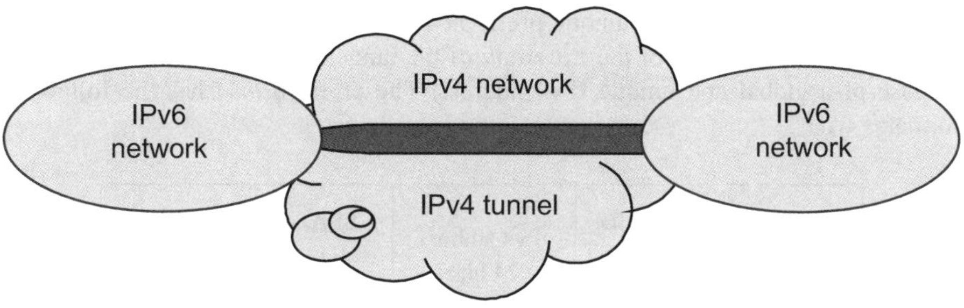

Figure 9.3. *IPv6 data transfer by encapsulation*

Encapsulation mechanisms all consist of the encapsulation of IPv6 packets from the emitting IPv6 network into IPv4 packets, which are then transmitted over the IPv4 Internet then decapsulated at the IPv6 network receiver. Encapsulation techniques can be classified into two categories:

– explicit encapsulation, which requires equipment configuration. In this case configured tunnels are put into place;

– implicit encapsulation, which does not require preliminary configuration. Automatic tunnel techniques can be used, from *6 to 4*, as can the ISATAP technique.

9.5.2.1. *Configured tunnels*

In order to connect each IPv6 network to another network, a tunnel is configured between each pair of routers in each network. These tunnels are manually configured so that each participating router knows the address of the extremities of the tunnels.

Routing protocols process tunnels as a single hop. These configured tunnels are often used with *tunnel broker* mechanisms. These are dedicated servers that automatically manage tunnel requests sent by users.

9.5.2.2. *Automatic tunnels*

Automatic tunnels use compatible IPv4 addresses (for example, "::137.37.17.53"). A compatible address is created by adding zeros to an IPv4 address to get 128 bits. When a packet is sent to a compatible address, the equipment is capable of automatically encapsulating by converting the IPv6-compatible IPv4 address toward an IPv4 address. At the other end of the tunnel, the IPv4 header is removed and the IPv6 packet is extracted.

9.5.2.3. *The "6 to 4"*

The goal of *6 to 4* is to connect IPv6 networks through IPv4 domains without explicit tunnels. In *6 to 4*, a unique prefix is attributed to each site and each prefix contains the IPv4 address of the extremity of the tunnel. Client sites are supposed to dispose of a global and unique IPv4 address. The prefix *6 to 4* has the following format:

FP (3 bits) 001	TLA ID (13 bits) 0x0002	IPv4 address 24 bits	SLA ID 16 bits	Interface ID 64 bits

Prefix format: 001 (3 bits).

TLA (top level aggregator): 0x0002 (13 bits).

SLA (site level aggregation).

The *6 to 4* prefix is broadcast within the *6 to 4* site, thus enabling the equipment to automatically configure their *6 to 4* addresses. IPv6 packets exiting a *6 to 4* site are encapsulated into IPv4 packets. The destination IPv4 address is the address of the extremity of the tunnel, which is a *6 to 4* router. IPv4 packets received are decapsulated and IPv6 packets are transmitted within the network.

The *6 to 4* mechanism is mainly implemented in access routers without modifying the terminals.

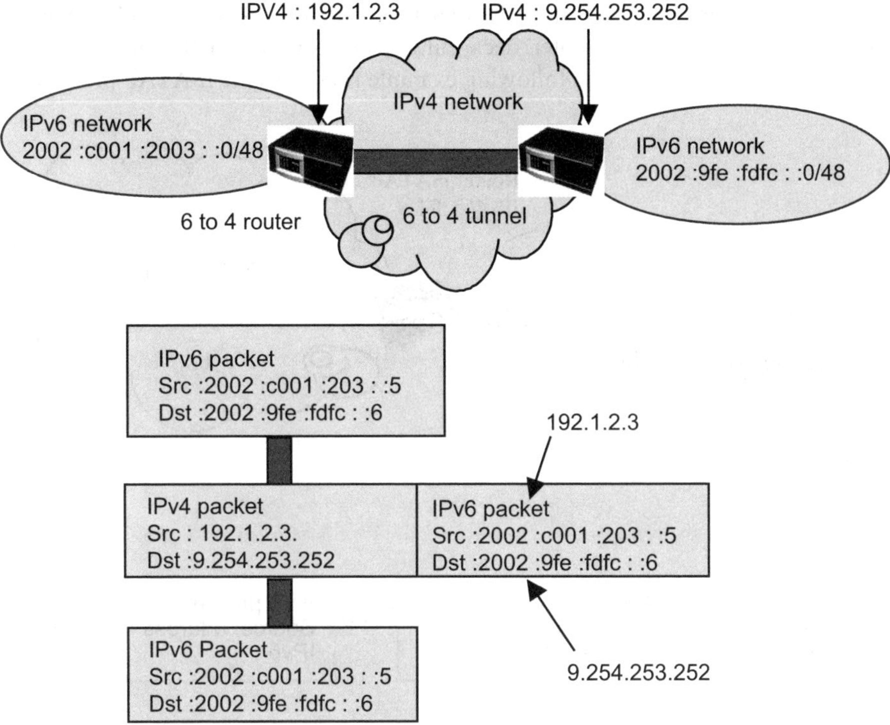

Figure 9.4. *6 to 4 functioning*

One of the major advantages of *6 to 4* is that the client can use IPv6 without recourse to a global prefix (allocated by one ISP or another). Another advantage of *6 to 4* is that it requires less configuring than configured tunnels do. One of the drawbacks of *6 to 4* is that it can result in an asymmetric routing path.

9.5.2.4. *ISATAP and 6 to 4*

The ISATAP technique was defined with the goal of offering an IPv4 connectivity to terminals and IPv4 routers, the objective being to deploy IPv6 while considering IPv4 as a level 2 layer of IPv6. ISATAP uses an address format that includes the IPv4 address in the interface identifier as follows:

Standard IPv6 prefix	0 16 bits	5FFE 16 bits	IPv4 address 24bits

Terminals and routers transmit IPv6 packets via an automatic tunnel within the site (LAN). The ISATAP router decapsulates the packet and transfers the IPv6 packet over Internet IPv6. The following example illustrates the ISATAP process:

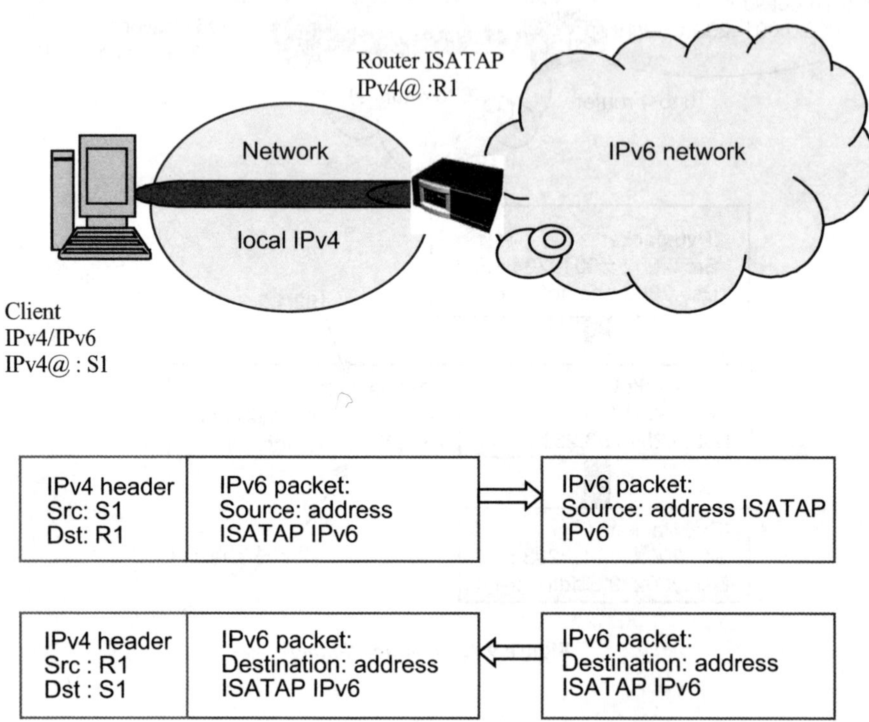

IPv4 header Src: S1 Dst: R1	IPv6 packet: Source: address ISATAP IPv6	⇒	IPv6 packet: Source: address ISATAP IPv6
IPv4 header Src : R1 Dst : S1	IPv6 packet: Destination: address ISATAP IPv6	⇐	IPv6 packet: Destination: address ISATAP IPv6

Figure 9.5. *ISATAP and 6 to 4*

9.5.2.5. *DSTM (dual stack transition mechanism)*

The DSTM mechanism resolves the problem of IPv6 terminals that are in IPv6 networks and need access to IPv4 services.

Based on the use of IPv4 tunnels in IPv6 and of a temporary allocation of a global IPv4 address, the DSTM offers a solution to the cited problem.

By means of the DSTM server, the terminal obtains a temporary IPv4 address that enables it to access the IPv4 services.

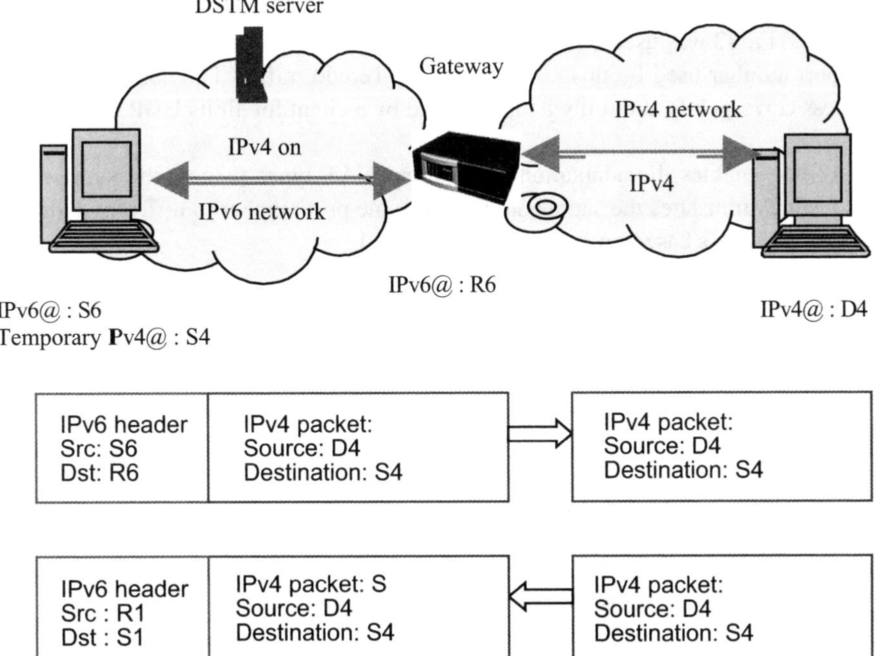

Figure 9.6. *DSTM*

9.5.2.6. *Teredo*

The goal of Teredo is to enable the nodes behind one or more NAT to obtain an IPv6 connection by encapsulating IPv6 packets in UDP protocols. As with *6 to 4*, Teredo is an automatic encapsulation mechanism. It uses the following address format:

Teredo prefix 32 bits	IPv4 address Teredo server	Flags 16 bits	Obscured external port (16 bits)	Obscured external address (32 bits)

The Teredo prefix has not yet been assigned by the IANA, however, the prefix 3FFE:831F::/32 was used for the first deployments. The external port corresponds to the port number used by this client for all its Teredo traffic. The obscured external address corresponds to the IPv4 address used by a client for all its UDP traffic.

Teredo enables the adaptation to different NAT types (except the symmetrical NAT which translates the same address and same port number to different addresses and port numbers based on destination addresses).

Many scenarios exist and a typical scenario is described in this section.

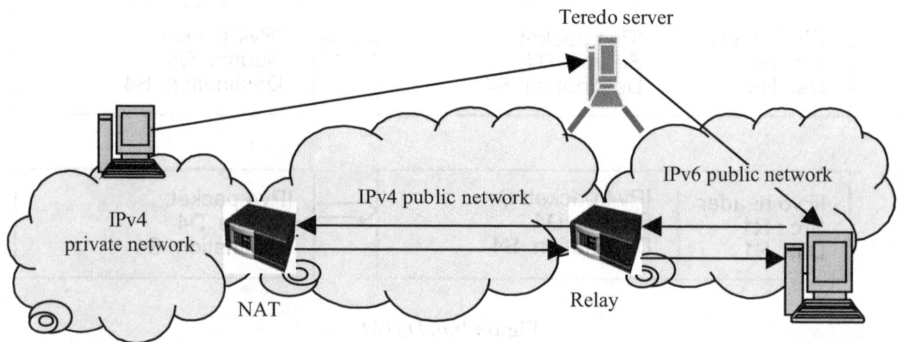

Figure 9.7. *Teredo*

To send a packet from a Teredo client to an IPv6 correspondent, the following process is used:

– the Teredo client must determine the IPv4 address and the UDP port number of the Teredo relay nearest to the correspondent IPv6. It sends an *echo request* ICMPv6 message to the correspondent via its own Teredo server;

– the Teredo server transfers the message to the correspondent;

– the correspondent responds with an *echo reply* ICMPv6 message to the client's Teredo server. Based on the Internet routing infrastructure, the Teredo packet is routed to the nearest relay;

– the relay encapsulates the message and sends it to the Teredo client. Based on the type of NAT used, the message is either directly sent or sent through the client's Teredo server;

– the client determines the correspondent's Teredo relay address and port number. A communication initialization packet is then sent from the Teredo client to the relay. The relay removes the IPv4 and UDP headers and transfers the packet to the correspondent.

Teredo is an efficient mechanism which enables the equipment that is behind NATs to access Internet IPv6. However, Teredo does not work with symmetrical NATs, which represent 10% of the total number of NATs deployed on the Internet.

9.5.3. *NAT-PT translation mechanisms*

Translation mechanisms are very different from encapsulating mechanisms. The former enable the communication between IPv4 nodes and IPv6 nodes, whereas the latter enables the communication between IPv6 nodes via IPv4 infrastructures.

The better-known translation mechanism is NAT-PT. It enables the communication between nodes located in IPv6 domains with nodes located in IPv4 domains. NAT-PT is based on a combination of an NAT (IP address translation) and an IPv4/IPv6 translation protocol. It relies also on the use of DNS. The following scenario details the NAT-PT process.

As with all translation mechanisms, NAT-PT suffers several limitations. The first is in its use of a state maintenance mechanism with regards to NAT-PT equipment, which induces a difficulty in scaling. The second critical point concerns the incompatibility of NAT-PT with IPSec and with applications that require address and port number storing.

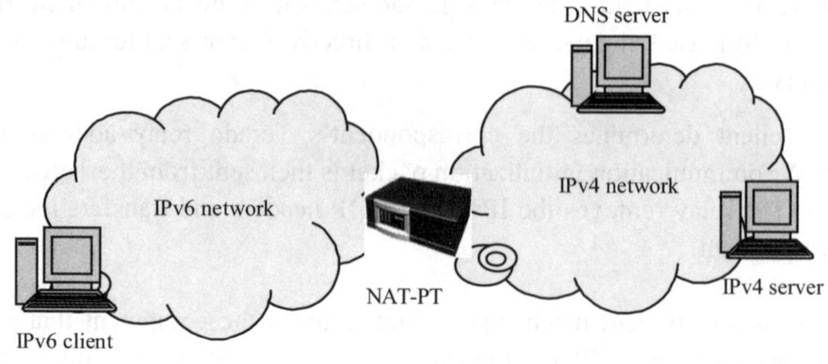

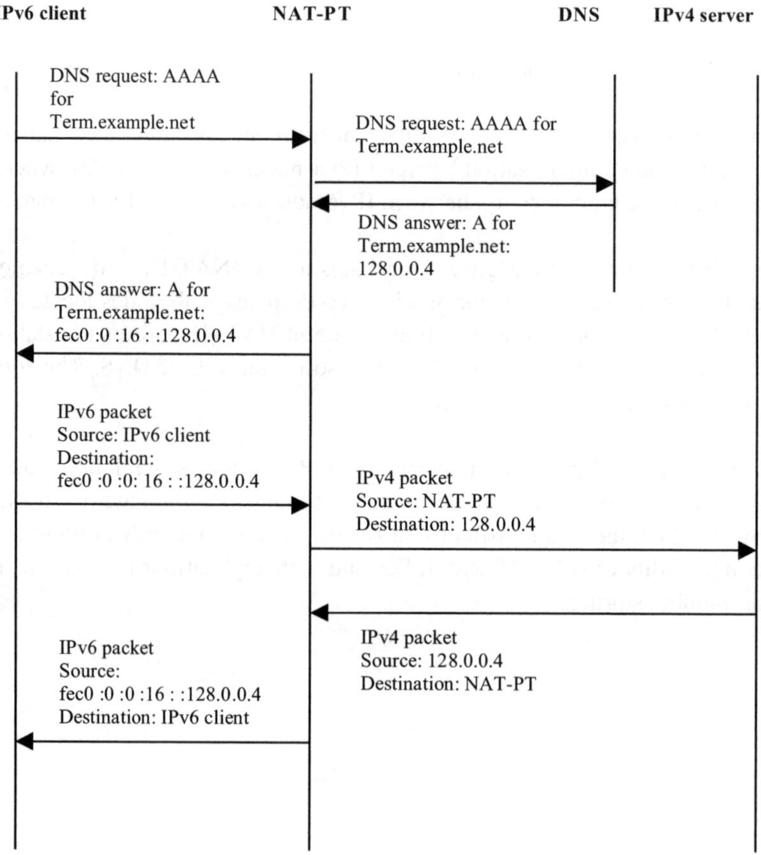

Figure 9.8. *NAT-PT translation*

9.5.4. *Conclusion*

With the goal of answering the user's needs in terms of service, the IETF has conceived the new Internet IPv6 protocol as a replacement of the basic IPv4 protocol. With its large addressing space, its native security support, its quality of service and its mobility, IPv6 enables simple, effective and flexible solutions for the new uses of the Internet. In order to integrate it into the current Internet, transition mechanisms have also been conceived with the goal of enabling a co-existence and an easy migration from IPv4 to IPv6.

All information concerning IPv6 standards can be found on the IETF's IPv6 workgroup page, http://www.ietf.org/html.charters/ipv6-charter.html.

Chapter 10

Addressing in IP Networks

10.1. Introduction

The Internet makes it possible to interconnect different heterogenous networks thanks to a common protocol: the IP protocol. The link layer frame is transmitted from point to point, whereas the IP packet it contains is sent on the network due to a routing technique. Thus, all the network machines are able to exchange data, without involving the link protocol. However, to be reachable, each machine must be identified with at least one address that must be unique on the IP network. The first addressing problem arises: how can IP addresses be efficiently allocated from the whole of the addressing space?

The value of the IP address is all the more important as it determines the routing of packets. An IP address contains location information with regards to the network to which it belongs. This information makes it possible for the router to determine the destination network among all the networks it knows. The addressing scheme of an IP network therefore affects routing and its performances. A second problem then emerges: how can addresses be allocated such that the routing technique is as efficient as possible?

At the beginning of the 1980s, a first IP protocol version was standardized to address these problems, but this standard has not stopped evolving. This chapter will endeavor to retrace the main evolutions in IPv4 network addressing and then those for IPv6, which is the new version of Internet Protocol.

Chapter written by Julien ROTROU and Julien RIDOUX.

10.2. IPv4 addressing

RFC 791 definitively standardizes addressing rules for an IPv4 network interface. No matter what interface it is, it must have a unique 32-bit IP address. The network elements, whose tasks are, for example, to route IP packets have at least as many IP addresses as they have interfaces. This first RFC standardizes IP network addressing by defining several address classes, which are detailed later on.

10.2.1. *Classful IP addressing*

This notion of class is designed to bring flexibility in IP address assignments. Each class has a different number of addresses and is composed of block addresses of different sizes assigned to an organization according to its needs.

10.2.1.1. *Two-level address structure*

This first specification defines a logical splitting of IP addresses into two levels: the network and the host. These two parts are distinguishable due to the analysis of the first high bits of the IP address. The first part, called the network number or the network prefix, identifies the network on which the host resides. The second part, the host number, identifies the host on this network. We therefore have a two-level addressing illustrated by Figure 10.1.

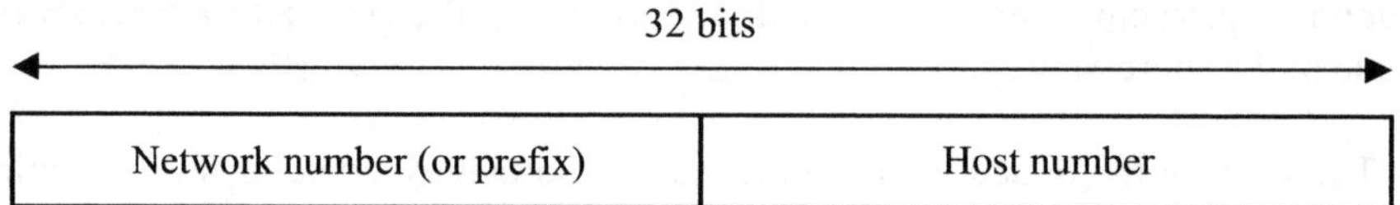

Figure 10.1. *Two-level address structure*

10.2.1.2. *IP address classes*

The number of possible networks in a class and the maximum number of network hosts are defined by the length of each part. Many classes were proposed in order to define different sized networks: classes A, B, C, D and E. The first three are destined to unicast addressing, in other words, addresses that are routable on the entire Internet. Class D is used for multicast addressing and class E is reserved for experimental uses. These last two classes are used only marginally and will not be addressed in this part.

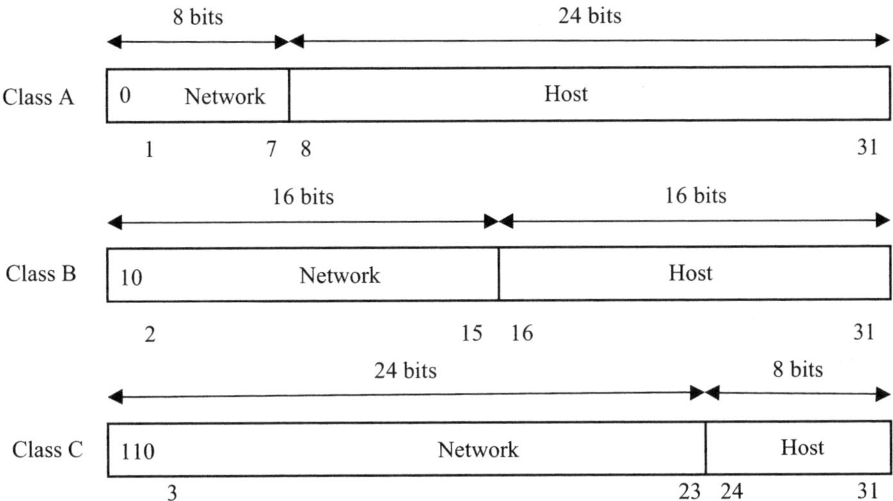

Figure 10.2. *Different classes of addresses*

In order to know to which class an address belongs, only the first high bits are needed. A router will therefore be able to determine where the network prefix ends, thus making it possible to search for a route in its routing table. Table 10.1 gives the number of networks and the number of hosts corresponding for each unicast addressing class. This table also gives the proportion of total space usage for each of them. The calculation takes into account the fixed part of the network prefix as well as the cutting off of certain special addresses, mentioned in the next section.

Class	First byte	Network prefix	Host number	Number of networks/hosts	Fraction of the addressing space
A	0xxxxxxx	8 bits	24 bits	126/16,777,214	1/2
B	10xxxxxx	16 bits	16 bits	16,384/65,534	1/4
C	110xxxxx	24 bits	8 bits	2,097,152/254	1/8

Table 10.1. *Class comparisons*

The dotted-decimal notation was specified to ease the reading of IP addresses. The series of bits is represented in the form of four integers, corresponding to the four bytes of the address, separated by dots.

Therefore, an address whose value is 10000100 11100011 01101110 11001111 is a class B address identified by its network prefix starting with the 10 bit string. This address can be written as 132.227.110.207 in decimal notation.

10.2.1.3. *Special addresses*

Precise uses have been specified for special addresses, among which the most common ones are cited here as examples:

– 127.0.0.0: this class A address is reserved for the local loop, in other words, the address of the machine itself;

– 0.0.0.0: is solely used as source address. This reserved class is used by a host when the network number is unknown and must be interpreted as "this host machine on this network";

– 255.255.255.255: is solely used as destination address. This special address corresponds to a limited broadcast address and therefore makes it possible to reach all the machines in the local sub-network;

– 192.168.0.0: this class B address is reserved for addressing machines on a private network. Addresses of this class are not unique, and can be assigned without constraint if they are not used on the Internet;

– RFC 3330, the most recent update, entirely redefines these special addresses.

10.2.1.4. *Classful addressing limitations*

This first IP specification, defined in RFC 791, introduces the use of classes for the IP protocol. It has been really successful since this standard is used worldwide, tending to become the sole level 3 protocol. However, its success has also quickly shown the weaknesses in its addressing system: in fact, it obliges a network administrator to ask for a new addressing class for each new network installed, no matter its size. This being the case, a large number of addresses have been wasted due to biased distribution. In addition, the size of the mail routers' routing tables has greatly increased.

A simple example shows the inadaptability of this addressing system. The distribution of class B to numerous companies that need just over 254 addresses has quickly shown that:

– each attribution of this type has caused a large waste of IP addresses;

– the number of available class B addresses has rapidly and worryingly decreased, and the size of Internet routing tables has proportionately increased.

In addition, companies legitimately requesting class B addresses have had to face a shortage and content themselves with many class C addresses. With the result of making the administration of such a network more complex, the size of routing tables has even more quickly increased and has reached an alarming level. In this context, a new specification has been made necessary in order to optimize the attribution and use of address classes.

10.2.2. *Subnetting*

The subnetting technique was specified in 1985 in RFC 950 to improve two-level addressing techniques in IP networks. This technique consists of adding a third hierarchical level and is destined to be used internally by an organization.

10.2.2.1. *Three-level addressing structure and use*

A third level of hierarchy divides the host number into two sub-parts: the sub-network number and the host number on this sub-network.

32 bits

Network number		Sub-network number	Host number
11111111	11111111	11111111	00000000

Figure 10.3. *Three-level address and its mask*

These three parts of the address are used in different ways according to whether they are used at the global Internet level or in an organization's internal network. In the first case, the routing technique remains the same and each router identifies a network from the network number in the address. Within the organization, the extended prefix is used: the concatenation of the network number and the sub-network number. This extended prefix makes it possible to define many sub-networks based on the internal needs of the organization.

In order to know where the border between the sub-network number and the host number lays, an extra piece of information is necessary: the sub-network mask. A mask is constructed in the same 32-bit format. The bits set to 1 in the mask indicate the length of the extended prefix. It must be noted that in this technique, the sub-network mask is longer than the network number. This makes it possible to identify different sub-networks in the address class attributed to the organization.

The mask is written in dotted decimal notation. Thus, the mask 255.255.255.0 will have its first three bytes set to 1 and the last to 0. However, this notation is not ideal to determine the length of the extended prefix. A new notation is used: the notation based on the / (slash). This makes it possible to give an IP address followed by the size of its mask: an address 132.227.110.207 with a mask 255.255.255.0 will be written as 132.227.110.207/24, 24 being the number of 1 bits in the mask. Figure 8.3 illustrates this example.

10.2.2.2. *Advantages and limitations of the subnetting techniques*

The three-level addressing model guarantees the invisibility of the splitting into sub-networks with regards to the external world of the organization. Whichever the granularity of the split, each Internet router only has one route to the group of defined sub-networks: the one defined by the attributed class that aggregates all the internally created sub-networks. It is therefore possible to create new networks, in reality sub-networks, without adding any entry in the routing tables of the Internet routers.

A current use of subnetting consists of creating, from a class B address, a multitude of sub-networks differentiated by the third byte of the address.

In addition, when creating a sub-network, the administrator does not need to request new IP addresses. Subnetting makes it possible to reuse previously allocated addresses to create a new network.

Beyond the economy of addresses and the struggle against the growth of routing tables, subnetting also makes it possible to avoid the propagation of routing table modifications. This phenomenon called route flapping is encountered when an administrator modifies the structure of the network. In the traditional addressing scheme, each modification of the smallest network must be reflected in all of the Internet routers. Scaling this mechanism is worrying in a global network in permanent extension. Subnetting makes it possible to limit route flapping because the administrators can modify, add and remove sub-networks without affecting their global visibility.

This mechanism requires modifications in the internal infrastructure where it is implemented. In fact, not all routing protocols enable the transport of sub-network masks. Nonetheless, current routing protocol versions exchange the network number and corresponding mask with each other.

However, this subnetting technique quickly shows its limitations: all the created sub-networks must have the same size. In fact, in subnetting, a network's addressing space is divided into a determined number of equally sized sub-networks. This leads to unoptimized addressing schemes where addresses are wasted. If an organization needs to divide its network into sub-networks of different sizes, the smaller sub-networks will have the same number of available addresses as the larger ones. A large number of addresses are therefore wasted.

10.2.3. *VLSM and CIDR*

In 1987, RFC 1009 defined the use of sub-network masks of varying lengths, in order to optimize the allocation of sub-networks. This technique is called variable length subnet mask (VLSM) as it is based on a variable length of the extended prefix.

10.2.3.1. *VLSM*

This specification thus mitigates the problem of same-size sub-networks. It also introduces a new possibility in the allocation of the addressing plan of different sub-networks: the recursive splitting of addressing space makes it possible to aggregate routes. A network is split into many sub-networks of the same size, which are themselves split into other sub-networks, and so on. A single route announcing the network of a given level suffices to route toward all the lower-level sub-networks.

An organization with a class A written as 13.0.0.0 can therefore divide itself into sub-networks whose addresses are 13.X.0.0/16. Due to the specification proposed by VLSM, each of these sub-networks can be split into sub-networks of different sizes. Figure 10.4 illustrates the possible splitting of an addressing space.

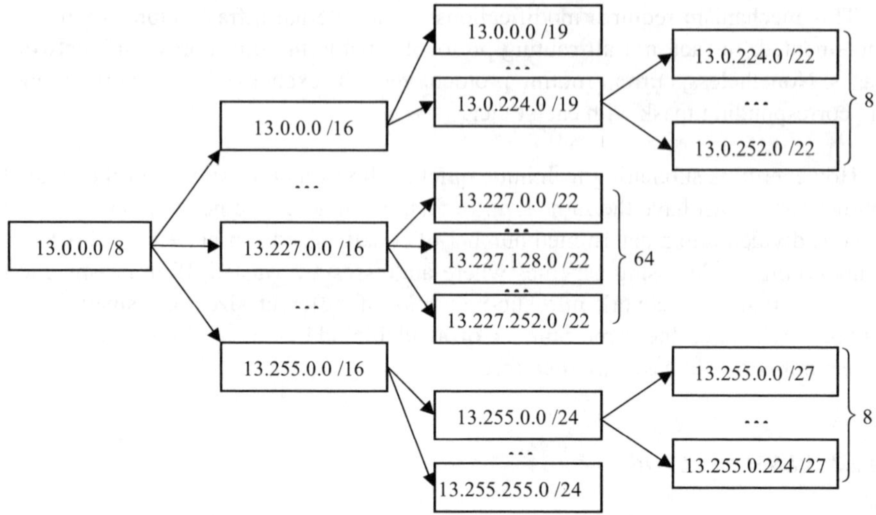

Figure 10.4. *Recursive splitting and aggregation*

In this example, the class A address is split into 256 sub-networks/16. Three of these are split into sub-networks of different sizes: 8 sub-networks/19, 64 sub-networks/22 and 256 sub-networks/24. Finally, two of these sub-networks are each split into two groups of 8 sub-networks.

From the point of view of the routers within the Internet, a single route is necessary to announce all the sub-networks created. Internally, the announcement of route 13.255.0.0/16 makes it possible, for example, to route 264 sub-networks. The aggregation of routes makes it possible to create a large quantity of sub-networks without increasing the size of the routing tables in higher levels.

So that this technique can be used in an organization's internal network, it must respect certain constraints. Like subnetting, a routing protocol that transports sub-network masks, such as OSPF or RIPv2, is indispensable.

The routers must also implement an adapted routing algorithm. In fact, in the example illustrated in Figure 10.4, sub-networks can have the same address but different masks. A router looks for the most adequate route for a packet among a group of identical prefixes but of different masks. With a classic routing algorithm, the first route will be chosen, whereas the IP packet can be destined to a different sub-network.

A "longest prefix-matching" algorithm is therefore required to implement a correct routing strategy. This algorithm consists of applying the concerned sub-network masks by the packet in order to choose the sub-network with the longest extended prefix. By following the principle of aggregation, the route with the longest extended prefix is more specific than one with a shorter prefix. The router therefore selects the more specific route to forward a packet to its destination.

One last point remains to be respected in order to use VLSM while aggregating the optimal routes. The splitting of the addressing space must follow the network's topology so that a route represents the maximum of sub-networks. If the network's topology is totally respected by the hierarchical splitting of the addressing space, a single and unique route will suffice for routing toward all the lower level sub-networks.

A larger flexibility is therefore obtained due to VLSM, which makes it possible to recursively split the addressing space according to the organization's needs. Moreover, by respecting the network's topology, we get an optimal aggregation of routes. However, these advantages can only be applied within the organizations. The attribution of classes A, B or C remains the same and routing based on them does not profit from route aggregation introduced by VLSM.

10.2.3.2. *CIDR*

In 1992, a lack of addresses was foreseen and B class addresses were almost all attributed. In addition, the size of routing tables within the Internet increased dramatically. A series of RFCs (RFC 1517, RFC 1518, RFC 1519, RFC 1520) was therefore defined in order to reclaim the advantages introduced by large-scale VLSM: this is the CIDR specification (classless inter-domain routing).

This specification proposes a radically different routing mechanism within the Internet. CIDR is called classless because the use of classes A, B and C is abandoned for an address attribution mode that better suits the needs. This new specification reclaims the functioning mode of VLSM by applying it to the entire Internet. We therefore benefit from the advantages of VLSM: the allocation of addresses is more efficient and the aggregation of routes makes it possible to relieve the core routers.

The use of network prefixes is now generalized throughout the Internet. All network addresses are therefore announced with their network prefix and it is possible to attribute blocks of addresses of arbitrary sizes. The first three bits of an address are no longer taken into account by the routers that are based on the announced network's prefix.

The principle of varying size prefix is used to optimize the use of addresses. An Internet access provider is no longer constrained to waste a number of addresses by attributing the number of necessary classes to cover a company's needs. In fact, it could not previously attribute anything but classes A, B or C, in other words blocks of addresses with /8, /16 or /24 prefixes. With CIDR, the access provider can split its addressing space as it wants and assigns, for example, an /11 or /19 address. The allocation is therefore more optimized and the size of routing tables is reduced: one route is enough to announce a network, whereas before it took as many routes as there were classes assigned to an organization (often more class Cs to meet the needs of an organization).

An organization that needs to attribute a certain address can also split it to give blocks of addresses of varying sizes, which can also be split. This results in a recursive splitting on the scale of the Internet.

As with VLSM, this recursive splitting makes it possible to aggregate routes. At each level, we only take into account the allocated blocks of addresses. Each route corresponds to a network and suffices to announce all the networks issued from it. Thus, an access provider only announces, on the Internet, one route per contiguous address space block that it administers, which suffices to announce all its networks. There again, the size of routing tables is reduced and makes it possible to have better performances and easier network administration.

The deployment of CIDR requires the same functionalities as VLSM: a routing protocol that transports network prefixes and implements a longest prefix matching algorithm. However, since CIDR no longer takes address classes into account, classless routers are indispensable. The first three bits no longer indicate the network prefix and must no longer be taken into account by the routers. Finally, the allocation of network addresses must follow the network's topology so that route aggregation is as efficient as possible.

Due to CIDR deployment, addressing space is more efficiently used and the size of routing tables is reduced. But its efficiency is linked to the topology, which can become difficult in certain cases. When an organization wants to change its access provider, it does not necessarily want to change its entire addressing plan, since renumbering can be costly. In this case, a routing exception indicating the route to this company's network is therefore necessary to indicate the path to follow. This exception makes it possible to reach this organization's network and not the network of their former access provider. This type of exception is more and more prominent and increases the size of routing tables.

Nonetheless, a highly hierarchical organization of address space is the most adapted solution for a network as extended as the Internet. The routers that form the network route the packets at their hierarchical level. They mainly know the network addresses along the same hierarchical level due to route aggregation. This considerably reduces the size of routing tables and enables an acceptable performance.

The new IP protocol specification, IPv6, is also based on a hierarchical network organization and reuses the basic CIDR principles. The use of network prefixes of varying sizes and the aggregation of routes possible by a strong hierarchy have been retained for this new protocol. In the following section, we will detail this protocol's principal innovations and notably the hierarchical structure imposed for the construction of an IPv6 address.

10.3. The future version of the IP protocol: IPv6

In order to adapt to the needs of new packet networks, a series of points has been modified in the IPv6 standard in relation to its IPv4 version. In this section, we will endeavor to describe the standard IPv6 addressing.

10.3.1. *Differences and comparisons between IPv4 and IPv6*

10.3.1.1. *A larger addressing space*

The IPv6 communication protocol defines and uses 128-bit (16 bytes) addresses. Although the size of the IPv6 packet's *address* field enables 3.4×10^{38} combinations, the addressing space was defined in order to enable several network hierarchy levels. Thus, this network hierarchy makes it possible to define sub-networks, from the largest Internet backbones to any sub-network in an organization.

10.3.1.2. *A totally hierarchical infrastructure*

The IPv6 addressing structure includes the last improvements made to that of IPv4. IPv6 addressing is thus defined hierarchically with variable length masks. This makes it possible to optimize the size of routing tables within the routers.

In addition, IPv6 defines multi-level address organization in order to guarantee a total hierarchy. Due to the implementation of different levels of service providers within the addressing structure, the routing table structure corresponds to the real physical organization of different level ISPs.

10.3.1.3. *Stateful and stateless address configuration*

In order to simplify host configuration, the IPv6 protocol supports both a stateful and a stateless configuration. Thus, the nodes that have an IPv6 stack will be able to receive an address from a stateful configuration through, for example, a DHCP server. The IPv6 nodes will also benefit from a stateless configuration by auto-configuring themselves.

In the case of stateless auto-configuration, the nodes are capable of defining their local address to communicate with their neighbors on the link and also their global address (routable over the Internet) derived from the sub-network prefix announced by the routers.

10.3.2. *IPv6 address representation*

As was mentioned, the size of IPv6 addresses is 128 bits, which enables approximately 3.4×10^{38} combinations. With IPv6, it becomes truly difficult to imagine that a new lack of addresses could occur. In fact, to put these values into perspective, the number of possible combinations given by IPv6 makes it possible to allocate 6.5×10^{23} addresses for each square meter of the earth's surface.

Of course, the choice of the value of 128 bits for the size of IPv6 addresses was not made according to the number of possible addresses per terrestrial square meter. This value was chosen in order to enable the organization of a better hierarchical sub-network structure. In addition this value of 128 bits enables a greater flexibility in the organization of different operator networks.

10.3.2.1. *Address distribution and formation*

Similarly to IPv4, IPv6 address spacing is divided into different uses. This distinction between different address types is based on the high bits. Their value is fixed and is often referred to with the term format prefix (FP). This allocation of address types is detailed in Table 10.2.

Allocation	Format prefix (FP)	Fraction of addressing space
Reserved	0000 0000	1/256
Unassigned	0000 0001	1/256
Reserved for NSAP allocation	0000 001	1/128
Unassigned	0000 010	1/128
Unassigned	0000 011	1/128
Unassigned	0000 1	1/32
Unassigned	0001	1/16
Unicast and aggregatable addresses	001	1/8
Unassigned	010	1/8
Unassigned	011	1/8
Unassigned	100	1/8
Unassigned	101	1/8
Unassigned	110	1/8
Unassigned	1110	1/16
Unassigned	1111 0	1/32
Unassigned	1111 10	1/64
Unassigned	1111 110	1/128
Unassigned	1111 1110 0	1/512
Local link unicast addresses	1111 1110 10	1/1024
Local site unicast addresses	1111 1110 11	1/1024
Multicast addresses	1111 1111	1/256

Table 10.2. *Distribution of addressing space*

All the unicast addresses can be used by IPv6 nodes that have a scope. This scope corresponds to the maximum range of the address within the network. The different defined scopes are the local link scope, local site scope and global scope. It should be noted that local site type addresses have just been abandoned by the various workgroups around IPv6 because their implementation complexity is too great.

10.3.2.1.1. From binary to hexadecimal

In IPv6, the 128 bits of the address are divided into 16-bit blocks. Each of these blocks is converted into a 4-digit hexadecimal number. Each of the 16-bit blocks is separated by a colon (:).

Therefore, taking the following 128 bit address:

0010000111011010100000001101001100000000000000000010111100111011

0000001000101010000000001111111111111110001010001001110001111010

We divide the address into 16-bit blocks:

0010000111011010 1000000011010011 0000000000000000 0010111100111011

0000001000101010 0000000011111111 1111111000101000 1001110001111010

Each of the 16-bit blocks is then transcribed into hexadecimal notation separated by ":":

21DA:80D3:0000:2F3B:022A:00FF:FE28:9C7A

The representation of IPv6 addresses can be simplified by the elimination of the first zeros in each of the 16-bit blocks. Nonetheless, each block must have at least one digit. This leads to the following representation:

21DA:80D3:0:2F3B:22A:FF:FE28:9C7A

10.3.2.1.2. Zero compression

Certain types of IPv6 addresses carry a large number of zeros. In order to further simplify the representation of addresses, a sequence of 16-bit blocks whose value is zero can be compressed in the form "::".

For example, the address FE80:0:0:0:25C3:FF:FE9A:BA13 can be compressed in the form FE80::25C3:FF:FE9A:BA13.

In the same manner, the multicast address FF02:0:0:0:0:0:0:2 can be compressed to FF02::2.

The compression of zeros can only be carried out in a single series of zeros within an IPv6 address. If ever we compressed two blocks of zeros in the same address, it would be impossible to know how many zeros are represented by the "::" symbol. Compressing zeros more than once would lead to address notation ambiguities.

10.3.2.2. *IPv6 prefixes*

As shown in Table 10.2, the format prefix defines the type of IPv6 address. This must not be mistaken for the IPv6 address prefix corresponding to the address' high bits, serving to identify the network. The IPv6 prefixes are represented in the same way as the CIDR representation (classless inter-domain routing) of IPv4 addresses. Thus, an IPv6 prefix is referred to as *network address/length of prefix*. For example, 21DA:D3::/48 is an IPv6 prefix.

10.3.3. *IPv6 address types*

There are three types of IPv6 addresses: unicast, multicast and anycast.

A unicast address identifies a single and unique interface. Due to an appropriate unicast routing infrastructure, packets destined to a unicast address are delivered to a single interface.

A multicast address identifies many interfaces. With the appropriate routing structure, the packets addressed to a multicast address are sent to several interfaces.

An anycast address identifies several interfaces. Due to an appropriate routing infrastructure, packets destined to an anycast address are delivered to a single interface. This interface is the nearest having been identified by the anycast address in the direction of routing. The nearest interface is defined in terms of routing distance (number of skips, for example). An anycast address corresponds to a "one toward many" communication, thus delivering packets to a single interface.

Only unicast addresses can be used as source addresses identifying a node in the course of communication.

10.3.3.1. *Unicast IPv6 addresses*

Unicast IPv6 addresses are divided into three categories: local link addresses, global addresses and special addresses. Local site addresses have been abandoned.

10.3.3.1.1. Global unicast addresses

Global unicast addresses are identified by FP 001. They are equivalent to public IPv4 addresses. These addresses are globally routable and therefore reachable on the IPv6 part of the Internet. They are made to be aggregated, which will make it possible to have a greater efficiency in the routing infrastructure. IPv4 is actually a combination of flat and hierarchical routing due to the routing exceptions it contains. IPv6 was completely hierarchically conceived from the beginning. The scope, or the region, in which the global addresses are unique is the whole of the IPv6 Internet network.

As we have specified, IPv6 routing is based on the notion of prefix. Nonetheless, in order to guarantee the coherence and hierarchy of network addressing, address splitting was introduced. This splitting, described in Figure 10.5, only serves for the attribution of addresses to different ISPs and has no influence on routing.

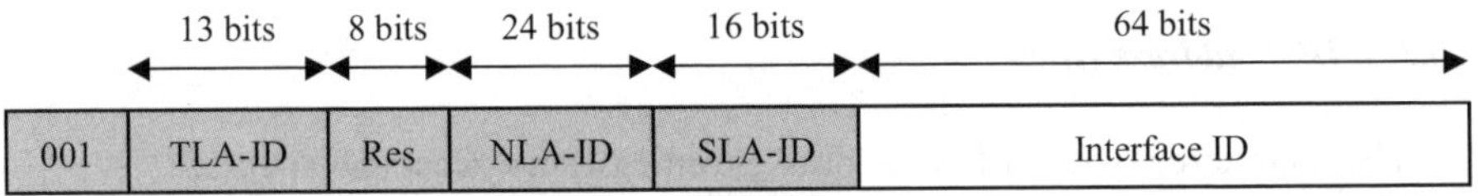

Figure 10.5. *Global unicast address definition*

TLA-ID: top level aggregation identifier. This 13-bit field identifies the highest level in the routing hierarchy. The TLA-IDs are administered by the IANA and attributed to local attribution organisms that attribute their TLA-IDs to high-level Internet access providers. The 13-bit field enables 8,192 combinations. The highest level routers in the Internet hierarchy therefore have no default route but have the whole of the entries corresponding to the TLA-IDs.

Res: these are bits used for future use in order to increase the size of the TLA-ID field or the NLA-ID field as needed. This field is 8 bits long.

NLA-ID: next level aggregator identifier. This field is used to identify a client site. This field is 24 bits long. The NLA-ID serves to create different addressing levels within a network in order to organize and route toward the lowest level access providers. The structure of these Internet access provider networks is not visible by the highest level routers. The combination of FP 001, of the TLA-ID and of the NLA-ID forms a 48-bit prefix attributed to an organization connected to the IPv6 portion of the Internet.

SLA-ID: site level aggregation identifier. The SLA-ID is used by an organization to identify its different sub-networks. This field is 16 bits long, which makes it possible for the organization to create 65,536 sub-networks or different hierarchical addressing levels. This makes it possible to implement an efficient routing infrastructure. Due to the 16-bit SLA-ID, assigning a prefix to an organization corresponds to attributing it a class A IPv4 address. The structure of the client's network is not visible by the access provider.

Interface-ID: makes it possible to uniquely identify an interface on a network. This field is 64 bits long.

The fields we have just described make it possible to define a three-level structure as shown in Figure 10.6. The public topology corresponds to high and low level Internet access providers. The site topology corresponds to the group of sub-networks of the organization. The interface identifier enables the addressing of the interfaces within each of the organization sub-networks.

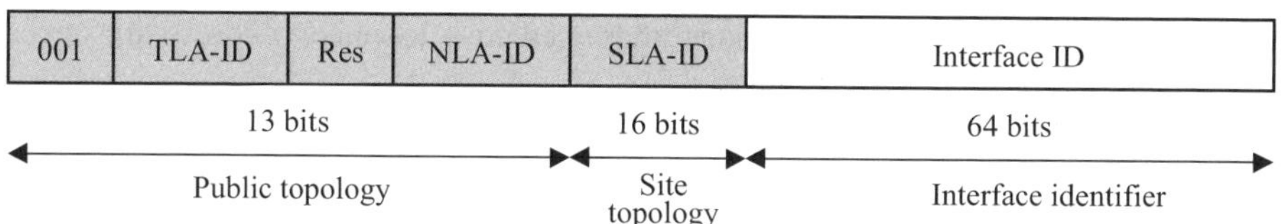

Figure 10.6. *Three-level hierarchical addressing*

10.3.3.1.2. Local link addresses

The local link addresses are identified by FP 1111 1110 10. These are used by the nodes that want to communicate with their neighbors on the same link. Thus, the hosts can simply communicate on an IPv6 link without a router. The scope of a local link address is the link on which the host resides.

A local link address is necessary for the communication between the nodes of a single link and is always automatically configured, even in the absence of all other unicast address types. Figure 10.7 describes the structure of a local link unicast address.

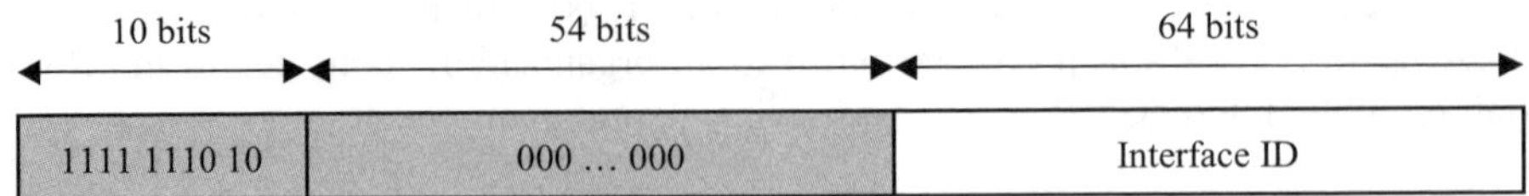

Figure 10.7. *Local link addresses*

Local link addresses are implemented by using FP FE80. With a 64-bit interface identifier, the local link address prefix is always FE80::/64. An IPv6 router never relays packets whose source address field contains the local link addresses from the links to which it is connected.

10.3.3.1.3. Special addresses

The unspecified address (0:0:0:0:0:0:0:0 or ::) is used to indicate the absence of an address. This address is typically used as source address during the configuration phase. This address is never attributed to an interface or used as a destination address.

The loopback address (0:0:0:0:0:0:0:1 or ::1) is used to identify a loopback interface that makes it possible for a node to send packets to itself. This is equivalent to the IPv4 address 127.0.0.1. Packets addressed to a loopback address are never sent on the link or relayed by a router.

10.3.3.2. *IPv6 multicast addresses*

In IPv4, multicast is a notion that is difficult to implement. In the case of IPv6, the multicast traffic was originally defined at the local link. Multicast is essential for the communication and auto-configuration of nodes belonging to a single link.

IPv6 multicast addresses are defined by FP 1111 1111. In the case of IPv6, multicast addresses are easily recognizable by the fact that they all start by FP FF. Multicast addresses cannot be used as source address in an IP packet.

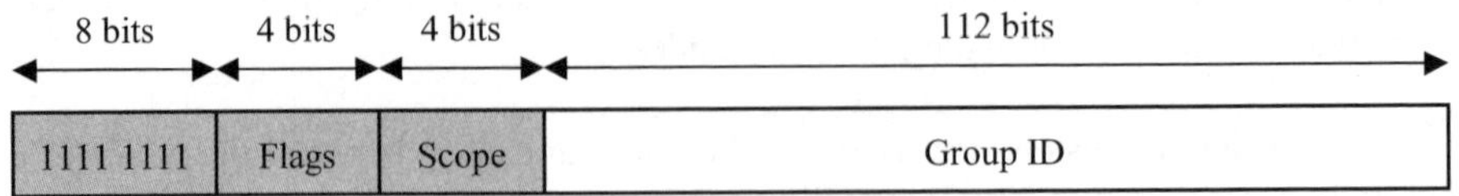

Figure 10.8. *Multicast addresses*

Putting aside the prefix used to identify them, multicast addresses have additional structures in order to identify their flag, their scope and their multicast group. Figure 10.8 describes the different fields used.

Flags: this field indicates the flag associated to the multicast address. This field is 4 bits long and makes it possible to differentiate addresses according to their life span. Currently, only two values have been defined. The T (transient) flag, when it is 0, indicates a permanently attributed address. On the contrary, when it is 1, it corresponds to a temporarily used address.

Scope: indicates the scope of the IPv6 address defining the corresponding multicast traffic. This field is 4 bits long. This field is notably used by multicast routers in order to know if they should relay packets or not. Table 10.3 defines the different scope values and their range within the Internet.

Value	**Scope**
0	Reserved
1	Local node scope
2	Local link scope
5	Local site scope
8	Local organization scope
E	Global scope
F	Reserved

Table 10.3. *Multicast scope values*

Thus, the traffic corresponding to the multicast address FF02::2 has a local link scope. An IPv6 router will never relay this traffic any further than the local link.

Group ID: this field enables the identification of the multicast group within the whole defined by the scope. Addresses between FF01:: and FF0F:: are reserved and defined for a particular use:

– FF01::1 all the nodes according to the local node scope;

– FF02::1 all the nodes according to the local link scope;

– FF01::2 all the routers according to the local node scope;

– FF02::2 all the routers according to the local link scope;

– FF05::2 all the routers according to the local site scope.

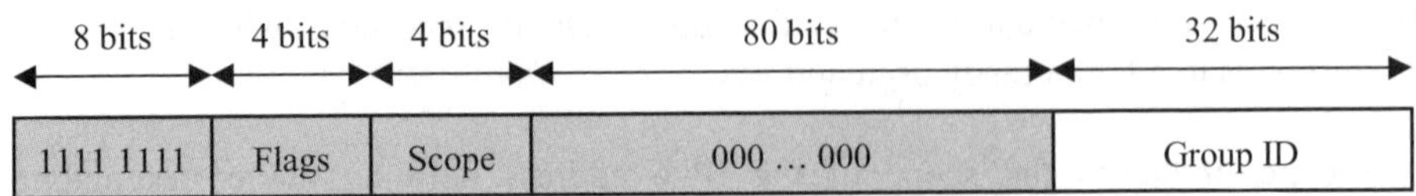

Figure 10.9. *Multicast addresses*

Due to the 112 bits in the Group ID field, it is possible to obtain 2^{12} different groups. Nonetheless, since the IPv6 multicast addresses are mapped on Ethernet addresses, it is recommended to only use the 32 low bits in the Group ID. Thus, each multicast group defined on the last 32 bits of the Group ID can be mapped on a unique Ethernet address. Figure 10.9 describes the fields of the modified multicast address.

10.3.3.3. *Anycast IPv6 addresses*

An anycast address is assigned to several interfaces. Packets intended for an anycast address are relayed by the routing infrastructure to the nearest interface among the ones to which the anycast address has been attributed. An anycast address makes it possible to designate a service by a well-known address; in this way, it is not necessary to interrogate a server to know the location of an equipment.

In order to complete the relaying of the packets, the routing infrastructure must know the interfaces to which the anycast address has been attributed and their distances in terms of routing metrics.

If the anycast concept is simple in its principle, its implementation is otherwise delicate. Moreover, it should be known that at the time of drafting the IPv6 specifications, this concept was only a research topic. Finally, another argument explains the prudence that should be conserved regarding this concept: there has been no real-life experience making it possible to test the functionalities of anycast addresses.

Anycast and unicast type addresses are allocated in the same addressing space. Anycast addresses are created by attributing a single unicast address to distinct nodes. Then it is the routing infrastructure that routes the packets to the closest next hop, therefore completing the anycast service implementation.

Currently, anycast addresses can only be used as destination addresses and are solely attributed to routers. Anycast addresses attributed to routers are constructed on the network prefix base. This prefix is created from the network prefix for a given interface. To construct the anycast address attributed to a router, the network prefix bits are fixed to the appropriate value and the leftover bits are fixed to 0. Figure 10.10 shows the construction of the anycast address for a router.

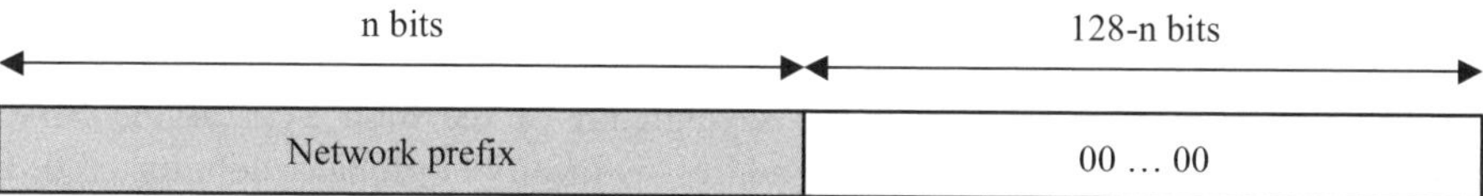

Figure 10.10. *Anycast address for a router*

10.3.4. *Interface identifier and IPv6 address*

The last 64 bits of an IPv6 address correspond to the interface identifier destined to be unique. There are different ways of determining the interface identifier:

– a 64-bit identifier derived from the extended unique identifier (EUI) of the MAC address;

– an interface identifier randomly and periodically generated to preserve anonymity;

– an identifier assigned during a stateful configuration (for example, by a DHCPv6 server).

Later on, we will only present the formation of IPv6 addresses due to the EUI-64 derived identifier. This can be derived from the interface's MAC (medium access control) address defined by the IEEE (Institute of Electrical and Electronic Engineers). We will quickly present MAC addresses and EUI-64 and then the formation of the last 64 bits of an IPv6 address.

10.3.4.1. *MAC address format*

Network interface addresses use a 48-bit sequence called the IEEE 802 address. These 48 bits are composed of the identifier of the company that produces the network interfaces represented on 24 bits and of 24 bits representing the extension of the network card.

The combination of the unique company identifier with the extension of the card produces a unique 48-bit address called the MAC address. Figure 10.11 illustrates the characteristics of a MAC address.

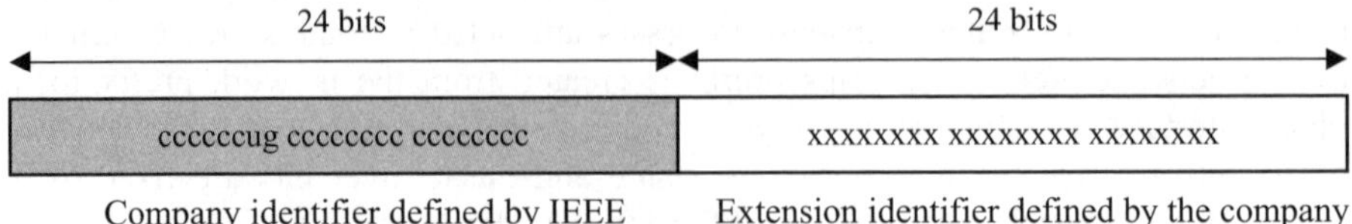

Figure 10.11. *IEEE 802 address*

The universal/local bit (U/L): the seventh bit of the high byte of the MAC address is used to indicate whether the address is locally or globally administered. If the U/L bit is 0, then the IEEE has administered the address. If the bit is 1, then the address is locally administered and there is no guarantee that the company identifier is unique.

The individual/group bit (I/G): the last bit of the high byte is used to determine whether the address is an individual address (unicast) or a group address (multicast). When this bit is 0, the address is a unicast address. When this bit is 1, the address designates a multicast address. A classic IEEE 802 address has U/L and I/G bits whose value is 0, which corresponds to a globally administered unique address.

10.3.4.2. *IEEE EUI-64 address format*

The IEEE EUI-64 addresses represent a new standard for network interface addressing. The company identifier is still represented on 24 bits but the extension that represents the card is represented on 40 bits. This makes it possible to increase the network card addressing space for manufacturers. This standard uses U/L and I/G bits similarly to the IEEE 802 addresses. Figure 10.12 describes the format of IEEE EUI-64 addresses.

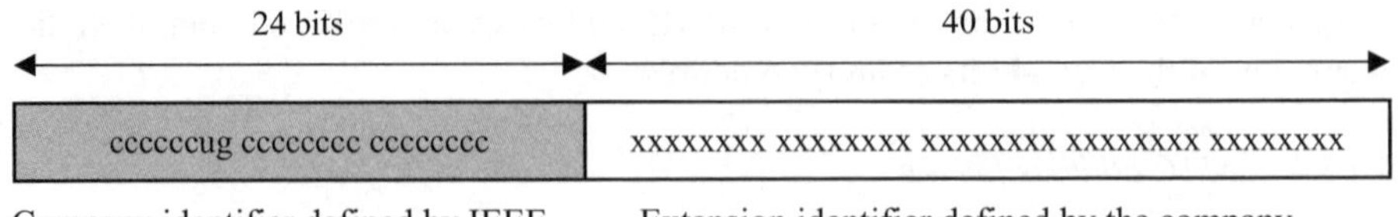

Figure 10.12. *IEEE EUI-64 identifier*

10.3.4.3. *IPv6 address creation*

All unicast IPv6 addresses must use a unique 64-bit interface identifier. In order to obtain this identifier, the U/L bit in the EUI-64 address is complemented. Thus, to create the interface identifier for an IPv6 address, first the EUI-64 must be formed, which is derived from the MAC address. For this, the 16 bits 11111111 11111110 (0xFFFE) are inserted in the IEEE 802 address between the company identifier and the card extension. The U/L bit in the EUI-64 address must then be complemented. Figure 10.13 illustrates these different steps.

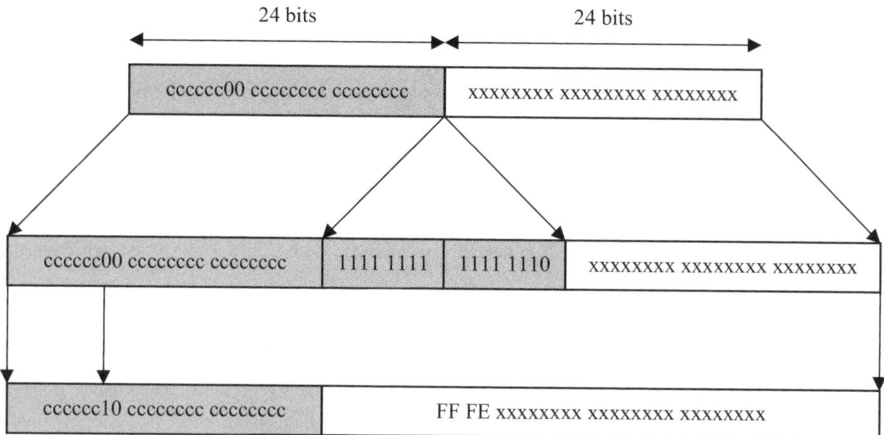

Figure 10.13. *IEEE EUI-64 identifier construction*

10.4. Conclusion

Throughout this chapter, the evolution of the IP network addressing technique is highlighted. The first specification of the IP protocol had not considered its scaling problem. Flexibility and performance problems have led to the specification of techniques more adapted to the general use of the Internet network. These successive modifications resulted in a radically different system. The use of fixed classes was finally abandoned for the use of network prefixes of varying sizes. Route aggregation benefiting from a multi-level hierarchy now makes it possible to generate performance problems due to a permanent extension of the network.

The new IP protocol specification reclaims these principles, while imposing a precise semantic for the attribution of IPv6 addresses. This organization makes it possible to obtain a completely hierarchical addressing and thus benefit from the previously cited advantages. In addition, IPv6 offers a much more vast addressing space, thus eliminating the problem of a lack of IP addresses. This point is not negligible in the face of the constant increase of IP terminals in the network. However, the migration toward IPv6 is not immediate and will surely be progressive, due notably to the coexistence within the network of the two versions of the IP protocol. In addition, the small quantity of addresses available for the number of countries that are still not well equipped will surely lead to a massive use of the IPv6 protocol in the near future.

Nonetheless, certain evolutions will carry unresolved problems, even in the framework of a fully operational IPv6 network. The major problem is induced by the dynamic topology of the network. To have optimal performances, the IP network hierarchy must reflect the physical topology of the network. When the topology changes, it is important to reflect these changes in the hierarchical splitting of the network, which can be very costly and therefore rarely possible. A simple change of access provider induces, for example, a complete renumbering. Another critical example concerns the ever-increasing mobility of terminals. This mobility provokes a changing network topology that can call into question the hierarchical organization of IP. This phenomenon introduces several problems, notably in terms of addressing and routing.

10.5. Bibliography

[RFC 791] "Internet Protocol, RFC 791", *IETF*, September 1981.

[RFC 950] MOGUL J., POSTEL J., "Internet Standard Subnetting Procedure, RFC 950", *IETF, Standards Track*, August 1985.

[RFC 1009] BRADEN R., POSTEL J., "Requirements for Internet Gateways, RFC 1009", *IETF*, June 1987.

[RFC 1517] HINDEN R., "Applicability Statement for the Implementation of Classless Inter-Domain Routing (CIDR), RFC 1517", *IETF Standard Track,* September 1993.

[RFC 1518] REKHTER Y., WATSON T.J., LI T., "An Architecture for IP Address Allocation with CIDR, RFC 1518", *IETF, Standards Track*, September 1993.

[RFC 1519] FULLER V., LI T., YU J., VARADHAN K., "Classless Inter-Domain Routing (CIDR): an Address Assignment and Aggregation Strategy, RFC 1519", *IETF, Standards Track*, September 1993.

[RFC 1520] REKHTER Y., WATSON T.J., TOPOLCIC C., "Exchanging Routing Information across Provider Boundaries in the CIDR Environment, RFC 1520", *IETF, Informational*, September 1993.

[RFC 1918] REKHTER Y., MOSKOWITZ B., KARRENBERG D., DE GROOT G.J., LEAR E., "Address Allocation for Private Internets, RFC 1918", *IETF, Best Current Practice*, February 1996.

[RFC 3330] IANA, "Special-Use IPv4 Addresses, RFC 3330", *IETF, Informational*, September 2002.

[RFC 3513] HINDEN R., DEERING S., "Internet Protocol Version 6 (IPv6) Addressing Architecture, RFC 3513", *IETF, Standards Track*, April 2003.

[RFC 3587] HINDEN R., DEERING S., NORDMARK E., "IPv6 Global unicast Address Format, RFC 3587", *IETF, Informational*, August 2003.

Chapter 11

SLA Driven Network Management

11.1. Introduction

Service Driven Network Management (SDNM) and Service Level Agreements (SLAs) have aroused great interest in the world network and telecom market in the past few years. We consider in this chapter SLAs regarding network services or those supported by a communication network such as the Internet. This will therefore exclude service level agreements relevant in other fields.

This chapter begins by describing the strong link between SLA-based network management on one hand and the policy management model and the QoS mechanisms studied in the preceding chapters on the other hand. It then proceeds to define the different components of an SLA. It is followed by a review of available SLA management systems and SLA modeling. The chapter ends with a description of a group of European projects which considered the provision of QoS based using SLAs.

11.2. Requirements for service driven management

The increasing impact of IT on the global performance of companies obliges IT managers to behave as true service providers. Also, telecommunication service

---

Chapter written by Issam AIB and Belkacem DAHEB.

providers and network operators competing in the current market are constrained to offer value-added services with tangible assurances to their customers. The differentiation of client requirements based on SLAs is then essential. An SLA is designed to establish a mutual agreement between the client (CL) and the service provider (SP), specifying the rights and responsibilities in which the CL and the SP engage regarding QoS in terms of negotiated price and duration. The CL-SP relationship founded in this contract requires that the provider has an exact knowledge of the level of service to be provided and a fine understanding of the users' subjective perception. It also requires transparent and regular communication to the client of the actual provided versus contracted service level.

11.2.1. *SLA and QoS*

QoS is a generic concept that covers several performance and statistical aspects. It is defined in [TMF 01] as "the collective effect of service performance that determines the degree of satisfaction of the user of the service" (for more details see Chapter 6). The final goal of an SLA is to specify the necessary parameters to ensure the client's desired QoS level.

11.2.2. *SLA and policy-based management (PBM)*

The set-up of an SLA between the CL and the SP implies the possibility of providing a QoS in a network that supports only best effort service. It therefore requires the installation of advanced mechanisms that make it possible to provide a more elaborated service than that provided by the IP architecture.

In this regard, work is being done by the IETF (Internet Engineering Task Force) in order to propose a model that ensures QoS enabled network control. This model, named PBM, consists of considering the network as a state machine on which actions take place. These actions emanate from policy rules resulting from the manner in which an SP wants to manage its network and therefore from the different SLAs with their clients. A design has been proposed for this model (Figure 11.1), more details of which can be found in Chapter 6.

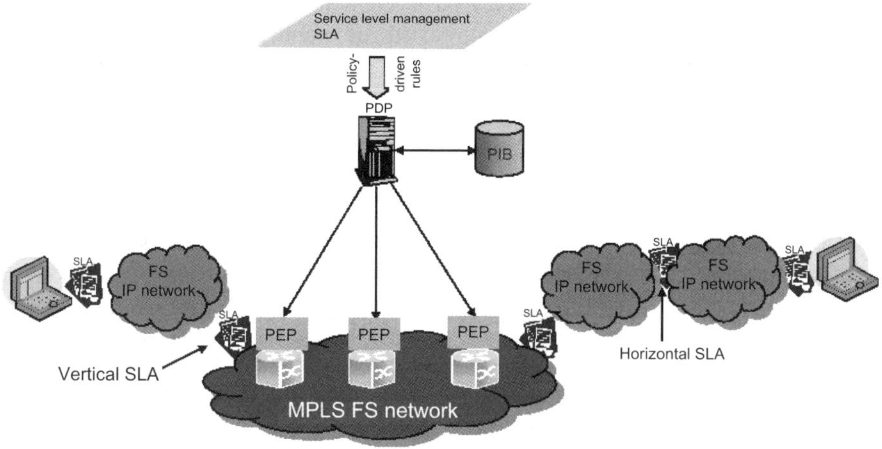

Figure 11.1. *Service driven management*

An SLA describes the high level business policy and is converted into Service Level Objectives (SLOs). An SLO consists of QoS parameters and their values. SLO values are decided after negotiations between the SP and the customer.

One way of providing the service agreed upon in the SLO is to derive a set of policy rules and manage the network according to these rules, [FAW 04]. The policy rules do not necessarily consist only of rules derived from the SLOs but they can have other sources such as the network administrator, etc.

11.3. The SLA

An SLA, to differentiate it from other types of contracts, is a formal measure between an SP and its CL which defines all the aspects to be provided by the service. Typically, an SLA covers availability, performance and invoicing, as well as specific measures in case of failure or malfunction of the service.

Consequently, an SLA constitutes a legal foundation, the basis on which the two participants (the customer and the SP) rely to plan their respective future affairs and relations. Eventually, the SLA serves as a means of judgment in case of dispute. This supposes the existence of a legal entity that certifies the contracts established between the clients and their SP.

An SLA can be composed of very basic elements that can be counted on the fingers of one hand, or can be composed of hundreds of lines of literary specifications surrounded by many ambiguities and complexities. A common error that many SLA providers and designers make is that they tend to think that the longer the SLA, the better it is.

Typically, an SLA contains the following components:

– a description of the contracting parties (the CL and the SP) as well as the legal party. This is called the WHO part of the SLA;

– a description of the service to be provided as well as the different components (the WHAT);

– the service scope: the duration of the validation of the contract and the geographical limits (the WHEN and the WHERE);

– the respective responsibilities of the contracting parties;

– the global QoS parameters of the expected service, as well as the QoS parameters of its components;

– the QoS parameters of an e-service concerning availability, performance, flow and rate of error;

– the procedure and method(s) of service billing;

– the procedures and mechanisms to use for the supervision of provided QoS and the manner in which the client receives a return to the QoS level;

– a detailed procedure concerning the maintenance and treatment of failures (rate of reactivity to the treatment of anomalies);

– a description of the exclusions in which the SLA is no longer valid;

– the consequences of disrespect of the contract:

- by the CL,

- by the SP.

In the following, we will explain in greater detail the components of an SLA. It is, however, true to say that there is not a global consensus among all service providers (network operators, Internet service providers, service provider applications, etc.) on what should be included in an SLA, nor on the semantic significance of an SLA and its components. This will be clarified in section 11.8.

11.3.1. *Designation of SLA contracting parties*

The designation of the CL and of the SP is used to identify the two parties as well as serving as a means of contact. Other attributes can be included, such as physical contact addresses, telephone and fax numbers, electronic addresses, etc.

The current tendency is to develop databases for user profiles. Profile databases make it possible to collect intelligent information on the client's use of services. Thus, the SP can propose a service adaptation according to its client's profile. In this case, an ID of the CL will be found in the identification field of the CL in the user profile database.

11.3.1.1. *Duration of the SLA*

This parameter tracks the global duration in which the SLA is considered valid by both parties. If the provided service is comprised of several sub-services, their respective periods of validity are included but are not necessarily equal to the global duration of the SLA.

11.3.1.2. *Service scope*

Considering that the SLA represents an assurance of QoS level, the SP must specify the physical extent of these guarantees. In fact, if an ISP can guarantee a level of QoS for their clients in the domain of its network operator, there is no way to guarantee the same QoS if the user flows travel through other domains, which is often the case. To remedy this difficulty, SLA chains with end-to-end guarantees have been introduced.

Nevertheless, there is a consensus on the fact that an SLA is a high level contract that describes the non-technical service parties. More details are given in the following section.

11.4. Specification of level of service (SLS)

The SLA is considered as the non-technical view of the service contract. The SLS represents the meticulous description of the technical parameters that constitute the service. The complete SLS is generally not accessible to the conventional client. This is principally due to the fact that it represents a network administrator view of the SLA. However, the client can receive a client SLS that contains technical parameters relevant to the user and make it possible for the client to confirm that the SP has adhered to the terms of the contract.

11.4.1. *Service availability guarantee*

Availability concerns more than just access to the service. It concerns a level of service on which the client can capitalize. For example, a website is considered accessible when the ping command sends an accessibility result. However, an IP telephony application considers that the communication service is inaccessible as soon as the available flow is clearly inferior to a predefined threshold.

For a composite service, an increase in its availability is, in general, a function of the increase in the availability of its components. However, it is important to know that the increase perceived by the client cannot correspond to that perceived by the network operator. Therefore, the semantics of service availability, as well as the SLS metrics used to verify the service, must be specified.

For example, considering the availability described in the IP service SLA offered to a company as proposed in [MAR 02]:

> The SLA specifies that the operator's network will be available 99.8% of the time evaluated over a period of a month and averaged with regard to the client company's n routers. The network is considered unavailable after 5 successive minutes of operation of failed pings; the frequency of pings is 20 sec with a 2 sec time-out.

It is clear that this offer of availability is very SP-oriented. In fact, in order to violate the availability metric, the network would have to be unavailable for approximately 14 hours/month. In fact, the network could well be "pingable" without really being available to the client and therefore be seen as unavailable by the client. Thus, an availability metric must consider the client's point of view rather than the SP's point of view to remain objective with regard to the SLA.

11.4.2. *Service performance guarantee*

The performance part of an SLS is principally the reason for its existence. The groups of precise parameters that determine the service's performance vary according to the service considered. We will see, in the section that addresses different SLA research projects, several typical examples of performance parameters.

For an IP network connectivity service, for example, a connection's performance is calculated with regard to four primary parameters: bandwidth, delay, jitter and rate of errors. Also, the manner in which these metrics are calculated must be specified as the supervision technique of the performance parameters considerably influences the final values.

[TMF 01] makes a distinction between performance parameters and performance events. An event is an instantaneous phenomenon that takes place within the network that is providing the service or in its environment and which affects the QoS provided to the CL. Examples of events are: errored seconds (ES), severe ES and severe errored periods.

Performance parameters are derived from performance event processing carried out during periods of defined measurements. The processing can consist of a calculation of an average, or of a comparison with a single threshold (for example: the maximum tolerated number of severe ES).

The exact determination of performance events and performance parameter calculation functions give the SP a practical means of determining and ensuring performance parameters proposed to its clients.

It can be noted that the supervision of performance parameters must be done with tools that the client trusts. If the client has no objective means of ensuring to himself of the reality and quality of provided services, he will not be able to ask for compensation in the case of a non-respected SLA. In certain cases, the SP and the CL call upon a neutral third party specialized in the measurements and which takes care of the supervision of performance parameter so that each party is more assured with regard to the validity of the performance reports.

11.4.3. *Billing methods*

There are multiple ways of billing a service. For example, an ISP can bill its customers at a fixed rate, by volume, by duration of use, by prepaid cards, or according to QoS requirements, etc. The CL must be well-informed on the billing method used in order to plan a budget and know how to select and manage the use of the service offered.

A billing system can be static or it can take into consideration the client's evolution and loyalty. However, the implementation of a complex billing structure is often costly and not justifiable unless the proposed services generate sufficient added value to ensure the perennial nature of the SP's system.

If we take, for example, the ISP market, the following tendency is shaped according to the importance of the ISP:

– For an ISP with few clients, notably during its implementation and as long as its client base does not exceed a few thousand, fixed billing is applied in most cases.

This does not require any particular billing software and does not incur excessive fees for the company.

– Beyond several thousand clients (say, between 2,000 and 10,000 clients), a billing software, preferably low-priced, becomes possible. These would generally be programs developed on the basis of Excel or File Maker Pro files, packed with sufficient functionalities to start proposing interesting service offers.

– For several tens of thousands of clients, a software package such as INFRANET from Portal becomes unavoidable. The cost of the licenses is justified in this case as it is compensated by the number of subscribers. These complete solutions have the necessary infrastructure to address the numerous evolving offers and the necessary developments for integration in the global ISP system. This would be a made-to-measure package.

11.4.4. *Addressing anomalies*

The SLS must contain clear and precise clauses regarding how service function anomalies are taken into consideration. Anomalies can originate with the SP (malfunction in the offered QoS) or with the CL (abuse of service use, non-conforming use of the service). The SLS also contains clauses that specify conditions that put the contract out of the field of the SP's QoS engagements.

In a policy management environment, anomalies are handled with the help of behavior-based rules that define the actions to take, often on the part of the SP in the case of malfunction, degradation of the level of service or poor service used by the CL. For example, in the case of excess bandwidth usage by an IP network CL, the SP can implement many policies to regulate the CL's usage. It can ignore excess packets, or mark them as un-guaranteed packets that can possibly be isolated in case of congestion. It can also alert the client of the excess, or simply bill the client for the excess packets. However, the rules for the management of anomalies of use and of service functions must be properly specified in the SLS so that neither party is surprised by the behavior of the other.

11.5. Service contract chains

The federation of services is an unavoidable consequence in a competitive environment, in which the best solution for the SPs' survival is to specialize and innovate in a precise (sub-)domain. Thus, in order to achieve a Final Client (FC) service, a chain of several CL-SP can take place, in which SPs become CLs for other SPs. For example, an ISP is an SP for the FCs, whereas it is a CL for its network operator. Similarly, a VoD (Video on Demand) server is a CL for its ISP. A CL can be a CL for its ISP for Internet access and during a period of time can also be a CL for a VoD service. This is how the problem of SLA chains and end-to-end SLA management arises.

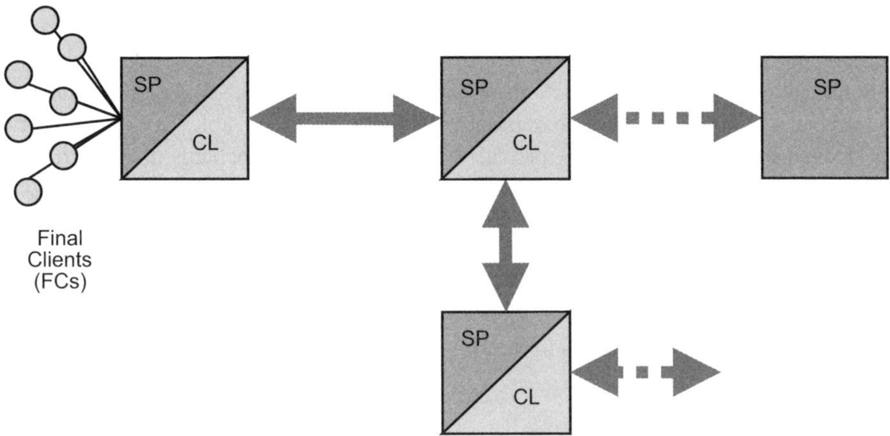

Figure 11.2. *SLA chains*

11.6. SLA types

The classification of SLAs can be done in various possible ways. We have already seen the differentiation of CL-SP and SP-SP SLAs. In other works, it was agreed that two main SLA types should be differentiated. The differentiation criteria concerns the OSI layers that are in question in the SLA.

11.6.1. *Horizontal SLA*

A horizontal SLA is an SLA between two SPs within the same OSI layer. In this category, for example, is an SLA between two IP domains or two request content providers.

11.6.2. *Vertical SLA*

A vertical SLA is an SLA between two SPs on two different OSI layers. In this category, for example, is an SLA between an MPLS network domain and an optic transport network (OTN), or an SLA between a VoD server and its ISP.

11.7. SLA management (SLM)

An SLA on its own is not sufficient if it has not been respected and properly managed. SLM represents the most difficult part of the Service-Oriented Architecture (SOA) which we proposed to study in this chapter. In fact, the challenge presented by an SP when it decides to adopt this architecture is to implement a management system that offers the contracted level of service to each client, as well as all the mechanisms and the gauges that make it possible to supervise the different service parameters, in order to be able to report on the performance and the reality of provided services. These mechanisms also make it possible to detect service failures. In short, an SLM system must manage [MAR 02] the individual life cycle of each client SLA, ensure the coherence of the SLA database as well as the coordination of end-to-end SLAs, as well as the observation of its business' global objectives.

We detail these three points in the following sections.

11.7.1. *An SLA life cycle*

An SP's SLM tool must support an SLA life cycle management. As with all modeling techniques that undergo several cycles, an SLA also goes through a group of development phases before its implementation. [TMF 01] identifies five principal phases: service development, negotiation and sale, implementation, execution and evaluation. Other divisions are possible according to the considered SLA's vision. [DAN 03] proposes four development phases: creation, deployment and provisioning, implementation and supervision, cancellation. Moreover, according to the economical scenario, there could be a varying number of sub-phases in each phase of the SLAs life cycle. Principally, this SLA life cycle contains the following phases [TMF 01].

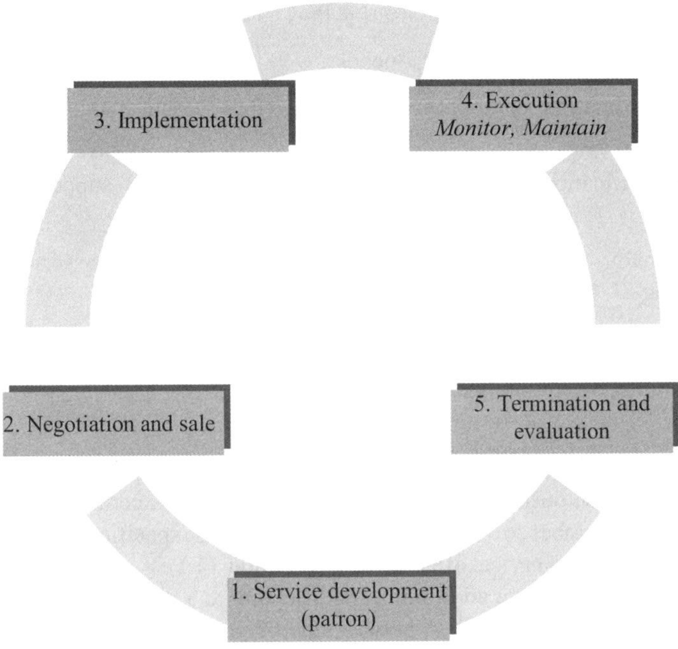

Figure 11.3. *SLA life cycle [TMF 01]*

11.7.1.1. *Service development phase*

The development of a new service offer begins with several external and internal sources.

The need for the development of new applications is especially caused by the aggressive competition on the current network service provider market (ISP, network operators, content providers, etc.). Also, the emergence of new communication technology implies the development of new services and new applications to support them. A last factor, which is just as important, is the group of client requirements for the development of new services, which are sometimes aimed at the specific needs of clients who are prepared to pay the corresponding costs.

In addition, the results of the evaluations of services and existing SLAs, of customer service and of the internal measures and reviews carried out on the revenues and existing service management efficiency contribute to the conception of new solutions to improve the image and quality of the services offered to the customers.

The development of the service itself is thus based on four principal components:

– the identification of the clients' needs;

– the identification of appropriate service characteristics: parameters, values, level of service;

– the identification of resource capabilities available to support the desired service;

– and finally, the development of a basic model for the service and the corresponding SLA.

The final result of this phase is therefore the description of the new service with its corresponding SLA template.

11.7.1.2. *Negotiation and sale*

This is the phase during which the client subscribes to a certain offer of service with a varying number of the concerned parameters specified. The client can subscribe to a global service offer of long duration (a year, for example) and he solicits from time to time an addition or modification to a new sub-service. The final result of this phase is a signed SLA between the SP and their CL.

11.7.1.3. *Deployment and implementation*

This is the phase during which the service is truly deployed (for example, the deployment of specific policies for the CL traffic), tested and activated.

11.7.1.4. *Execution and supervision*

This phase encompasses the agreed-upon duration of the validity of the SLA. It consists principally of tracking the proper functioning of the service, the real-time follow up of activity reports and service levels, as well as the detection and handling of eventual anomalies.

11.7.1.5. *Termination and evaluation*

This phase is composed of two principal parts. The first concerns evaluating the CL's level of satisfaction and identifying the evolution of the CL's needs. The latter represents the SP's internal evaluation to measure the overall satisfaction level of its clients, to group the performance results of the resources that were put into play to assure the SLAs, to compare the obtained performance results with regard to the SP's general goals and finally to propose improvements that will serve as a link to return to the first phase of the SLA life cycle.

11.7.2. *End-to-end SLM*

An SLA is considered an interface between a CL and an SP. In practice, there can be one or more domains and therefore several SPs between the final CL and the actual SP. Each SP domain is supposed to be independently and proprietarily managed.

To reduce the service deployment time, it is necessary to establish a standard for SLA representation and negotiation. Chapter 7 will handle more precisely the problems encountered in the establishment of these SLAs.

11.7.3. *Observing the SP's global objectives*

The observance of the SP's global objectives begins by respecting each SLA concluded with its clients. The terminology used for this is FAB (Fulfillment, Assurance, Billing), which constitutes the key components to ensure a successful management.

An SP must be able to provide, following the different concluded SLAs, a reliable provisioning of resources (routers, servers, processors, etc.). The existence of tools that make it possible to deduct the necessary provisioning operations from a group of SLAs is primordial and constitutes a major challenge to the SOA architecture.

The development of techniques that enable the automatic translation of SLAs into policy rules for the provisioning and management of resources but also the maintenance of QoS supervision and billing is the current solution in network and connectivity service management.

11.8. SLA modeling and representation

After the study we have just carried out on SLAs, we now understand that the starting point of an SOA solution that lends itself well to portability, life cycle management automation and SLA interoperability is that of a standard SLA model.

The standardization task is difficult enough, due to the complexity of the subject, and there currently is no unanimity for a precise SLA model.

We describe here the group of TMF recommendations [TMF 01] for the conception of an SLA. TMF considers that it is very difficult to define a generic model that encompasses all SLA types because it is not always easy to translate client QoS needs in terms that the SP uses in the SLA.

From the view of the TMF, an SP proposes a group of products and services to their clients. This group consists of a number of commercial offers. A commercial offer represents a pack of services offered by the SP to the client and can be a single service (for example, an ATM PVC) or a group of services (for example, an xDSL access with email and Web access).

This commercial offer is composed of a group of service elements. Each service element can be associated to a service class (for example, Gold or Silver). A service element models the capabilities specific to a particular technology (for example, IP connectivity, xDSL connectivity) or operational capabilities (for example, customer service).

Service Resources are the basic blocks that compose basic service elements and are often invisible to the user. Figure 11.4 shows such a service package [TMF 01].

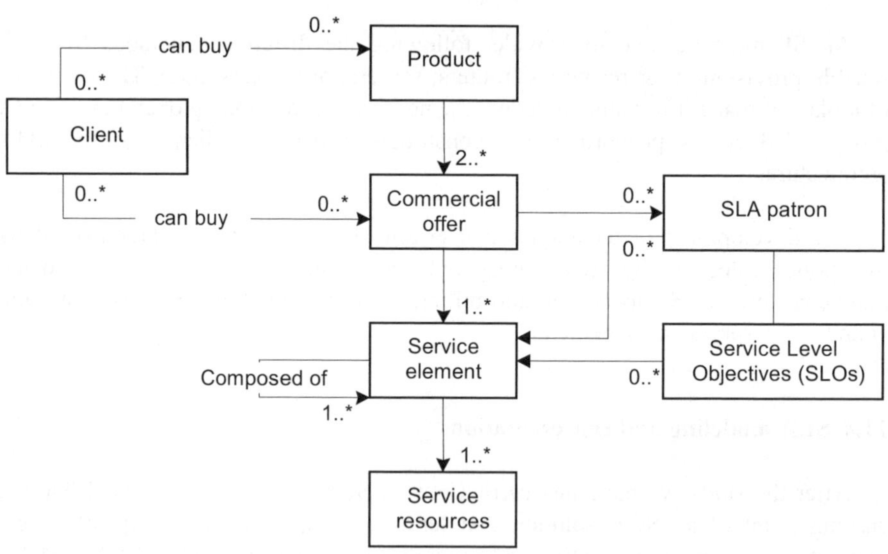

Figure 11.4. *Service composition [TMF 01]*

For example, a residential Internet connectivity service offer can comprise two SLA template instances: basic email and residential access. The basic email template uses the Service Email service element and the residential access template uses the

Access IP service element. Finally, these service elements rely on basic resources for their implementation, such as an email server, an authentication server, a DHCP server, access routers or an access modem.

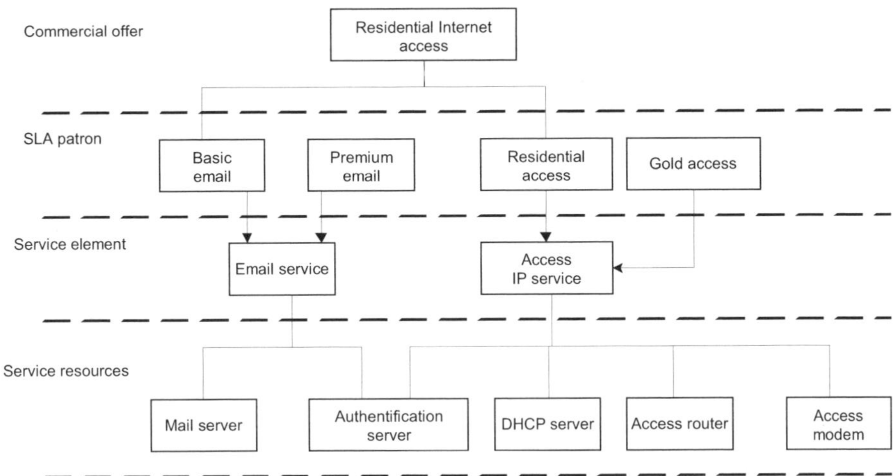

Figure 11.5. *Example of a service package*

An intrinsic part of the development of a service is the production of an SLA template. The role of an SLA template is to capture a group of SLOs for a certain service (see Figure 11.4). The objective of service level is the representation of that guaranteed level of service offered in the SLA.

11.9. Research projects and activities

This section gives a general overview of the principal projects that have addressed QoS and the field of defining SLAs.

There are many projects whose objective is to provide recommendations for constructing the future generation of data communication networks.

11.9.1. *The TEQUILA project*

The objective of the TEQUILA (Traffic Engineering for Quality of Service in the Internet, at Large Scale) project is to study, specify, implement and validate a group of service definitions and traffic engineering tools to obtain end-to-end QoS as required in the Internet [GOD 01].

The TEQUILA group differentiates between a qualitative and a quantitative SLS:

– a qualitative SLS is used to specify a service with relative QoS indicators, such as the packet loss rate;

– a quantitative SLS is used to represent a service for which the QoS indicators are measured exactly, for example, time of transit.

The group does not stop there. It has also defined different SLS attributes. These are described as follows:

– the span: uniquely identifies the geographical or topological region in which the QoS applies;

– flux identifier: identifies a flow of data with specific characteristics;

– conformity and traffic envelope: describes IP traffic flow characteristics;

– excess handling: describes how the service provider will handle excess traffic, in other words, traffic outside the profile;

– performance guarantee: the performance parameters describe guarantees of the services offered by the network to the user for profile traffic only;

– calendar of service: indicates the start and end times of service, for example, when the service is available;

– availability: indicates the average service breakdown allowed per year and the maximum time allowed for service re-establishment after breakdown.

This is, by and large, the SLS format defined by the TEQUILA consortium.

11.9.2. *The AQUILA project*

The AQUILA (Adaptive Resource Control for Quality of Service Using an IP-based Layered Architecture) consortium's objective is to define and implement a layered architecture for QoS support in IP networks [SAL 00]. To accomplish this, the AQUILA consortium has worked on defining the SLS based on the works of the TEQUILA project.

The consortium is in agreement with the need to have a formalized standard SLS representation between the client and the network. This SLS representation should be very general and able to express all the possible service offers based on the

DiffServ model [BLA 98]. The AQUILA consortium has signaled the need for a mechanism to simplify the generic SLS description. This has led to the definition of predefined SLS types [SAL 00].

From the application point of view, a predefined SLS can simplify the interaction between the client and the network. It makes it possible to support a range of applications with the same communication behavior and therefore similar requirements to QoS, such as delay, packet loss, jitter, etc. From the operator's point of view, this simplifies the network management and enables an efficient flow aggregation [SAL 00].

The SLS structure defined by AQUILA is composed of the following attributes:

– SLS type;

– span;

– flow identification;

– traffic description and conformity test;

– performance guarantees;

– service calendar.

As we can see, AQUILA's SLS omits some of the TEQUILA's SLS entries (excess handling, availability) while including a new attribute called the SLS type that distinguishes between a client SLS and a predefined SLS. The predefined SLS values proposed by AQUILA are:

– PCBR – Premium CBR;

– PVBR – Premium VBR;

– PMM – Premium Multimedia;

– PMC – Premium Mission Critical.

11.9.3. *The CADENUS project*

The goal of the CADENUS (Creation and Deployment of End-User Services in Premium IP Networks) project is to provide dynamic creation and configuration across the components linked to the user. This includes the authorization and registration to service components linked to the network along with the proper QoS and compatibility control [ROM 00].

The CADENUS project introduces the notion of dynamic service creation (DSC). This is obtained from communication with many IP service functions and components. These components can be firewalls, proxy servers, SMTP relays, etc. They define CADENUS' premium IP layer, the service creation layer, which manages entities and resources, negotiations and dynamic SLA creation [CAM 01].

CADENUS uses SLA and service templates during the negotiation phase and considers at least two different types of dynamic behaviors:

– user requirements that vary with time;

– network conditions that vary with time (the user is updated with regard to these changes via feedbacks).

Thus, four scenarios can be created by combining each of the types described above [ROM 00]:

– no user requirement variations/no network condition variations;

– user requirement variations/network condition variations;

– no user requirement variations/network condition variations;

– user requirement variations/no network condition variations.

Each of these scenarios needs a different SLA type. The first case needs a static SLA and the other cases need renegotiable SLAs.

11.9.4. *The SEQUIN project*

The objective of the SEQUIN (Service Quality across Independently Managed Networks) project is to define and implement an end-to-end QoS approach that will operate across multiple management fields and will exploit a combination of the two technologies, IP and ATM [CAM 01].

SEQUIN's work focuses on the development of a parameter definition of QoS. Four parameters are adopted:

– delay;

– IP packet delay variation;

– capacity;

– packet loss.

Different QoS classes have been defined based on these parameters.

With the goal of developing the QoS parameter definition cited above, SEQUIN used the works of the TEQUILA, CADENUS and AQUILA projects previously presented and has contributed to the analysis that follows in relation to SLAs.

Multi-domain SLA support requires a group of standardized semantics for SLSs negotiated in different locations [CAM 01]:

– between the client and the service provider;

– within an administrative domain (intra-domain SLS negotiation);

– between administrative domains (inter-domain SLS negotiation).

CADENUS approach attempts to contribute to the support of multi-domain SLAs by defining, initially, detailed and global SLA terms. The detailed SLA (DSLA) refers to the contract between a final user and a service provider. The global SLA (GSLA) refers to the inter-network contract of multi-domain scenarios created by service providers, with the purpose of supporting their final client SLAs. An GSLA takes into account traffic aggregates passing from one domain to another. In general, there is no direct relationship between DSLA and GSLA [CAM 01].

TEQUILA approach handles the support of multi-domain SLAs by supposing that the service negotiation process is composed of a service subscription and a service invocation phase. The following scenarios are possible in the negotiation of inter-domain SLS [CAM 01]:

– hop-by-hop SLS negotiation (where a jump is an autonomous system – AS);

– end-to-end SLS negotiation (with the pre-establishment of interAS conduits);

– local SLS negotiation.

Consequently, the SEQUIN project defines a simple SLA and an SLS architecture that conforms to its own end-to-end QoS development needs across multiple management domains.

The SLS can contain the following:

– two types of predefined SLS as described in the AQUILA approach:

- premium IP,

- IP+,

– a group of attributes as defined in TEQUILA and AQUILA;

– a group of predefined values for each of these attributes (by using the IETF's IP performance measures as metrics the values of which will be presented to the users as an assurance for the network's conformity with the current SLA);

– use of implementation directed by the data (flow) from a recursive service level negotiation between the transportation sub-domains, for the support of inter-domain SLAs, as described by CADENUS.

Finally, the SEQUIN project mentions that by comparing TEQUILA, AQUILA and CADENUS approaches to SLAs, we get the impression that they evolve from AQUILA's predefined static SLS to the entirely specified SLS creation architecture in TEQUILA and the most dynamic CADENUS aspect, thus making DSC possible [CAM 01].

11.9.5. *The EURESCOM project*

Founded in 1991 by 20 European network operators, EURESCOM (European Institute for Research and Strategic Studies in Telecommunications) is an institution leader in collaborative telecommunications R&D [EUR 01]. Some of the projects developed by EURESCOM have great importance in defining SLAs. Some of these projects are mentioned below.

11.9.5.1. *The QUASIMODO project*

Indicated as QoS MethODOlogies and solutions within the service framework: measuring managing and charging QoS.

The QUASIMODO project started in 1998 with a first objective of proposing a QoS model that considers user level QoS classes and network-level network performance parameters. Its second objective was to execute tests with a few significant services and applications with the purpose of finding the correlation between QoS classes and performance parameters [KRA 01].

Among the most pertinent results of this project, we can consider the development of a QoS model. The principal characteristics of this mode are summarized as follows [KRA 01]:

– a group of application categories is considered (this makes the model conscious of the applications);

– the users can choose a quality class. These classes are created such that a user will always see the difference between one class and another;

– a group of network performance parameters is to be managed in order to provide the required quality class to the user for a certain application category.

Laboratory tests (with experts) and acceptability tests (with real users) were carried out on certain significant applications belonging to different application categories (principally VoIP, Streaming and e-commerce on the Web). These tests were carried out to find the correlation between quality classes and network performance levels. The results of these tests show that, for each application category, it is possible to create two quality classes such that they present different network performance values (in terms of delay, jitter and loss) and are perceived differently by the user [KRA 01].

The project's results do not directly contribute to the definition of SLA, but it is interesting and useful to consider the conclusions concerning different quality classes being effectively perceived by the users.

11.9.5.2. *The "interoperator interfaces for ensuring end-to-end IP QoS" project*

This project started in 2000 with the intention of supporting the interests of European operations in the managed interaction of networks and IP services, especially in the framework of QoS. In the goal of perceiving QoS, end-to-end QoS must be supported. If the service is offered through many operator domains, these operators must cooperate to ensure that the client requirements are achieved. The processes of inter-domain management, interfaces and models are necessary for the support of this cooperation [BRU 01].

The main objectives of this project are [BRU 01]:

– to understand the new management requirements presented by the end-to-end IP services and to link these requirements to existing works and standards;

– to produce specifications for management processes, models and interfaces required to support the assurance of end-to-end IP QoS;

– to detect network performance monitoring requirements and parameter measuring of services required for the support of these processes, interfaces and models.

The results of this project are published in a group of documents, which are very precious for the project's objectives. A document called *Measurement of Performance Metrics and Service Events* is certainly interesting for defining and implementing SLAs. The goal of this document is to identify key-service performance and event metrics that need to be controlled for supporting IP QoS inter-domain management and to identify the different abstraction views and levels required by the users and receivers of this information [BRU 01].

The document skips through the traffic profile and the service performance metrics that the vendor and the buyer of quality IP services can use to verify the accomplishment of the SLA. Many primary metrics have been described by using the IETF's and ITU-T's contemporary definition. The metrics are derived from transit and end-to-end services, including the definitions of QoS availability.

11.9.6. *The Internet2 – Qbone project*

Internet2 – Qbone is a consortium that receives a network of QoS test in the Internet. This group defines the concept of Bandwidth Broker. From the point of view of Internet2 – Qbone, policy control, policy admission control, AAA functions, network management functions, intra- and inter-domain routing are affected by or affect the bandwidth broker.

The project's goal is not to define a general SLA or a standard detailed SLS. Its first goal is to attempt to provide Premium Service based on the EF PHB of the DiffServ architecture.

Among the parameters agreed upon by the provider and its customer for a Premium Service, we include:

– the start and end time of the service;

– the source and the destination;

– the size of the MTU;

– the peak flow.

The Premium Service guarantees are:

– no loss caused by congestion;

– the delay is not guaranteed;

– the IP Delay Variation is guaranteed on the condition that the IP route is not changed.

Premium Service is unidirectional and the excess traffic is rejected.

In this project, we suppose that the SLA/SLS between two adjacent domains (peer domains) is statically established and configured. Then, the bandwidth broker receives requests that require a Premium service, called resource allocation requests. Based on these requests, the definition of Premium service, and the SLA/SLS between two domains, the bandwidth broker will deduct the necessary PHB and the traffic conditioning mechanism, and will configure the boarding routers.

More specifically, the SLS declares that the traffic of a certain class must follow certain traffic polishing conditions, enter the domain by a specific link, be handled with certain PHB(s) and whether the traffic's destination is in the domain that receives the traffic.

However, the TCS (Traffic Conditioning Specification) is defined as an entity that specifies the classifier rules, the traffic profiles, the marking, reject and modeling rules that are applicable to a traffic selected by a classifier. Here are some of the parameters that can be included in a TCS:

– performance: flow, reject probability, delay;

– span: domain entrance and exit points;

– traffic profile: token bucket parameters;

– how to handle excess traffic;

– marking;

– modeling;

– mapping to a well-known DSCP service.

Internet2 – QBON distinguishes between SLA/SLS and resource reservation. SLA/SLS is a means of making it possible for a company to have a number of resources and resource reservation is the means for company members to obtain resources within the limit of available company resources. This idea is illustrated in Figure 11.6.

An SIBBS (Simple Inter Bandwidth Broker Signaling) protocol enables the discussion between two bandwidth brokers to fulfill dynamic resource reservation. SLS and resource reservation are both based on well-known services in the DiffServ environment.

11.9.7. *The ARCADE project (ARchitecture for Adaptive Control of IP Environment)*

The ARCADE project has reunited academics and R&D centers (LIP6, INRIA) and industries (France Télécom R&D, Thomson-CSF Communications, QoSMIC).

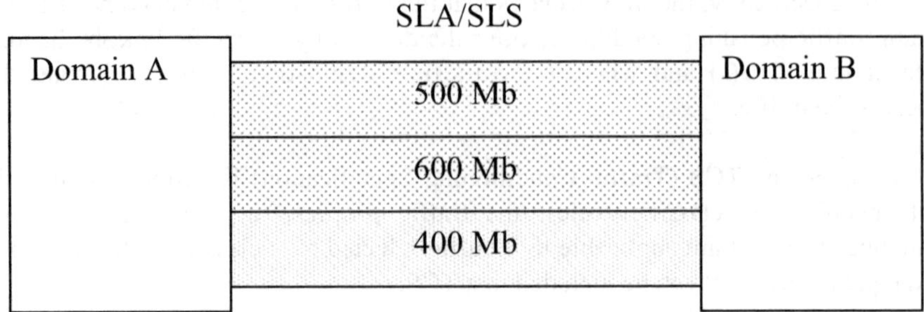

(a) SLA/SLS makes it possible for domain A to have:
500 Mb of type Resource 1
600 Mb of type Resource 2
400 Mb of type Resource 3
and domain A pays domain B for these resources

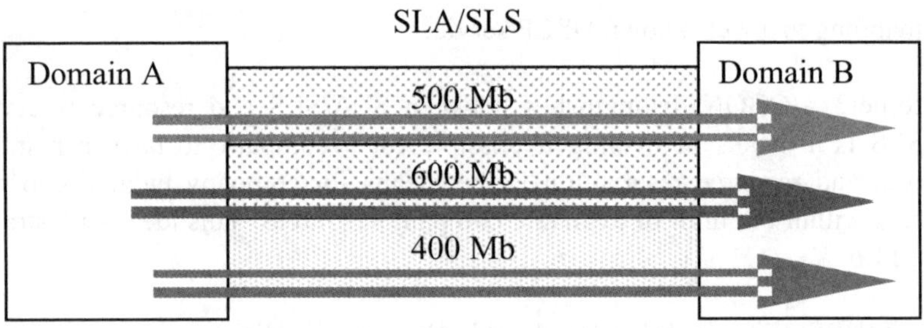

(b) Resource reservation makes it possible for domain A to dynamically assign its resources to its traffic

Figure 11.6. *SLA/SLS and resource reservation*

The goal of the ARCADE project [ARC 01] is to draw up a general model that makes it possible to control IP networks. This control is based on the determination of a profile for each user and client, with the aim of allowing it to communicate adapted resources. These resources can by dynamically allocated. The control is carried out on the security, mobility and QoS. The part of the architecture that was conceived and developed in this project concerns the policy servers and the definition of an intelligent interface (an extension of the COPS protocol) between the policy server and the IP network nodes. This protocol is called COPS-SLS.

COPS-SLS is a protocol defined by LIP6, ENST and Alcatel [NGU 02] for the dynamic negotiation of a level of service. The idea of COPS-SLS is to define a means of communication between the client and the network and also between networks to establish an end-to-end QoS through a policy management system. The result of the SLS negotiation will be implemented by PDPs (Policy Decision Points) that take care of resource allocation in the concerned domains.

COPS-SLS can be used to dynamically provide an intra- or inter-domain SLA while dynamically establishing an end-to-end QoS.

The SLS parameters used in COPS-SLS are based on the parameters defined in TEQUILA, apart from the generally static parameters such as reliability. In addition, COPS-SLS makes it possible to negotiate predefined services (for example, well-known services) as well as unpredefined services. The rest of this part presents defined and potentially defined parameters in COPS-SLS.

The QoS level in a data flow is expressed in terms of:

– service ID: this parameter can be used for a well-known service or a defined service in the predefined SLS mode. When the Service ID is used, it is not necessary to specify duration, jitter and packet loss parameters in the message;

– delay: the duration required in the negotiation phase;

– jitter: the jitter required in the negotiation phase;

– packet loss: the packet loss rate required in the negotiation phase.

These parameters can be gathered from the applications that are QoS conscious. For example, the size of the playback memory in streaming applications can define duration and jitter needs.

Traffic is characterized by different types of rates and parameters according to the conformance algorithm and the QoS level required by the data flow:

– conformance algorithm: simple token bucket, tri-color token bucket, etc.;

– peak flow;

– average flow;

– maximum flow;

– maximum burst size;

– excess handling.

The extent of negotiations identifies the starting point and the end of the QoS guarantee. It makes it possible for a domain to know if it is final or not.

– Starting point: [IP address(es)] the point(s) from which traffic needs a service level guarantee.

– End point: [IP address(es)] the point(s) where the service level guarantee ends.

The identification of flows defines parameters for recognizing the flow at a domain's entrance. These parameters depend on the type of data flow for which the client wants to negotiate a service: IP address, source sub-network's mask length, port numbers, protocol number, DSCP, MPLS label, IPv6 flow label.

The service calendar specifies the start and end time of the SLS: hour, date, etc.

Each COPS message can transport more than one SLS. Each SLS is identified by an SLS ID. The client-handle value is used to identify different requests sent by the SLS client.

11.10. Conclusion

Significant steps in SLA design have been completed by TEQUILA, AQUILA, CADENUS, SEQUIN and ARCADE. Their results represent a good basis for defining any type of network SLA.

It is currently possible to define flexible and more dynamic SLAs addressing certain changes in one or more parts, notably the SLS. The SLA no longer concerns a single domain but several domains. This is how the inter-domain SLA was born, which created an end-to-end reflection involving a chain of SLAs. Each can be dynamically negotiated via specific negotiation protocols.

11.11. Abbreviations and acronyms

AAA	Authentication, Authorization, Accounting
CBR	Constant Bit Rate
CL	Client
COPS	Common Open Policy Service
COPS-SLS	Common Open Policy Server – Service Level Specification
DiffServ	Differentiated Services
DSC	Dynamic Service Creation
DSCP	Differentiated Service Code Point

DSLA	Detailed SLA
ENST	Ecole National Supérieure des Télécommunications
e-Service	Electronic Service
FC	Final Client (end user)
GSLA	Global SLA
ID	Identifier
IETF	Internet Engineering Task Force
IP	Internet Protocol
ISP	Internet Service Provider
ITU-T	International Telecommunication Union – Telecommunication Standardization Sector
LIP6	Informatics Laboratory, University Paris 6
MPLS	Multi-Protocol Label Switching
MTU	Maximum Transport Unit
PBM	Policy-Based Management
PHB	Per Hop Behavior
QoS	Quality of Service
SDNM	Service Driven Network Management
SLA	Service Level Agreement
SLO	Service Level Objective
SLS	Service Level Specifications
SP	Service Provider
TCS	Traffic Conditioning Specification
VBR	Variable Bit Rate
VoD	Video on Demand
VoIP	Voice over IP

11.12. Bibliography

[ARC 01] http://www-rp.lip6.fr/arcade.

[BLA 98] BLAKE S., BLACK D., CARLSON M., DAVIES E., WANG Z., WEISS W., "An Architecture for Differentiated Services", *Request for Comments: 2475*, December 1998.

[BRU 01] BRÜGGEMANN H., "Inter-operator Interfaces for Ensuring End to End IP QoS", EURESCOM P1008 project, http://www.eurescom.de/public/projects/P1000-series/p1008/default.asp.

[CAM 01] CAMPANELLA M., CHIVALIER P., SEVASTI A., SIMAR N., SEQUIN "Deliverable 2.1-Quality of Service Definition", http://www.dante.net/sequin/QoS-def-Apr01.pdf.

[CIM 99] Common Information Model (CIM) Specification, version 2.2, DMTF, June 1999.

[DAN 03] DAN A., LUDWIG H., PACIFICI G., "Web Services Differentiation with Service Level Agreements", *IBM Software Group Web Services Web site*, 2003.

[EUR 01] EURESCOM home page, http://www.eurescom.de.

[FAW 04] FAWAZ W., DAHEB B., AUDOUIN O., BERDE B., VIGOUREUX M., PUJOLLE G., "Service Level Agreement and Provisioning in Hybrid Photonic Networks", *IEEE Communications Magazine Special issue on "Management Of Optical Networks"*, January 2004.

[GOD 01] GODERIS D., T'JOENS Y., JACQUENET C., MEMENIOS G., PAVLOU G., EGAN R., GRIFFIN D., GEORGATOS P., GEORGIADIS L., VAN HEUVEN P., "Service Level Specification Semantics and Parameters", *Internet Draft*, *draft-tequila-sls-01.txt,* work in progress, June 2001.

[KRA 01] KRAMPELL M., "QUASIMODO – QUAlity of ServIce MethODOlogies and Solutions Within the Service Framework: Measuring, Managing and Charging QoS", EURESCOM P906 project, http://www.eurescom.de/public/projects/P900-series/p906/default.asp.

[MAR 02] MARILLY E. *et al.*, "Requirements for Service Level Agreement Management", *IEEE*, 2002.

[MAR 02] MARTIN J., NILSSON A., "On Service Level Agreements for IP Networks", *IEEE INFOCOM 2002*, 2002.

[MEL 01] MELIN J. L., "*Qualité de Service sur IP*", Editions Eyrolles, 2001.

[NGU 02] NGUYEN T. M. T. *et al.*, "COPS Usage for SLS Negotiation", *Internet Draft*, <draft-nguyen-rap-cops-sls-03.txt>, July 2002.

[RAI 03] RÄISÄNEN V., *Implementing Service Quality in IP Networks*, John Wiley & Sons, 2003.

[ROM 00] ROMANO S. P., ESPOSITO M., VENTRE G., CORTESE G., "Service Level Agreements for Premium IP Networks", *Internet Draft*, draft-cadenus-sla-00.txt, November 2000.

[SAL 00] SALSANO S., RICCIATO F., WINTER M., EICHLER G., THOMAS A., FUENFSTUECK F., ZIEGLER T., BRANDAUER C., "Definition and Usage of SLSs in the AQUILA Consortium", *Internet Draft*, draft-salsano-aquila-sls-00.txt, November 2000.

[TMF 01] TMF, "SLA Management Handbook", *Public Evaluation 1.5*, June 2001.

Chapter 12

New Approaches for the Management and Control of IP Networks

12.1. Introduction

The IP packet transfer network is currently migrating from a simple network with a best effort guarantee to a network that includes Quality of Service (QoS) and security. Given the broad range of devices that can compose an IP network, integrating QoS and security will only exponentially increase the complexity of its management. Moreover, this management is not common to all the organizations that make up the big picture (Internet operators, access networks, company networks, etc.). In order to facilitate this management, which appears to be very complex and dependent upon the organization that manages the IP network, the policy-based management approach has recently been proposed and its different components are being standardized.

The main goal of this approach is to integrate the management of all the network's components into a single management system. This management system will then need to enable the application of a global management strategy (a policy) appropriate for the organization that is managing the IP network. Such a system will enable the introduction of organization management policy rules and will apply these rules to adequate areas of the network. This succeeds in causing:

– a reduction in cost that has been unceasingly increases as related to management;

Chapter written by Yacine GHAMRI-DOUDANE.

– a reduction in the complexity of management and control issues;

– a centralization of management information which makes it possible to ensure the integrity of the global management strategy.

This technique can be used in an operator network or in a company network, as well as in an access network.

12.2. Network management policies

In this section, we will attempt to answer two questions:

– What is a policy in the context of networks?

– What task will a policy have within the framework of network management?

12.2.1. *Definition of the term "policy"*

As the term "policy" [RAJ 99] is widely used, it is imperative to give a clear definition in the context of networks. As a starting point, the term policy denotes a unified rule for accessing resources and network services based on administrative criteria. Figure 12.1 denotes different levels to which such a rule can be expressed and exerted.

Let us take, for example, the case of managing the QoS on the Internet. The network view of a policy is expressed in terms of end-to-end performance, of connectivity and of dynamic network states; it will also take topologic criteria into account.

This view is composed of several nodal views. These views correspond to the objectives and the needs of the policy at the level of different nodes. These are composed of policy rules, which must be seen as atomic injunctions through which different network nodes are controlled. As each node possesses specific resource allocation mechanisms[1], each nodal view must finally be translated into specific instructions to the node's hardware devices, in other words, into a hardware device view.

---

1 Specific in a non-standard sense: each industry has its own mechanisms and implementations installed in the network devices (nodes) it produces.

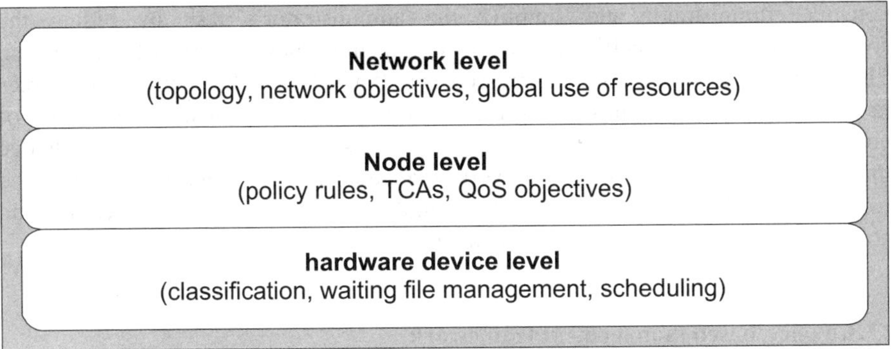

Figure 12.1. *Conceptual policy hierarchy*

12.2.2. *Objective of policy-based network management*

A policy [VER 00] is a report that defines what traffic must be processed by the different network elements and how this traffic will be processed. This is defined by an administrator and is applicable to the IP network he administers. We therefore speak about administrative domains.

The network administrator must specify the rules that are related to the handling of different types of network traffic. An example of a policy can be as follows: "all traffic between the accounting and finance departments in the company must be encrypted", or "all IP telephony traffic must be classified as *premium*". The policies are therefore a means of representing different actions that a network element can apply. The application of these actions can also be conditional. In fact, the preceding policies can be written respectively in the form of rules, as follows:

– **If** ((dpt_src == 'acct') & (dpt_dest == 'finance')) **THEN** encryption;

– **If** (traf_type == 'VoIP') **THEN** classification ('premium').

Policies which are defined above are called business level policies. As illustrated in this example, these can be easily introduced by the administrator. Following this, they must be translated to network level and then to lower levels as described in the previous section. Thus, the business level represents a fourth possible view of a policy. Polices which are defined are therefore seen as a means to simplify the management of different network element types which can deploy very complex technologies.

To sum up, a policy has the goal of enabling the definition of high level objectives for the operation of the network. These rules are established by the network administrator and simplify the administrator's task by hiding the configuration details specific to network equipment or more generally to the system. These functional objectives are expressed as policy rules for the control of resources in the network. These rules are defined actions that must be active when a set of conditions are met. The policies provide a way to specify and automatically change the operational strategies for network management and control. Policy-based management has many assets for implementing efficient, coherent and comprehensible network management and control systems.

12.3. Policy-based management framework

For a fast deployment of a new business-level strategy, the network manager needs a set of tools that enable him to define and implement high level decisions in the network. This implementation must be flexible and without specific knowledge of the characteristics of the equipment deployed in the network [RAP 04].

The Internet Engineering Task Force (IETF) *rap* working group [RAP 04] and the Desktop Management Task Force (DMTF) *policy* working group [POL 04] have specified a general framework that makes it possible to take into account the scaling factor for the policy definition and administration. This general framework introduces a group of components that enable the definition, the representation and the realization of policy rules. There are four such components:

– a policy administration console;

– a policy repository;

– a policy decision point (PDP);

– policy enforcement points (PEP).

Figure 12.2 illustrates the interactions between these components.

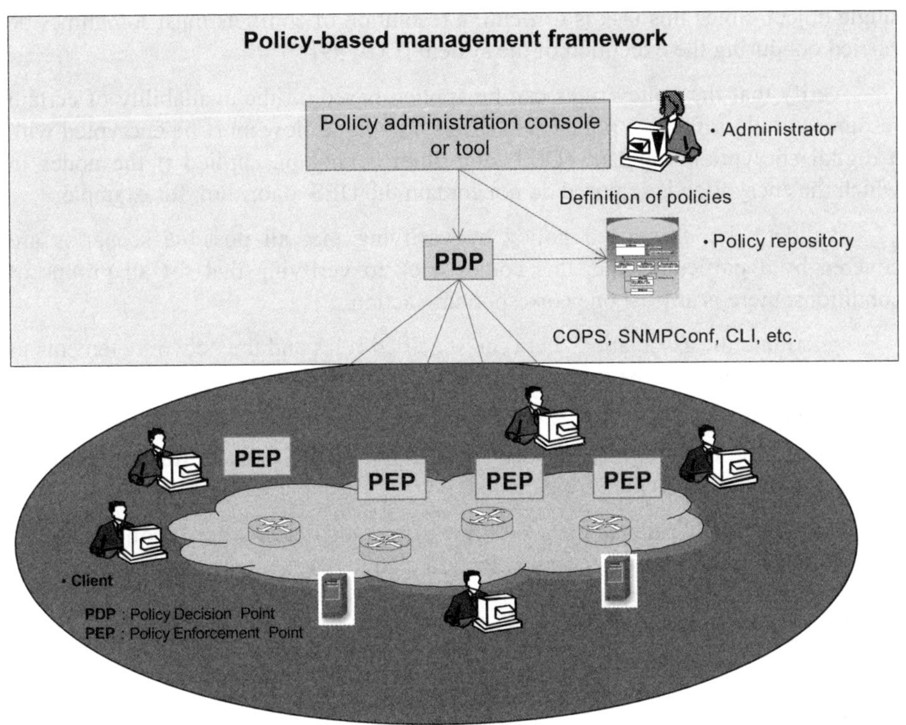

Figure 12.2. *Policy-based management framework and its components*

12.3.1. *The policy administration console or tool*

The policy administration console is the tool that enables the network administrator to interact with the management system. In fact, thanks to this console, the administrator can define and introduce policy rules to be applied to a network. This station must fulfill the following tasks, in the order in which they are listed:

– present an interface from which the administrator can introduce the rules: the structure of the introduced business-level rules must be very simple (human language);

– enable the verification of the syntax and semantic integrity of the introduced rules: this includes type-checking and the verification of the values specified for the parameters that constitute the policy. Conformity, independent or not of relationships between the parameters, must also be verified;

– make it possible to verify that the introduced rules are consistent between each other: rules are inconsistent if at a given moment conflicting actions are applied on a single object. Since this task is difficult, a resolution of conflicts must sometimes be carried out during the execution of the system [LUP 99];

– verify that the policy rules can be applied based on the availability of certain resources or algorithms: a policy rule specifying that a flow must be encrypted with a digital encryption standard (DES) algorithm cannot be applied if the nodes in which the encryption is required do not contain the DES algorithm, for example;

– validate the introduced policy by verifying that all possible scenarios are covered by a particular rule; this comes back to verifying that for all groups of conditions, there is at least one corresponding action;

– determine the associations between specified rules and the network elements to which these rules must be applied; identify the places where each introduced rule must be applied;

– correspond each high-level policy rule to one or more low-level rule based on the group of network elements concerned;

– store the policy rules in the *policy repository* and detect when a policy rule has been changed in the repository so that the affected network elements can be notified. This notification is made via the PDP which makes adequate decisions at each modification, such as the introduction of a new policy.

12.3.2. *The policy repository*

The policy repository, or the directory of policy rules, is the place where policy rules are stored by the administration console. These policy rules are then used by the decision-making element, the PDP. This directory offers a central point from which all the network management activity is carried out. The policy repository can as easily contain the business-level policy rules as the lower-level policy rules in the policy hierarchy (see Figure 12.1).

The LDAP directory [HAD 02] is an example of a policy repository; we can even say that in most cases we would use an LDAP directory to store policy rules. In fact, the storage method for policy rules in this type of directory concentrated a maximum of effort regarding standardization. This does not necessarily mean that we cannot use other technologies for storing policy rules; on the contrary, a simple database or a web server can be used to this effect.

Besides the problem of where to store policy rules, the most important thing is to know in which format these rules will be stored. Indeed, representation conventions of the information contained in the specification of policy rules must be defined. These conventions are called information models. More details on this part of the information modeling for policy rules are given in section 12.6.

12.3.3. *PDP*

The PDP is the entity responsible for decision-making based on stored policy rules in the policy repository. Decisions made from this base are actions to be applied to network elements. These are therefore seen as being enforcement points for policy rules, or PEP. The PDP is therefore responsible for the following tasks:

– identify the group of rules applicable to the different PEPs under its responsibility. It accesses this group of rules in the policy repository;

– transform this group of rules into a format or syntax that is comprehensible by PEP functions. This refers more often to the notion of *policy information base* (PIB) [MCC 01];

– regularly verify the current state of the network in order to validate the required conditions for the enforcement of a policy;

– notice changes in policy rules and take into account these changes. This can be done by the continuous supervision of the policy repository or via a notification from the administration console.

12.3.4. *PEP*

The PEP is responsible, on the one hand, for applying the PDP decisions; it is therefore concerned with the enforcement of the action part of the policy rules. On the other hand, the PEP must supervise and collect a certain number of statistics and information that can modify the functioning of the network. Therefore, it eventually has the task of transmitting these in the form of a report to the PDP. This enables the PDP to verify the proper unfolding and installation of PEP decisions as well as eventually initiating the application of other policy rules, since their condition part has changed state.

As previously described, the PEPs are often located along the path followed by the traffic of data. Based on the type of policy rules, it can be located in the routers, firewalls, terminals or other proxies. Each of these entities responds to a specific function that can be managed via a policy rule.

The policy-based management architecture as described here can be seen as a client/server architecture. In this view, the PEP plays the role of a COPS client, whereas the PDP plays the role of the COPS server [BOY 00], the COPS protocol being the means of communication between the PEP and the PDP.

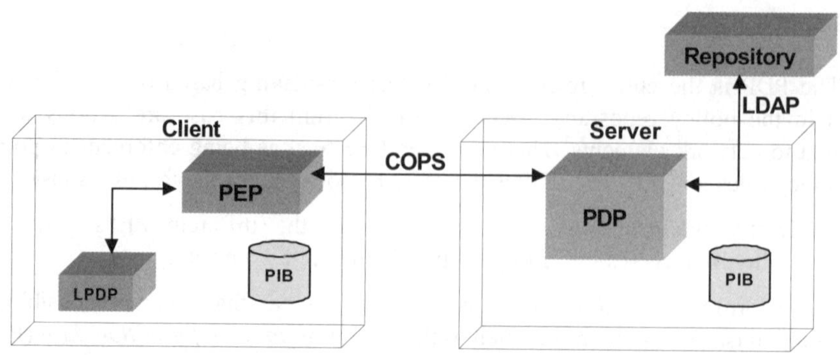

Figure 12.3. *Client/server architecture of the COPS protocol*

12.4. COPS protocol

The COPS protocol [BOY 00] describes a client/server model for the support of alerts between a PDP and a PEP (see Figure 12.3). This model makes no supposition on PDP functionalities and methods; it is based, however, on the fact that the policy server, the PDP, returns decisions to PEP (client) requests. It is therefore a request/response type protocol.

For reasons of reliability of the transport of policy requests and responses, COPS uses the Transport Control Protocol (TCP). The protocol also offers authentication, data protection and integrity control mechanisms.

The COPS model always uses the directories for storing information linked to policy rules and it also introduces the PIB notion for the storing of information relating to low-level policies and relative to a particular type of client. In fact, the COPS protocol presents a generic management model that is flexible in the sense that it can be applied to several policy domains (in other words, QoS provisioning policies, access network control policies, etc.). In this case, it is a question of defining an extension of the COPS protocol by defining suitable objects and a value for the type of client. This value is indicated in the *client-type* field in the COPS message header. Based on this value, the following objects are translated appropriately. As examples of this type of client, we cite the two that are standardized by the IETF *rap* working group, which are:

– COPS-RSVP [HER 00a]: the first proposed extension of the COPS protocol (client-type = 1). Its objects transport policies for the admission control of RSVP message [BRA 97];

– COPS-PR [CHA 01]: another extension of the COPS protocol. Its objects transport rules for router configuration. The configuration of QoS parameters in DiffServ routers [BLA 98, NIC 99] is one use case handled by COPS-PR.

The COPS protocol is adapted to the different management strategies. In fact, this protocol supports the two management models: the *outsourcing* model and the *provisioning* model. In the outsourcing model (in other words, the case of COPS-RSVP), a request is sent by the PEP when there is a request for a resource (such as an RSVP message arriving at an RSVP router). The sent request is therefore an explicit decision request relative to this event. Upon receiving the decision, the policy is applied to this event (accept/reject the reservation request). In the provisioning model (the case of COPS-PR for DiffServ), the idea is not the same. In this model, the PEP notifies its presence to the PDP at the onset. This notification has the role of a request. In fact, from that precise moment, the PDP continuously checks if it has configuration rules to apply to this PEP and applies them if necessary. When the resource request event arrives at the PEP (a DiffServ packet arrives at a DiffServ router), the PEP does not send a request to the PDP to ask for the decision relative to this event, but applies the corresponding policy rules that are already installed (in other words, marks the packet based on the rule that applies to it and puts it in the corresponding queue, etc.).

12.4.1. *Segment and COPS message format*

As illustrated in Figure 12.4, a COPS segment is composed of a header and of data. The header contains information making it possible to identify the message.

The most important fields in this header are the operation or the sent message (OP-CODE) type as well as the client-type. The data includes the objects and information concerned by the COPS message. Each of these objects identifies a specific type of data. We are solely interested here in messages that can be exchanged between the PEP and the PDP.

These messages are illustrated in Table 12.1. This table's objective is to facilitate the understanding of the global functioning of the protocol.

Figure 12.4. *COPS segment structure within a TCP/IP segment*

OP-CODE	Message	OP-CODE	Message
1	REQ (Request)	6	OPN (Client-Open)
2	DEC (Decision)	7	CAT (Client-Accept)
3	RPT (Report State)	8	CC (Client-Close)
4	DRQ (Delete Req. State)	9	KA (Keep-Alive)
5	SSQ (Sync. State req.)	10	SSC (Sync. Complete)

Table 12.1. *COPS messages and associated OP-CODEs*

OPN, CAT and CC messages are used to establish connections or sessions. These messages are exchanged between the PEP and its serving PDP according to the sequence chart presented in Figure 12.5. The KA message is periodically sent by the PEP to the PDP to check if the TCP connection is still active. This message therefore enables the connection between the PEP and the PDP to remain active as long as it is required. The PEP also transmits PDP requests by using the REQ message and creates reports on the proper enforcement of PDP decisions via the RPT message. The RPT message can contain, apart from information on the success or failure of the installation of a policy, statistic information related to the functioning of the network element on which the PEP is installed. This supervision information represents a continuous update of network state information, which is stored by the PDP in the policy repository. The DEC decision message is sent by the PDP in response to the received REQ message. There should always be a DEC message and only one in response to an REQ message. However, this DEC message can, based on the context and the type of client, convey several objects. Among these objects, the Error object can appear to notify the PEP that an error was encountered during the message formatting or that one of the objects sent in the request is unknown.

The DRQ message is sent by the PEP to notify the PDP that its state is not available or is not up to date. The PDP then makes provisions for the use of objects sent in the DRQ message by the PEP. The SSQ message enables the PDP to explicitly ask the PEP for its internal status information. The SSC message is sent by the PEP in response to the PDP's SSQ message for the synchronization of these internal statuses. Safeguarding of respective states between the PEP and the PDP consequently renders the COPS protocol a stateful protocol.

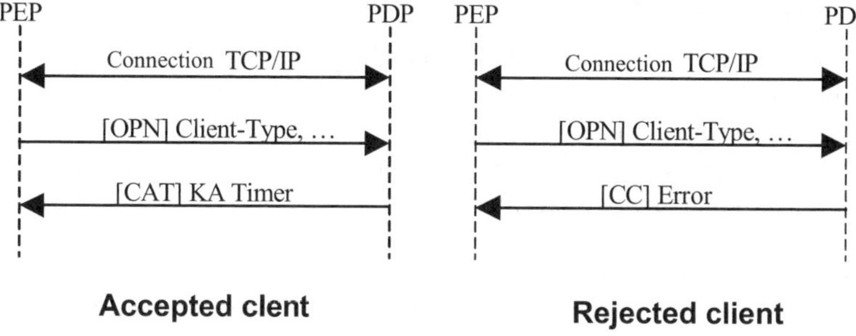

Figure 12.5. *Exchanges between PEP and PDP*

12.5. Policy domains

The notion of management policy, as introduced in the preceding sections, can be applied to the management of different sectors in the domain of IP networks. This notion can also be applied to other types of networks or in all other domain which is not related to networks. Each sector to which a policy-based management can be applied will be denoted under the name policy specialty [VER 00].

The two sectors that currently present the most interest in the network field are QoS and security. In each of these two sectors, several techniques are in competition to achieve a certain number of service objectives. In the following two sub-sections, we will describe some of these techniques and we will introduce a way of using the policies for their management.

12.5.1. *QoS*

Three main technologies have emerged during the last decade to offer QoS in the framework of IP networks:

– traffic shaping;

– differentiated services, or DiffServ;

– integrated services, or IntServ.

Each of these techniques has its own policy to offer or guarantee a certain QoS level and to consequently define its own needs in terms of management. These can be addressed by a global management policy.

12.5.1.1. *Traffic shaping*

This very simple technique makes it possible to reserve the bandwidth for flows traversing a link that can be subjected to congestions. Typically, this mechanism is installed in the output link of company or campus routers in order to enable an efficient sharing of the bandwidth and to reserve more bandwidth for priority traffic, whether for the company or the campus.

In fact, the bottleneck is in these access links to the Internet. The technique of traffic shaping makes it possible to control the bandwidth use by different network applications and to attribute adequate shares of the bandwidth to these applications.

The resulting questions are the following:

– What are the highest priority applications?

– What is the exact portion of bandwidth to be given to each class of applications?

– What will be done if a priority traffic class immediately needs more bandwidth than was attributed to it?

– Under which conditions shall unused bandwidth be distributed and how will the distribution be made?

There is no single response to each of these questions. In fact, each organization can have its own policy for answering these questions. A policy-based management can therefore serve to translate the responses to its questions (business-level policy) in a group of configuration policies for this element (low-level group of rules).

12.5.1.2. *Integrated service*

The second technology is IntServ technology [SHE 97]. The idea here is that each flow must reserve the resources it needs all along the path it will travel. In order to do this, the reservation uses the RSVP signaling protocol [BRA 97, WRO 97]. The packets belonging to traffics that have made these reservations must then obtain a better service, in terms of end-to-end delay and bandwidth, than the packets of those that have not made any reservations. These will have a best effort service like the one currently offered by the Internet.

The role of policies in the IntServ model is to respond to the following two questions:

– Who is allowed to request resource reservations using RSVP?

– Which requests must be honored by a router and which must be rejected?

For the first question, we an answer with: only those organizations having signed a service level agreement (SLA) contract with the network operator can claim the use of services specified by this service contract. This contract includes a certain number of information which makes it possible to answer to the second question. In fact, following the contents of the contract (maximum threshold of reserved resources, maximum duration of service use, etc.), the decision to accept or decline a new request can be made.

In order to achieve these objectives, the IETF *rap* working group has defined a particular client-type for the support of RSVP messages by the COPS protocol [HER 00a]. This group also defined RSVP protocol extensions to enable its interoperability with a policy-based management system [HER 00b].

12.5.1.3. *Differentiated service*

DiffServ technology [BLA 98, NIC 99] relies on the capacity of network elements to classify passing IP packets. This classification implies marking the packets based on their respective classes. The marking process is carried out by the terminals or network access routers using the DiffServ technology. These terminals or network access routers are part of the so called DiffServ domain. When the packet is correctly marked, it will be processed by the routers it will meet on its path. This processing will consist of giving the adequate degree of priority to the packet for resource access.

The DiffServ architecture defines two types of network elements: access routers and network core routers. Each of these has a group of specific tasks: classification and marking of IP packets based on the contents of their header for the access router, and the differentiated processing based on the previous marking for the core network router.

The access router can also control and limit the allocated flow rate of a particular flow. Policy rules therefore enable the access router:

– to correspond each flow, as a result of its identifier[2], to a particular service class via an appropriate marking;

– to apply future limitations in bandwidth for each flow and/or aggregate.

The core network routers must apply a unique processing on a traffic aggregate, and not flux by flux, thus simplifying their management task. This processing includes, among others, attributing priority parameters between different aggregates

2 The flowId field in IPv6, or the five-part field <addr_src, addr_dest, port_src, port_dest, num_proto> in IPv4.

thereby enabling differentiated processing. The configuration of this group of priorities can then be carried out by policy [CHA 03].

12.5.2. *IP security*

Access and communication securitization issues require a lot of attention. Among the currently used techniques for introducing security in Internet networks, the following can benefit from policy-based management:

– the installation of packet filters in the firewalls aiming at protecting a local network from external attacks;

– the use of secure socket layer (SSL) [RES 01] for encrypting applicable data before transmitting within the network; this solution aims at ensuring confidentiality and flow integrity when they travel through the Internet;

– the use of IPSec (IP security) tunnels [IPS 04] to encrypt network data; this network security, contrary to application security brought about by SSL, is carried out independently of the wish of the application to encrypt this data transparently to this one. This technique enables the encryption of all of the traffic between two points (routers or terminals) on the network.

The use of policy rules makes it possible, in these three cases, to facilitate the deployment of a new security policy, for example, or to configure parameters and algorithms for the introduction of a certain level of security.

12.6. Information modeling

One of the big standardization challenges in policy-based management is the creation of a common information model for the policies. This model will then be extended to the different possible cases.

During the design of any large software system, we often need to store data somewhere: a database, a data directory, files or other. An information model (or data model) is therefore a simple abstraction which enables the description of the structures and the type of data or information that must be stored. It is often more practical and especially more flexible to use an object-oriented representation for the specification of these information models. This results in a representation that is easier to understand and carry out. The information model must identify object types and their attributes as well as the relationships between objects (for example, inheritance relationships).

When a real system is developed, the defined information model must then be carried out by using a particular technology. In fact, if the real system uses an LDAP directory, for example, information model objects, defined independently from the technology used, must then be translated into an LDAP schema.

From the perspective of network and system management, the DMTF policy group has defined a common information model for policy-based management. This model is called common information model (CIM) [CIM 03]. This model is common and extensible; it must in fact be extended for each particular policy domain. The IETF policy and IPSP groups, for example, are proposing extensions to this model for QoS policy management [MOO 01, MOO 03a, MOO 03b, SNI 03] and for security [JAS 03] in IP networks.

In the following sections, we will briefly describe CIM functioning as well as its extensions. For more details on CIM operation, its extensions or its realization under LDAP, see [C2L 00, CIM 03, JAS 03, MOO 01, MOO 03a, MOO 03b, SNI 03].

12.6.1. *The CIM model*

The first version of the CIM, proposed by the DMTF, appeared in April 1997. It was created for the purpose of standardizing a company's technological information representation. It has not ceased to evolve since then [CIM 03] and includes, in its latest versions, an end-to-end model for company information management or network service management. This management model is called end-to-end because it includes material aspects as well as software and service aspects. The goal of CIM is to offer a comprehensive and consistent base model for system and network management.

This CIM model defines a conceptual information model describing the entities to be managed, their composition and the relationships between them. The model is called basic or common because it was not developed for a domain or a particular problem, or for a particular implementation. Its sole purpose is to address end-to-end management. CIM therefore defines the content and the semantics of manageable entities in a company or in an IP network.

The management model is composed of a core model and common models that extend the core model. These common models were defined for all technological domains with information to manage: from networks and operating systems to distributed applications and databases. Other common models were developed for security support, event management and management infrastructure.

This model was conceived to make the distinction between logical and physical aspects of the manageable entities. The physical model describes how the material components are physically configured. It thus defines the information model to be used for the group of low-level policy rules (for hardware devices). The logical aspects of the CIM model describe the high-level policy rules (for example, network policies). The logical model is more abstract. The functional or behavioral aspects of the managed entities are defined by the logical model. The core model and the common models that extend it also cover a significant breadth of manageable elements. This includes low-level devices to applications, as illustrated in Figure 12.6.

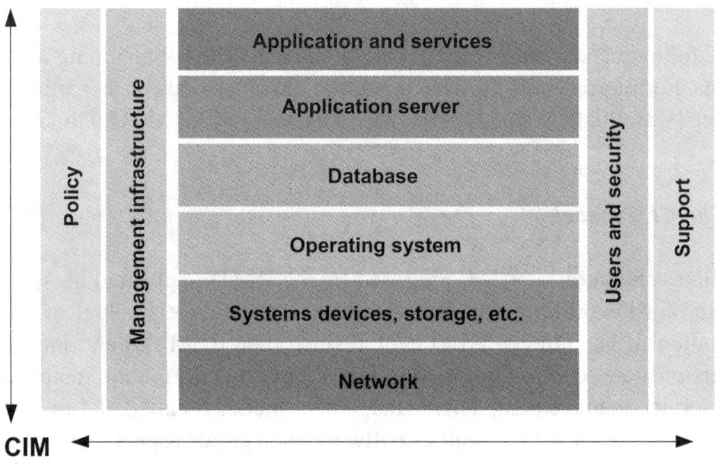

Figure 12.6. *CIM core model extension and common models*

Each of the common models defines appropriate entities that need to be managed for a particular technological center. They also define the way they can interact with other global model entities.

The richness of the CIM model lies in its completeness and its object-oriented design. This design enables it to be easily reusable, in part or completely, or to be easily extensible. Note that reusing a standard and valid model is always easier than to conceive a new one from scratch.

A new service or system can therefore benefit from all parts of the CIM model and extend it to the new entities it introduces. This extension will be done most often by inheritance of object classes previously defined by the CIM. The CIM model can therefore be specialized for a particular issue.

12.6.2. *CIM extensions for IP networks*

The generalization and extensibility of the CIM model have allowed it to be reused by IETF *policy* and IPSP working groups to define information models regarding policy-based QoS and security management. Standard models standardized by the IETF, based on CIM, are:

– policy core information model (PCIM) [MOO 01] and PCIM extensions (PCIMe) [MOO 03]: these are basic models governing the representation of network management policy information and the control information of this policy. The PCIMe model was introduced to mitigate certain lack of PCIM identified during PCIM extension to IP QoS management and IP security. The two models, PCIM and PCIMe, are directly derived from CIM;

– QoS policy information model (QPIM) [SNI 03]: is an extension of the PCIM model. It makes it possible to model the information used for the administration, management and control of the access to QoS resources making it possible to manage a network's QoS. More specifically, it enables the definition of policy rules for resource access in terms of QoS in the DiffServ and IntServ models;

– QoS device datapath information model (QDDIM) [MOO 04]: QDDIM is also a PCIM extension. This model makes it possible to model low-level policy rule information (for hardware devices) concerning network and terminal elements. It concerns the configuration of mechanisms that enable traffic shaping along the paths borrowed by the data. This model is also defined so as to function with the two IETF-defined QoS models: DiffServ and IntServ. It must, however, be used in conjunction with QPIM, which it complements;

– IP security policy (IPSP) [JAS 03]: this model is also a PCIM and PCIMe extension. It was conceived to facilitate the configuration of IPSec protocol parameters [IPS 04], for the implementation of secure tunnels, as well as the Internet key exchange (IKE) protocol [HAR 98], for the exchange of encryption keys.

12.7. Conclusion

In this chapter, we have introduced the new approach recommended for the management and control of the next generation of IP networks. In fact, this new generation, which should incorporate new mechanisms to provide QoS and security, introduces a significant complexity with regard to control and management. The recommended approach for this control and management, whether for standardizing organisms or for research initiatives, is the policy-based management approach. In fact, this has already demonstrated its attributes in terms of flexibility and ease of management of complex systems. This is achieved by the definition of several points of view of a policy. This approach also defines the mechanisms, components, models and protocols which enable the translation of a business-level policy (in human language) into the configuration of a given equipment. This chapter also makes it possible to appreciate the advancements linked to this new management approach.

12.8. Bibliography

[BRA 97] BRADEN R., ZHANG L., BERSON S., HERZOG S., JAMIN S., "Resource Reservation Protocol (RSVP) – Functional Specification", *RFC*, 2205, September 1997.

[BLA 98] BLAKE S. *et al.*, "An Architecture for Differentiated Services", *RFC*, 2475, December 1998.

[BOY 00] BOYLE J., COHEN R., DURHAM D., HERZOG S., RAJA R., SASTRY A., "The COPS (Common Open Policy Service) Protocol", *RFC*, 2748, January 2000.

[C2L 00] CIM-TO-LDAP MAPPING SPECIFICATIONS, "Guidelines for CIM-to-LDAP Directories Mapping", DEN Ad Hoc WG, May 2000.

[CHA 01] CHAN K. *et al.*, "COPS Usage for Policy Provisioning (COPS-PR)", *RFC*, 3084, March 2001.

[CHA 03] CHAN K., SAHITA R., HAHN S., MCCLOGHRIE K., "Differentiated Services Quality of Service Policy Information Base", *RFC*, 3317, March 2003.

[CIM 03] CIM SPECIFICATIONS, "Common Information Model (CIM) Specification Version 2.8", *DMTF Policy WG*, August 2003.

[HAR 98] HARKINS D., CARREL D., "The Internet Key Exchange (IKE)", *RFC*, 2409, November 1998.

[HER 00a] HERZOG S. *et al.*, "COPS usage for RSVP", *RFC*, 2749, January 2000.

[HER 00b] HERZOG S., "RSVP Extensions for Policy Control", *RFC*, 2750, January 2000.

[HOD 02] HODGES J., MORGAN R., "Lightweight Directory Access Protocol (v3): Technical Specification", *RFC*, 3377, September 2002.

[IPS 04] GROUPE DE TRAVAIL IPSEC DE L'IETF, http://www.ietf.org/html.charters/ipsec-charter.html.

[JAS 03] JASON J., RAFALOW L., VYNCKE E., "IPsec Configuration Policy Information Model", *RFC*, 3585, August 2003.

[LUP 99] LUPU E., SLOMAN M., "Conflicts in Policy-Based Distributed Systems Management", *IEEE Transactions on Software Engineering, Special Issue on Inconsistency Management*, 25(6), p. 852-869, November/December 1999.

[MCC 01] MCCLOGHRIE K. *et al.*, "Structure of Policy Provisioning Information", *RFC*, 3159, August 2001.

[MOO 01] MOORE B., ELLESSON E., STRASSNER J., WESTERINEN A., "Policy Core Information Model – Version 1 Specification", *RFC*, 3060, February 2001.

[MOO 03] MOORE B. (ed.), "Policy Core Information Model (PCIM) Extensions", *RFC*, 3460, January 2003.

[MOO 04] MOORE B., DURHAM D., STRASSNER J., WESTERINEN A., WEISS W., "Information Model for Describing Network Device QoS Datapath Mechanisms", *RFC*, 3670, January 2004.

[NIC 99] NICHOLS K., JACOBSON V., ZHANG L., "A Two-bit Differentiated Services Architecture for the Internet", *RFC*, 2638, July 1999.

[POL 04] GROUPE DE TRAVAIL *POLICY* DU DMTF, http://www.dmtf.org/download/about/workgroups/slaWGCharter.pdf.

[RAJ 99] RAJAN R., VERMA D.C., KAMAT S., FELSTAINE E., HERZOG S., "A policy framework for integrated and differentiated services in the Internet", *IEEE Network Magazine*, vol. 13, no. 5, p. 36-41, September/October 1999.

[RAP 04] GROUPE DE TRAVAIL RAP DE L'IETF, http://www.ietf.org/html.charters/rap-charter.html.

[RES 01] RESCORLA E., *SSL and TLS: Designing and Building Secure Systems*, Addison-Wesley Professional Publishing, 2001.

[SHE 97] SHENKER S., WROCLAWSKI J., "General Characterization Parameters for Integrated Service Network Elements", *RFC*, 2215, September 1997.

[SNI 03] SNIR Y. *et al.*, "Policy Quality of Service (QoS) Information Model", *RFC*, 3644, November 2003.

[VER 00] VERMA D.C., *Policy-Based Networking – Architecture and Algorithms*, New Riders Publishing, Indianapolis, 2000.

[WRO 97] WROCLAWSKI J., "The Use of RSVP with IETF Integrated Services", *RFC*, 2210, September 1997.

Chapter 13

Internet Security

13.1. Introduction

The importance of security has greatly increased following attacks against certain large-scale sites such as Amazon or Yahoo. These attacks, along with increasing fears of cyber-terrorism since September 11th, 2001, have encouraged researchers to find means and methods of protection for users and machines.

The aim of this chapter is to give a view of the state of the art of security in the Internet. The first part addresses a few elements of security. Next, we will present Internet security according to its field of application. Firstly, we will address the security of user data. Then, we will discuss the security of the Internet infrastructure. Finally, we will look at protecting user installations.

13.2. Elements of security

The international NS standard ISO 7498-2 [NF 90] defines security services and the associated mechanisms. Firstly, we will give the definitions of some of the security services. Then, we will give a few elements of cryptography before citing a few security mechanisms cited in the ISO document. Finally, we will broach the problem of key management in the Internet.

Chapter written by Vedat YILMAZ.

13.2.1. *Security services*

Authentication makes it possible to identify a communicating entity and/or the source of data.

Confidentiality ensures the protection of data against all unauthorized disclosures.

Integrity services neutralize the modification of the message between its source and its destination.

Access control ensures protection against all unauthorized use of resources accessible via the OSI architecture. These resources can be OSI or non-OSI resources, reachable via OSI protocols. In the context of the Internet, the definitions linked to OSI are extended to TCP/IP protocols and the resources accessible via these protocols. These resources can be TCP/IP protocols or the resources accessible via these protocols.

The objective of *non-repudiation* is to provide the receiver with a means to disallow emissions from the emitter, or to the emitter a means that disallows the receiver to deny reception.

13.2.2. *Cryptography*

The principle of cryptography is to transform an illegible message for all into a message readable only by authorized people. The basic principle is that cryptography is bi-directional: it enables the encryption and decryption of a message by restoring its original form.

Symmetrical cryptography requires the source and the destination key information, used for encrypting and decrypting. The main issue with this method is that the message sender must communicate its key to the receiver via a secure channel.

In the case of asymmetric cryptography, the sender and receiver each have a private key and a public key, both determined by mathematical methods. This method guarantees that data encrypted with a private key can be decrypted with the corresponding public key.

13.2.3. *Security mechanisms*

Encrypting can ensure confidentiality and play a role in a number of other security mechanisms.

The process of digital signing implies either the encryption of a unit of data, or the creation of a cryptographic control value for the unit of data by using the signer's private information as a private key. The signature can only be created by using the private information of the signer. Therefore, when the signature is verified, we can prove that only the unique holder of the private information can have produced the signature. The signature ensures non-repudiation.

Access control can use the authenticated identity of an entity or information relative to the entity to determine and apply access rights of the entity. If the entity tries to use an unauthorized resource, the access control will reject this attempt.

The determining of unique data integrity implies two processes, one for the emitter and the other for the receiver. The emitting entity adds a value to a data unit, the value being a function of the data. This value can be supplementary information, such as a block control code or a cryptographic control value, and can also be encrypted. The destination entity generates a corresponding value and compares it to the received value to determine if the data was modified during transit.

The following are some of the techniques that can be applied to authentication exchanges:

– the use of authentication information, such as a password (provided by an emitting entity and controlled by the destination entity);

– cryptographic techniques;

– the use of characteristics and/or information unique to the entity.

Properties which are relative to data communicated between two or more entities, such as their integrity, their origin, their date and their destination, can be guaranteed by providing a notarization mechanism. The guarantee is provided by a notary (third party) that is trusted by the communicating entities and holds the necessary information for providing the required guarantee in a verifiable manner. When calling on this notarization mechanism, the data is communicated between the communicating entities via the protected communication instances and the notary.

13.2.4. *Key management issue*

We have seen that asymmetrical cryptography does not require the encryption key, which means it is particularly adapted to applications that use networks (such as the Internet). In fact, it is only the public key that is transmitted via a directory and it is only used during encryption.

However, nothing guarantees that the public key is really the one associated with the user. A hacker can corrupt the public key by replacing it with his own public key. The hacker is then in a position to decrypt all messages having been encrypted with the current directory key. It is for these reasons that the notion of a certificate was implemented. A certificate enables the association of a public key to an entity (a person, a machine, etc.) to ensure validity. The certificate is in some way the identity card of the public key. It is delivered by a trusted third party (TTP) called a certification authority (CA). We speak then of public key infrastructure (PKI) [PKI 04].

The private key must always be protected in a fixed and safe location. These days, the users are more and more driven to mobility. They can be called to change their work tools. It is both risky and complicated to ask users to carry their own private keys. With the technological advances in the chip card industry, a private key is being associated with it more and more. This guarantees that the private key cannot be copied and remains in the control of the subject.

The use of biometrics presents a great interest in security. Identifying a person uniquely by their physiological characteristics (retina, iris, digital imprints, voice, etc.) can have different applications. For example, imagine its use in PKI, where users would have a chip card with digital imprint identification.

13.3. User data security

Securing data on the Internet requires an analysis in order to clearly define the security needs. What data needs to be secured? Between points A and B, does the entire stream need to be secured or just a critical sub-set? Also, must the protected stream be completely protected or only partially so? In other words, must we apply security mechanisms to the application by adding a protective field? Or can we simply ensure a protection in the transfer? Unless we do not want to get to the IP level to implement a VPN IPSec [IPS 04] from end to end or only on a portion of the network. The Internet, although built on TCP/IP protocols, requires IP packets to be transported via frames (level 2) that can take various forms between a source and a destination. So even if the highest security is available at the level of IP and higher, what should de done if level 2 is poorly protected? We will think notably of the problems linked to the transmission method in Ethernet networks and wireless network broadcasting, such as Wi-Fi whose use is increasing more and more. Finally, the Internet relies also on physical infrastructures. Despite available security protocols for level 2 and higher, how will we protect ourselves against failures or malicious use of property linked to the physical environment?

We will present an aspect of security at each level and will attempt to answer these few questions.

13.3.1. *Application level security*

Applications, in the framework of the Internet, include the flow that uses the services of sub-layers starting with the transport layer (or the network layer in the case of raw IP packets). The traditional network applications are concerned, such as databases or network games (often using UDP for transportation); also concerned are the network services that make the Internet a success, such as FTP for file transfer, SMTP and POP for messaging, DNS for machine naming and others.

We will see forms of security brought to this level. Firstly, we will see application security with two examples: PGP and tattooing. Secondly, we will see security extensions applied to an application protocol, in this case RSVP.

13.3.1.1. *Application security*

PGP (pretty good privacy) [IPG 04] is a cryptography software in the public domain. It is based on asymmetric cryptography (RSA) and on symmetric cryptography (IDEA at 128 bits). Its most current use is in the protection of emails. It can also offer confidentiality by encryption and digital signature authentication. PGP uses IDEA for the text with a secret random key that is different for every session. This key is encrypted with the destination public key and transmitted with the message. Upon reception, PGP decrypts the secret key with the private RSA key. PGP decrypts the data with the previously obtained secret IDEA key.

Before speaking of tattooing, let us recall steganography. Steganography enables the hiding of an image in a document. One of its interesting applications is watermarking. It no longer hides an image within a document, but marks it indelibly. The main purpose is the protection of copyright. Information relating to the author is inscribed in the document such that nobody can appropriate it. This is applicable to images as well as to software. Then, each copy of the document contains the same mark, that of the legal owner.

In these two examples, we use cryptographic properties in the interest of applying mechanisms to application flows. The security services can be considered as a processing layer between the initial information and the TCP/IP layers.

13.3.1.2. *Security extensions*

RSVP [BRA 97] is a protocol that requires resource reservations in the network. It is an indication whose objective is to ensure QoS for data streams. RSVP messages rely on UDP/TCP.

Take two users: Alice and Bob. Alice wants to have a videoconference with a certain QoS with Bob; Alice initiates an RSVP signal in the network so that the adequate resources are allocated.

Here, various needs linked to security appear. First there is the authentication of the entity or of the user. We have to make sure that it is really Alice's entity who is making the request and that this entity communicates with the network entities she trusts. If this is not the case, another entity can usurp the identity of Alice's machine and use the service for which Alice pays. Also, the request of Alice's entity can be redirected to a malicious node for perverse reasons. User authentication can be added to entity authentication to ensure the identification of the user of the improved service.

In RSVP, extensions were brought for services other than authentication, namely integrity and anti-replay. In fact, a usurper could modify Alice's request so that it was never accepted by the network, for example, by excessively modifying the QoS parameters. Also, this same usurper could, after hearing Alice's messages, decide to replay the same messages to attempt to access the improved services for which Alice is paying.

To counter such risks, the IETF decided to enrich the structure of RSVP messages. A new RSVP object was created to ensure the authentication of entities (origin of data) but also to arm itself with possible modifications (integrity) of other RSVP objects, as well as the replay of RSVP messages: this is the Integrity object [BAK 00].

Here is an illustration of the contents of this object, followed by a brief description of the attributes of interest.

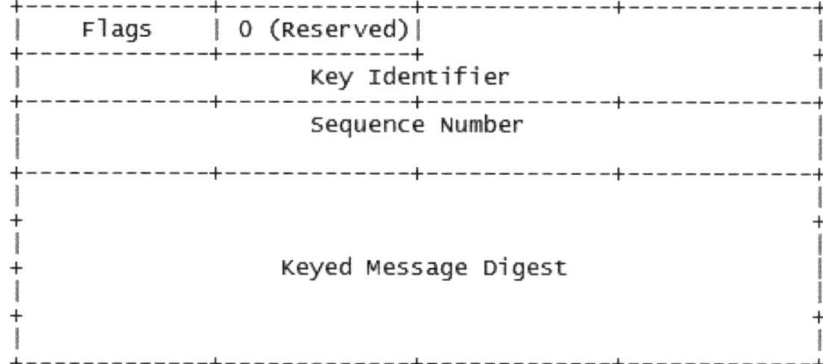

Figure 13.1. *RSVP Object Integrity*

The *keyed message digest* is used to ensure the integrity of indication messages by calculating a condensed version of the entire RSVP message with the help of a key. The *sequence number* is used by the receiver to distinguish between a new and a replayed message.

We note that the use of a condensed version is not proper to the RSVP Integrity object. In fact, this technique is largely used to address integrity and authentication protection of the origin of the data at any level.

13.3.2. *Transport level security*

In this section, we will give an overview of transport-level security. The most widespread protocol is called *security socket layer* (SSL) [FRE 96], proposed in 1994 by Netscape.

SSL is a client/server protocol that enables the authentication of the origin of the data, confidentiality and integrity of exchanged data. It is independent of communication protocols. It is an intermediate layer between application protocols like HTTP (access to www servers), FTP, Telnet, etc., and the TCP layer. SSL is composed of a key generator, of chopping functions, of RC4, MD5, SHA and DSS encryption algorithms, of negotiation and session management protocols whose main one is the *handshake protocol*, and of X509 certificates.

SSL is especially used by the HTTP application (called HTTPS). Its success is notably due to its ease of use and to its integration in all the browsers on the market[1].

---

1 If the server that sends the information uses SSL, you will see at the bottom left of the browser a small key or lock that automatically appears.

Many commercial companies have found here a means of communicating securely with their clients, notably to obtain payment for services rendered. Typical uses are the transfer of credit card codes for online sales sites and viewing bank account information.

The better-known implementations in HTTP are in Netscape and Apache Web servers, which use, in France, 40-bit RSA keys[2]. However, today SSL is known for its vulnerability to brute force attacks (exhaustive) when using 40-bit keys. It is therefore recommended that 128-bit keys are used instead.

Today, a workgroup called *transport layer security* (TLS) [TLS 04] is active within the IETF to ensure the proper functioning, standardization and the eventual evolution of the SSL protocol. The last evolution is currently known as TLS, which takes SSL and improves it, but is not compatible.

The security of SSL/TLS distinguishes itself by its capacity to adapt to an application and does not need additional application security functions.

In terms of security services, we see once again the use of a message authentication code ensuring the authentication of the origin and the integrity; we also have the possibility of making the data confidential.

13.3.3. *Network level security*

SSL or TSL is not a layer that secures all applications. Moreover, this protocol was conceived to ensure generally punctual transactions. These are neither regular nor periodic. Although this security addresses current Internet needs, securing the entire stream, and permanently so, is sometimes necessary. Also, it would be preferable to act without modifying the application itself[3]. This is proposed in the continuation[4] of *IP security* or IPSec protocols [IPS 04].

IPSec provides security services to the IP layer by enabling a selection of security protocols (AH [KEN 98a] or ESP [KEN 98b]), algorithm determination (DES, 3DES, SHA1, MD5, etc.) and keys to be used for these services (through the IKE negotiation protocol). IPSec offers access control services, connection-less data integrity, data origin authentication, protection against packet loss, confidentiality.

2 RSA keys are part of the asymmetrical cryptography domain and are especially used during the entity authentication phase.
3 The use of SSL/TLS requires secure socket manipulation functions.
4 IPSec appears to be a unique protocol, but it is really a series of protocols.

These services offered in IP are independent from higher-layer protocols (TCP, UDP, ICMP, etc.).

There exists the notion of mode in IPSec, with transport mode and tunnel mode. In transport mode, security services are applied to the data of higher-layer protocols and in the tunnel mode, security services are applied to IP packets that are encapsulated into IP packets.

Among the notable defaults in IPSec are parameter setting, key management and "too-full" security.

Parameter setting in IPSec is quite complex given the multitude of data to be managed between two IPSec nodes A and B. Among these are the authentication method, the mode, the protocols, the algorithms and the lifespan of security associations. Today, even if two IPSec stock editors are meant to support some of these mandatory parameters according to IETF standards, there are always interoperability problems, either because the standards are not applied to the letter, or due to an incorrect configuration of IPSec nodes.

The key management problem touches not only IPSec but also the entire field of data security in general. Whichever the authentication method is used by the nodes, a secret must be implemented on each node before establishing an IPSec security association. If we consider the X509 certificate that is widely used these days, following the example of SSL/TLS, a request will first have to be lodged with a CA to obtain a signed certificate. The difference relative to a current use of SSL/TLS is that the client/server mode no longer exists: each of the IPSec nodes will have to obtain its certificate. On most Web servers using SSL/TLS, the latter possess their certificate and the client that connects obtains the server's certificate to authenticate it. The "medium" client will not have to worry about creating his certificate with a CA. However, with IPSec, he will have to manage this task. IPSec ensures a mutual authentication, whereas in most cases with SSL/TLS, only the server will be authenticated by the client.

Up until now, we have had an overview of the available security in the TCP/IP layers. Even if for many the Internet can be restricted to these principal layers, the proper functioning relies also on a network's link and physical layers.

13.3.4. *Link level security*

Today, a user who connects to the Internet often does it through a local network, a traditional network (Ethernet) or a wireless network (Wi-Fi). Evidently, there exist many other level 2 methods and technologies to access the network of networks. The goal here is solely to put forward certain existing menaces on level 2 technologies currently used.

The Ethernet (10 Mbps, Fast or Giga) is known for its broadcast method by which information sent from a machine A to a machine B is visible by all other machines on the same Ethernet branch. From here, if the encapsulated information in the Ethernet block of data is not already secured, a malevolent user can read everything that is going on. For example, if the integrity is ensured to whichever higher level, the information is always readable. To ensure confidentiality, all PCs must ensure encryption.

In traditional Ethernet technologies, anybody can come and plug their PC on the network to found out what is going on. From the moment where some information such as the MAC address, the IP address and even the user's DNS name are known, the game is up. Domain control enables user authentication and ensures access security in both Windows and Linux environments. Having said that, there is a plethora of protocol analysts (sniffers) who can read what is going on in an Ethernet network, without connecting to the domain[5].

For wireless access, the Wi-Fi standard (*wireless fidelity* or IEEE 802.11b) generalizes as much for companies as for private networks. It presents numerous security gaps. Above all, Wi-Fi communications can be intercepted by an external user. If there are no security policies, an intruder can access network resources by using the same wireless equipment. A first level of protection is available by using WEP (wired equivalent privacy). However, WEP works with 64-bit encryption keys (128-bit keys are optional) that can be too easily hacked. Also, there is neither encryption key distribution mechanism nor veritable user authentication. In conclusion, WEP only offers weak security.

To improve WLAN security, the IEEE is aiming toward an encryption that reuses point-to-point protocol techniques. Many solutions are considered to improve security in these networks. In the short term, the use of a protocol such as SSL, SSH and IPSec is recommended. The Wi-Fi Alliance proposes a standard based on 802.1x/EAP authentication mechanisms and an improved encryption (TKIP, temporary key integrity protocol) whose group is called WPA (wireless protected access). The IEEE 802.11i workgroup has standardized this environment by proposing an extension using the AES (WPA2) encryption algorithm.

5 We can also listen to a network without being connected.

13.3.5. *Physical level security*

All securities reviewed up until now were of a protocol nature. In this section, we will see defaults linked to the physical environment. They apply to the Internet, as well as to networks that do not enforce the use of TCP/IP protocols. It is not an exhaustive list of defaults logged in the physical layer of the Internet, but a relatively comprehensive example.

A fault in the networks, as well as in the electronic and data processing fields in general, is the electromagnetic radiation. Radiation presents the opportunity for external attacks. We also refer to *side channel attack*. Globally, the Internet is composed of PCs linked by physical links through routers. Each of these components is susceptible of producing electromagnetic radiation. An interesting example is the PC screen. As stated by Quisquater and Smyde at SECI02, "it is important to note that the light emitted from a screen contains useful information that is nothing more than the video signal, etc. it is possible to reconstruct a video image from the luminosity of a distant screen".

In the middle of the 1950s, the American military initiated the Tempest project [TEM 04], whose goal was to develop a technology capable of detecting, measuring and analyzing electromagnetic radiation emitted by a computer to intercept data from a distance. Thus, Tempest makes it possible to capture a screen's electromagnetic emissions and recreate them into an image. This is distance image duplication. Everything that the spied upon computer displays on its screen, Tempest captures and reconstructs the image in real-time. Today, the NSA could read a screen through walls and up to a kilometer away. It is no longer necessary to spy on TCP/IP packets to understand what the victim is doing.

To reduce the security distance from a kilometer to a few meters, we can make sure that all the peripheries are class B (class A, which is the most common class, offers hardly any protection) and that our cables are shielded (especially between the screen and the UC); RJ45 cables can serve as antennae if they are poorly protected.

13.4. Internet infrastructure security

This section attempts to give answers to the primordial question: is the Internet reliable and secure? Today, we know that Internet infrastructure attacks can lead to considerable damage, from the moment where principal components, such as DNS, routers and communication links have implicit confidence relationships between each other.

[CHA 02a] gives an overview of the faults in the Internet structure and the responses put forward. It presents the attacks against the Internet infrastructure in four categories: DNS hacking, poisoning of the routing table, poor packet processing and service denial.

13.4.1. *DNS*

The *domain name system* (DNS) is the global hierarchical distributed directory that translates machine/domain names into numerical IP addresses in the Internet. Its faculties have made the DNS a critical Internet component. Thus, a DNS attack can affect a large portion of the Internet. Its distributed nature is a synonym of robustness but also of different types of vulnerabilities.

To reduce a request's response time, DNS servers store information in a cache. If the DNS server stores false information, the result is a poisoned cache. In this case, the aggressor can redirect traffic to a site under his control. Another possibility is a DNS server controlled by an adversary (malicious server) which will modify the data sent to users. The malicious servers are used to poison the cache or to commit DoS on another server. There is also the case of an aggressor that passes for a DNS server and responds to the client with false and/or potentially malicious information. This attack can redirect traffic to a site under his control and/or launch a DoS attack on the client himself.

To answer these possible attacks, the IETF has added security extensions communally known under the term DNSsec [DNS 04]. DNSsec provides authentication and integrity to DNS updates. All attacks mentioned are attenuated by the addition of the authentication of the data source and the authentication of transactions and requests. The authentications are provided by the use of digital signature. The receiver can verify the digital signature with the received data. To make DNSsec viable, secure server and secure client environments must be created.

13.4.2. *Routing table*

Routing tables are used to route packets in the Internet. They are managed by information exchanges between routers. Poisoning attacks correspond to the malicious modification of routing tables. This can be done by modifying routing protocol update packets, creating bad data in the table.

The work on routing protocols in the Internet has been principally directed into two directions: distance vector protocols (for example, RIP) and link state protocols (for example, OSPF). These two types of protocol present different characteristics with regard to state information exchange and routing calculation. In a distance vector protocol, each node *frequently* sends its routing distances *to its neighbors.* A neighbor, upon reception of a distance vector packet, updates its routing table if necessary. In a link state protocol, each node *periodically* inundates the state of its links *to all network nodes*. Upon reception of the link state updates (called *link state advertisement* or LSA, in OSPF), each router calculates the shortest path tree (SPT) with itself as the root of the tree. Distance vector protocols consume more bandwidth than link state protocols. Also, differently from link state protocols, they suffer a lack of complete topology information at each node. This lack of knowledge encourages a variety of attacks that are not possible in link state protocols.

The poisoning of the routing table can be done by link and router attacks. Link attacks, differently from router attacks, are similar for both types of routing protocol.

13.4.2.1. *Link attacks*

Link attacks appear when the adversary has access to the link. The routing information can be intercepted by an adversary, without being propagated any further. However, the interruption is not efficient in practice. There is generally more than one path between two nodes[6]. Therefore, the victim can always obtain the information from other sources. Most routing protocols use updates that use acknowledgments. Due to this, interruptions are detected. However, if the links are selectively interrupted, it is possible to have asynchronous routing tables throughout the network, which can create buckling and DoS. Asynchronous routing tables can also be created if a router suppresses its updates, but sends an acknowledgment.

Routing information packets can also be modified or fabricated by an adversary who has access to a network link. Digital signatures are used for the integrity and authenticity of messages. In the case of digital signatures, the emitter signs the packets with his private key and all nodes can verify the signature by basing themselves on the emitter's public key. Routing updates increase by the size of the signature (typically between 128 and 1,024 bits). It is a viable solution in link state routing protocols because messages are not frequently transmitted. This is also proposed for distance vector protocols. However, these consume an excessive amount of bandwidth. Therefore, adding a header in the form of a digital signature is not very much appreciated by researchers.

---

6 The average degree of each node is relatively high (around 3.7).

The poisoning of the routing table can also be done by replicating old messages, where a malicious adversary holds onto routing updates and replays them later. This type of attack cannot be resolved by using digital signatures because the updates are valid, only postponed. A sequence information is used to prevent this attack. The sequence information can be in the form of sequence numbers or time stamps. An update is accepted if the sequence number in the packet is greater than or equal to the sequence number of the previously received update in the same router. This resolves the replication problem; however, the packets in the same period of time can be replayed if the time stamp is used as sequencing information. No remedy has been found for this problem. However, it has limited effects because it can only be used if a router sends multiple updates within the same time period.

13.4.2.2. *Router attacks*

A router can be compromised and would then be called *malicious*. Router attacks differ according to the nature of the routing protocol.

In the case of a link state routing protocol, the malicious router can send incorrect updates regarding its neighbors, or remain quiet if the link state of the neighbor has indeed changed. Solutions are the detection of an intrusion and techniques added to the protocol. In the detection of an intrusion, a central attack analysis module detects attacks based on a possible sequence of alarm events. However, using such a module is not a solution that can be applied on a large scale. The other solution is to integrate the capacity of detection in the routing protocol itself. This was proposed in SLIP (secure link state protocol) [CHA 02b]. A router believes in an update on the condition as long as it also receives a "confirmation" update of the link state of the node that supports the suspicious link. However, this solution also proved to be incomplete.

In the case of distance vector protocols, the malicious router can send false or dangerous updates concerning any node in the network because the nodes do not have the complete topology of the network. If a malicious router creates a bad distance vector and sends it to all his neighbors, the neighbors accept the update because there is no way to validate it. Since the router is malicious, standard techniques such as digital signature do not work. [SMI 97] proposes a validation by adding old information (predecessor) in the distance vector update. Although it performs for the detection of incoherence, the algorithm presents a few faults. It is incapable of detecting router attacks when a malicious router changes the updates in an intelligent way.

13.4.3. *Poor packet processing*

In this type of attack, the malicious router poorly manipulates the packets, thus generating congestions, DoS or a reduction in connection bandwidth. The problem becomes difficult (or untreatable) if the router selectively interrupts or routes packets poorly, thus causing loop routing. This type of attack is very difficult to detect.

The adversaries can retain real data packets and mistreat them. This is an attack during the data transmission phase, different from the poisoning attack. Poor packet processing attacks have a limited efficiency compared to router table poisoning and to DoS attacks. In fact, these attacks are limited to a portion of the network, whereas poisoning attacks can affect the entire network. However, this type of attack is possible and very difficult to detect.

In a similar fashion to poisoning attacks, an adversary can perform a link attack or a router attack.

13.4.3.1. *Link attacks*

An adversary, after accessing a link, can interrupt, modify/fabricate or replicate data packets. The interruption of TCP packets could reduce the global bandwidth of the network. The source that feels congestion reduces the transmission window provoking a reduction in connection bandwidth.

In [ZHA 98], the authors have firstly shown that the selective suppression in a small number can largely degrade TCP performance. The authors used packet suppression profiles and intrusion detection profiles to identify the attacks. These are the only types of solutions attempted to detect these types of attacks. However, there are still questions regarding scaling the intrusion detection techniques throughout the Internet. Similarly to routing updates, data packets can be modified or fabricated by adversaries. IPSec, the standard series of protocols for adding security characteristics to the Internet IP layer, provides authentication and encryption for data packets throughout the Internet. To counter replay attacks, IPSec integrates a little protocol called anti-replay window protocol. This protocol can provide an anti-reply service by including a sequence number in each IPSec message and by using a sliding window.

13.4.3.2. *Router attacks*

Malicious routers can cause all the link attacks. Moreover, they can poorly route packets. Malicious packet routing can provoke routing congestion toward heavily loaded links, or can even be used as DoS attacks by directing an incontrollable number of packets toward a victim. This attack is cited in Cisco's *white papers* [CIS 00], where packets received and emitted by the same interface of a router are suppressed. This simple filtering scheme can avoid a naïve bad routing attack. However, a malicious router can create loop routing, which remains an open problem.

13.4.4. *Denial of service (DoS)*

DoS are destined to specific machines with the intention of breaking the system or provoking a denial of service. These are done by individuals or groups, often for personal notoriety. They become extremely dangerous and difficult to avoid if a group of aggressors coordinate the DoS. This type of attack is called the distributed DoS (DDoS). It should be noted that a DoS can be the consequence of routing table poisoning and/or poor packet processing.

In general, DoS attacks can be of two types: ordinary and distributed. In an ordinary DoS attack, an aggressor uses a tool to send packets to the target system. These packets are created to put the target system out of service or to quash it, often forcing a *reboot*. Often, the source address of these packets is spoofed, making it difficult to locate the real source of the attack. In the DDoS attack, there can always be just one aggressor, but the effect of the attack is greatly multiplied by the use of attack servers known as agents[7].

Several DoS attacks are well known. The *flooding UDP* makes it possible to send UDP packets with spoofed return addresses. A hacker links the character generation service of a UDP system to the UDP echo service of another system. In a *TCP/SYN flood*, the hacker sends a large quantity of SYN packets to a victim. The packet return addresses are spoofed. Thus, the victim puts SYN-ACK's in queue, but cannot continue to send them because it never receives ACK from the spoofed addresses. Finally, in an *ICMP/Smurf attack*, the hacker broadcasts ICMP *ping* requests with a spoofed return address toward a large group of machines on a network. The machines send their response to the victim whose system is submerged and cannot provide any service.

The proposed solutions for DoS attacks fall into the preventive and reactive categories. All preventive techniques look for the detection of DoS and to do so they base themselves on older information making it possible to ensure filtration. Known techniques use the verification of inverse *unicast* path, the control of the flow of

---

7 To have an idea of the range of a DDos, consider that more than 1,000 systems were used at different times in a concerted attack on a unique server at the University of Minnesota. The attack not only put this server out of service but also denied access to a large network of universities.

SYN packets and the verification of incoming and outgoing interfaces. Reactive techniques attempt to identify the adversary after the attack was carried out.

This is an active field of research because the current identification techniques are completely manual and can be spread out over several months. Current solutions consist of testing the link until reaching the source, logging data packets in the key routers (very loaded solution), ICMP traceback and IP traceback. In ICMP traceback, each router stores a packet with a low probability (1/20,000). When a packet is stored, the router sends an ICMP traceback message to the destination. In IP traceback, each router marks a packet with a reliable probability. In both cases, if there is a DoS, the destination can retrace the packets up to the source, based on ICMP packets or on information in the marked packet.

13.5. Internet access infrastructure security

We have seen the security that can be implemented for user data and Internet infrastructure. In this last section, we see the risks incurred when an entity connects to the Internet. The term entity can include a PC connected with a traditional V.92 modem, a wireless PC linked to a hot-spot, or a company network linked through an access router.

The open nature of the Internet is a quality and a defect. Once connected, whether they are ill-intentioned or not people can relatively easily access our software and material resources. If no protection is considered regarding access control, a hacker could see, consult and in the worst case destroy the content of our entity. To counter these access types, firewalls have been created. These systems have evolved to detect intrusions more intelligently with IDS. Also, with the increase of "always connected" entities, the virus plague requires more and more serious consideration[8].

13.5.1. *Access control by firewall*

The firewall is a hardware or software concept that makes it possible to protect an internal network (or a PC) from external attacks. The firewall is a unique and mandatory point of passage between the internal network and the Internet.

8 The defaults in the physical layer also apply to the Internet access infrastructure. Tempest attacks can be taken into account in this section. We will not return to this subject as it was already introduced in the preceding section.

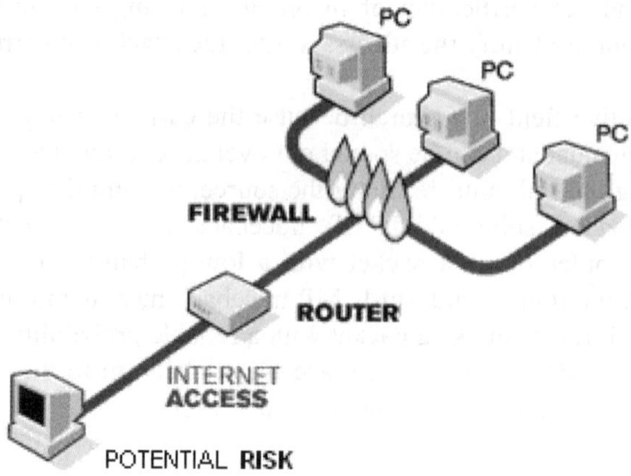

Figure 13.2. *Firewall*

At the beginning, firewalls behaved as authorized address managers and then evolved to control application accesses. The following section deals with access control in general. Today, firewalls are adapted to other security needs of which we will give an overview later on.

13.5.1.1. *Access control*

Access control makes it possible to accept or reject connection requests, but also to examine the nature of the traffic and validate its content, due to filtering mechanisms. There is static filtering and dynamic filtering.

Static filtering verifies IP packets (header and data) in order to extract the quadruplet "source address, source port, destination address, destination port", which identifies the current session. This solution makes it possible to determine the nature of the requested service and to define whether the IP packet should be accepted or rejected. The configuration of a filter is done through an access control list (ACL) which is constituted by the concatenation of the rules to follow. For example, a first rule indicates that all the machines can connect to the Web server on port 80 and the following rule authorizes the Web server to response to all the clients of the service (on a port greater than 1,024). These rules enable all the machines to access the Web. At the end of the list, there must be a rule specifying that all communication is prohibited in the group of services and machines at the entrance or exit of the protected network. The principle consists of prohibiting everything that is not authorized. This example is of a TCP service. With TCP, the differentiation between incoming call and outgoing call relies on the ACK bit in the header, which characterizes an established connection. This distinction does not exist for UDP where differentiating a valid packet from an attack attempt on the service is not possible. We also see this problem with certain applications that answer clients' requests on dynamically allocated ports (for example, FTP). It is generally impossible to satisfactorily manage this type of protocol without opening the access to a larger number of ports and therefore make the network even more vulnerable.

Dynamic filtering follows the principle of static filtering. Its efficiency extends to the quasi-totality of currently used protocols (TCP, UDP, etc.), due to the implementation of state tables for each established connection. To manage the absence of a UDP circuit and the dynamic allocation of ports implemented by service, dynamic filtering examines in detail the information up to the application layer. It interprets the particularities linked to the application and dynamically creates rules for the duration of the session.

13.5.1.2. *Other functionalities*

The proxies (application relays) are now integrated in the firewalls. A proxy is interposed between clients and servers for a specific surveillance and isolates the network from the exterior. All data is routed to the network and it decides which action to take. It verifies the integrity of the data and authenticates the exchanges. The authentication of the users is indispensable for carrying out reliable access control relating to user identity. The most currently used mechanism consists of associating a password to a person's identifier. However, this password is often sent visibly. On the Internet, this situation is not acceptable. To remedy this, firewalls propose solutions based on encryption. Virtual private networks (VPN) use the Internet to securely interconnect different affiliate and/or partner company networks. To mitigate the lack of confidentiality on the Internet and to secure the exchanges, solutions based on the SSL protocol or on IPSec are integrated in the firewall. Thus, societies can interconnect their different sites and can provide an access control to their mobile employees.

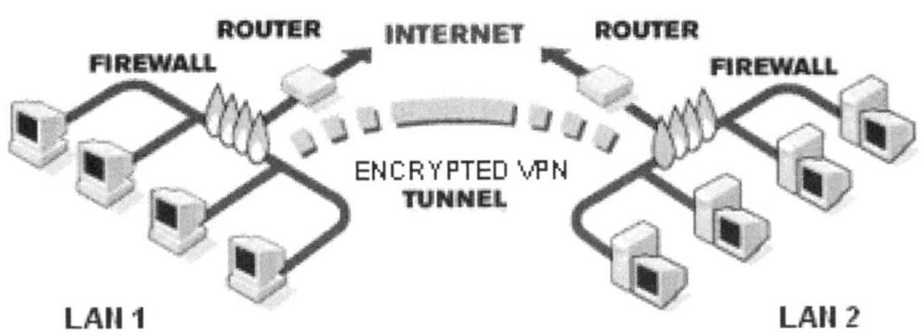

Figure 13.3. *VPN firewall type*

13.5.2. *Intrusion detection*

At the entrance of a network there is often a simple firewall that blocks unused access roads. However, a firewall does not sufficiently filter requests that pass through open accesses. For example, hackers often use open port 80 (HTTP) to infiltrate poorly protected systems. A veritable surveillance system that permanently controls the identity of network requests must be considered. The role of the IDS is to locate the intruder in the traffic transiting in the ports left open by the firewall. The administrators do not have enough knowledge of the problem and the company networks are still under-equipped.

13.5.2.1. *IDS types*

13.5.2.1.1. IDS network

The most traditional method is the IDS network. Its role is to immobilize each request, to analyze it and to let it continue along its path only if it does not correspond to an attack referenced in a database.

A quality IDS network has an exhaustive file of attack signatures. This file is centralized and updated by the constructor. The last update must be regularly uploaded. The other important point is the placement of the IDS network. A probe placed at a bad location can be inefficient. It is common that we cannot content ourselves with a single filtering system: the more complex a network, the more vulnerabilities it presents. It is logically more difficult to protect it. However, each added IDS network is expensive: they are machines greedy for resources. The flows are very heavy and it is necessary to dedicate very powerful machines to the IDS network, or specialized equipment to support them. The quality of the detection system varies from one system to another. Since the first versions, the technology has progressed and the products are becoming more performing and more pertinent. Performances have improved by targeting detection methods better. Today, each block of data is no longer analyzed from top to bottom: IDS networks have learned to target strategic points. Regarding the pertinence of intrusion detection systems, things have also changed. An IDS network does not signal all the dangerous character strings, but only those that are in a position to be exploited. If a security expert sends mail to a colleague containing an attack code, the IDS network will not initiate an alert. However, if the code is contained in a request whose goal is to saturate a server's memory, the IDS network will signal it right away.

13.5.2.1.2. The IDS host

The IDS network does not guarantee a 100% security level on its own. To get to this, another system needs to be implemented that will observe the behavior of each block of data and will signal everything that seems unusual. This is the role of the IDS host, a probe that is placed on each system that needs to be protected.

The IDS host detects known and unknown anomalies: if an attack makes its way through the meshes of the net, the probe will locate it. The IDS host takes a snapshot of the system at a given time. It defines everything that is legitimate. Everything outside of the frame and the system's habits is considered an attack. For example, a modification of the registry will be blocked and will raise an alert. An IDS host probe is less onerous than an IDS network but must be placed on each machine to supervise. They are often reserved for well-protected machines. They are less greedy for resources than an IDS network: they are found in the form of software and are not integrated in a server or specialized equipment like IDS networks.

13.5.2.2. *Administration problems*

The level of maturity of intrusion detection systems is far from being optimal. Administrators are easily drowned under a mass of inevitably irrelevant alerts. We end up seeing IDS that are no longer administered and have fallen into disuse. This reality is difficult to accept when we know that the price of entry is around tens of thousands of Euros. To avoid such a loss, we must predict that an IDS requests daily administration. It is pointless to invest in it if there is no possibility of taking the time to attend to it daily. The IDS needs regular supervision, which is not always the case. Signatures are not always updated and detections are not always investigated. What makes this task easier to bear is to filter the blocks of data to the maximum before they even get to the IDS. The detection system must be the last layer of the anti-intrusion filter. Other layers must stop benign attacks. To do this, optimally configured routers and firewalls must be carefully implemented. Thus, the rate of alerts that merit signaling will be greatly reduced.

13.5.3. *Viruses*

13.5.3.1. *Terminology*

A virus is a program that exerts a damaging action such as the modification or destruction of files, the erasing of the hard drive, the lengthening of processing time, generally worrisome visual or audio manifestations, etc. This action can be continuous, sporadic, periodic, or take place on a precise date[9]. Some variations are the worm, the logic bomb, the Trojan horse or the macro virus. A worm is a network virus. A logic bomb is a device that contains a launch condition, such as a system date, for example. A Trojan horse presents itself as an executable file (utility, game, etc.) that contains an insidious feature capable of causing damage. Macro viruses are

---

9 For example, the Michelangelo virus only launches on March 6th.

linked to Office series software. These have macro commands whose initial goal is to automate repetitive tasks. This goal was diverted to destroy or modify files on the infected machine[10].

13.5.3.2. *Reconnaissance*

Many symptoms can reveal an infection: for a file, it can be a change in size or creation date or checksum, for a program, it can be a slow loading time, a different screen aspect, surprising results, etc. The size of available memory can be reduced with respect to what we normally see. The regular user will need to use specialized software capable of a fine and complex analysis of the contents of the memory and of the hard drive.

A category of virus detectors operates on a collection of signatures. The simpler viruses comprise a series of instructions unique to themselves but identifiable and called signature. We can establish a catalog that will grow every time a new virus appears. The programs that exploit this method are called scanners. They give very few false alarms[11]. The inconvenience of this method is the need for periodic catalog updating, which imposes on the user a paid subscription with the anti-virus editor.

Another method exists that has the advantage of not needing any updates. It is based on heuristic algorithms that suspect, in certain successions of instructions, the possibility of a virus. The probability of false alarms is greater than with scanners but the efficiency is permanent, at least until the appearance of a new general form of attack.

13.5.3.3. *Prevention*

To protect ourselves against viruses, recommended solutions are the control of new installed applications, the locking of storage mediums when they do not need to be overwritten and having an updated anti-virus.

Myths, hoaxes and urban legends are falsely presented as a security alert. Their goal is to create confusion around security and to provoke cascading email transmission to overload email in-boxes (spamming)[12]. Manipulations of this type have brought organisms to sign their messages. Also, it is easy to verify that a virus

---

10 A known macro virus, I love you, led Microsoft to modify Outlook 98.

11 Scanners are inefficient against polymorphic viruses because these viruses can modify their appearance.

12 This practice was inaugurated with the false announcement of the Good Times virus, presented as very dangerous. This false alert continues, several years after its inception, to circulate over the Internet, sometimes with variations (Penpal Greetings, AOL4FREE, PKZIP300, etc.).

alert is a hoax by consulting specialized sites[13]. In general only authenticated sources should be trusted. In all cases, the information should be verified before propagating the message, especially to a distribution list. If we are credulous, we knowingly participate in the malevolence.

The platform that consists of the most macro viruses is Microsoft Word for Windows. Viruses propagate easily because .doc files contain both text and all the associated macros. The first action to carry forward is to deactivate the execution of macro commands when receiving a Word document (or any other Office series document), especially if it is from an unknown source. The fabrication of a macro virus is available to any neophyte. A large quantity of new macro viruses is created every day. The implementation of signature files on the antivirus must take place at least once a month.

13.6. Summary and conclusion

The Internet is historically of an open nature. At its beginnings, the goal was to link a few computers to each other. Mutual confidence was implicit.

Today, millions of machines use the Internet. Mutual confidence between all communicating parties is no longer present. Hacking is more and more common, with goals of making a profit and even achieving and even personal notoriety. Security therefore comes into play.

First and foremost, security is not a service but a notion that groups various services. Among the most commonly used are integrity, confidentiality and non-repudiation. Therefore, the more fashionable mechanisms are message authentication code, encryption and digital signature.

For a medium user, unless using applications preconceived for supporting a secured traffic, protecting data demands an analysis in order to define his security needs. This is not available to everybody. There is the exception of SSL, where in most cases, the user does not have to do anything. IPSec is standardized today but its use and configuration are still complex. Underneath the TCP/IP layers, the type of transport network also has its own problems. We must also not forget the flaws in the physical layer. These can put into question all the security services grafted on the higher levels, which are derived from engineering protocols.

---

13 For example, Hoaxbuster, http://www.hoaxbuster.com.

The Internet infrastructure is not infallible. A secured DNS is far from applicable to the entire Internet. Routing protocols present enough defaults to make routing table poisoning attacks possible. In this field, router attacks demand research attention because very few efforts have been made in this direction. Poor packet processing from a malicious router represents an attack that is very difficult to detect. The interruption of packets still presents a problem. With regard to poor routing for packet loop routing, the problem is still open. Certain DoS attacks are well-known and detectable. However, the DoS still presents problems in finding the aggressor. In fact, IP address spoofing is a major inconvenience when trying to locate it.

Finally, we saw that installation security is still in a maturation phase. Firewalls are necessary but not sufficient. The IDS are efficient and offer an optimal level of protection in company networks. However, there is still progress to be made. The big problem with the IDS is their administration. With regard to viruses, present before the advent of the Internet, they do not stop evolving and renewing every day to the point where a monthly update of antivirus software can be insufficient these days.

13.7. Bibliography

[BAK 00] BAKER F., LINDELL B., TALWAR M., "RSVP Cryptographic Authentication", *IETF RFC*, 2747, January 2000.

[BRA 97] BRADEN R., ZHANG L., BERSON S., HERZOG S., JAMIN S., "Resource Reservation Protocol (RSVP) – Version 1 Functional Specification", *IETF RFC*, 2205, September 1997.

[CHA 02a] CHAKRABATI A., MANIMARAN G., "Internet Infrastructure Security: A Taxonomy", *IEEE Network*, p. 13-21, November/December 2002.

[CHA 02b] CHAKRABATI A., MANIMARAN G., "Secure Link state Routing Protocol", *EcpE*, Iowa State University, 2002.

[CIS 00] CISCO, Strategies to protect against Distributed Denial Of Service Attacks (DDoS), DoS Cisco White Paper, February 2000.

[DNS 04] DOMAIN NAME SYSTEM SECURITY WORKING GROUP, IETF, http://www.ietf.org/html.charters/old/dnssec-charter.html.

[FRE 96] FREIER A.O., KARLTON P., KOCHER P.C., *The SSL Protocol Version 3.0*, Netscape Communications, 18th November 1996.

[IPG 04] *International PGP Homepage*, http://www.pgpi.org.

[IPS 04] IP SECURITY PROTOCOL WORKING GROUP, IETF, http://www.ietf.org/html.charters/ipsec-charter.html.

[KEN 98a] KENT S., ATKINSON R., "IP Authentication Header", *IETF RFC*, 2402, November 1998.

[KEN 98b] KENT S., ATKINSON R., "IP Encapsulating Security Payload (ESP)", *IETF RFC*, 2406, November 1998.

[NF 90] FRENCH STANDARD, Systèmes de traitement de l'information – Interconnexion de systèmes ouverts – Modèle de référence de base – Part 2 : Architecture de sécurité, *ISO 7498-2*, September 1990.

[PKI 04] PUBLIC KEY INFRASTRUCTURE WORKING GROUP (X.509), IETF, http://www.ietf.org/ html.charters/pkix-charter.html.

[SMI 97] SMITH B.R., MURTHY S., GARCIA-LUNA-ACEVES J.J., "Securing Distance Vector Routing Protocols", *Proc. SNDSS*, February 1997.

[TEM 04] *The complete, unofficial TEMPEST information page*, http://www.eskimo.com/ ~joelm/tempest.html.

[TLS 04] TRANSPORT LAYER SECURITY WORKING GROUP, IETF http://www.ietf.org/html.charters/tls-charter.html.

[ZHA 98] ZHANG X. *et al.*, "Malicious Packet Dropping: how it might impact the TCP performance and how we can detect it", *Symposium Security Privacy*, May 1998.

Chapter 14

Security Protocols for the Internet

14.1. Introduction

IP's flexibility and simplicity have met the needs in data processing network matters; however, its starting goal was not to ensure secure communication, which is a reason behind a strong absence of functionalities. The ease of attacks and the will to use IP for sensitive applications have incited the development of security solutions such as filtering routers, firewalls and secure protocols and applications. IPSec is a standard that defines a security extension for the IP protocol to make it possible to secure the IP layer and the higher layers. It was developed, at first, for the new generation IPv6 protocol and was then carried over to IPv4. The security services provided by IPSec are confidentiality, authentication and data integrity, anti-replay protection and access control.

The rapid evolution and deployment of wireless networks based on the IEEE 802.11 standard for Internet access present new security problems that did not exist previously in telegraphic networks. The WEP (Wired Equivalent Privacy) protocol was developed with the intention of bringing security solutions to wireless networks. However, this WEP protocol proved not to be adapted as it contains security weaknesses and presents key distribution problems. Then, there was the development of a security architecture called RSN (Robust Security Network). This architecture is based on the IEEE 802.1X standard, which offers strong authentication and authorization mechanisms and dynamic key distribution mechanisms for communication encryption in wireless links. In this chapter, we will detail the two security architectures IPSec and IEEE 802.1x.

Chapter written by Idir FODIL.

14.2. IPSec

IPSec is a group of data security mechanisms that is integrated in TCP/IP to process each emitted or received IP packet. This processing can be the rejection of the packet, the application of security mechanisms or the permission to pass. Its integration in TCP/IP makes IPSec exploitable by higher layers by offering a unique security mechanism for all of the applications.

14.2.1. *Security services*

IPSec aims at preventing various possible attacks on the IP networks by preventing in particular the spying of the data circulating on the network or being passed off as something else. For this, IPSec can provide the following security services:

– *Confidentiality*: the data that is transported can only be read by the emitter and the receiver. No information circulates unencrypted on the network. It is also possible to encrypt the IP packet headers.

– *Data authenticity and access control*: this consists of two services, which are the data source authentication and the integrity of the data.

– Authentication guarantees that the received data come from the declared expediter.

– Integrity guarantees that the data has not been modified.

– Authenticity of each received packet makes it possible to implement access control, particularly to protect access to resources or private data.

– *Anti-replay protection*: this makes it possible to detect an attack attempt consisting of resending a valid packet previously intercepted on the network. The whole of these services are based on modern cryptographic mechanisms which give them a high level of security.

14.2.2. *Protection modes*

IPSec uses two protocols, AH and ESP, to ensure communication authentication, integrity and confidentiality. IPSec can be used to secure either the entire IP packet, or the layers above it. The appropriate modes are called tunnel mode and transport mode.

In transport mode only the content of the IP packet is protected. This mode is useable on terminal equipment (clients, servers).

In tunnel mode: the whole content of the IP packet is protected by its encapsulation into a new packet. This mode is used by network equipment (routers, firewalls, etc.).

The AH (Authentication Header) and ESP (Encapsulating Security Payload) protocols can be used either separately or combined.

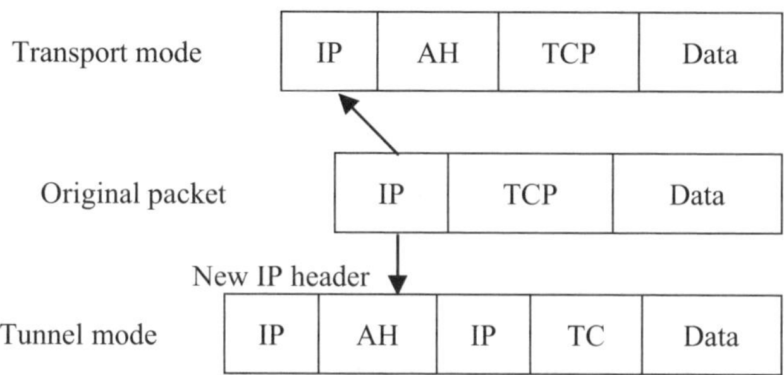

Figure 14.1.

14.2.3. *AH (Authentication Header) protocol*

The AH protocol ensures the authenticity of non-confidential IP packets (without encryption). The principle of this protocol is to add a supplementary field to the IP packet, which makes it possible, upon receipt, to verify the authenticity of the data. A sequence number makes it possible to detect replay attempts. AH calculates an HMAC based on the secret key, the packet payload and the consistent parts of the IP header. Then, the AH header is added to the packet. The AH header is illustrated in Figure 14.2.

The AH header is 24 bytes long. The *next header* field is the same as the one in the IP header. In tunnel mode, the next header field is equal to "4" because it is an encapsulation of the IP packet within another IP packet. In the transport mode, if we encapsulate TCP, the *next header* field is 6. Given that AH protects IP by including the immutable parts in the IP header, it therefore does not cross the NATs.

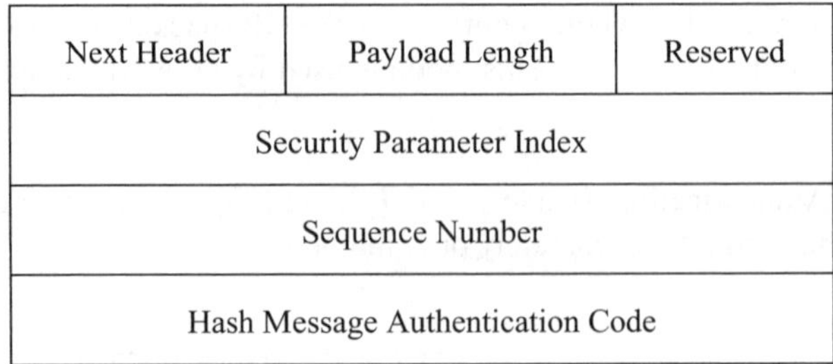

Figure 14.2. *AH protocol header*

14.2.4. *ESP protocol (Encapsulated Security Payload)*

This protocol guarantees data confidentiality and can thus guarantee authenticity. The ESP principle is to generate, from an IP packet, a new packet in which the data (possibly the header) is encrypted. It can also ensure authenticity by adding an authentication block and anti-replay protection by using a sequence number. After packet encryption and HMAC calculation, the ESP header is generated and added to the packet. The ESP header consists of two parts, as illustrated in Figure 14.3.

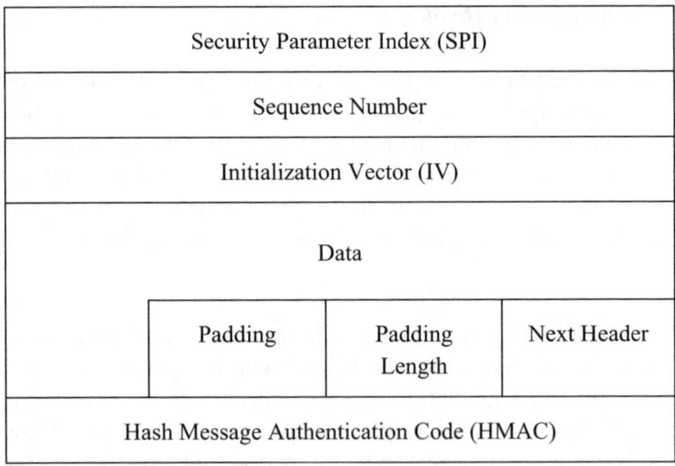

Figure 14.3. *ESP header*

The initialization vector (IV) is used in the encryption process. Symmetric encryption algorithms are susceptible to attacks if the IV is not used. The initialization vector ensures that the encryption of two identical pieces of information (payload) gives two different pieces of information. IPSec uses block encryption, which means using stuffing in the case where the length of the encrypted data is not a multiple of the size of the block. The HMAC is calculated on the basis of useful information in the packet and does not take the IP header into consideration.

14.2.5. *Security parameter management*

To protect packet integrity, IPSec uses HMACs. To derive the HMAC, IPSec uses hash algorithms such as MD5 and SHA in order to calculate the hash based on the secret key and the IP packet contents. This HMAC is then integrated into the IPSec header and the receiver of the packet can verify the HMAC and see if it has the secret key.

To guarantee IP packet confidentiality, the IPSec protocol uses symmetrical encryption algorithms. The IPSec standard imposes the implementation of NULL and DES. Currently, more powerful algorithms are used such as 3DES, AES and Blowfish.

To ensure protection against service denial attacks, IPSec uses a sliding window. A sequence number is assigned to each packet and the packet is only accepted if the number is within the window. Old packets are immediately suppressed. This solution protects against replay.

14.2.5.1. *Security association*

To encapsulate and decapsulate IPSec packets, the extremities need to store the participants' secret keys, algorithms and IP addresses. All these necessary IPSec parameters are stored in security associations (SA). These SAs are stored in an SAD (Security Association Database).

An SA is composed of the following parameters:

– source and destination IP addresses of the extremities that are protecting the IP packets;

– IPSec protocol (AH or ESP);

– the algorithm and the secret key used by IPSec;

– index (SPI). A 32-bit number identifying the SA;

– IPSec mode: tunnel or transport;

– size of the anti-replay protection window;

– lifetime of the SA.

A security association can only protect one single direction of traffic. To protect the two directions, IPSec needs two security associations. Other information stored in the SP (Security Policy) is necessary to define what type of traffic needs protection.

An SP specifies the following parameters:

– source and destination packet addresses. In transport mode, these are the same addresses as in the SA;

– the protocol (and port) to be protected;

– the SAs to be used.

14.2.5.2. *Key exchange*

The secret keys and the algorithms must be shared between all the participants of a secured session. This can be carried out by a manual approach and an automatic approach, both of which are defined as follows:

– *Manual*: a first approach for key exchange is the manual management, which consists of enabling the administrator to manually configure each piece of equipment using IPSec with the appropriate parameters. This approach is practical in a small-sized and static environment; on the other hand, it is not at all convenient in a large-scale environment. In addition, it is based on a static definition of the parameters without renewal.

– *Automatic*: the second approach is the automatic management by means of an appropriate protocol. The parameter management protocol relative to IPSec is IKE (Internet Key Exchange), which is not only related to keys. This protocol does not manage (negotiate, update, suppress) all the security parameters. IKE is a high-level protocol whose role is to open and manage a pseudo-connection on top of IP (AH and ESP directly act on the IP level). IKE includes, at the beginning of the negotiation, a mutual authentication of the participants that can be based either on a shared secret or on public keys. The exchange of public keys used by IKE can be done manually directly in the framework of IKE by the exchange of online certificates, or by means of a public key infrastructure (PKI).

14.2.5.3. *IPSec configuration*

The security services offered by IPSec are based on choices defined by the network administrator by means of security policies. These policies are stored in a Security Policy Database (SPD). An SPD is composed of a list of rules, where each rule carries a number of criteria that make it possible to determine what part of the traffic is concerned.

Consulting the SPD makes it possible to decide, for each packet, if it should follow security services and if it will be allowed to pass or, on the contrary, be blocked. This basis indicates to IKE what SA it must negotiate and in particular which secure tunnels it must establish.

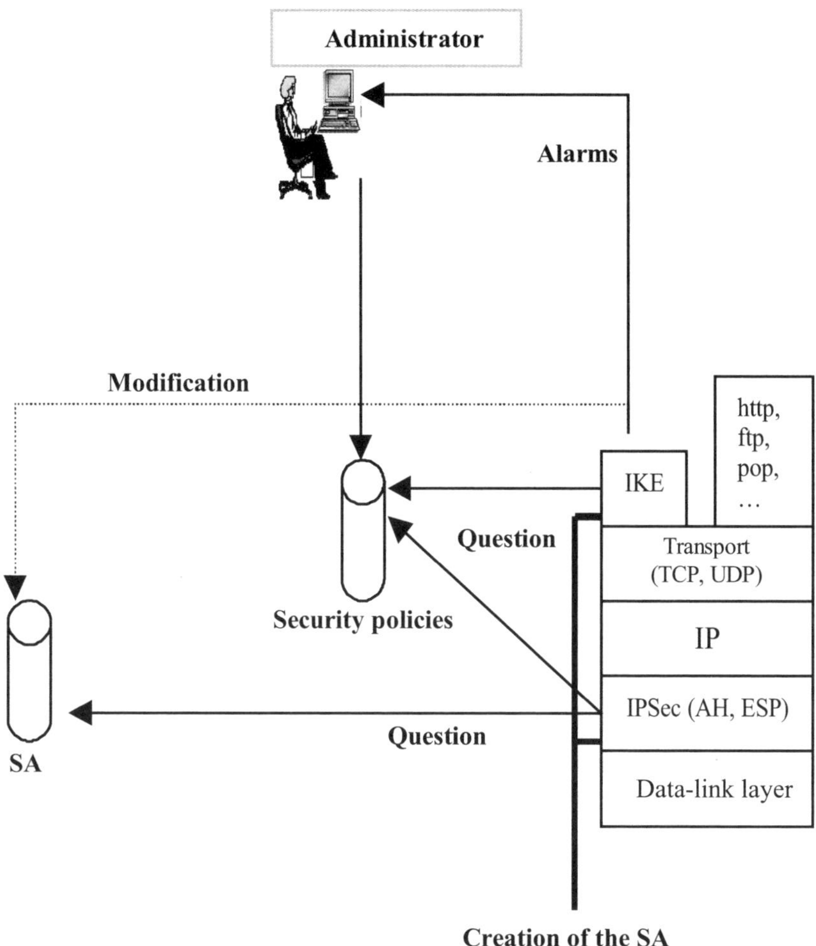

Figure 14.4. *IPSec configuration*

All the details concerning IPSec can be viewed on the IPSec workgroup page of IETF at http://www.ietf.org/html.charters/ipsec-charter.html.

14.3. IEEE 802.1x

The IEEE 802.1x standard (Port-Based Network Access Control) defines a mechanism that makes it possible to control access to resources on a local IEEE 802 network that can be wired or wireless. It offers mechanisms that make it possible to authenticate and authorize users connecting point-to-point on a local network's port.

The 802.1x standard was conceived to be used in wireless networks but it is now widely used in 802.11-type wired networks. Thus it offers the possibility to access points to control network access and to distribute or obtain key information on the part of authenticated users. The development of the 802.1x standard was subscribed from the following needs:

– Rapid 802.11 network deployment in public and private locations.

– A wireless network operator wishes to control the users that are accessing it. Currently, we can connect easily enough on any 802.11 network.

– The need to use AAA. AAA technology such as the Radius protocol is widely used to control the access of users on the networks. 802.1x can use the existing infrastructures to offer AAA functionality to the users.

– The need to dynamically distribute the keys. The WEP (Wired Equivalent Privacy) protocol was created to offer security on a wireless network by using symmetrical encryption keys. One of the limitations of this approach is the management and distribution of keys.

The 802.1x standard makes it possible to distribute WEP keys to users and to access points.

It should be noted that WEP technology was completed by taking into account IEEE 802.1x authentication and by the regular modification of keys due to TKIP technology. This new standard is called WPA (Wireless Protected Access). This WPA technology was completed in June 2004 by the 802.11i standard, which still includes IEEE 802.1x authentication and incorporates a new encryption algorithm, AES, much more robust than RC4, which is in WPA. This new standard is called WPA2.

14.3.1. *Architecture*

There are three entities in the 802.1x protocol architecture:

– The client, who represents a user wanting to connect to the network.

– The authentication system, which:

- communicates with the client and submits client information to the appropriate authentication server,

- enables the user, if authenticated, to use the network,

- acts independently from the authentication method used as it acts as relay between the authentication server and the client.

– An authentication server which contains users' information as well as their network use rights. The most widely used and deployed today is the Radius server.

The 802.1x architecture offers, to the authentication system, the possibility of creating two network connection points for users. The first connection point, called the uncontrolled port, makes it possible to exchange blocks of EAP data between the client and the authentication system. The second connection point, called the control port, makes it possible to transfer client traffic if it has been authorized.

14.3.2. *EAP protocol*

The 802.1x standard relies on the use of the EAP (Extensible Authentication Protocol) for user authentication. The EAP was developed as an extension of the PPP protocol in order to make it possible to deploy and use authentication mechanisms in a given network. The EAP was created to enable the addition of authentication modules at the level of client access and authentication server.

EAP extendibility is ensured by:

– the installation of an EAP library on the client and on the server; different types of EAP protocols can be supported;

– the modification of the EAP type can be carried out at any moment;

– the support of authentication schemes such as:

- generic Token Card,

- OTP (One-Time Password),

- TLS (Transport Level Security),

- chip cards,

- certificates,

- future authentication technologies.

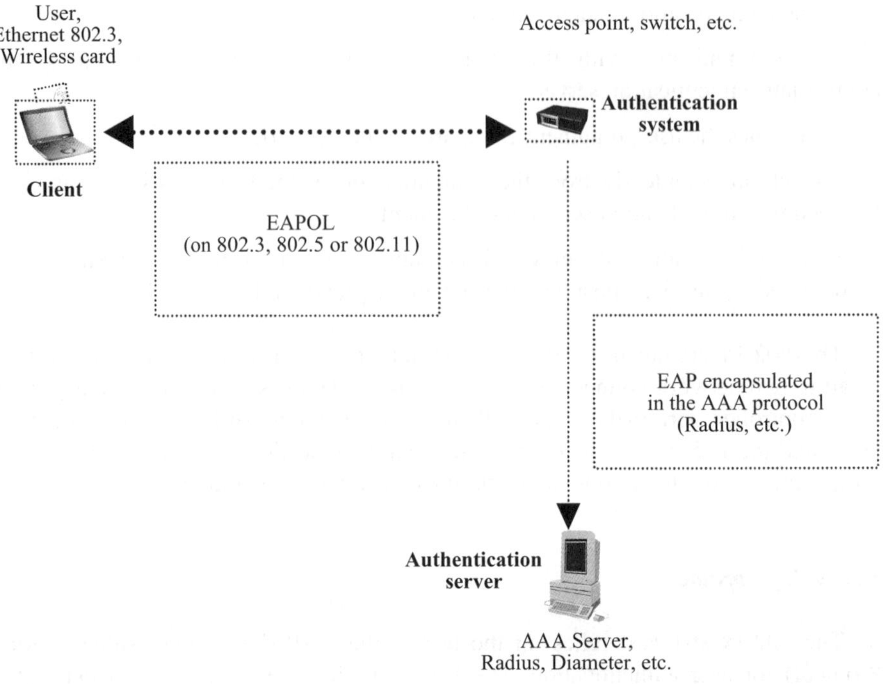

Figure 14.5. *IEEE 802.1x architecture*

An EAP session unfolds as follows:

– Establishment of a link.

– No choice in authentication protocol.

– Each PPP pair negotiates to execute EAP during the authentication phase.

– Authentication phase.

– Negotiation of the EAP authentication scheme to use as EAP type.

– Choice of EAP type (finished).

– Message exchange between the client (user) and the authentication server.

– Conversation, which consists of authentication information requests and responses.

The EAPOL protocol (EAP over LAN) is an EAP message encapsulation protocol on an LAN network. It makes it possible to notify EAP sessions by beginning and end session messages.

14.3.3. *IEEE 802.1x protocol functioning*

The 802.1x standard enables:

– *Logic access*. In a wireless 802.11 network, a station must create an association with the access point before using the network. The protocol that creates the association enables the client and the access point to exchange their MAC addresses. This creates a logic access that the station can use to communicate with the access point and thus dynamically derive the available WEP keys. Once the association is established, the client can then be authenticated by using the EAP protocol.

– *WEP key management*. The 802.1x standard does not exclude and does not require the WEP protocol or any other encryption protocol. It offers a mechanism that makes it possible to distribute encrypted information from an access point to a client using the EAPOL-Key message. This can be done in a session. Thus, if an ill-intentioned person obtains a WEP key, it will not be of any use after the user ends the session.

14.3.3.1. *EAP/Radius authentication*

The diagram in Figure 14.6 describes the EAP authentication phase. As previously indicated, a client must first be associated with the access point before exchanging EAP messages.

The exchanged EAP messages between client and access point are transported by the EAPOL protocol. EAP messages are then transmitted by the Radius protocol between the access point and the Radius server. This makes it possible to create an authentication session between the client and the authentication server. Once the user is authenticated, the EAP-Key message is sent with the goal of exchanging key information between the access point and the client.

14.3.4. *Conclusion*

The IEEE 802.1x standard makes it possible to implement security mechanisms on local 802 networks and particularly on wireless IEEE 802.11 networks. In addition, the fact that access points use the EAP protocol for authentication makes it possible to support multiple authentication methods such as certificates, chip cards, etc. Currently, the 802.1x is the most deployed security standard in wireless networks and is supported by the majority of access point equipment and by computer operating systems.

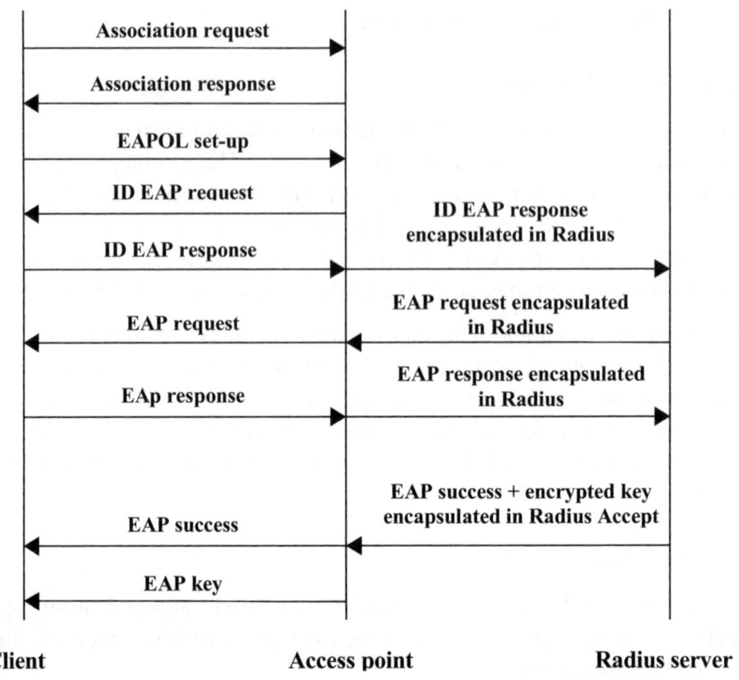

Figure 14.6. *802.1x protocol unfolding*

14.4. Bibliography

[ATK 98] KENT S., ATKINSON R., "IP Encapsulating Security Payload (ESP)", *RFC 2406*, November 1998.

[BLU 98] BLUNK L., VOLLBRECHT J., "RFC 2284: PPP Extensible Authentication Protocol (EAP)", *IETF*, March 1998.

[GRA 02] GRAHAM J. W. II, "Authenticating Public Access networking", *SIGUCCS'02*, Providence, Rhode Island, USA, 20-23 November, 2002.

[IPS 98] KENT S., ATKINSON R., "RFC 2401: Security Architecture for the Internet Protocol", *IETF*, November 1998.

[JAS 01] JASON J., RAFALOW L., VYNCKE E., "IPsec Configuration Policy Model", *Internet Draft*, November 2001.

[KEN 98] KENT S., ATKINSON R., "IP Authentication Header", *RFC 2402*, November 1998.

[KRA 97] KRAWCZYK K., BELLARE M., CANETTI R., "HMAC: Keyed-Hashing for Message Authentication", *RFC 2104*, February 1997.

[MAD 98] MADSON C., GLENN R., "The Use of HMAC-MD5 within ESP and AH", *RFC 2403*, November 1998.

[MAL 02] MALLADI R., AGRAWAL D. P., "Current and Future Applications of Mobile and Wireless Networks", *Communications of the ACM*, vol. 45, no. 10, October 2002.

[NIK 02] NIKANDER P., "Authorization and Charging in Public WLANs using FreeBSD and 802.1x", *USENIX annual technical conference*, 10-15 June 2002.

[P8021.X] "Standard for Port-Based Network Access Control", *IEEE Draft P802.1X/D11*, LAN MAN Standards Committee of the IEEE Computer Society, 27 March 2001.

[RIG 00] RIGNEY C., WILLENS S., RUBENS A., SIMPSON W., "RFC 2865: Remote Authentication Dial in User Service (Radius)", *IETF*, June 2000.

[SCH 03] SCHMIDT T., TOWNSEND A., "Why WI-FI Wants to Be Free", *Communications of the ACM*, vol. 46, no. 5, May 2003.

[VAR 00] VARSHNEY U., VETTER R., "Emerging Mobile and Wireless Networks", *Communications of the ACM*, vol. 43, no. 6, June 2000.

[WLAN 99] IEEE. 802.11b/d3.0 Wireless LAN Medium Access Control (MAC) and Physical Layer (PHY) Specification, August 1999.

[ZHA 02] ZHANG J. *et al.*, "Virtual Operator-Based AAA in Wireless LAN Hot Spots with Ad-Hoc Networking Support", *Mobile Computing and Communications Review*, vol. 6, no. 3, 2002.

Chapter 15

Secured Infrastructure for Ambient Virtual Offices

15.1. Introduction to ambient Internet and to its needs in terms of security

Pervasive computing, *ambient computing* [CIA 02a] or *ubiquitous computing* [WEI 91, WEI 93a, WEI 93b, WEI 96] consists of creating computerized and networked environments that will help the users accomplish their daily professional and personal tasks. It is at the convergence of four great fields of computer science: networks, applications, hidden systems and man-machine interfaces. There exist different ways of naming this computing tendency, by combining names and adjectives that point to two notions of intelligence/capacity for calculation in our daily universe (see Table 15.1). In Europe, the term *ambient intelligence* [ERC 01, IST 01] seems to dominate. These new universes will need permanent connectivity that will be made possible by an ambient Internet and ambient networks [DUD 02]. However, the users, apart from the fact of being permanently able to access their data, will want to be assured of a high level of security (in terms of confidentiality but also of reliability). Reciprocally, the services will call upon authentication mechanisms vis-à-vis the users.

15.1.1. *Ambient Internet and security*

Ambient Internet is a recent concept: it proposes an Internet accessible everywhere and all the time, in other words without spatial or temporal constraints,

Chapter written by Laurent CIARLETTA and Abderrahim BENSLIMANE.

with as many multi-media data transfer streams as possible. Therefore, if Internet access must be omnipresent, this supposes remaining connected to a network while being mobile. Wireless networks and more particularly those based on Wi-Fi technology or IEEE 802.11 [IEE 03] will be at the core of our issue.

Nouns	Adjectives
Computing	Ambient
Intelligence	Pervasive
Internet	Ubiquitous
Networks	Smart
Space	Disappearing
Environment	

Table 15.1. *Pervasive computing vocabulary*

Every day, we find more and more wireless Internet access points, whether they are *hot-spots* in public locations or company Wi-Fi access points. It is also possible to use mobile telephony technology to have access to service requests and data (and notably personal and professional data). In addition, research in the field of ad hoc networks, or networks without infrastructure, will enable the dynamic and sometimes punctual expansion of this coverage in response to user needs and to the variations in fixed infrastructure loads.

Security is a very delicate field in computer science. This is even truer for systems such as ambient Internet. We will find ourselves in dynamic wireless environments, which are, in other words, all the more easy to listen, modify and difficult to manage. Thus, we can use cryptographic technologies to ensure communication security by providing confidentiality, authentication of both sources and destinations, and the integrity of data exchanged on the channel. However, the problems of management and implementation of infrastructures that enable the deployment of solutions is still unresolved. pervasive computing intrinsically raises a large number of questions regarding this aspect, all the more as it will be at the service of "Mr. Everybody" and that therefore its users will not often be network administrators or security specialists. The increased use of wireless communication technologies, user mobility, the presence of these technologies in our daily environment and network dynamism reinforce the security needs in the structures that are currently being implemented.

15.1.2. *Chapter organization*

Firstly, we will present the framework in which we are currently, that of ambient Internet and the proposed security infrastructure (it can be noted that this infrastructure also takes into account aspects of QoS associated to security). Next, we will detail the implementation of various elements of this infrastructure in prototypes. Finally, we will propose an evolution of our works toward an architecture that is even more modular and adaptive.

15.2. Virtual Private Smart Spaces (VPSS)

The design of Virtual Private Smart Spaces (VPSS) [CIA 02] addresses the control and security of access to the resources in an ambient computing environment. They rely on RBAC (Role-Based Access Control [FER 01]) philosophies stipulating that resource access must be authorized or declined depending on the roles that the user is fulfilling at the given moment. VPSS use the notion of virtual private network (VPN) to make it possible to create environments in which each VPN corresponds to a role. It is a question of differentiating access to resources present on the network: this will make it possible to control the access but will also give different priorities and dynamics to network flows. To summarize:

VPSS = RBAC + VPN

Roles constitute key components to define access policies for the proposed services. On the one hand, the equipment must be identified and communication must be secure, but on the other hand the users are central in these environments and must consequently also be identified. To go even further, we consider that security policies based on the identity of users must take precedence over those that use equipment authentication.

15.2.1. *VPSS context and motivation*

15.2.1.1. *Ambient Internet security problems*

Ambient computing corresponds to strongly but discretely computing in the fabric of our daily environment. In the near future, wherever you may be, you will probably be close by and consequently have access to ambient computing equipment and networks. They will provide you and other users in this location with an access to present services.

You will carry with you some of this equipment and they will help you fulfill tasks such as Web navigation, time management, environment adaptation to specific needs, the management and supervision of various other activities, etc. This could become quite useful. These technologies could, for example, bring an improvement to monitoring the health of the elderly, or a contextual aid to disabled people, or will enable better supervision of young children.

There are, however, many obstacles and a certain number of dangers when it comes to public acceptance of pervasive computing technology. Among these are security, respect of privacy and reliability. We can mention:

– eavesdropping: first and foremost, there will always be those who will attempt to hear your communication. Of course, this can also happen by accident, that is someone "listening" to your conversation without meaning to, but this is still a crucial security and privacy issue;

– impersonating: in order to obtain a number of private data or unauthorized access to some of your resources, users or malevolent services can try to take users' and valid services' identities (for example, a false online banking service could get you to divulge your credit card number);

– loss of control to a third party, confidentiality and privacy: your neighbor's son who wants to practice his hacking skills might try to get your intelligent refrigerator to stop running, from a distance, while you're away on vacation. More and more people could be tempted to gain access to your personal information and gain control of your goods (your electronic and digital goods being of course at the top of their list). This becomes a real problem when wireless access points become cheap and widespread. An unsatisfied passerby, a professional hacker, an industrial spy or a government agent are but a few of the possible intruders or eavesdroppers.

Other less obvious threats also exist: the "blue screen of death" and other *kernel panic* syndromes or current interruption during a storm can be a danger in an environment where everything is controlled by microprocessors and software: your hospital bed, your stove, or your car. Simply imagine the last time you were cornered in an elevator, knowing this technology is quite old, and think about what could happen if it was connected to the Web.

Finally, users will no doubt accept that these machines integrate themselves to their environment but not so if they become involved or invading. They want to maintain global control of the situation unless something exceptional occurs.

Let us hypothesize that a good number of other technical problems have been resolved, such as the knowledge of context, the coexistence and best use (of clients' interests and needs, in terms of cost, flow, mobility support, etc.) of wireless communication protocols, microcontroller and processor electrical consumption, standardization and compatibility of technologies of discovered service. Consequently, the principal security and reliability problems in a perfect world of ubiquitous computing can be organized as follows:

– Authentication: you want to be able to identify the person or machine with whom you will be in communication and vice versa, and be sure of the integrity of the content of exchanged messages.

– Confidentiality and privacy: you will want, no doubt, your communications to be automatically encrypted with a level of security adapted to the type of activity you're carrying out. Law enforcement or your company's security agents will also want to be able to access a part of your communication flow when they have the authorization to do so. In case of danger or immediate need, your identity could be revealed or hidden to the environment in which you are evolving (services and users).

– Safety: these machines and services should function in such a way that they cannot become a danger to your health and that of other users (the example of HAL in *2001, A Space Odyssey*, or the dangers presented by Bill Joy of the research into nanotechnology and the possibility of reproduction of the robots [JOY 00]).

– Reliability: you want these services and equipment to be universally available at all times and that they function as expected.

– Access control based on technical competencies: if you are not a specialist, your rights must be limited. It is a question of avoiding your actions or those of others present being dangerous to the system or to your health. The services should be able to adapt to your level of qualification.

– Access control based on the situation: when you find yourself in a situation that requires that someone take control (for example, a person who is performing an important operation just has a heart attack, someone or something needs to ask for help and (limited) control will be passed on to a third party in order to maintain system stability). In other words, access control will have to be based on the identity and the role of the person who is using the services and equipment of the environment and can be delegated.

– Hierarchical access: according to the current users and their role, an access hierarchy policy will need to be implemented.

However, and contrary to the norm, security must be taken into account from now on because it is one of the deployment conditions for ambient computing. So that it can be accepted by the broadest public, these questions will need to be considered: the users wish no doubt for a great ease of use but they also need to have confidence in these environments.

In the case of 802.11 networks, security was not ensured at the time of the works presented in the following discussion. The weakness of the WEP protocol, notably, was often made obvious. The works of the 802.11i group make it possible to provide a personalized response to security problems in Wi-Fi environments, but our objective in the development of our infrastructure is to propose generic solutions that are easily transposable to other wireless network technologies and that make it possible to take into account pieces of equipment that are already in place. In addition, we would like to associate the notion of security and of QoS as characteristics of the network resource that will be variables in terms of roles.

We note that the standardization of ambient intelligence technology will certainly be a pre-requisite for a larger audience, but it will also enable industrial proceedings to be more powerful. This will surely be particularly true for security.

15.2.1.2. *The central role of the users*

We are convinced of the central role of the user. The services are provided to the users. The user can be one person, a group of individuals or of other services, but generally a human being should always be taken into account from the specification step and in the whole process of service conception that the environment is supposed to obtain.

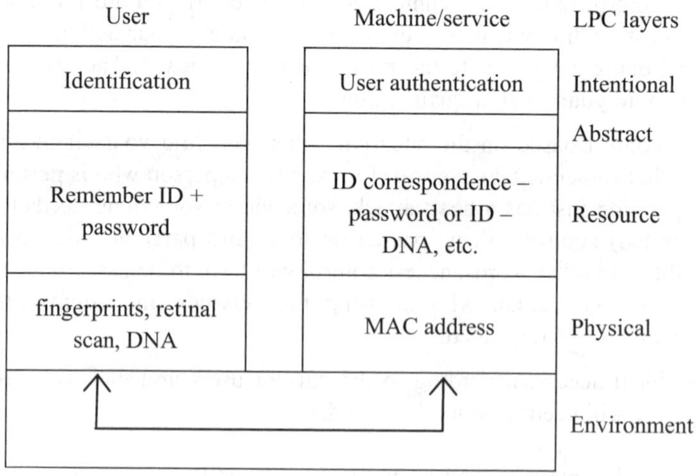

Figure 15.1. *Security and LPC model*

Figure 15.1 shows the Layered model of Pervasive Computing (LPC) [CIA 00] that we can apply. As the physical layer of the user's model corresponds to the physical part of the objects, we can evoke the use of fingerprints or retinal scans, or again our DNA signature as a method of user authentication in parallel with MAC address or a serial number of an electronic machine. At the level of the resource layer, the capability of the user to remember a password and a unique identifier corresponds to the capability of applications to verify correspondence, the validity of this combination, to encrypt or decrypt information. Our model shows that the users and the machines must be identified, and that the choices of creation will depend on the balanced combination of physical identification (the recognition of the machine or the individual by their characteristics) and abstract identification (the use of a challenge, a combination of user name and password and a PIN code).

In our societies, interactivity relies essentially on the identity and roles (social status, parent-child relations, professional hierarchy, company organization, etc.). Access control of ubiquitous computing services must duplicate the way we organize our society: it must be based on the roles played by the users, the services and objects of the ubiquitous computing environments.

The VPSS architecture does not resolve particularly low level security problems linked to wireless technology, but corresponds rather to a high level methodology, independent of the communication medium used (although functioning on IP, since it is, for us, the central protocol for Internet and ambient computing), that relies on the associated security roles and policies.

15.2.2. *RBAC*

"Role-Based Access Control" was developed to provide a high level of security but simple to use in the framework of large scale organizations. It is an access control method of data processing resources in multi-user environments. The decision to allow or refuse this access is based on the roles that the individuals play in the organization that implements RBAC security policies. The roles are defined after a meticulous analysis of the operation mechanisms of the organization in question. Once the roles are defined, they are then virtualized into an electronic security scheme.

Access rights are grouped by role and only those who perform this role can of course access the resource.

RBAC uses the notion of least privilege, which means that only the rights needed to complete the actions depending on a given role are granted.

Among the functionalities that make it a powerful architecture, we find a hierarchy of roles. A user can have many roles that are not mutually exclusive. Consequently, when the roles and access rights overlap, the hierarchical organization gives the optimum access rights.

RBAC is implemented in the health care field, an environment very sensitive to questions of security, privacy and professional confidentiality.

Finally, it is possible to assimilate an RBAC operation to the concept of object technology method. This enables its use in modern applications and operating systems.

15.2.3. *VPSS scenarios*

VPSS are destined to address the following scenarios:

– In an intelligent meeting room such as the Aroma space [DIM 00, CIA 02], people gather for a conference. Some of them are employees who work in the building, others are also employees but from another site, others still are invited guests, and among all of these some will be orators. We want the people present to be able to access the services at their disposal in the room. As the organizer of this meeting, you have access to your company's private network. If you are an orator, you want to be able to use the projector and have the possibility, due to secured channels or mechanisms similar to SSH, to access your personal files on the company's private network. Moreover, as a member of the audience, you only have access to the orators' slides and electronic visitor cards.

– In a hospital, when moving from one patient's room to the next, a doctor can have only his personal wireless electronic assistant at his disposal. He can thus obtain the files of the present patients or share this information with other doctors who accompany him. In the same way, a nurse who is giving care can get information on his portable computer. If an unpredicted event takes place and the doctor needs access to other resources (such as a videoconference kit with portable computer), he is limited by the capabilities of his PDA. He can then use the nurse's computer by identifying himself as a doctor, do what he needs to do then disconnect, returning the computer to its owner, all the while maintaining the confidentiality of the exchanged information.

– In the street, there is an accident. We need to find the nearest doctor or police officer, even if they are not in service. Although they are not currently fulfilling their roles at this precise moment, the environment is able to send messages and call them to help. The personal electronic assistants or portable telephones that these people have with them receive these messages because they are in this specific zone, which enables a rapid first reaction before the intervention of emergency services.

VPSS must enable an ad hoc access, based on the roles, to the services in a ubiquitous computing environment.

15.3. An infrastructure secured by ambient virtual offices

15.3.1. *The application of virtual office*

The application of virtual office can be described as the universal presence and the request of data, applications and services of a user, wherever the user may be, to which we add the possibility of using specific services depending on the geographic location, on the network, on the context and on the role played at a given moment.

We will use VPSS, in other words an authentication method based on the users' roles at the core of the infrastructure, to control access to virtual office resources. More precisely, a user or group of users having identical roles within an organization will have the same work environment, with the same security and QoS, wherever they may be in the platform.

15.3.2. *General architecture*

We have developed the VPSS infrastructure by using private virtual networks.

We use IPSec to provide subjacent security mechanisms, particularly in order to secure communication channels.

The alias IP mechanism in Linux makes it possible to give many IP addresses per interface to each node.

Each role corresponds to a VPN. Each VPN is a virtual sub-network protected by IPSec. In Figure 15.2, each link (horizontal line) is protected by IPSec and corresponds to a role (Role 0 = sub-network 0, etc.).

The first role/sub-network is particular as it is the point of entry where all users must register with the access control server (ACS).

The ACS sends information concerning physical and logical configurations of other accessible roles/sub-networks (such as IPSec keys, network configurations). The first sub-network is generic and has a very limited role called the invited role, or role 0. This role is only an entry point to the network and to the security policy server and access rights that provide configuration information corresponding to roles that can be assumed by the users at a given moment.

In a given sub-network, users (rectangles) can only access the registered services (ovals) for the corresponding role. A service can be available for several roles (as shown vertically) and the users can have several roles at the same time.

In the current state of our work, this architecture is used as proof of concept and to establish the schedule of conditions of our prototype implementation.

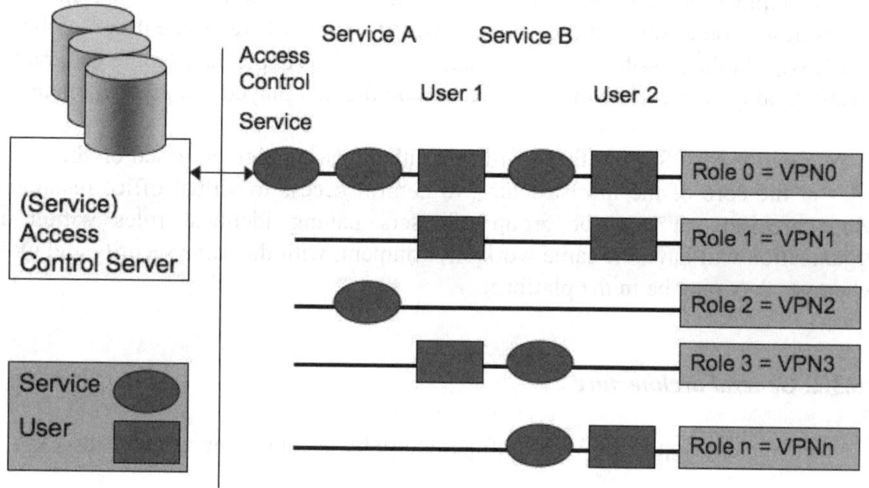

Figure 15.2. *VPSS architecture*

15.3.3. *An implementation of network interface access control*

Our prototype implementation was developed in Linux. This choice was guided by the availability of the core source code and the open source software for this platform.

In order to test the Linux model developments, the ACS and the connection clients (*log-in*) were implemented by using Java and a simple database created for the occasion. The different roles are saved with a short description of their properties, the configuration information and the different members, as well as their profiles.

We will use group management tools in Linux (the chgroup commands, the files/etc./group and /etc./passwd) to define special groups that will be the equivalent of roles. For example, the groups with group numbers (GID) 1001 to 1005 represent 5 roles in our test environment.

For each role, an alias-IP address is provided. This can either be given by a DHCP server or by an ACS server. This address corresponds to the sub-network of the corresponding role.

A Linux mode, once loaded, [BEC 97] makes it possible to capture certain core system network calls. According to the group to which you are connected, your communication is sent to the corresponding IP socket and then to the network interface of the corresponding VPN.

We want to make scenarios such as the one where a doctor needs to connect quickly in another user's session possible. We have used XNest [XNE 03] to make different and personalized environments possible during these temporary sessions.

The choice of IP for our prototype architecture was guided by its omnipresence at the heart of private and Internet networks and by the personal knowledge of its programming as well as that of IPSec in the Linux core (bearing of the referenced IPSec of the NIST of IP to IPv6 under Linux [CIA 98], which is a standard presence in the new Linux node (2.6)). However, the use of VLANs (Virtual LANs, virtual local networks) could also be considered.

15.3.4. *Prototype infrastructure of network control access*

Scenario

We want to authenticate a user and assign certain privileges according to the roles for which he is registered, and notably the role he is playing at the current moment. These privileges are notably access to certain services, to certain network elements as well as a certain priority in these accesses and an associated QoS.

This infrastructure makes it possible to have a permanent virtual office and the essential of the configuration (network) is automatically regulated at a low level. However, profiles more precise than the roles must be considered in the management of the intelligent office. Thus, a higher-level management application could sort everything that is available for a given role and manage the role changes according to preferences linked to the individual and to a large context (external events, user interaction, etc.).

It can be noted that the impossibility under certain operating systems of having virtual network interfaces and using IP aliases can force series of disconnection and reconnection to change roles if these prototypes are related to the aforementioned systems.

In the framework of advanced development of a prototype infrastructure, we based ourselves on the following scenario.

A Wi-Fi equipped university wants to provide a controlled ambient Internet access to all intervening people in their center, in other words to make is possible to use virtual office applications.

People accessing the ambient network can be of diverse origins and play different roles: administrators, local teachers/researchers, external consultant, administrative personnel, students, visitors.

The first problem to address is the initial connection. Thus, a person arriving for the first time on the university campus with their portable computer will first have to identify themselves on the access server which will provide the rights according to a profile corresponding to the roles they are authorized to play. Initially, this person being unknown, his rights are limited to a basic Internet access with limited bandwidth which enable him simply to navigate and check emails. If this person had a particular identity (and therefore particular roles) he would then get supplementary rights once the registered profile is validated.

Once this step is completed, we can find ourselves in different situations: an external consultant, for example, occasionally giving lectures in the university or participating in research projects partially carried out locally, will already have registered and when reconnecting, his profile will already be known and his rights will be granted automatically. The necessary data are stored in a file (certificate) on his computer. He will therefore have Internet access, with a flow enabling the upload of medium-sized files (for uploading his lectures, for example), and also access to the entire university network (except the administration) to store files on a server in he personal space and/or to check his university address emails.

A student will have the same type of rights, except where university research network access is concerned.

However, a local teacher/researcher who is regularly connected to the university network already has rights granted by the administrator. He has access to the research lab network and the university network. He can store files on different servers and has restricted rights of access to the administration server to store messages, for example. He has ADSL Internet access to enable videoconferencing with other research laboratories. He can, for example, have priority in terms of access and QoS over other roles.

However, administrative personnel connect regularly, even daily, to the network. These rights include, of course, complete access to the administrative network and to medium-flow Internet. However, they also have access to the university server to store files (announcements).

Administrative personnel can also be a teacher/researcher so for this person, the rights will be that of administrative personnel and those of the local teacher/researcher mentioned above.

Finally, an administrator often has limited rights with regard to user management. However, this person also has other roles and rights corresponding to their status/profile in the university and to the activities they are currently carrying out.

The granting of roles is done by a role administrator, whereas the choice of implemented roles rests with the user. Therefore, it is the use of RBAC dedicated to the access and use of the network. The real complexity is in the installation, the configuration and the maintenance of services on good virtual networks.

15.3.5. *Implementation of the solution*

15.3.5.1. *Technologies used*

With regard to secured access to 802-type networks (802.3: Ethernet, 802.11, Wi-Fi), we use 802.1x technology which makes a strict authentication of the station or active equipment possible.

This authentication architecture is comprised of three entities:

– supplicant or client: terminal with a network card (wireless) and wanting to connect to the local network;

– authenticator: intermediate element generally serving as access point (Wi-Fi access point here) that relays information provided by the client, who is not yet authorized, which is necessary to his authentication, to the network authentication server. Once the authentication succeeds, it then opens access to traffic apart from that to which he is already linked;

– authentication server: verifies the validity of the access request made by the supplicant (we have used a RADIUS (Remote Authentication Dial-In User Service) server) and confirms or not to the authenticator.

During the authentication phase, the following protocols are implemented:

– EAP (Extensible Authentication Protocol): this PPP (Point-to-Point Protocol) extension is a generic authentication protocol that supports multiple authentication mechanisms (described further);

– EAPOL (EAP Over Lan – local network): the access point only allows messages to the authentication server, until the client is authenticated;

– RADIUS: this is a specific protocol used between the access point and the authentication server, in which EAP packets are encapsulated.

The access point therefore serves as intermediary and as a filter (see Figure 15.3): on the one hand it only opens the port for EAPOL and then relays with EAP into RADIUS. Only the authentication protocols circulate, while the client is still not recognized.

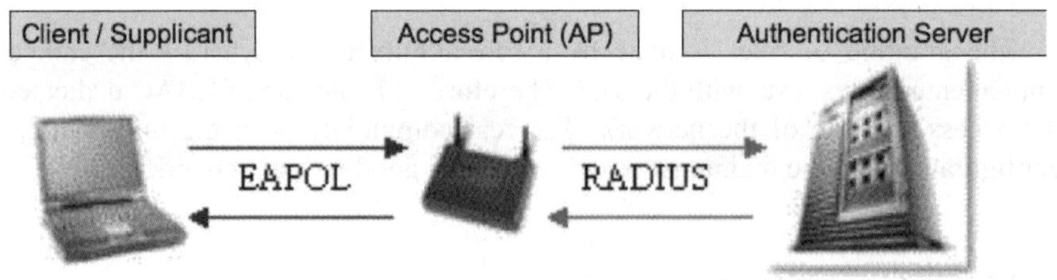

Figure 15.3. *Different protocols in an authentication phase*

The EAP protocol can be transported by 802 systems (.3 or .11, for example) or encapsulated in RADIUS systems. It supports several authentication methods (of which at least 50 are described in IETF drafts) of which the following 2 are the most widely used and which we have used:

– EAP/MD5: this method is based on the user name/password pair. Each client has such a pair registered on the RADIUS server which must be communicated to the server. The server then verifies the validity of this information to decide whether or not to authorize the client to connect to the network. This authentication, however, is not reciprocal because only the client's identity is verified. The insertion of a false access point/RADIUS server pair makes it possible to gain authentication information without the client's knowledge (man in the middle). The attacker is then able to connect instead of the client.

– EAP/TLS: this method makes it possible to address this problem because authentication is symmetrical: the client verifies the server's identity before authenticating himself due to a Public Key Infrastructure (PKI). This then supposes the generation of certificates for the RADIUS server and for each client, and asymmetrical public/private keys. These make it possible to verify the identity of communicating entities by using their public keys and of the signature of the exchanged information.

However, this reliability of the EAP/TLS method is ensured at the cost of increased complexity, that of the PKI model and the software and servers to be implemented, as well as the dependence with regard to its implementation in operating systems, access points and RADIUS servers.

15.3.5.2. *Concrete realization*

When a client is within the zone of coverage of an access point, there is an association between the network card and the Wi-Fi access point. This enables an exchange of authentication messages between the client and the RADIUS server via the access point. The access point will then validate or reject the request.

Two cases can then arise:

– If this is a first connection (Figure 15.4), the request is made to the authentication server by using a default login corresponding to an absence of authorization. A message is sent to the client specifying to identify himself with an *iup* login and *iup* password, which makes it possible for him to have access to a DHCP server that then provides him with an IP address in a very limited field: this address simply makes it possible to view a Web page that then gives instructions on how to proceed. From here, he can create a profile on the system and request associated rights and roles. To do this, he must complete a form that will or will not be validated by an administrator. The generated certificate will be a function of the roles associated with the user's identity within the organization. We use LDAP (standardized directory system, known for its speed of information restitution) to store user profiles, but a database would also suffice at this stage. The script that enables network connection must then be uploaded.

A person who has already been connected and taken into account in the system already has a certificate, knows his user name/password pair and can initiate the connection script. This script makes it possible, from the client's connection certificate, to know his role and establish a configuration file specific to the user that will be used by the supplicant (connection 802.1x software client).

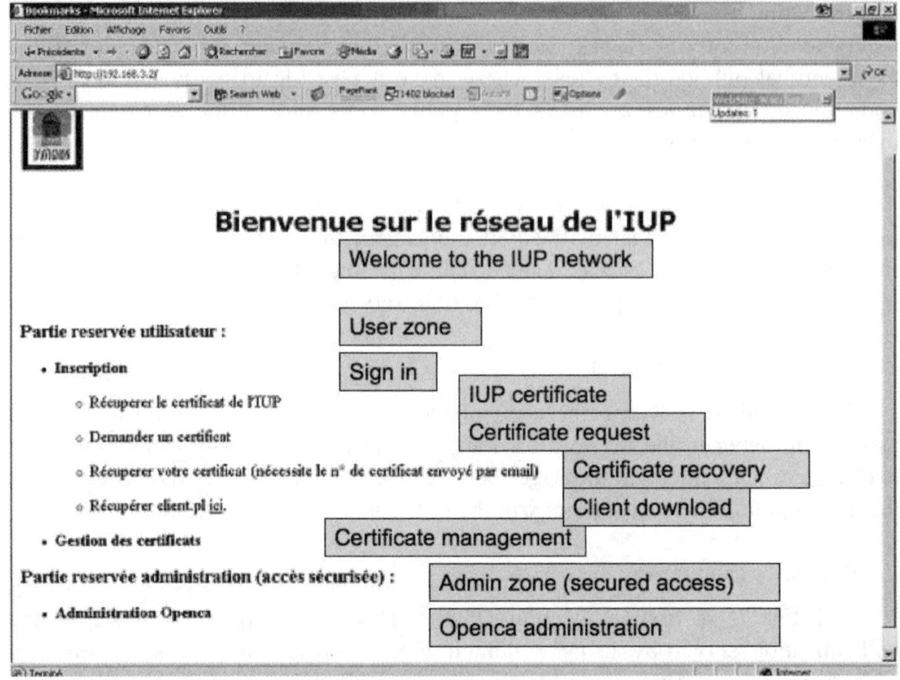

Figure 15.4. *Welcome page during initial connection*

– If the authentication succeeds, the access point will open a port on the machine guaranteeing WLAN access. Then, due to the information contained in his certificate, a request is sent to the DHCP server corresponding to the role of the user.

The addressing plan was preset and blocks of specific addresses are assigned by role. There are therefore many DHCP servers on the authentication server, each of which manages a block of addresses and therefore a role. The role/IPaddress pair makes it possible to simply manage network resource access (printing, saving, etc.) and priorities (making it possible to give a certain QoS and a certain flow), due to filtering in the network routers/firewalls. IP address changes are transparent for access points that manage access according to machine MAC addresses.

The security of the different streams linked to roles is ensured by the use of a VPN (Virtual Private Network) using the same certificates as established during authentication. The traffic can be encrypted to ensure confidentiality and signed to be sure of the identity of the communicating machines. The use of a VPN makes it possible to be freed of subjacent technology (Wi-Fi) and of security protocols such as WEP and its evolutions (WPA). This can also lighten the load in access points, but implies an overload on client machines (encryption cost) that could be penalizing for light clients (such as PDA).

This infrastructure and notably certificate management makes it possible to address the following situations:

– A single machine can be used by different people because each will have their own user name/password pair, their own certificate and their own script in their directory (/home/login under Linux). With adequate rights, this information remains confined to the user's personal directory. This information can simply be carried on a USB key, or any other mobile data storing format, which enables the use of other machines as well.

– The same person can have several roles in the university, such as a teacher fulfilling administrative personnel and role administration functions. The VPSS administrator must organize these according to a precise hierarchy. The use of the least privilege rule in RBAC must apply here. According to the role played, the user must both authenticate himself with the server each time and reload the corresponding certificate, or when possible (such as, for example, by using the modified Linux operating system in the previous section), have several available certificates and use the one corresponding to the role he wants to fulfill.

15.3.5.3. *Secure platform implementation*

Authentication

To complete network authentication we have used Freeradius, which is an open-source implementation of RADIUS. After a test phase in EAP/MD5 mode, we were able to deploy a more reliable solution using PKI and TLS.

Client information

We use OpenLADP, an LDAP open-source implementation to store information.

Addressing and DHCP

Access control is carried out by managing blocks of IP addresses according to user roles. A script makes it possible, by using certificate information, to identify the client's role and, after his authentication, to orient him to a DHCP server (on a good port) which is working on a virtual surface in the authentication server and listening to his role on a specific port. The IP address thus provided is in a block corresponding to the role in question and service access will be established according to the filters on it.

Confidentiality

The use of VPNs enables a combination of:

– authentication of intervening elements in the network. Only authorized people have access to the data;

– confidentiality of the flows.

The initialization of the tunnels between servers and clients requires that the former know the IP address of the latter. At the moment a client connects, Freeradius initiates a script that will scan the files called role_name.leases (for example, prof.leases or student.lease) in which are located the IPs granted by the server. We use the Freeswan version of IPSec for Linux.

Firewall

According to the roles and therefore the granted IP addresses, routers/firewalls filter the flows. The rules, which must be updated according to the blocks of addresses and the implemented IPSec channels, make it possible to let through or restrict the flows.

15.4. Conclusion and perspectives

VPSS is a high level security architecture that must make it possible to control access to ambient Internet, to secure the flows and to differentiate them.

15.4.1. *Limitations and future developments*

The currently developed prototypes introduce a certain number of limiting factors.

15.4.1.1. *Complexity of implementation*

Therefore, whether it is the Linux network interface access model, the supplicant elements or the VPN clients, they suppose the pre-installation of the corresponding software. Our solution, although easy to use, only functions in a certain type of environment and notably in the Linux operating system. It supposes in part a modification of the Linux core. However, with the arrival of the new 2.6 core, the integration of IPSec and the multiplication of netfilter/iptables hooks [NET 03], we can more discretely carry out the redirection of network flows within the operating system (we could filter the flow according to the group or user identifier and redirect the flow to the virtual network interface of its choice).

Client and service administration requires planning in order to properly define the roles and therefore the resource access policy. However, these environments must have a great reactivity with regard to the dynamic aspect of clients and services, and the design as is presented here seems rather heavy, and it would be interesting to study its scaling capabilities.

Moreover, this solution must be able to be used by anybody, in a perpetually moving context and sometimes very dynamic. We are therefore currently looking at automatic management aspects of dynamic environments such as VPSS. Policy management and dynamic configuration/deployment solutions are possible to simplify the implementation of VPSS and their administration.

15.4.1.2. *Ambient office and profiles*

The application of ambient offices will also need to implement more precise elements with regard to profiles. In fact, for a single role, it is important that each user is able to define preferences regarding available services he wants to access permanently, for example. This could be done above the VPSS which will take care of managing rights and access quality transparently for applications and users.

Similarly, managing the different user roles and profiles requires the development of a boarding table that makes it possible to easily navigate displacements and ambient Internet access.

15.4.1.3. *Interoperability*

Ambient Internet supposes a use of technology which the question of interoperability addresses. Such a solution must function across the different operating systems and network access methods and on equipment with variable technical characteristics (mobile telephone, top of the line portable, communication gadget, etc.).

15.4.1.4. *Virtual networks and QoS*

The aspects linked to QoS have not yet been implemented, but they are also available in the Linux core. Automatic VPN deployment can also come with the implementation of reservation mechanisms, for example. The use of VPN IPSec is a surcharge in terms of resource consumption for clients and servers, and of bandwidth that will need to be studied. The use of level 2 VLAN [VLA 03] for flow differentiation is also a path we are considering.

Ambient Internet security will be as essential as the omnipresence of high-flow access to its data, its services and the locally present services. Security has of course a considerable cost, but that is all the more true when safety mechanisms are added after the fact. We must therefore think now about solutions such as VPSS, which take into account the popular, interactive and dynamic character and the commercial stakes associated with ambient intelligence. In fact, the possibility of differentiating the flows and therefore QoS is an element to consider in commercial deployment and an offer of ambient Internet service.

15.4.1.5. *IPv6*

The prototypes described here were developed for IPv4 but IPv6, the new generation IP, has a certain number of interesting characteristics. The large number of addresses and addressing diagrams will enable an easier administration of VPNs. In addition, IPSec and autoconfiguration mechanisms are included.

15.4.2. *Conclusion*

VPSS are yet only an outline of a possible solution that seeks to meet the needs for identification and resource access security of intelligent ambient environments. A total solution will have to use the narrow association of media access security mechanisms, of user identification, of dynamic management adaptable to the needs and will be controlled by a rights and role management architecture of the highest level.

15.5. Bibliography

[BEC 97] BECK. M., BOHME H., DZIADZKA M., KUNITZ U., MAGNUS R., VERWORNER D., *Linux Kernel Internals*, 2nd edition, Addison-Wesley, 1997.

[CIA 00] CIARLETTA L., DIMA A., "A Conceptual Model for Pervasive Computing", *Proceedings of the 2000 International Conference on Parallel Processing Workshops*, Las Vegas, August 2000.

[CIA 02a] CIARLETTA L., Evaluation des technologies de l'informatique ambiante, PhD Thesis, Henri Poincaré University, Nancy, November 2002.

[CIA 02b] CIARLETTA L., CHRISMENT I., "Espace Virtuels Privés Intelligents", SAR'2002.

[CIA 98] CIARLETTA L., Vers une approche sécurisée des mécanismes d'autoconfiguration d'IPv6, DEA Report, 1998.

[DIM 00] DIMA A., CIARLETTA L., MILLS K., "Aroma-Wireless Jini Connection Technology-Based Adapters for Smart Space Devices", *Technical Session 622 of the Java One Conference*, San Francisco, 2000.

[DUD 02] DUDA A., "Ambient Networks", *Proceedings of the tutorials, Réseaux Haut-Débit et Multimédia* (RHDM'2002), Autrans, 2002.

[ERC 01] AOLA J., "Ambient Intelligence", *ERCIM News, Special Theme: Ambient Intelligence*, volume 47, October 2001.

[FER 01] FERRAIOLO D. F., SANDHU R., GAVRILA S., KUHN D. R., CHANDRAMOULI R., "Proposed NIST standard for role-based access control", *ACM Transactions on Information and System Security (TISSEC)*, p. 224-274, ACM Press, 2001.

[IEE 03] IEEE, "802.11, Home Page", http://grouper.ieee.org/groups/802/11.

[IST 01] ISTAG, "Scenarios for Ambient Intelligence in 2010, Final Report", February 2001.

[JOY 00] JOY B., "Why the future doesn't need us", *Wired*, 2000.

[NET 03] NETFILTER, "the netfilter/iptables project", http://www.netfilter.org.

[VLA 03] IEEE, "802.1Q, Virtual LANs", http://www.IEEE 802.org/1/pages/802.1Q.html.

[WEI 91] WEISER M., "The computer for the 21st century", *Scientific American*, September 1991.

[WEI 93a] WEISER M., "Some computer science issues in ubiquitous computing", *Communications of the ACM*, p. 75-83, 1993.

[WEI 93b] WEISER M., "Hot Topics: Ubiquitous Computing", *IEEE Computer*, 1993.

[WEI 96] WEISER M., BROWN J. S., "The Coming Age of Calm Technology", *technical report*, Xerox PARC, October 1996.

[XNES 03] XNEST, "XNEST Manual Page", http://www.xfree86.org/4.2.

Chapter 16

Smart Card Security

16.1. Introduction

The inherent lack of security in the world of wireless data networks pushes users to secure their digital environment. The network can no longer be considered a trustworthy environment. It is necessary from now on to use safe mechanisms that make it possible to use networks without risk.

The smart card was initially designed with the purpose of securing banking transactions. It has also put forward its security advantages in second-generation mobile telephone networks, within the GSM standard. Smart card technology has been tried and tested and is standardized by the ISO 7816 standard. Its mobile character along with its capability of handling information makes it a perfect candidate for securing data networks.

In this chapter, we will present the microprocessor smart card's different characteristics that could make this technology the cornerstone of Internet security.

We will also present the microprocessor contact smart card, while keeping in mind that its contactless counterpart benefits from the same possibilities, as well as those characteristics of a contactless interface.

Chapter written by Vincent GUYOT.

16.2. History

The smart card is not a new technology. In fact, since the end of the 1960s, European, American and Japanese researchers have been considering it [RAN 00, JUR 03].

Many people around the world have contributed to the implementation of this technology, but it is the French who gave rise to the current smart card. In 1974, the first patent on the memory card was registered. In 1978, a French engineer invented the microprocessor card with the SPOM patent[1].

Its information processing capabilities equips it with a certain intelligence[2]. The smart card will not undergo any major changes in its functioning principle[3].

Since the beginning of the 1980s, industrial applications have been appearing. In France, the use of the smart card as a method of payment in public phone booths and its use in the banking industry caused a new industry to emerge [TOW 02].

16.3. Different parts of a smart card

A smart card is created in a series of steps relating to its different parts: the chip, the plastic support and the external module. All these parts are standardized, which makes a segmented industrialization possible.

16.3.1. *The chip*

The smart card can be considered a microscopic computer. In fact, it is composed of the same functional elements as computers. It integrates a microprocessor, often a second microprocessor specialized in the acceleration of cryptographic operations, memory, a communication port and functions with an operating system. The power of current cards is often compared to that of the first microprocessor, the IBM-PC[4]. The fundamental difference with a true computer is

---

1 Self-Programmable One-chip Microcomputer [UGO 92].

2 Hence its English name of *smart card* or *smartcard*.

3 The SPOM patent is still used by microprocessor smart card makers.

4 Which appeared in 1981.

that all these elements are on a single silicone substrate, on an electronic chip[5], whereas they are separated into several electronic components in a computer.

Figure 16.1 presents the different components on the chip as well as their interactions. The microprocessor (CPU) through which all information flows plays a preponderant role.

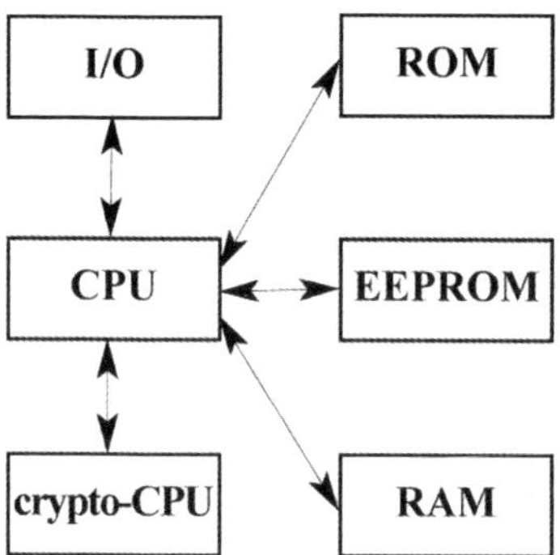

Figure 16.1. *Interactions between a smart card's components*

Smart card processors are most often based on an 8-bit design and are operated at a speed of 3.3 MHz. The new generation of 32-bit processors achieves a speed of 33 MHz.

A smart card's memory is generally less than 64 Ko. New types of memory, such as Flash or FeRAM, make it possible to achieve a laboratory memory capacity close to a megabyte. In a few years, it is possible that this capacity will be in the order of gigabytes [VER 00].

Today, it is nevertheless possible to simulate larger storage capacity in a smart card.

For example, there are solutions such as [BIG 98], which combine encrypted data storage on the network and the use of a personal smart card that makes it possible to restrict access to this information to only the card carrier.

Typically, the viewing of sensitive images such as x-rays in the medical field is therefore limited to the doctor who is treating the concerned patient.

5 Hence the name smart card.

16.3.2. *The plastic card*

This electronic chip is inserted in a plastic card that serves as transportation support. The format of the plastic card is similar to that of old magnetic cards whose format is ID-I[6]. However, the ID-000 format, made popular by the use of SIM cards in GSM terminals, is much smaller, as shown in Figure 16.2.

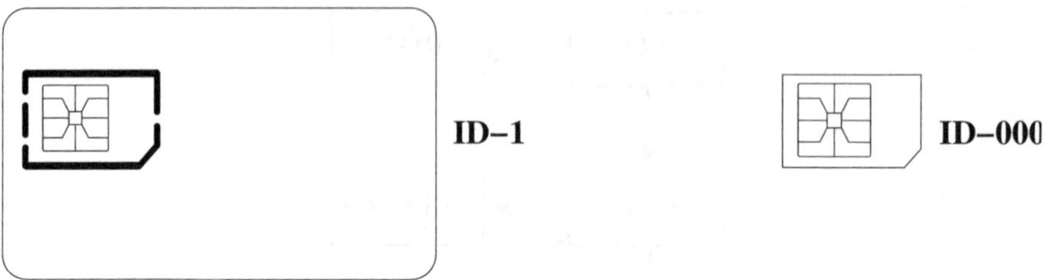

Figure 16.2. *Different smart card formats*

16.3.3. *The external module*

As it is naturally very fragile, the chip must be protected. For this reason, it is covered with a module that ensures a physical protection and serves as an interface with the external environment. It feeds the chip with an electronic current, provides it with a clock and allows it to communicate. This visible module[7] is what physically differentiates a smart card from a magnetic card. It is composed of eight contact zones, as indicated in Figure 16.3.

C1	C5
C2	C6
C3	C7
C4	C8

Figure 16.3. *The placement of a smart card's contacts*

6 Standardized in the ISO 7810 standard.
7 The form of this module makes it possible to instantly identify its manufacturer.

Each contact's role is described below:

– C1: electric feed;

– C2: re-initialization;

– C3: clock;

– C4: unused;

– C5: electric mass;

– C6: no longer used (initially programming tension);

– C7: serial communication;

– C8: unused.

We note that on these eight standard contacts, only five are used. Some manufacturers use contacts C4 and C8, reserved for future use, in conformance with the ISO 7816 standard, to add supplementary functions to the card, such as the possibility of communicating via USB[8] instead of the standard serial link.

16.4. Smart card communication

To use a smart card, it must be inserted into an active electronic device capable of interacting with it[9]. Such a reader exchanges standardized messages with the card. These messages authorize the reading and writing in the card's memory. Despite its denomination, a smart card reader therefore makes it possible to write data into a card.

16.4.1. *A low-level communication protocol*

A smart card can be compared to a micro-server that uses a half-duplex communication protocol.

8 Universal Serial Bus, which is a standard used to connect external peripheries to microcomputers.

9 To ensure the card carrier's legitimacy, all exchange generally begins by PIN code verification or a biometric test that can definitely block the card in case of an error.

Smart card communication can be done by the intermediary of APDU messages[10]. There are two types[11]: command APDUs and response APDUs (see Figure 16.4).

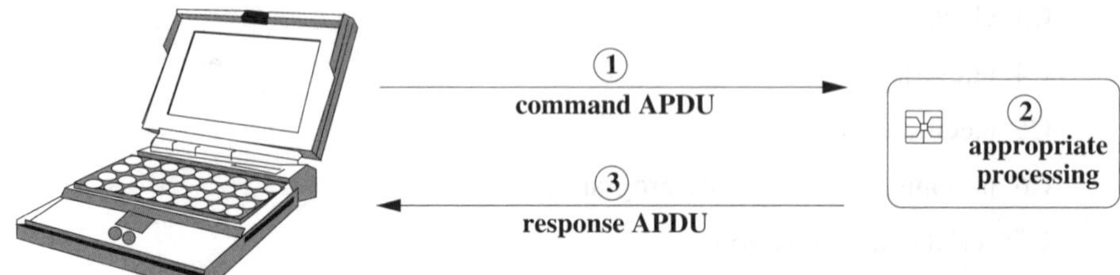

Figure 16.4. *Communication by APDU*

Command APDUs are always sent by the entity[12] that wishes to use the card, whereas response APDUs are sent by the card in response to command APDUs. Figures 16.5 and 16.6 show the different fields[13] that make up the command APDUs and the response APDUs.

Compulsory header				Optional part		
CLA	INS	P1	P2	Lc	Data	Le

Figure 16.5. *Command APDU*

10 Application Protocol Data Units.
11 Defined by the ISO 7816-4 standard.
12 That always initiates the communication.
13 Each is a byte in size, except the data field that has a variable size.

Compulsory header	Optional part	
Data	SW1	SW2

Figure 16.6. *Response APDU*

Significance of the initials:

– CLA: the class of the instruction to be sent (ISO, proprietary, etc.);

– INS: instruction to be executed by the card;

– P1: first instruction parameter;

– P2: second instruction parameter;

– Lc: length of the optional data field;

– Le: length of the expected APDU response;

– SW1 and SW2: card status indicators.

There can be several types of command APDUs.

When the optional part is non-existent, the command sent to the card is to execute an instruction without entrance or exit data (for example, a command to destroy data stored on the card). The card's response indicates if there were any processing errors (0x9000 indicates success).

If the optional part contains a Le field, the card also returns the data.

In the case of an LC field and a data field without an Le field, the data will be sent and processed by the card but no data will be sent back.

16.4.2. *Higher level communication protocols*

A communication protocol common to all smart cards has been defined: ISO 7816-4. This is an archaic protocol based on message exchange. It only defines reading and writing operations for raw data without defining the high level interface such as data flows or files, which slows the development of applications that use a smart card.

To hide this protocol, several approaches propose an interaction with the card with the help of higher abstraction level protocols.

In [ITO 99], the author proposes accessing the smart card file system transparently by integrating it into UNIX type system file structure in the manner of NFS [SAN 85]. The operating system's automatic handling of the complexity of the ISO 7816 communication protocol greatly facilitates the development of applications.

There exist different approaches to access smart cards via the HTTP protocol, whose use is widespread in the World Wide Web (WWW).

[BAR 99] proposes accessing the card by URLs that are relayed by an HTTP proxy. This proxy, working on the computer on which the card is connected, translates the card's destination URLs into a series of specific APDUs.

The authors of [REE 00] followed a very different road by implementing an HTTP server directly into a smart card. They implemented a TCP/IP communication specific stack between the card and the computer to which it is connected in order to link the network card.

To open the world of GSM networks to the Internet, the authors of [GUT 00] propose implementing an HTTP server in SIM cards by using the SIM Toolkit[14]. To function with the current mobile GSM terminals, the authors recommend the GSM operator's use of HTTP proxies. These encapsulate the HTTP requests into SMSs[15] that they send to the mobile terminal.

Java's RMI method is used in [MIC 01] to describe the smart card's resources. Two APDUs are successively used to choose and invoke the smart card's service. This method is limited to the use in the Java environment.

16.4.3. *Different readers*

The evolution of smart card readers followed that of microcomputer communication ports. The oldest readers communicate through serial or parallel microcomputer ports. These ports, which were not electronically powered, required an external source of electricity[16]. In a marginal way, certain keyboard

14 Development platform for standardized GSM 02.19 and GSM 03.19 applications.

15 Short Message System.

16 By the PS/2 port, an electric transformer or batteries.

manufacturers integrate a reader into their products. Another attempt at the democratization of the use of smart cards relies on a card reader introduced in floppy disk drives already integrated in machines. Portable microcomputers also have readers that are dedicated to the PCCARD[17] format. From now on, most new smart card readers use the USB port for all of its advantages[18]. Certain microcomputer makers deliver certain models with a reader directly integrated into the machine.

In [ITO 01b], the authors propose the use of a remote smart card, which would make it possible to benefit from the advantages of smart cards on machines that do not have a reader. Their solution consists of keeping the smart card in a safe place, where it would be permanently connected to a computer linked to the network. The traditional services relying on local cryptographic resources, such as SSH [YLO 96] or PGP [ZIM 95], must be modified to use the remote smart card's cryptographic resources by using the Simple Password Exponential Key Exchange (SPEKE) protocol [JAB 96].

16.4.4. *An autonomous card*

Recent technological advances in this matter have led us to think that readers will soon disappear, the cards having integrated communication port management. In fact, electronic chip manufacturers have succeeded in integrating the complete USB protocol management into the chip. Thus, the simple fact of physically linking such a card to a microcomputer's USB port makes it possible for the terminal to use it without a reader. The suppression of the reader, which is a costly device, causes a reduction in cost. In addition, the use of smart cards becomes easier as they can now be used in terminals that have no readers. This major technological advance facilitates the adoption of smart cards in networks.

16.4.5. *The software*

A smart card reader needs a specific driver to be recognized and used by the operating system to which it is attached. For a long time, there were no standards. To write an application that used a smart card, the reader was programmed in a proprietary manner. Changing the reader meant re-writing the application so that it could have a dialogue with the new reader. To write such an application, it was necessary to work in narrow collaboration with the manufacturers of smart card readers to have information on the particular functioning of each reader destined to

---

17 Formerly PCMCIA.

18 Electric powered for peripheries and increased communication speed.

be used by the application. These difficulties have considerably slowed the use of smart cards in data networks.

16.4.5.1. *PC/SC*

In the 1990s, the various players on the smart card and micro-processing market met within the "PC/SC[19] Workgroup" to standardize a design whose goal was to efficiently integrate the smart card in microcomputers. The PC/SC design was published at the end of 1996. Due to this standard, applications are able to use the PC/SC API without worrying about the type of reader they use.

The PC/SC design was conceived independently of all operating systems. In fact, the only official implementation is integrated in Microsoft systems[20]. Today, smart card readers are all equipped with their PC/SC system driver.

16.4.5.2. *OCF*

At the same time, another organization was formed, the OpenCard consortium, to provide a purely Java design destined to use smart cards in a multiplatform environment. In 1997, the OpenCard Framework (OCF) was published. Its use makes it possible to write a Java application (or an applet) that functions in any operating system equipped with a virtual Java machine. The readers' OCF driver must also be written in Java for portability reasons.

Unfortunately, there are very few smart card readers that are delivered with their OCF drivers. The reader manufacturers prefer providing their product with a single driver, the purely Windows PC/SC driver. Fortunately, there is a gateway between PC/SC and OCF, which makes it possible to write Java programs that use OCF while benefiting from the PC/SC driver provided with the readers.

16.4.5.3. *PCSC Lite*

Since 1998, the MUSCLE project[21] has developed a free implementation[22] of the PC/SC design for UNIX systems, called PCSC Lite, which is interoperable with the Microsoft version[23] with the goal of facilitating the use of existing Windows applications in UNIX systems. Today, PCSC Lite functions in the most current UNIX systems and with OCF programs.

---

19 Personal Computer/Smart Card.

20 The PC/SC design is included in this system since Windows 2000.

21 http://www.musclecard.com.

22 In the Open-Source sense.

23 As it uses the same API.

16.5. A secure component

The development of the smart card was done with the goal of maximizing security. It contains specific defense mechanisms.

16.5.1. *A single entity*

The electronic chip, which is the basis of all smart cards, is a component that was designed with the single goal of providing strong physical security to the data that it contains. The fact that the memory and the microprocessor are part of a single physical component confers to the smart card an as yet unequalled level of security.

It is in fact quite complex to intercept the exchanges between the microprocessor and the memory if these two entities are physically unified within a single electronic component.

16.5.2. *Physical attacks*

Since their conception in the 1970s, corruption attempts on smart cards have multiplied. The people doing that do not lack in imagination to attempt to get around the security mechanisms.

Attempts in modifying the microprocessor's behavior are based on altering the card's power supply or its clock. It can also be cut out to make it possible to directly read the memory cells. Exposure to various rays is another possible form of attack.

16.5.3. *Countermeasures*

Different conception particularities secure the smart card against all types of attack. Memory management is much more complex, such that all direct reading attempts will fail. Captors placed within the chip have the goal of detecting abnormal external behavior, such as suspicious variations in the clock or in the power supply, or even an attempt of extracting the chip. It can react differently according to information uploaded by its sensors, by interrupting all current operations or by self-destruction.

Recently produced smart cards are also protected against passive attacks by monitoring their electric consumption. This in fact reveals information on the chip's behavior.

16.6. Smart card alternatives

Many types of portable devices that are easy to use from one computer to another are sometimes designed as alternative solutions to the smart card. We will show that these solutions do not guarantee the same services as those offered by the microprocessor smart card.

16.6.1. *Magnetic cards*

In existence long before smart cards, and having the same physical format in common, although not quite secure, the magnetic card is still used throughout the world as a credit card.

16.6.1.1. *Advantages*

The magnetic card is a tried and tested technology, which has been standardized[24] for a long time. Due to its inherent technological simplicity, it can be mass-produced at a low cost. This type of card is based on banking network security.

16.6.1.2. *Disadvantages*

A magnetic card has no great data storing capacity. It is not possible to record more than a million bits on it.

In addition, the magnetic card has no means of protection against reading or writing on its magnetic stripes, thus making it possible to copy such cards and accidentally altering the data when passing in a magnetic field[25].

As a result, there are security problems with regard to confidentiality and authenticity of stored data. In addition, the card only serves as data storing support and has no processing capacity.

16.6.2. *Cryptographic tokens*

With the adoption of standard USB interfaces, new available transmission speeds[26] coupled with the possibility of directly providing power supply to the

24 ISO 7811-1 to ISO 7811-6.
25 Magnets, CRT monitors, etc.
26 Port speeds: serial 0.1 Mbit/s, parallel 0.8 Mbit/s, USB 12 Mbit/s, USB2 480 Mbit/s.

external device have made it possible to develop removable active cryptographic modules.

16.6.2.1. *Advantages*

Equipped with a cryptographic processor and a memory capable of storing a private key, a cryptographic token makes it possible to deploy security solutions based on asymmetric cryptography, of PKI type.

16.6.2.2. *Disadvantages*

In addition to the fact that there is very little available memory for storing information, the security module is dissociated from the EEPROM. This makes it possible to extract secret information contained in the cryptographic token [KIN 00]. The retrieval of keys is therefore possible and annihilates the system's security.

16.6.3. *USB memory cards*

With the reduction in price of electronic components in the last few years, and notably of Flash memory, a new kind of mobile data device has appeared.

16.6.3.1. *Advantages*

Flash memory is equipped with a very large information storage density, which explains the existence of USB keys with storage capacities reaching several gigabytes. Based on the standard "USB storage", these new devices function directly in all recent microcomputer operation systems without requiring a specific intermediary reader.

16.6.3.2. *Disadvantages*

Even if some models attempt to control access via a password, data is readable in Flash memory. Such a device also does not have any processing capabilities. It serves solely as support for information storage.

16.7. Smart card security advantages

Data processing security often calls upon cryptographic algorithms. After recalling the various types of available mathematical tools, we will see how the smart card, due to its advanced cryptographic capabilities, is particularly adapted to reinforcing security in networks.

16.7.1. *Cryptographic concepts*

The different available tools in cryptography rely on two concepts: data encryption and hashing.

16.7.1.1. *Encryption*

The current encryption algorithms follow the Kerckhoff principal [KER 83a, KER 83b]: an encryption algorithm's security should never rely on the algorithm's confidentiality but rather on a secret key. The encryption algorithms that have not respected this rule have always ended up broken, like those used in the GSM [GOL 97], DVD [STE 99] or Adobe's electronic book [SKL 01]. To reinforce the security of an algorithm that follows this principle, it is enough to increase the size of the key used.

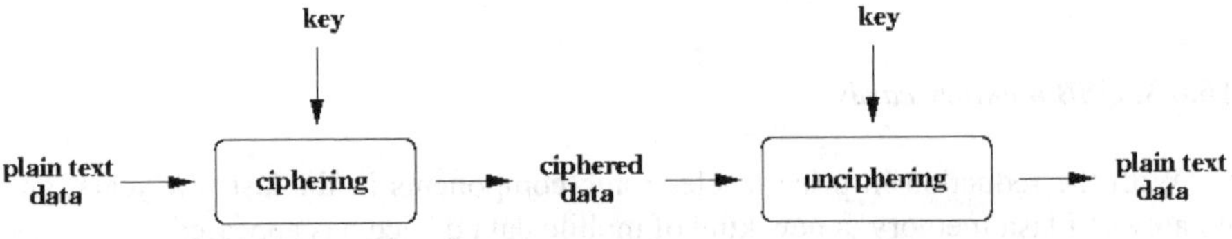

Figure 16.7. *The principle of symmetrical cryptography*

Secret key encryption algorithms function as indicated in Figure 16.7. A key makes it possible to encrypt and decrypt the data. The conception of the algorithms is such that it should never be possible to deduce the original data from the encrypted data without having the key. This principle forces the emitter and the receiver to keep the key secret, which can create key distribution problems in a system where many players participate in the data exchange. In fact, key compromise signifies the compromise of the whole system. On the other hand, encryption and decryption operations with secret keys are very fast.

The principal secret key encryption algorithm used is the Data Encryption Standard (DES). It is being replaced by its successor, the Advanced Encryption Standard (AES).

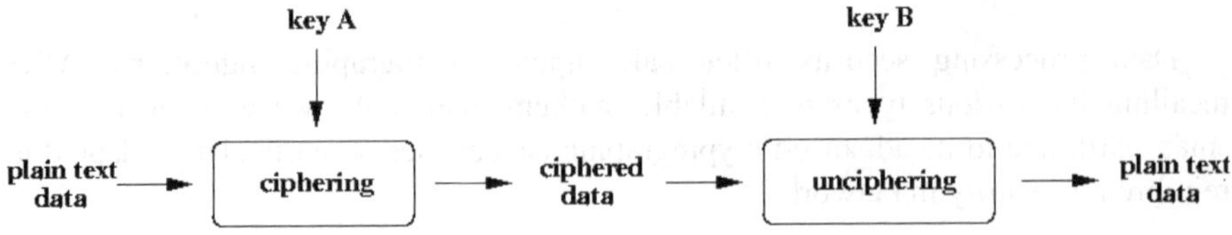

Figure 16.8. *Asymmetrical cryptography principle*

At the same time that the smart card appeared, the concept of asymmetrical cryptography saw the light of day through a new cryptographic principle that implemented two particular keys [DIF 76, MER 78]. As shown in Figure 16.8, encryption and decryption actions are dissociated and each implements a different key. These keys are calculated in such a way that knowing one key does not make it possible to deduce the other key. This new principle implies keeping one of the keys secret and distributing the second. The flexibility of asymmetrical algorithms is counterbalanced by slow encryption and decryption operations, much slower than those of symmetrical cryptography, which limits these algorithms to processing fewer data.

The most frequently used public key encryption algorithm is RSA[27].

16.7.1.2. *Hashing*

Hashing algorithms are functions that apply to data of all sizes and that calculate a fixed-size message digest, as shown in Figure 16.9. A hashing function must be made in such a way that it is very difficult to find two messages with the same message digest.

The most widely used hashing algorithms are Message Digest 5 (MD5) and the Secure Hash Algorithm (SHA).

Figure 16.9. *Hashing*

16.7.2. *Authentication*

When we communicate with a remote entity, we must be sure that the remote entity is the one we think it is. In our everyday lives we can, for example, authenticate a person by their telephone number or by their address. The guarantee that the messages that I send to a person really get to the person in question lies in my confidence in the phone operator or the postal service to contact the right person. This confidence does not exist in the Internet.

---

27 From the name of those who conceived it, Rivest, Shamir and Adleman.

In the digital world, authentication is made possible through the use of cryptographic tools.

The most current method consists of verifying that the remote entity has an authenticating secret in its possession.

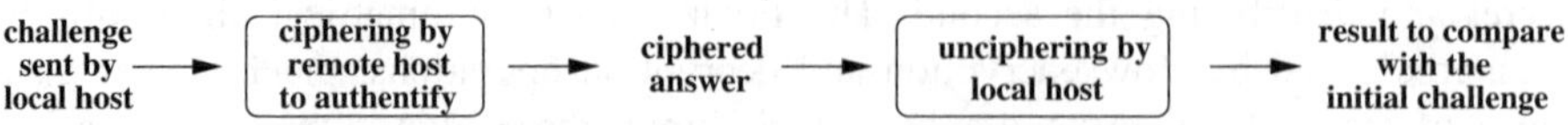

Figure 16.10. *Authentication*

Symmetrical cryptographic algorithms offer a solution to this problem, as shown in Figure 16.10.

The wary entity sends a random message to the entity to be authenticated. The latter encrypts the message with its secret key and sends it back. The wary entity only has to decrypt the received message and compare the result with the random message. If they are the same, this proves that the remote entity is in possession of the secret key and the authentication is a success.

Smart cards are well adapted to this scheme. In fact, they are conceived to keep information safe, a key for example, and can contain a cryptographic processor that accelerates private key encryption algorithms such as DES or AES.

However, there is still the problem of secret key distribution when several entities communicate. The compromise of this key means the compromise of the entire system.

Asymmetrical cryptographic algorithms make it possible to address the authentication problem in a manner that is more satisfying than symmetrical algorithms.

As in the case of symmetrical cryptography, the wary entity will send a random message to the entity to be authenticated, which will encrypt it with its private key and send it back. The wary entity will try to decrypt the response with the public key, which is known by the entity to be authenticated.

If the decrypted message corresponds to the initial random message, this proves that the remote entity is in possession of the private key of the entity its claims to be and the authentication is successful. This scheme is more satisfying than the case of symmetrical cryptography, as the implicated entities do not have to share any secrets.

Once again, the smart card is adapted to this situation. In fact, a smart card, along with being appropriate for the secure storage of a key and capable of implementing algorithms such as RSA, makes it possible to generate the keys necessary for asymmetrical cryptographic algorithms.

This considerably reinforces the confidentiality of the secret key that will never leave the card.

16.7.3. *Confidentiality*

To ensure the confidentiality of sensitive data in a risky environment, they must be made unusable to all non-authorized people.

The symmetrical and asymmetrical cryptographic algorithms seen previously fulfill this task, as in Figure 16.11, where we see the process of data delivery.

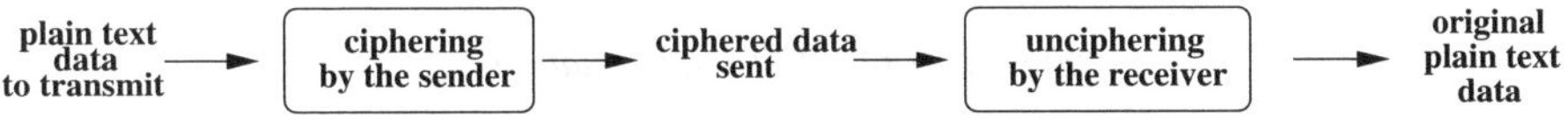

Figure 16.11. *Confidentiality*

Current smart cards, equipped with cryptographic processors, are particularly adapted to the confidential transportation of data.

The current processing power of smart cards does not enable the encryption or decryption of large quantities of data in a reasonable time frame.

However, technological advances such as changes in communication interfaces for faster interfaces, new types of memory, as well as faster processors, tend to reduce this processing time.

In [BLA 96], the author proposes a solution to this problem of encryption speed in the current smart cards. It exposes a new protocol, the Remotely Keyed Encryption Protocol (RKEP), which makes it possible for a smart card to symmetrically encrypt a rapid flow of data by using the power of the computer to which it is linked.

The strength of this protocol is that it makes it possible to use a computer in which the smart card has no confidence because there is no transmission of secret information from the chip. The RKEP establishes a low flow information exchange between the smart card and the computer to which it is connected, providing to the computer enough information to be able to encrypt or decrypt the rapid flow of data.

16.7.4. *Integrity*

More than ever with the advent of wireless networks, the information exchanged in the networks is susceptible to alterations, whether voluntary attack or involuntary (technical failure). There is therefore a need to verify if the received information really corresponds to the information that was sent. This verification is carried out by means of an electronic signature.

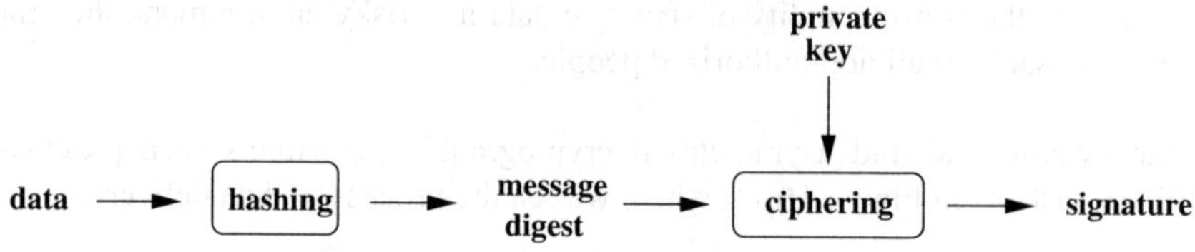

Figure 16.12. *Electronic signature*

For this, we combine asymmetrical encryption and the principle of hashing, as in Figure 16.12.

The sender calculates the message digest of the data to be sent, then encrypts this message digest with its private key to create a signature that will be sent with the data. The receiver will receive the data with the signature. They will decrypt the signature with the help of the sender's known public key and will verify if the resulting message digest corresponds to the message digest obtained when it hashes the received data itself. If they are equal, this will prove the integrity of the received data.

Once again, the smart cards contain the cryptographic material necessary for this task.

16.7.5. *Random numbers*

In cryptography, random numbers play a very important role. It is primordial to have a good random number generator to ensure a proper running of the cryptographic algorithms.

Smart cards contain a material random number generator. They are the ideal complement for all microcomputer that wants to make the most of cryptographic algorithms.

In [ITO 01a], the author proposes reinforcing the security of the contents of a hard drive by encrypting each of the files with a random password. It uses a smart card to carry this out. From a secret in the card and relative to the files to be encrypted or decrypted, the card will generate a different password for each file. The random number generator it contains will ensure the security of these files.

16.8. Network security

Data networks carry many security problems. We will study different solutions, based on the use of the smart card, that improve network security.

16.8.1. *Authenticated data*

[SCH 99] addresses the problem of authentication of data transmitted over the Internet to personal services, such as the use of licensed software. The author proposes identifying the user by sharing a secret key between a personal smart card and the application.

The software that requires user information sends the information request to the card along with a random number. The card first verifies the identity of its carrier: it asks for a PIN code. Then, the card sends the required information, signed, to the application with the secret key shared.

The application that receives the message recalculates the signature from the sent information and on the known secret key. If the signature, so generated, corresponds to the one sent, the data sent by the user are authenticated.

16.8.2. *High level communication protocols*

[MIL 88] defines the Kerberos protocol. This protocol makes it possible to authenticate a user on a remote server through communication encrypted with a secret key. The weakness in this protocol lies in the security of the password storage as well as the choice of user password. The user generally chooses a low security password that does not resist dictionary attacks.

In [BOL 91], the authors warn against the use of the Kerberos protocol without the help of specific material.

One solution to secure Kerberos with the smart card is presented in [ITO 98]. It consists of using a randomly generated password within the card. The password stays in the card and then all cryptographic calculations are carried out in the card on the password that never leaves the card.

16.8.3. *Program execution in hostile environments*

In a network environment, to access remote services a user can find himself using a potentially hostile computer. One possibility would be downloading the services into a smart card to execute them in a secure environment. The current limitations of smart cards do not make it possible to implement this principle.

[LOU 02] addresses this problem by installing mathematical bases to secure the execution of mobile code with a smart card.

The execution of a small part of the calculations by the card extends the smart card security to the entire environment.

16.8.4. *Wi-Fi network authentication*

Wireless networks have profited from the low cost of Wi-Fi technology[28] to become widely used. Unfortunately, serious security breaches [BOR 01] were discovered in the WEP[29] protocol, the protocol destined to control access and ensure confidentiality in Wi-Fi networks. These failures make it possible to get round the measures implemented by the WEP protocol.

The IEEE 802.1X security architecture makes it possible to remedy this problem. It involves three parties:

– the supplicant;

– the authenticator;

– the authentication server.

The user's station, the supplication, wants to connect to the network[30]. The authenticator is the network point of entry[31]. The authentication server is linked to

28 IEEE 802.11 standard.

29 Wired Equivalent Privacy.

30 Typically a laptop equipped with a Wi-Fi interface.

31 Typically a Wi-Fi access point.

an AAA[32] infrastructure that contains the users' network profiles and their authentication parameters.

The exchanges necessary for the connection use the EAP[33] protocol. They use keys specific to the user's station.

In [URI 03], the authors propose the use of a smart card that implements the EAP protocol in which the supplicant's authentication keys are stored. Executing cryptographic operations within a smart card prevents the duplication of authentication keys and consequently excludes illicit connections.

16.9. Conclusion

The smart card was conceived with the purpose of reinforcing security. In the world of data networks and notably that of wireless networks where security is an important issue, the smart card is capable of responding to this security need. Although the addition of a smart card to a system renders it more secure, the security of the network protocol that uses it is fundamental.

16.10. Bibliography

[BAR 99] BARBER J., "The Smart Card URL Programming Interface", *Gemplus Developper Conference (GDC'99)*, Paris, France, 1999.

[BIG 98] BIGET P., VANDEWALLE J.-J., "Extended Memory Card", *European Multimedia Microprocessor Systems and Electronic Commerce Conference (EMMSEC)*, Bordeaux, France, 1998.

[BLA 96] BLAZE M., "High-Bandwidth Encryption with Low-Bandwidth Smart cards", Springer, *Fast Software Encryption*, 33-40, 1996.

[BOL 91] BOLLEVIN S. M., MERITT M., "Limitations of the Kerberos Authentication System", *USENIX Winter 1991 Technical Conference*, 253-268, Dallas, USA, 1991.

[BOR 01] BORISOV N., GOLDBERG I., WAGNER D., "Intercepting Mobile Communications: The Insecurity of 802.11", *7th Annual International Conference on Mobile Computing and Networking*, Rome, Italy, 2001.

[DIF 76] DIFFIE W. HELLMAN M. E., "New Directions in Cryptography", in *IEEE Transactions on Information Theory*, 644-654, 1976.

32 Authentication, Authorization, Accounting.
33 Extensible Authentication Protocol.

[GOL 97] GOLIC J. D., "Cryptanalysis of Alleged A5 Stream Cipher", *Advances in Cryptology – EUROCRYPT'97*, Konstanz, Germany, 1997.

[GUT 00] GUTHERY S., KEHR R., POSEGGA J., VOGT H., "GSM SIMs as Web Servers", *7th International Conference on Intelligence in Services and Networks (IS&N'2000)*, Athens, Greece, 2000.

[ITO 98] ITOI N., HONEYMAN P., ARBOR A., "Smart card Integration with Kerberos V5", *7 CITI Technical Report 98*, 51-62, 1998.

[ITO 99] ITOI N., HONEYMAN P., REES J. ARBOR A., SCFS: A UNIX Filesystem for Smart cards, *USENIX Workshop on Smart Card Technology*, 107-118, Chicago, USA, 1999.

[ITO 01a] ITOI N., SC-CFS: Smart Card Secured Cryptographic File System, *10th USENIX security Symposium*, Washington DC, USA, 271–280, 2001.

[ITO 01b] ITOI N., FUKUZAWA T., HONEYMAN P., Secure Internet Smart Cards, *Lecture Notes in Computer Science*, 2041, 73-89, 2001.

[JAB 96] JABLON D. P., "Strong password-only authenticated key exchange", *ACM Computer Communications Review*, 1996.

[JUR 03] JURGENSEN T. M., GUTHERY S. B., *Smart Cards the Developer's Toolkit*, 2-3, Prentice Hall, 2003.

[KER 83a] KERCKHOFF A., "La Cryptographie Militaire", 5-38, *Journal des sciences militaires*, 1883.

[KER 83b] KERCKHOFF A., "La Cryptographie Militaire", 161-191, *Journal des sciences militaires*, 1883.

[KIN 00] KINGPIN, "Attacks on and Countermeasures for USB Hardware Token Devices", *5th Nordic Workshop on Secure IT Systems (NordSec 2000)*, 35-57, Reykjavik, Iceland, Reykjavik University, 2000.

[LOU 01] LOUREIRO S., Mobile Code Protection, PhD Thesis, EURECOM Institute, 2001.

[LOU 02] LOUREIRO S., BUSSARD L., ROUDIER Y., "Extending Tamper-Proof Hardware Security to Untrusted Execution Environments", *5th Smart Card Research and Advanced Application Conference*, San Jose, USA, 2002.

[MER 78] MERKLE R., Secure Communication Over Insecure Channels, *Communications of the ACM*, 294-299, 1978.

[MIC 01] MICROSYSTEMS S., JavaCard 2.2 Remote Method Invocation Design Draft 1 Revision 1.1, 2001.

[MIL 88] MILLER S., NEUMAN C., SCHILLER J., SALTZER J., "Kerberos Authentication and Authorization System", *Project Athena Technical Plan*, Section E.2.1, 1988.

[RAN 00] RANKL W., EFFING W., *Smart Card Handbook*, 2-5, John Wiley & Sons, 2000.

[REE 00] REES J., HONEYMAN P., "Webcard: a Java Card Web Server", *4th Working Conference on Smart Card Research and Advanced Applications, CARDIS 2000*, Bristol, UK, 197-208, 2000.

[SAN 85] SANDBERG R., GOLDBERG D., KLEIMAN S., WALSH D., LYON B., Design and Implementation of the Sun Network Filesystem, *USENIX Conference*, 119-130, Portland, USA, 1985.

[SCH 99] SCHOLTEN I., BAKKER J., "Authentication on WWW using smart cards", *Information Technology Shaping European Universities (EUNIS'99)*, Helsinki, Finland, 1999.

[SKL 01] SKLYAROV D., MALYSHEV A., "eBooks security – theory and practice", *conference DEF CON 9*, Las Vegas, USA, 2001.

[STE 99] STEVENSON F. A., *Cryptanalysis of Contents Scrambling System*, 1999.

[TOW 02] TOWNEND R. C., "Finance: History, Development & Market Overview, Smart Card News", 2002, http://www.smartcard.co.uk/resources/articles/finance.html.

[UGO 92] UGON M., GUILLOU L., QUISUATER J., "The Smart Card: A standardized security device dedicated to public cryptology", 1992.

[URI 03] URIEN P., LOUTREL M., "The EAP smart card. A tamper resistant device dedicated to 802.11 wireless networks", *Third Workshop on Applications and Services in Wireless Networks*, Berne, Switzerland, 2003.

[VER 00] VERTIGER P., DESPONT P., DRECHSLER U., DÜRIG U., HÄBERLE W., LUTWYCHE M., ROTHUIZEN H., STUTZ R., WIDMER R., BINNIG G., "The Millipede – More than one thousand tips for future AFM data storage", *IBM Journal of Research and Development*, 44, 3, 2000.

[YLO 96] YLONEN T., "SSH – Secure login connections over the internet", *6th USENIX Security Symposium*, San Jose, USA, 37-42, 1996.

[ZIM 95] ZIMMERMANN P. R., *The Official PGP User's Guide*, 1995.

Chapter 17

Video Adaptation on the Internet

17.1. Introduction

One of the greatest challenges of the decade is large scale video broadcasting. Dedicated networks offering large bandwidths have seen the light of day as an answer to these requirements. However, with the explosion of the Internet, new perspectives have emerged, namely the integration of video broadcasting to this new media. This integration offers many advantages including a large cover, a great flexibility and a reduced cost. In fact, video transmission over the Internet can currently touch tens of millions of people who can use their personal computer to receive multimedia, whether in real-time or not.

Unfortunately, due to the current Internet design, many problems have been identified, namely the total lack of guarantees on the Quality of Service (QoS). In fact, the Internet network is a packet-oriented data transmission network in which the data is routed in various completely independent ways. The principal objective is to be able to send this data into the network as efficiently as possible. This type of service is known as best effort service. It is therefore not possible to guarantee a certain level of QoS for a video stream on a particular path connecting a source to a destination. One of the direct consequences of this type of service is that the network can suffer from saturations, better known as congestions.

This phenomenon is directly linked to the amount of information transmitted in the network, which can sometimes greatly exceed the network's capacity. In such a case, the network performance decreases, resulting in a brutal increase in the loss

---

Chapter written by Nadjib ACHIR.

rate and of delays from end-to-end. To avoid these problems, transmission control and error control mechanisms are indispensable in limiting the resource usage rate in the network and in enabling a certain video quality.

The most common mechanisms defined by the IETF (Internet Engineering Task Force) to introduce QoS in the Internet world, are differentiated services (DiffServ) [GLA 95] and integrated services (IntServ/RSVP) [BRA 97]. The DiffServ model consists of introducing several service classes, each offering a different quality. Each traffic flow is allotted a suitable class of service. This classification of the traffic is carried out in the network's periphery, directly at the source or on an access router. Each packet is marked with a code that indicates the class of traffic assigned to it. Routers in the network's core use this code to determine the QoS required by the packet.

All packets having the same code receive the same treatment, even if they belong to different flows. Packet classification criteria must reflect the real needs of the information they are transporting in terms of flow, sensitivity to packet losses, delay and jitter.

IntServ is a model that proposes a resources reservation in intermediate nodes before data transmission within the network. Contrary to the DiffServ model, each application is free to request a specific QoS. Routers on the transmission path verify whether they can grant this QoS or not and they consequently accept or refuse the reservation of the application. If the reservation is accepted, the application is then ensured to obtain guarantees for the transfer of data. The main inconvenience in this approach is the scalability (in other words the increase of flows crossing a node). Moreover, if only one router along the transmission path does not offer IntServ service, the path is no longer a good one. However, these two approaches require changes in the design of the communication network, which is not an easy task.

If there is no guarantee on the QoS at the network level (which is currently the case for the Internet), the application must manage this quality itself and ensure the best service possible by using control or adaptation mechanisms. These mechanisms can be implemented at the source, at the destination or within the network. Their objective is to adjust the application's parameters such as the type of encoding, the level of protection, the transmission rate, in order to eliminate the impact from the losses, delays and bandwidth variations on the video quality. The main advantage in this type of control is that if the application tracks the specificity of flows that it generates, it can then offer a much more efficient control for this data.

In this current chapter we present the principal solutions introduced in the literature which enable a video adaptation on best effort networks such as the Internet. In addition, we will approach an example of video adaptation.

17.2. Error control

The objective of error control is to increase the resistance against losses of flows transmitted on the network by offering possibilities of error corrections. In the following sections, we present the most common techniques used in other works for video protection. We describe in detail one technique in particular (error correction by anticipation) since it is used in this chapter for one of our propositions.

17.2.1. *ARQ approach*

The ARQ (automatic repeat request) approach [LIN 84] consists of the retransmission from the source of the packets that were not properly received by the receiver. The source must have access to the sequence numbers of the lost packets via the reception of positive or negative acknowledgment. However, in the case of real-time video transmission, this type of mechanism is not acceptable due to penalizing delays introduced by retransmissions. Moreover, this type of mechanism cannot be extended to multicast environments due to the number of receivers, which can be rather high. The emission of acknowledgment by these receivers can single-handedly congest the entire network.

17.2.2. *The forward error correction approach*

The FEC (forward error correction) approach has been one of the most common error correction techniques in multimedia communication systems in the last 30 years.

The main idea is to add redundancy to data streams sent via any transmission system that produces errors or losses. Redundancy is used to evaluate transmission losses and to correct errors. This additional data is called parity. Thus, there is no need to retransmit the data.

There are two main families of FEC mechanisms. The first perfectly reconstructs the information contained in lost packets and the second is simply a downgraded version of the first. In this chapter, we will concentrate on the first family of FEC mechanisms. This redundancy will enable the receiver to determine the value of each lost packet on the condition that a sufficient amount of data was successfully transmitted.

Using this last mechanism, each packet containing L bits is assimilated into an alphabetic symbol of size $2^L(X)$. In this case, a block of k data packets represents a vector of X^k. This encoding consists of multiplying the vector by a generating matrix. The result is a vector of size X^n called a code word. This vector represents n data packets. The encoding mechanism has added m (m = n – k) packets to the initial data stream (the relationship $\delta = m/n$ is called the rate of redundancy). These n packets will then be transmitted over the network. At the receiver's end, a certain number of packets are received and this number can be less than n based on the network's loss rate. The received packets form a linear equation system and the compensation for loss consists of the resolution of this system if a sufficient amount of data packets was received.

As shown in Figure 17.1, an optimal encoding diagram is a coding for which any portion of code equal in size to the transmitted message (of size k) is sufficient for the reconstruction, at destination, of the entire original message. These diagrams are called MDS (maximal distance separable) code.

Such mechanisms are available and can be practically applied. Reed-Solomon codes [MAC 84] and codes based on the Cauchy matrix [BLO 95] are the two most common practical examples. The use of these diagrams was also standardized by the IETF [ROS 98].

In such mechanisms and in a memoryless channel, where the loss probability is equal to ε, for a packet to really be lost at least n – k – 1 more packets must also be lost. In this case, the loss probabilities after decoding at destination, $\varepsilon_{receiver}$, can be calculated as follows:

$$\varepsilon_{receiver} = \varepsilon\left(1 - \sum_{j=0}^{n-k-1}\binom{n-1}{j}\varepsilon^{j}(1-\varepsilon)^{n-j-1}\right) \quad [17.1]$$

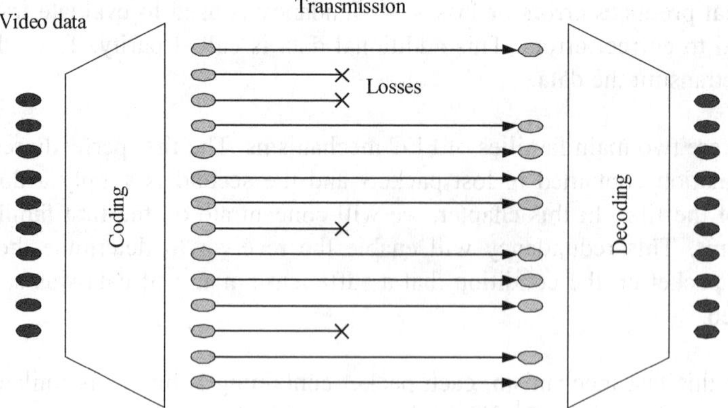

Figure 17.1. *Optimal encoding diagrams*

In order to see the impact of this type of mechanism on the loss rate, we show, in Figure 17.2 the evolution of the loss rate perceptible at the destination for varying values of the redundancy rate δ. We consider the size of the original message to be 100 packets and we vary the loss rate in the network from 1% to 10% and the redundancy rate from 1% to 20%.

The results show that the loss rate at the receiver end, $\varepsilon_{receiver}$, can be significantly reduced despite the fact that the real loss rate in the network may be significantly greater.

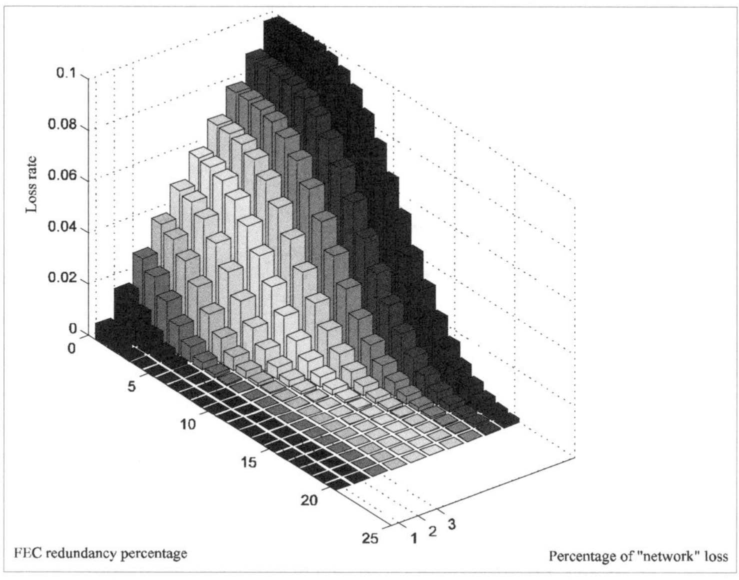

Figure 17.2. *Loss Rate evolution for different protection levels*

17.2.3. *The hierarchical or priority FEC approach*

The hierarchical FEC approach takes into consideration the hierarchical nature of the video stream (multi-rates) in order to grant a greater protection for streams of higher importance or priority regarding video quality. This approach can be used in the point-to-point case, but has a large field of application for multicast [TAN 99] in which video multi-rates coding is predominant. In fact, a protection adapted to each receiver can be offered against the receivers' loss rate. An example of this diagram is shown in Figure 17.3.

In this example, the source generates a video composed of three video layers (a basic layer and two improvement layers). Moreover, it generates two levels of hierarchical FEC protection. The ability to receive these two layers offers an optimal protection of the video stream. Receivers r1 and r2 uniquely receive the three video layers, since they have a low loss rate. However, receiver r3 receives the three video layers as well as an FEC layer since it has an average packet loss rate. Finally, receiver r4 receives all the layers (video and FEC) because its loss rate is high.

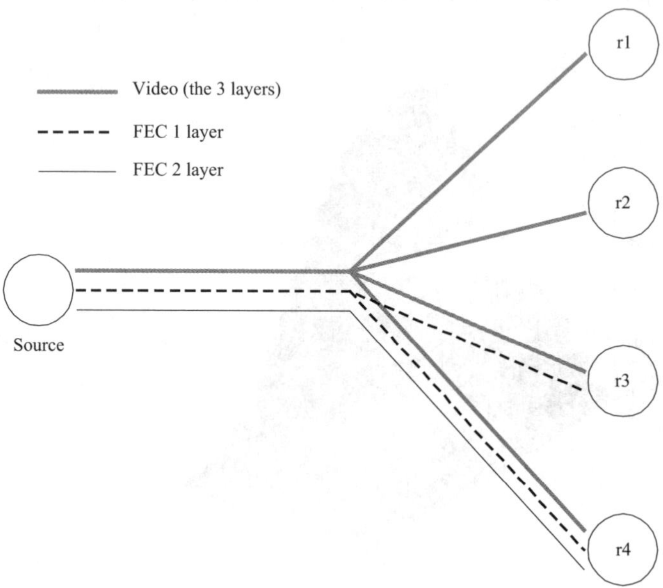

Figure 17.3. *Multicast hierarchical FEC transmission*

17.2.4. *The hybrid FEC/ARQ approach*

In order to ensure a zero loss rate for a real-time video transmission, the FEC approach is often associated with the ARQ approach. We then refer to hybrid-ARQ diagrams [NON 98]. The goal is to limit, as much as possible, the number of lost packets via the FEC approach, thereby reducing the number of retransmitted packets by the source due to losses. A first approach consists of requesting the retransmission of packets that could not be corrected via the FEC approach. A second approach, more adapted to multicast transmission, consists of informing the source of the number of losses detected by each receiver. In so doing, the source can potentially resend redundancy data useful to several receivers.

17.3. Point-to-point congestion and flow control

Knowing that the Internet network does not offer a guarantee on the QoS, each application chooses, on the basis of its characteristics, the most adapted protocol to improve its performances. HTTP, FTP and Telnet protocols use TCP because it enables the source to control congestion and limit the transmission rate during congestion periods. However, for video transmission applications, such as video-conferencing, UDP is the chosen protocol.

This raises an important issue regarding the fact that UDP does not use any congestion control mechanism. In fact, in the case of many TCP and UDP connections traversing one congested node, TCP sources reduce their transmission rates, whereas UDP sources continue to transmit without control and risk maintaining a much higher rate than what is available in the network. This can provoke large video streaming losses that use UDP without the source even realizing it. It is obvious that the source needs to implement congestion control mechanisms to adjust the video streaming according to the bandwidth available in the network.

Moreover, in order to ensure an equitable use of the bandwidth between different streams transiting in the network, it would be equally advantageous that these streams, which are not using TCP, share resources equally with other TCP flows (the dominant protocol on the Internet). With this intention, they must adjust their flows according to the TCP behavior. By making this flow adjustment at the source, network resources can be equitably shared between the different existing connections.

We can outline three main congestion control algorithm families:

1. The first family regroups protocols controlled by windows. These are the approaches that most closely resemble TCP from an operation point of view. In these approaches, flow control mechanisms maintain a congestion window, like TCP does, which is used directly or indirectly to control packet transmission. One of the first video stream control algorithms using this approach is [JAC 99]. In this proposition, the authors directly use the TCP congestion control algorithm to regulate the transmission rate. The size of the TCP window is converted to encoding rate for the encoder.

Another TEAR (TCP emulation at the receivers) approach [RHE 00] relies on the receiver to regulate the rate at the source. The receiver indirectly estimates the TCP flow from the RTT and the size of the window. This algorithm can be used in point-to-point cases as well as multicast cases.

2. The second family of congestion control protocols is an approach that adjusts its transmission flow similarly to TCP, in other words by additive increase and multiplicative decrease, but without using a congestion window. This approach is called AIMD (additive increase multiplicative decrease). The LDA (loss delay adjustment) approach [SIS 98, SIS 00] emulates the mechanisms AIMD of TCP via the use of the RTP/RTCP (real-time transport protocol/real-time transport control protocol) couple [SCH 96]. After the RTCP report, if no losses were detected, the source increases its transmission rate by a ratio of the current rate and if losses are detected the rate is exponentially reduced according to the number of lost packets.

3. The third family regroups protocols controlled by models. These diagrams use a mathematical equation derived from an analytical model that models TCP behavior to approximate the appropriate transmission flow. One of the very first models was proposed in [MAH 97]. TCP response is given by the following equation:

$$R_{TCP} \approx \sqrt{1.5}\frac{MTU}{RTT\sqrt{p}} \qquad [17.2]$$

In this case, flow is limited according to the RTT, the loss rate and the MTU (maximum transfer unit). [TAN 99] presents one of the first approaches based on this analytical model for point-to-point video transmission. In this proposition, the receiver measures the loss rate and the RTT, then calculates based on the preceding equation the flow of transmission and the associated video codec encoding.

Thereafter, a second model was proposed, this time taking into consideration the impact of delays [PAD 98]. The formulation of the TCP model is given by the following equation:

$$R_{TCP} \approx \frac{MTU}{RTT\sqrt{\frac{2p}{3}} + T_0\left(3\sqrt{\frac{3p}{8}}\right)p\left(- + 32p^2\right)} \qquad [17.3]$$

The preceding equation gives us a maximum limit, R_{TCP}, on the transmission flow in bits/second according to the size of the MTU packet in bits, to the return RTT travel time, to the probability of packet loss p and finally to the time-out value T_0 of TCP retransmission.

One of the most interesting approaches, using this second model, is the TFRC (TCP-friendly rate control) protocol [HAN 01]. In addition to congestion control, TFRC enables a better rate stability compared to TCP connections sharing the same path. TFRC connections react better to network conditions (congestions) than the standard TCP connections. One TFRC source estimates the state of the network by exchanging control packets with the final system (the destination) to collect information or feedback on the state of the network. Upon receiving these control packets, the destination sends back, in a return packet, information necessary to calculate the RTT and the loss probability. With these details, the source estimates the flow of a TCP connection competing for bandwidth with a TFRC connection, using the same data path.

In addition to network congestion controls, other adaptations have been coupled with the preceding protocols, taking into account video coding characteristics (the logic structure of the video stream). One of the first approaches consists of the maintenance of several versions of the same video with many qualities at the source. If changes in available bandwidth are perceived by the network, the server then toggles between the different versions, thereby improving or down-grading the quality. This stream control method is adapted for video on demand applications in a more or less restricted environment. It requires, at the source, a large amount of disk space to enable the storage of all versions of the video.

An alternative to this first approach was given by Amir *et al.* [AMI 95]. In this approach, a design enabling video transcoding by the network was proposed. This design is based on the principle of video gateway. This video gateway enables video streaming transcoding on the fly depending on the amount of available bandwidth in the network. Moreover, this video gateway uses an RTP protocol, which enables it to easily interact with video tools used currently on the Internet. However, the strong point in this design, on the fly transcoding, can also be viewed as a weakness due to the length of time required for calculations leading to changes in the coding.

Another approach, largely addressed in other works, consists of the use of multi-layer video. Rejai *et al.* [REJ 99] proposed a method that enables point-to-point congestion control through the use of hierarchical multi-layer video. A semi-optimal allocation is proposed to allocate available bandwidth to different video layers. This study concentrates on first generation multi-layer video standards.

Wakamiya *et al.* [WAK 01] proposed a flow control method for MPEG-4 video using a similar TCP protocol. However, in this proposition, the entire video scene is considered as a single object. The compression method used is an FGS (fine granular scalability). In FGS, video is compressed into two video streams, the first representing the base layer (BL) and the second flow representing the enhancement layer (EL). The adaptation of FGS video flow is carried out by eliminating portions of data in the enhancement layer.

Finally, many commercial video transmission applications are currently spread out on the Internet, such as Windows Media Player [WIN 04] or RealOne [REA 04]. Unfortunately, there is no any performance analysis information available on these.

17.4. Multicast congestion and flow control

Up until now, we have considered point-to-point communication environments. However, many video transmission applications on the Internet (video-conference, television, distance education, etc.) use one or more sources towards several destinations. This type of transmission is considered multicast transmission. Multicast transmission of multimedia data over the Internet relies on the use of IP-multicast technology. This technology consists of simultaneously broadcasting packets to all subscribed clients. This process is possible on a portion of the Internet called the multicast backbone, or MBone. The problem with such technology is that the current best effort Internet network is characterized, not only by an unguaranteed QoS and variable in time, but also by its heterogenity.

This heterogenity results in a reception capability and in a processing capacity relative to each receiver. Receivers are connected to the network by irregular bandwidth capacity connections. Consequently, a multicast broadcasting of video data with a uniform rate is not desirable as it can be the source of congestion for a less privileged receiver if the rate exceeds its capacity. This being the case, one of the key factors in the deployment of multicast video transmission is the capability to adapt to the heterogenity of the receivers and to adapt to the state of the network. Proposed solutions for video transmission in point-to-point cases are not applicable in multicast cases. To handle this new problem, different approaches have been proposed to enable an adaptation to the transmission rate according to the state of network connections. These approaches are primarily based on the multi-rate aspect (scalability) associated with video encoding.

From the point of view of video sourcing, traditional multi-rate video streams can be produced in one of two ways. In the first way, the replication of data, the source generates many streams for the same video but with varying flow rates. Each stream can serve a subset of receivers within the same bandwidth. In the second approach, data breakdown (or multi-layer approach), the video is broken down into a set of video streams or layers. At the receiving end, if a single layer is decoded then the video quality is weak but can be significantly improved if several layers are decoded. There are two diagrams for the multi-layer approach: cumulative and non-cumulative. In the following we will detail each approach by quoting several studies proposed in other works for each of them.

17.4.1. *Multicast transmission by information replication*

The information replication approach can be seen as a compromise between a single-flow multicast connection and many point-to-point connections. This approach is justified in a multicast environment where the receivers' bandwidth follows a certain pre-determined distribution, for example, by using standards such as an ISDN network at 128 kbps, an ADSL network at 2 Mbps or a local network at 10 Mbps. So a limited number of flows can be used to approach the threshold of streams available to users on each network. Figure 17.4 shows an example of this type of multicast transmission.

In this figure, it is noticeable that the source transmits the same video in three streams (each stream represents one level of quality). Group G_1 applies to the first stream (the stream with the highest quality) seeing as all the receivers from this group have a significant bandwidth. However, group G_3 receives the last stream (the stream with the weakest quality) because its bandwidth is limited.

One of the protocols that follows this approach is DSG (destination set grouping) proposed by Cheung *et al.* [CHE 96, JIA 00, LI 96]. In DSG, the source attempts to satisfy the heterogenity constraints in the bandwidth by offering a small number of completely independent video streams, each stream representing the same video but at different rates of flow. The streams are transmitted to different receiver groups. There are as many multicast groups as there are streams. Each receiver chooses the group that matches the closest to its bandwidth. It can also change groups when its bandwidth changes. DSG protocol has two main components: an intra-flow protocol used by the receivers of a same group to adjust the transmission flow according to certain pre-established limits and an inter-flow protocol used by the receivers for changes in multicast groups. DSG is a relatively scalable method able to respond to a relatively high number of receivers, while offering evenness in the regulation of flow. However, one of the biggest problems in this approach is that the transmission of independent video streams to various rates leads to a poor utilization of the bandwidth due to the large amount of redundancy of the information in the links shared by many groups of receivers.

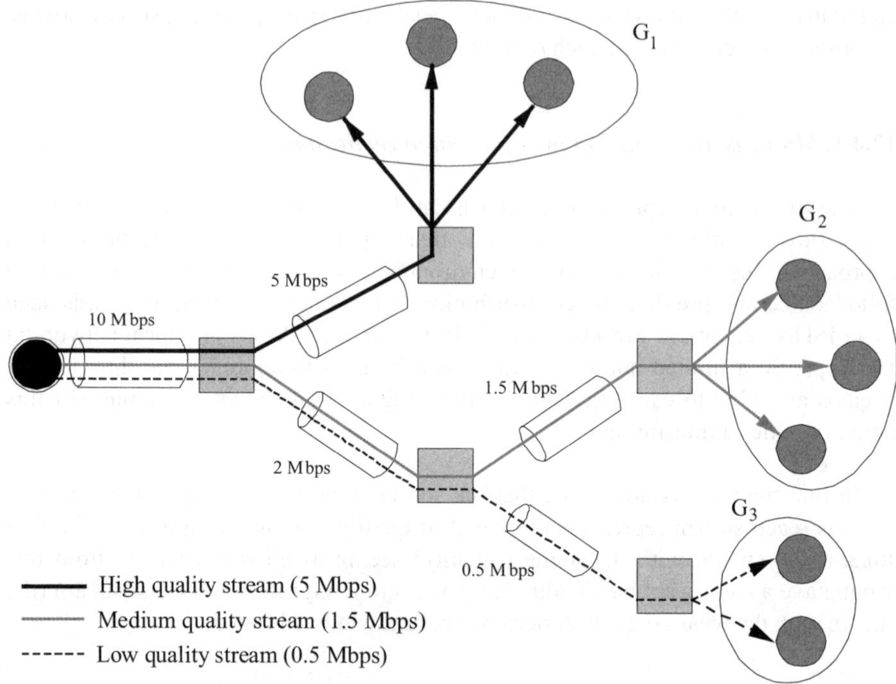

Figure 17.4. *Multicast transmission by information replication*

17.4.2. *Cumulative multi-layer multicast transmission*

The use of cumulative multi-layer video transmission and the transmission by priority is still one of the most attractive approaches for multicast transmission. The greatest advantage of this representation, with regard to information replication, is the absence of redundancy of transmitted information. The multi-layer video coder compresses the video sequence in one or many layers of different priorities. The layer with the highest priority, called the base layer, contains data representing the most important information in the video sequence; however, the additional layers, called enhancement layers, contain data that refines the quality of flow of the base layer.

The source generates a flow of data packets for each layer and assigns a unique elimination priority to each of these flows. The highest priority is assigned to the base layer packets and the enhancement layer packets receive a progressively lower priority. In the case of a network congestion notification, the lower priority packets are automatically eliminated, thereby assisting the network nodes in preventing possible losses in the base layer or in the more important video enhancement layers. However, this type of transmission imposes important changes in the network and in the video coding: the source must be able to generate a cumulative multi-layer video, network nodes must carry out a separation between packets from differing video layers according to their priority and finally the ability to eliminate packets according to their importance in times of congestion. A representation of this multicast transmission is shown in Figure 17.5.

We note in this figure that the source transmits three layers (a base layer and two enhancement layers). Group G_1 is associated to the three layers since all the receivers in this group have a high bandwidth. However, group G_3 solely receives the base layer because it is all its bandwidth enables it to receive.

McCanne *et al.* proposed the first complete adaptation algorithm for multicast video receivers [MCC 96, MCC 97]. In this algorithm, known as RLM (receiver-driven layered multicast), the video source generates a fixed number of video layers, each with a fixed rate of flow, and the receivers subscribe to one or more layers according to the available rate of flow. Congestion is managed by the receivers by observing the rate of packet loss. This approach has the advantage of using cumulative multi-layer video in order to mitigate the problem of heterogenity in the receivers' bandwidths. However, the users are limited in the number of layers that the source is supposed to provide and often the number of layers generated by the source does not optimize the utilization rate in the network and the video quality at the receivers' level. Moreover, RLM adapts rather slowly to changes in available network bandwidth. It suffers several behavior problems: transitory periods of congestion, instability and periodic loss in strength [MCC 97]. Recently, RLM, LVMR (layered video multicast with retransmission [LI 97]), TCCLD (TCP-like congestion control for layered data [VIC 98]) and PLM [LEG 00a, LEG 00b] extensions have been proposed to improve their weaknesses and to adapt to TCP behavior.

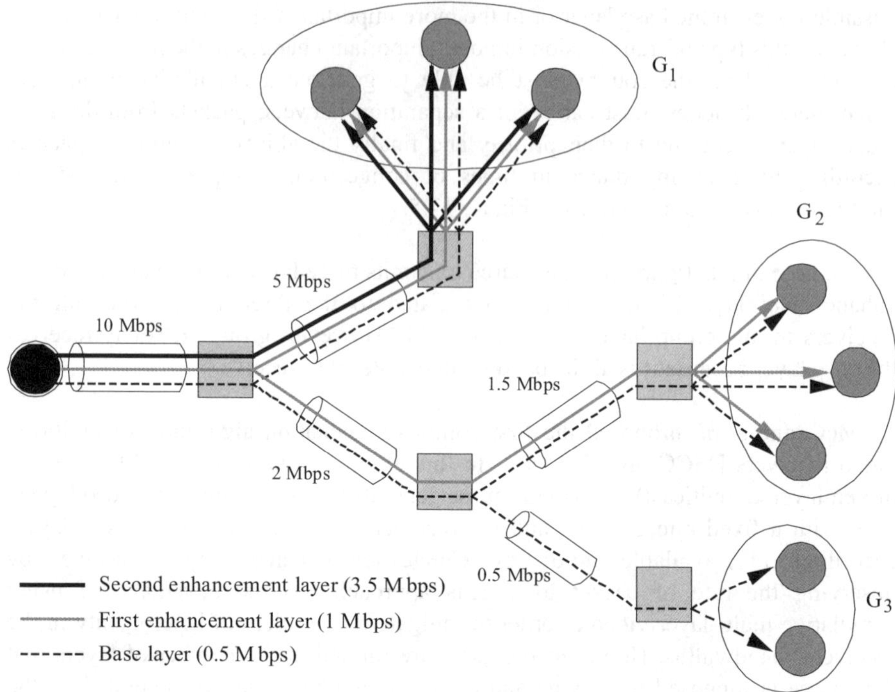

Figure 17.5. *Cumulative multi-layer multicast transmission*

17.4.3. *Non-cumulative multi-layer multicast transmission*

In this approach, video is coded in two or more completely independent video layers. Each layer can be decoded independently and offers a specific video quality. Each receiver can subscribe to any video layer sub-group and thereby improve video quality [COV 82, COY 01]. Non-cumulative multicast transmission was made possible by recent advances in multiple description coding [GOY 98] (MDC). An MD coder generates multiple layers, or descriptions, for a source signal. Figure 17.6 illustrates an example of this approach.

We note in this figure that the source transmits three layers (one layer represents one video description). Group G_1 corresponds to descriptions d_1 and d_2, since all the receivers in this group have a bandwidth equal to the total of these two descriptions. However, group G_3 receives descriptions d_2 and d_3 because its bandwidth is equal to the sum of the flows of these two descriptions.

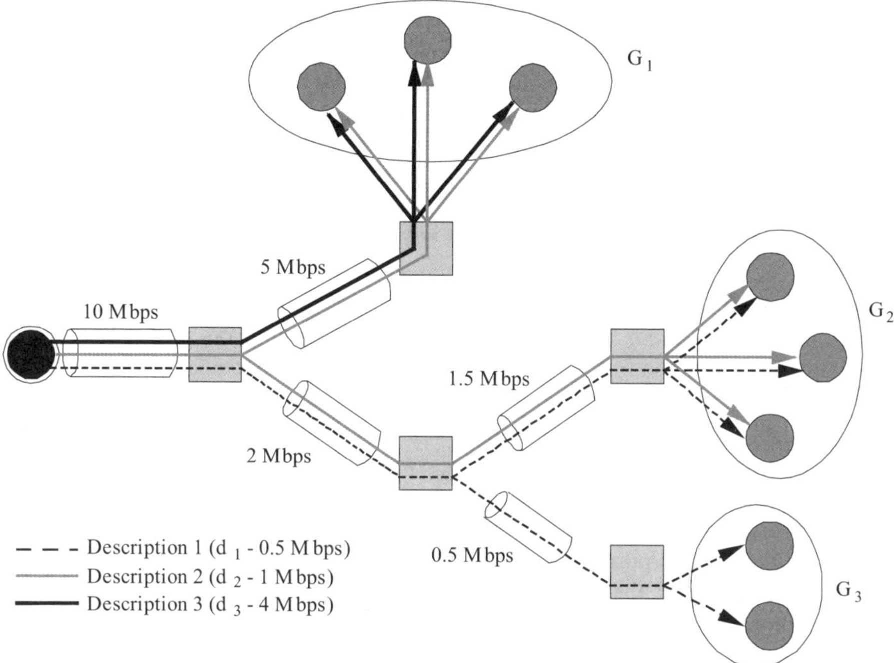

Figure 17.6. *Non-cumulative multi-layer multicast transmission*

17.4.4. *Other multicast video transmission methods*

Other alternative approaches have seen the light offering other network adaptation algorithms [AMI 95, AMI 98, ELA 96, IAN 99], for example, video bridging [AMI 95]. In the video bridging approach, the source generates a high quality video and the intermediate network nodes recode the video to a lower quality in case of bandwidth limitations. The main problem in this approach is the complexity and amount of time the recoding algorithms take in the intermediate network nodes.

17.5. An example of video adaptation: multi-object video transmission

Another consequence of the advent of video on the Internet is a significant evolution in video compression standards. In fact, for a long time, different proposed compression models considered, in their method, the video to be a sequence of fixed rectangular images one after another [MPE 93, MPE 96]. These models are called *first generation encoding diagrams*. In this case, user interaction is reduced to initiating or stopping the running of a video. However, new *second generation encoding diagrams* or object-oriented (multi-objects) have seen the light of day, such as MPEG-4 [MPE 98]. These models consider reasoning at the level of the semantic contents of the video and favor user interaction. With this reasoning, it is possible to carry out an independent manipulation of these semantic entities (objects) that constitute the video, such as is shown in Figure 17.7. This independent manipulation of video objects enables the assignment of relative importance to the objects, which can then be used, for example, to allocate more flow to the more important objects rather than to the less important objects such as a static background in the video. The case study which we will carry out in this last part of the chapter is within the framework of second generation video coding diagrams.

Previously, we saw that to mitigate the problem of packet loss, the source can use error protection mechanisms. We also saw that one of the most commonly used methods for this protection is FEC (forward error correction) which consists of adding redundancy to the video stream in order to increase the loss tolerance. For this, at source level, a portion of the bandwidth is used to transmit the redundancy. On the receivers' end, the client is then able to recover possibly lost packets, in a specific proportion, according to the portion of bandwidth allocated to this redundancy.

However, to our knowledge, all studies carried out up until now [NGU 02, BOL 96] exclusively concern first generation encoding diagrams and no second generation encoding diagrams. The main issue, in a first generation diagram, is finding the right rate of redundancy to allocate to the entire video to improve the quality. However, in a multi-object context, this issue becomes more complex due to the specific structure of this coding. In this case, the source must find the redundancy rate to be applied not for the whole video, but rather to each object that constitutes the video scene. This redundancy rate must match the importance of each object on the global quality of the video. So, a possible adaptation consists of the introduction of a multi-object unequal error protection algorithm, which enables an unequal protection between different video objects. Therefore, objects with the most influence on video quality are protected as much as possible, to the detriment of objects with a lesser impact on quality. To accomplish this, we use FEC techniques that have the ability to assign a certain level of protection to a data stream, with the goal of reducing the rate of noticeable losses at the receivers' level.

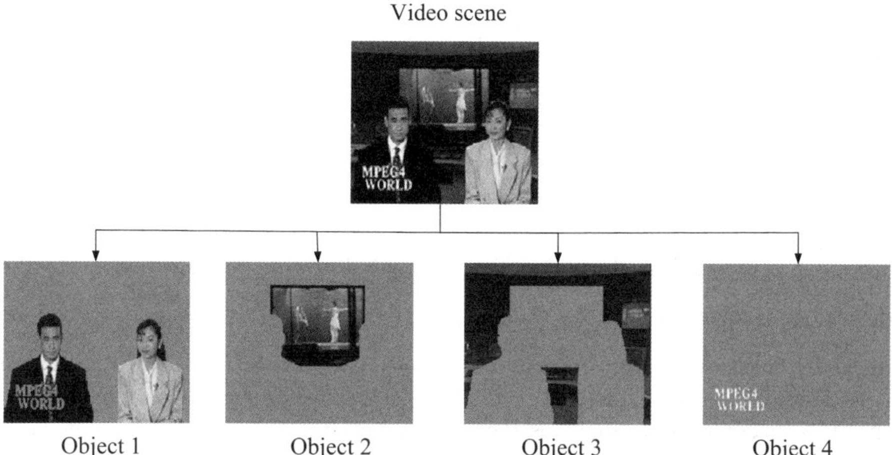

Figure 17.7. *Cutting of a multi-object video scene*

Given that the objective of any video transmission system is the improvement or the optimization of quality perceived by the receiver, we will start this latter part by proposing a quality model adapted to multi-object video transmission. Following this, we will propose an unequal protection algorithm.

17.5.1. *Modeling multi-object video quality*

17.5.1.1. *At the source*

In order to obtain a model representative of video quality, we will adhere to the usual supposition, which is that the quality of an image is measured by its coding distortion. Distortion is defined as the square root of the difference between the value of the original image pixel and the value of the pixel at the level of the decoder, for a given flow. This distortion is usually called MSE (mean squared error) and indicates the image degradation. The smaller the distortion, the greater the image quality.

In our given problem, the video scene comprises many objects and each object is characterized by its priority and by its flow/distortion. We can therefore imagine a function of global quality of the video scene representing the average distortion of the different objects within it.

So, if we suppose that the source generates a multi-object coded video, and that this video is made up of a group of objects $O = \{Oi\} i \in \{1, N\}$, then we can represent the quality function for any video scene as follows:

$$D_{scene} = \sum_{i=1}^{N} \alpha_i \times D_i \qquad [17.4]$$

where D_{scene} represents the video scene global distortion, D_i corresponds to the distortion of object O_i in the video scene ($\sum_{i=1}^{N} \alpha_i = 1$). This priority can be manually initiated by the user, or calculated based on heuristics that take into consideration a few characteristics relating to the object such as its position, its speed, or its complexity. So, by assigning a high priority to an object, the source can be indirectly informed to be more careful with this object than with less important objects.

In order to represent this flow/distortion function, several studies came to the conclusion that distortion evolves exponentially when the source encoding flow decreases [BAI 00, LIN 98]. So, the quality of each object, at source, can be represented by a flow/distortion function as introduced in [BER 71, COV 91, KAV 98]:

$$D_i\left(R_i^{(s)}\right) = D_i^{max} \times 2^{-\beta_i R_i^{(s)}} \qquad [17.5]$$

The parameter β_i is relative to the complexity of the object encoding O_i and $R_i^{(s)}$ represents the maximum encoding flow of the object at source. The choice of parameter β_i is directly linked to the coder used. In this flow/distortion model, to normalize the analysis, the distortion is limited by a maximum value of D_i^{max} ($D_i^{max} = 1$). In Figure 17.8, we represent the perceptible distortion trend at the receiver for different encoding flows at the source. Note that a small increase in source flow induces a large reduction of distortion.

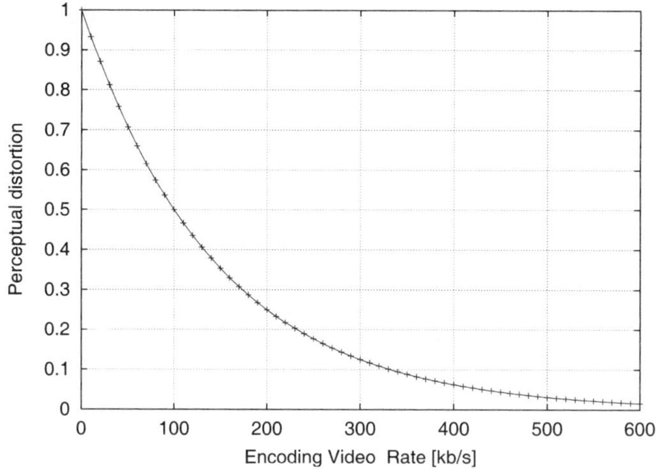

Figure 17.8. *Noticeable distortion as a function of video flow, where βi = 0.01 and R = 600 kbps*

17.5.1.2. *At destination*

The function calculated in equation [17.4] gives us the degree of quality at the source, in other words, just after coding. However, at the receiver this quality can potentially change according to network packet losses. In the following, we will propose a model that estimates the video scene quality at the receiver due to network loss rates.

Let R(t) be the available source rate at time t and for a period Δ(t). Let $R_S(t)$ and $R_{FEC}(t)$ respectively be the source video flow rate and the FEC flow rate used to protect the multi-object video against errors and losses in the network (R(t) = RS(t) + $R_{FEC}(t)$). For now, we suppose that the video encoding flow $R_S(t)$ is always less than the rate of flow $R_S(t)$. We focus solely on the problem of flow rate allocation $R_{FEC}(t)$ between video objects in order to increase the perceptible receiver quality.

In the following, we suppose that each image belonging to object O_i is subdivided into n_i packets, k_i different data packets and u_i FEC redundancy packets (ui = ni + ki). Therefore, the probability of correctly receiving an image is equal to the probability of receiving at least k_i packets among the n_i packets transmitted at the source. This probability can be written in the following format:

$$c_{O_i}(\delta_i) = \left[1 - \varepsilon_{O_i}(\delta_i)\right]^{k_i} \delta_i = \frac{u_i}{n_i} \qquad [17.6]$$

$\varepsilon_{Oi}(\delta_i)$ represents the perceivable loss rate for O_i at the receiver when we use a redundancy rate of δ_i. The calculation of $\varepsilon_{Oi}(\delta_i)$ can be done using equation [17.1]. Therefore, if k_i packets of an image are correctly received, then this image can be correctly decoded with a distortion equal to:

$$D_i^{max} 2^{-\beta_i R_i^{(s)}} c_{O_i}(\delta_i) \qquad [17.7]$$

However, if less than k_i packets among the n_i packets transmitted at the source are received by the receiver, we suppose that the perceptible distortion, for the object O_i at the source, is equal to D_i^{max}. With this supposition, we choose the worst case because in reality the loss of a part of an image description is not equal to the total encoding loss of the image. We can therefore represent the distortion of the object O_i if we use, for this object, an encoding flow rate of R_i and a redundancy rate equal to δ_i as follows:

$$D_i(\delta_i) = \left[D_i^{max}\left(1 - c_{O_i}(\delta_i)\right) + c_{O_i}(\delta_i) D_i^{max} \times 2^{-\beta_i R_i^{(s)}}\right] \qquad [17.8]$$

From equations [17.4] and [17.8], we can represent the global quality of a multi-object video by the receiver's distortion, as follows:

$$\begin{aligned} D_{scene}(R_{FEC}) &= \sum_{i=1}^{N} D_i(\delta_i) \\ &= \sum_{i=1}^{N}\left[D_i^{max}\left(1 - c_{O_i}(\delta_i)\right) + c_{O_i}(\delta_i) D_i^{max} \times 2^{-\beta_i R_i^{(s)}}\right] \end{aligned} \qquad [17.9]$$

where $q = \sum_{i=1}^{N} \delta_i$ represents an amount of FEC bandwidth available for objects. The distortion function given in equation [17.9] represents, for us, the metric we must take into consideration in the FEC bandwidth allocation mechanism. The minimization of this metric leads to an improvement in the video quality and vice versa; the closer this function is to 1, the more the video will be degraded.

17.5.2. *Unequal multi-object video protection*

In the previous section, we presented a model that represents the quality of a multi-object video for a particular FEC bandwidth. This model considers the characteristics of each object, namely its priority and its flow rate/distortion rate. We now consider the problem of optimal distribution of this FEC bandwidth (R_{FEC}) between the different objects that constitute the video scene, under which equation [17.9] is minimized.

With the goal of finding the optimal distribution, we transform our problem into one of resource allocation. In fact, we have a certain FEC budget represented by R_{FEC} and we want to distribute this budget among N entities representing the video objects. In this case, equation [17.9] represents a cost function that gives us the price to pay for a certain allocation. This cost function decreases according to the parameters. In order to resolve this allocation problem, we use a mathematical optimization method, namely dynamic programming. In the following, we will start by describing a dynamic programming optimization model. After that, we will propose an optimal FEC allocation algorithm for multi-object video.

In order to apply the optimization model introduced above, we will need to describe our problem as a dynamic programming model. For this, we must define our system's steps, the transfer function and the system cost function. We model our system as follows:

– we assume that the number of steps corresponds to the number of objects that constitute the video scene (N). The kth step corresponds to the kth object (O_k);

– we consider that the variable state of the system, x_k, corresponds to the available FEC budget for steps {k, k + 1, ..., N}, that is to say for objects $\{O_k, O_{k+1}, \ldots, O_N\}$. In this case, the quota of available global FEC bandwidth for all video objects is represented by x_1;

– we also consider that the decision variable u_k, which must be made at step k, corresponds to the FEC redundancy quota allocated to object O_k in order to improve the loss tolerances for this object.

In order to obtain a dynamic system in discrete time, we must make the whole of the decision variables discrete. To do this for C_k unit we have divided C_k into M small intervals, each of size R_{FEC}/M. Based on these assumptions, our discrete time dynamic system can be formulated as follows:

$$x_{k+1} = f_k(x_k, u_k) = x_k - u_k, \qquad k = 1, 2, \ldots, N \qquad [17.10]$$

This means that the FEC redundancy budget (R_{FEC}), available for objects $\{O_{k+1}, O_{k+2}, \ldots, O_N\}$, is equal to the available FEC redundancy budget for objects $\{O_k, O_{k+1}, O_{k+2}, \ldots, O_N\}$, from which we subtract the FEC redundancy quota which is uniquely allocated to the object O_k. In this case, the cost function that represents the cost to pay during step k is equal to the distortion introduced by the object O_k for the receiver, multiplied by the priority of this object in the scene (α_k). This cost function is written as follows:

$$g_k = \left(x_k, u_k\right) = \alpha_k \times D_k\left(\delta_k\right) \tag{17.11}$$

Note that the cost function that we chose for our system respects the condition of additionality required by dynamic programming, given that the global distortion of the video scene is equal to the sum of distortions of each object, multiplied by its priority. From what is written above, we can define our dynamic programming algorithm with an initial state equation, which is easily calculated for all possible values of u_k, in the following form:

$$J_k\left(x_k\right) = g_N\left(x_N\right) = \alpha_N \times D_N\left(\delta_N\right). \tag{17.12}$$

As well as a recursive function $J_k(x_k)$ in the following form:

$$J_k\left(x_k\right) = \min_{\pi \in \Pi} E\left[\alpha_k \times D_k\left(\delta_k\right) + J_{k+1}\left(x_k - u_k\right)\right], \tag{17.13}$$

where $\pi = \{u_1, u_2, \ldots, u_{N-1}\}$.

The recursive function J expresses the optimal cost obtained for periods k to N with an initial FEC redundancy budget equal to x_k. The function J_k is recursively calculated in time, starting from period $k = N - 1$ and ending at period $k = 1$. The value of $J_1(R_{FEC})$ represents the optimal cost when the global FEC redundancy budget for all video objects, at period 1, is equal to R_{FEC}. During this calculation, the optimal policy π^*, which represents the optimal FEC budget allocation, is simultaneously calculated by minimizing the right side of equation [17.13]. We can then obtain the optimal FEC redundancy rate relative to each object constituting the video scene; this optimal distribution is equal to:

$$S^* = \left(\frac{u_0^*}{n_0}, \frac{u_1^*}{n_1}, \frac{u_2^*}{n_2}, \ldots, \frac{u_N^*}{n_N}\right) \tag{17.14}$$

17.5.3. *Adaptation results and performances*

In the previous sections, we have studied a method of improving the quality of a multi-object video during a point-to-point transmission in an unreliable environment.

This study enabled us to propose an unequal protection mechanism between video objects according to the impact of each object on the global quality. This section is dedicated to the evaluation of the performances of the unequal protection algorithm. We will concentrate our study on a simple network model consisting of a source S and a receiver D. We suppose that S is a multi-object video source, generating a multi-object video stream and that the receiver D is capable of composing the video scene from the different received objects. We have modeled the source path as having a uniformly distributed loss rate between packets, equal to ε.

In our analysis, we suppose at first that the video scene is made up of three objects O_1, O_2 and O_3. A priority is associated to each object. This priority is equal to 0.6 for object O_1, to 0.3 for object O_2 and to 0.1 for object O_3. This means that the most important object is object O_1 and the least important object is object O_3. The assignment of object priority is arbitrary because the goal of our work is to exploit this priority in order to improve the transmission of this type of video coding. We assume that the maximum flow of each object is equal to 600 kbps for object O_1, to 400 kbps for O_2 and finally to 200 kbps for O_3.

To our knowledge, no unequal protection algorithm between multi-object video objects has previously been proposed. So we propose another algorithm which we call *naïve* (Algorithm 1) in order to compare it with our proposition. We want this naïve algorithm to behave as if the source did not take into account the priority of objects in order to better show our proposal's contribution. So this algorithm is an equitable FEC allocation between different objects of a video scene.

Algorithm 1: *naïve algorithm*

$Bw_{available} \Leftarrow R(t)$

$i \Leftarrow 1$

while $(Bw_{available} \neq 0)$ **do**

$$R_i \Leftarrow R_i \frac{R(t)}{M}$$

$$Bw_{available} \Leftarrow Bw_{available} - \frac{R(t)}{M}$$

if $i \neq N$ **then**

$i \Leftarrow i+1$

else

$i \Leftarrow 1$

end if

end while

In Figure 17.9, we represent the perceived distortion at destination for different FEC budget values. We represent this budget by the relationship between the amount of bandwidth used for FEC and the total amount of bandwidth used to transmit the video. We first use the naïve algorithm and then the unequal protection algorithm. In this simulation, we have assumed that the network loss is mainly due to congestion and the loss rate has been fixed at 10%.

It is clear from Figure 17.9 that the unequal protection algorithm gives better results than the naïve algorithm. We note, also, that the distortion curve of the unequal protection algorithm develops in three steps. Each step corresponds, in reality, to the protection of a video object. Therefore, in step 1 our algorithm starts by protection O_1 because this object has the greatest influence on the video's global quality. Then, when it judges that this object is sufficiently protected (in other words, it is not possible to improve the quality of the object, even if we add more FEC), it goes to O_2 and then to O_3.

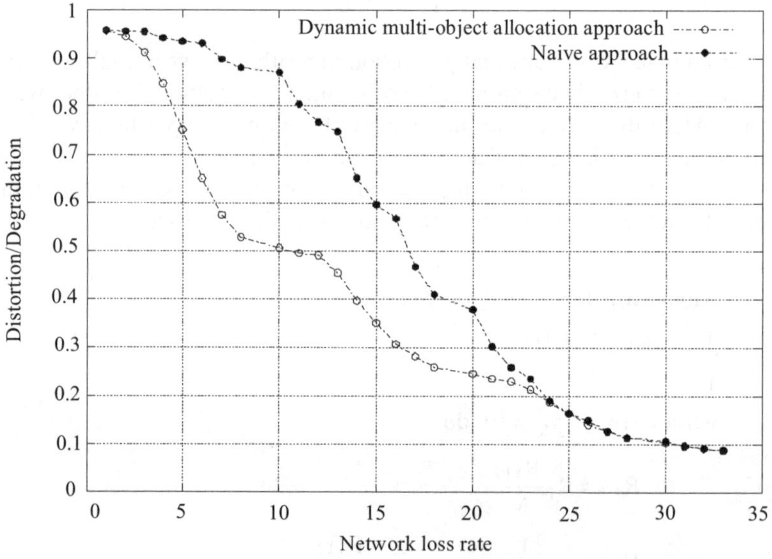

Figure 17.9. *Perceptible distortion trend at the receiver with regard to the rate of protection*

In Figure 17.10, we represent the percentage of FEC allocated to each object, according to the available FEC budget for video, using the unequal protection algorithm.

We note that there are three steps identified in the preceding figure. In fact, when the FEC budget is less than a particular threshold, 12%, our algorithm protects only the most important object O_1. Between the thresholds of 12% and 22%, the FEC budget is distributed between the objects O_1 and O_2. Finally, when the FEC budget exceeds 22%, all three objects are protected. These thresholds are calculated completely dynamically according to the characteristics of each object (priority and flow/distortion) and to the state of the network (losses).

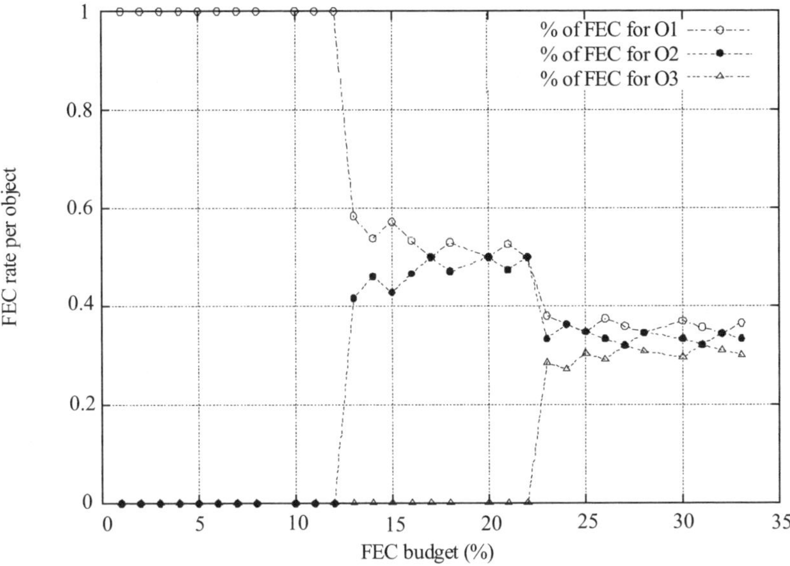

Figure 17.10. *Percentage of FEC allocated to each video object*

17.6. Conclusion

In this chapter, we described the principal approaches proposed in other works for a video adaptation in best effort networks, such as the Internet. The proposed approaches concern protection against errors and the rate control in video transmissions in point-to-point and multicast modes. These solutions offer an alternative for the improvement of video transmission over the Internet besides the guarantee designs for QoS. The greatest advantage is the consideration of transmitted video flow characteristics.

Following this, we presented an example of a video adaptation algorithm and we focused on the problem of protecting against multi-object video errors in a best effort environment. We proposed an unequal protection algorithm enabling an unequal protection between video objects using an FEC mechanism. More precisely, a different rate of protection is applied to each object following its impact on the global quality of the video. In order to do this, we have proposed a model that quantifies the video global quality according to the priority of the object in the video scene, to the function of flow/distortion that characterizes each object and finally to the loss rate in the network. We have considered the function resulting from this model as the objective function which we will need to minimize to obtain a minimal distortion and therefore a minimal degradation at the receiver.

The results obtained from the simulations show the efficiency of our proposition. In fact, the unequal protection algorithm enables the preservation of the quality of the objects with the highest impact on the global quality of the video. This induces a great reduction in distortion. This type of mechanism is very well-adapted to applications such as video-conferencing, where the objective is to protect as best as possible a part of the video, that is, the speaker.

17.7. Bibliography

[AMI 95] AMIR E., MCCANNE S., ZHANG H., "An Application Level Video Gateway", in *Proc. ACM Multimedia'95*, San Francisco, 1995.

[AMI 98] AMIR E., MCCANNE S., KATZ R. H., "An Active Service Framework and Its Application to Real-Time Multimedia Transcoding", in *Proceedings of Conference of the Special Interest Group on Data Communication, ACM SIGCOMM'99*, p. 178-189, Vancouver, Canada, September 1998.

[BAI 00] BAI J., LIAO Q., LIN X., "Hybrid models of the rate distortion characteristics for MPEG video coding", in *7th International Conference on Communication Technology*, August 2000.

[BER 71] BERGER T., *Rate Distortion Theory: A Mathematical Basis for Data Compression*, Prentice-Hall, Englewood Cliffs, 1971.

[BLA 95] BLAKE S., BLACK D., CARLSON M., DAVIES E., WANG Z., WANG W., "Architecture for Differentiated Services", *IETF RFC 247*, December 1995.

[BLO 95] BLOMER J., KALFANE M., KARP R., KARPINSKI M., LUBY M., ZUCKERMAN D., "Theory An xor-based erasure-resilient coding scheme", Technical report TR-95-048, International Computer Science Institute, Berkeley, August 1995.

[BOL 96] BOLOT J. C., TURLETTI T., "Error Control for Packet Video in the Internet", *Proc. CIP'96*, Lausanne, 1996.

[BRA 97] BRADEN R., ZHANG L., BERSON S., HERZOG S., JAMIN S., "Resource Reservation Protocol (RSVP) – Version 1 Functional Specification", *IETF RFC 2205*, December 1997.

[CHE 96] CHEUNG Y. S., AMMAR M. H., LI X., "On the Use of Destination Set Grouping to Improve Fairness in multicast Video Distribution", in *IEEE Infocomm '96*, p. 553-560, San Francisco, March 1996.

[COV 82] COVER T., EL GAMEL A., "Achievable rates for multiple descriptions", *IEEE Trans. Infocomm '82*, p. 851-857, 1982.

[COV 91] COVER T. M., THOMAS J. A., *Elements of Information Theory*, John Wiley and Sons, 1991.

[ELA 96] ELAN P., ASSUNÇÃO A. A., GHANBARI M., "Multicasting of MPEG-2 video with multiple bandwidth constraints", *7th International Workshop on Packet Video*, Brisbane, Australia, March 1996.

[GOY 98] GOYAL V. K., KOCEVIC J., AREAN R., VETTERLI M., "Multiple description transform coding of images", in *Proceedings of ICIP*, Chicago, October 1998.

[GOY 01] GOYAL V. K., KOCEVIC J., AREAN R., VETTERLI M., "Fine-Grained Layered multicast", Anchorage, April 2001.

[HAN 01] HANDLEY, PADHYE J., FLOYD S., "TCP Friendly Rate Control (TFRC): Protocol Specification" (Internet Draft: draft-ietf-tsvwg-tfrc-02.txt), 2001.

[IAN 99] IANNACCONE V. G., RIZZO L., "A layered video transcoder for videoconference applications", Technical report Mosaico Research (Report PI-DII/4/99), Pisa University, Italy, 1999.

[JAC 99] JACOBS S., ELEFTHERIADIS A., "Video services over networks without QoS guarantees", *WWW consortium Workshop on Real-time Multimedia and the Web, RTMW'96*, Sophia Antipolis, October 1999.

[JIA 00] JIANG T., AMMAR M. H. ZEGURA E. W., "On the use of destination set grouping to improve inter-receiver fairness for *multicast* ABR sessions", in *IEEE INFOCOM'00*, Tel-Aviv, Israel, March 2000.

[KAV 98] KAVE M. R. S., "Transmission multimédia fiable sur Internet", PhD Thesis, Paris XI University, 1998.

[LEG 00a] LEGOUT A., BIERSACK E. W., "Pathological Behaviors for RLM and RLC", in *Proceedings of International Conference on Network and Operating System Support for Digital Audio and Video NOSSDAV'00*, p. 164-172, Chapel Hill, North Carolina, USA, 2000.

[LEG 00b] LEGOUT A., BIERSACK E. W., "PLM: Fast Convergence for Cumulative Layered multicast Transmission Schemes", in *Proc. of ACM SIGMETRICS'2000*, p. 13-22, Santa Clara, 2000.

[LI 96] LI X., AMMAR M. H., "Bandwidth Control for Replicated-Stream multicast Video Distribution", in *HPDC'96*, Syracuse, USA, August 1996.

[LI 97] LI X., PAUL S., PANCHA P., AMMAR M., "Layered Video multicast with Retransmission (LVMR): Evaluation of Error Recovery Schemes", in *Proceedings of the Sixth International Workshop on Network and Operating System Support for Digital Audio and Video*, 1997.

[LIN 84] LIN S., COSTELO D. J., MILLER M., "Automatic Repeat Request Error Control Schemes", *IEEE Communications Magazine '84*, vol. 1, p. 5-17, 1995.

[LIN 98] LIN J., ORTEGA A., "Bit-Rate Control Using Piecewise Approximated Rate-Distortion Characteristics", in *Proc. of IEEE Transactions on Circuits and Systems for Video Technology*, vol. 8, no. 4, p. 446-459, August 1998.

[MAC 84] MACWILLIAMS F., SLOANE N., *Theory of Error-Correcting Codes*, North-Holland, 1977.

[MAH 97] MAHDAVI J., FLOYD S., "TCP-friendly *unicast* rate-based flow control", technical memo sent to the mailing list of end2end-interest, January 1997.

[MCC 96] MCCANNE S., JACOBSON V., VETTERLI M., "Receiver-driven Layered multicast", in *ACM Sigcomm'96*, p. 117-130, New York, August 1996.

[MCC 97] MCCANNE S., Multicast multilayer videoconferencing: enhancement of multilayer codec and implementation of the receiver driven layered multicast protocol, PhD Thesis, Texas, December 1997.

[MCC 97] MCCANNE S., "Scalable Compression and Transmission of Internet multicast Video", in *IEEE Infocomm'96*, 1997.

[MPE 93] MOTION PICTURE EXPERT GROUP (MPEG), "Information technology – coding of moving pictures and associated audio for digital storage media at up to about 1.5 Mbit/s – part 2: Video", Technical Report 11172-2 (MPEG-1) – ISO/IEC, 1993.

[MPE 96] MOTION PICTURE EXPERT GROUP (MPEG), "Information technology – generic coding of moving pictures and associated audio information: Video", Technical Report 13818-2 (MPEG-2) – ISO/IEC, 1996.

[MPE 98] MOTION PICTURE EXPERT GROUP (MPEG), ISO/IEC 14496-1, "Information Technology – Coding of Audio-visual Objects, Part 1: System", ISO/IECJT1/SC 29/WG 11 Draft International Standard, 1998.

[NGU 02] NGUYEN T., ZAKHOR A., "Distributed Video Streaming over the Internet", *Multimedia Computing and Networking (MMCN),* 2002.

[NON 98] NONNENMACHER J., BIERSACK E. W., TOWSLEY D., "Parity-based loss recovery for reliable multicast transmission", *IEEE\slash ACM Transactions on Networking*, vol. 6, p. 349-361, 1998.

[PAD 98] PADHYE J., FIROIU V., TOWSLEY D., KUROSE J., "Modeling TCP throughput: a simple model and its empirical validation", *Proceedings of Conference of the Special Interest Group on data Communication, ACM SIGCOMM'98*, p. 303-314, Columbia University, Vancouver, Canada, August 1998.

[REA 04] Real One Player, http://france.real.com/index.html.

[REJ 99] REJAIE R., HANDLEY M., ESTRIN D., "Quality Adaptation for Congestion Controlled Video Playback over the Internet", *ACM SIGCOMM'99*, p. 189-200, 1999.

[RHE 00] RHEE I., OZDEMIR V., YI Y., "TEAR: TCP Emulation at Receivers – Flow Control for Multimedia Streaming", Technical report, April 2000.

[ROS 98] ROSENBERG J., SCHULZRINNE H., "Theory An RTP payload format for Reed-Solomon codes", Technical report IETF AVT Working Group, November 1998.

[SCH 96] SCHULZRINNE H., CASNER, F. J., "RTP: A Transport Protocol for Real-Time Applications", *IETF RFC 1889*, 1996.

[SIS 98] SISALEM D., SCHULZRINNE H., "The Loss-Delay Based Adjustment Algorithm: A TCP-Friendly Adaptation Scheme", *International Conference on Network and Operating System Support for Digital Audio and Video NOSSDAV'98*, Cambridge, UK, July 1998.

[SIS 00] SISALEM D., WOLISZ A., "LDA+: A TCP-Friendly Adaptation Scheme for Multimedia Communication", *IEEE International Conference on Multimedia and Expo (III)*, p. 1619-1622, New York, August 2000.

[TAN 99] TAN W., ZAKHOR A., "Real-Time Internet Video Using Error Resilient Scalable Compression and TCP-Friendly Transport Protocol", *IEEE Transactions on Multimedia'99*, vol. 1, p. 172-186, 1999.

[VIC 98] VICISANO L., RIZZO L., CROWCROFT J., "TCP-Like Congestion Control for Layered multicast Data Transfer", in *IEEE INFOCOM*, p. 996-1003, 1998.

[WAK 01] WAKAMIYA N., MIYABAYASHI M., MURATA M., MIYAHARA H., "MPEG-4 Video Transfer with TCP-Friendly Rate Control", *MMNS'01*, 2001.

[WIN 04] Windows Media Player, http://windowsmedia.com/9series/download/download.asp.

Chapter 18

Voice over IP

18.1. Introduction

Voice over IP has two types: real-time and non-real-time. In the first case, we find all the applications that require a human interaction such as telephony of generally good quality. In the second case, we find all the voice applications that do not require real-time, in other words, that can be executed in differed time, such as remotely recovering a sound track. This second category essentially combines file transfers. In this chapter, we will especially be interested in telephony.

Telephonic application remains essential in the communication between people and is on the rise, as we will see, because the implementation cost of telephony is currently falling. This reduction in cost poses problems to the operators who can no longer claim to earn more money on this application.

Telephony has, for a long time, been in the hands of circuit switching networks, but a strong competition exists with packet transfer networks, essentially of IP type. The proportion of packet transfer types in France at the beginning of 2007 is 60% and the rest is of circuit switching type. Let us keep in mind that these are approximate values since more and more circuit-driven telephonic communications continue as packet transfers to return to circuit type before arriving at destination.

In this chapter, we will essentially be interested in packet IP telephony, called ToIP (telephony over IP). The switch over to IP telephony is inescapable, as it will enable the integration of data services and telephony in one network. In fact, many

---

Chapter written by Guy PUJOLLE.

companies are trying to integrate their telephonic environment into their packet transfer network to lower communication costs and to simplify their network maintenance by switching from managing two networks (telephony and data) to only one network (data).

The difficulty in telephony via packets lies in the very strong temporal constraint due to the interaction between individuals. The latency time must be less than 300 ms in order to maintain an acceptable human interaction. If we wish for a good quality of conversation, the latency must not exceed 150 ms. A more complex case arises when there is an echo, that is, a signal that comes back into the ear of the sender. The echo is produced when the signal meets an obstacle such as in the use of a headset. The echo that sets out again in the opposite direction is digitized by a codec and has no problem traveling through a digital network. The digitized value of the echo latency being 56 ms, so that the echo is not displeasing to the ear, the one-way travel time must not exceed 28 ms by assuming a symmetrical network requiring the same response on the way out as on the way back. Therefore, the software in the terminal equipment must be capable of managing delays and of re-synchronizing incoming bytes. In general, modems such as GSM terminals are equipped with echo suppressors that avoid this strong temporal constraint.

Another essential point in telephony is the need to alert the person being called by a ringing. Telephonic communication is therefore broken down into two phases: the first phase makes it possible to alert the call receiver and the second corresponds to the transport of the conversation itself. In fact, a third phase concerns ending the conversation when one of the two terminals terminates the communication. In fact, this third phase uses the same protocols as the first: a signaling protocol.

This chapter examines first the evolution in telephony toward Internet networks, the first phase of signaling and then the second phase of voice transportation. Finally, IP telephony in companies and for individuals is studied in the last part of this chapter.

18.2. Telephonic application in an IP context

As previously explained, the application of telephony is complex to deal with due to its interactive character and its strong synchronization. Three successive operations are necessary for voice digitization, whether telephonic or not:

– *sampling*: consists of taking points from the analog signal as it is unfolding. It is obvious that the importance of bandwidth increases as the need for per-second sampling increases. The sampling theorem gives the solution: it is necessary to sample an amount equal to at least twice the bandwidth;

– *quantification*: consists of representing a sample by a numeric value by way of a law of correspondence. This phase consists of finding the law of correspondence so that the signal value has the most significance possible;

– *coding*: consists of giving a numeric value to the samples. These values are transported in the digital signal.

The size of the analog telephonic voice band is 3,200 Hz. To correctly digitize this signal without loss of quality, since the quality is already relatively poor, it is necessary to sample at least 6,400 times per second. Standardization has opted for a sampling of 8,000 times per second. Quantification is carried out by semi-logarithmic laws. The maximum allowable amplitude is divided into 128 positive levels in the European version of PCM, to which we must add 128 negative levels, giving 256 values, which requires 7 or 8 bits of binary coding. The total value of the telephonic voice digitization flow is obtained by multiplying the number of samples by the number of levels, giving:

– 8,000 × 7 bits/sec = 56 kbps in North America and Japan;

– 8,000 × 8 bits/sec = 64 kbps in Europe.

Many other solutions have been developed with regard to quality and hearing defects:

– AD-PCM (adaptive differential-pulse code modulation);

– SBC (sub-band coding);

– LPC (linear predictive coding);

– CELP (code excited linear prediction).

The following section reviews the principal audio coders.

18.3. Audio coders

Audio coders associated with the different techniques previously cited are numerous. We find, namely, the traditional codecs but also new low-flow coders. Figure 18.1 illustrates the exit speeds for different telephonic voice coder norms based on a standard sampling of 8 kHz. The ordinate represents the quality of the sound in reception, which is obviously a subjective criterion. We have also represented in this figure coders used in mobile GSM networks and the regional norms.

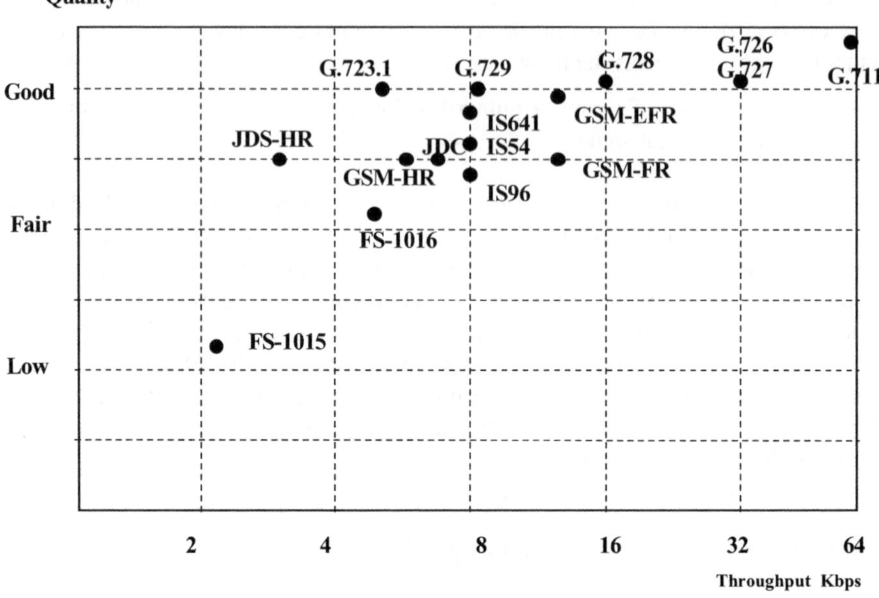

Figure 18.1. *Different audio coders*

For high-definition audio, a more significant bandwidth is considered because the human ear is sensitive to frequencies of 20 to 20,000 Hz. Sampling is carried out at 40 kHz and we have chosen 44.1 kHz. Coding carried out on a CD is at 16 bits per sample, which gives 705.6 kbps.

Among numerous proprietary coders on the market, we mention:

– FlowWorks at 8.5 kbps;

– VoxWare at 2.4 kbps with an RT24 coder;

– Microsoft at 5.3 kbps with a norm of G.723;

– VocalTec at 7.7 kbps.

The recommendation of G.711 corresponds to the traditional digitization of 64 kbps in Europe or 56 kbps in the USA.

G.723 is a voice compression used by many manufacturers, Microsoft among others, who use it in a Windows environment. The flow decreases to almost 5 kbps.

G.726 is the adopted norm for voice compression in adaptive differential coding in 16, 24, 32 or 40 kbps. In this case, instead of coding the entire sample, only the difference between the sample and the previous sample is sent, thus enabling a coding of a lot less binary elements.

G.727 also uses a differential coding, which complements the previous coding. This recommendation indicates how to change, in the course of digitization, the number of bits used to code the samples. This is particularly useful in the framework of networks that require the application to adapt according to the network load.

G.728 is a compression at 16 kbps using a prediction technique, which consists of coding the difference between the real value and an estimated value of the sample based on previous samples. We understand that this difference can be even less than in the differential technique. If the estimate is good, the value to be transported is always close to 0. Very few bits are necessary to transmit this difference. FS standards come from the American Ministry of Defense.

More recent coders are G.723.1, G.729 and G.729.A. Coder G.723.1 enables a total flow of between 5.3 and 6.4 kbps. The two G.729 coders enable a flow of 8 kbps, but the quality of communication is better. This codec was chosen to compress the voice in the UMTS.

Telephonic voice is a very constraining application, as we have seen many times in this book. The first constraint comes from the interactivity between two users, limiting the return trip time to 600 ms at the most. UIT-T norms carry this value to 800 ms. However, to have a good quality of communication, we need to go down to 300 ms for the return trip. Following subjacent protocols, many methods satisfying these constraints were developed at the end of the 1990s, which we will examine.

18.4. Telephony over IP

The issue of telephonic voice transport in IP environments is quite different since we are on an uncontrolled IP network, such as the Internet or on a network that enables the introduction of controls, such as the private Intranet network in a company or that of an ISP.

In first-generation Internet, the network must not be heavily loaded so that the 300 ms constraint is respected. On Intranet networks and ISP networks, and also those of operators, voice transfer is possible as long as the network is controlled so that the total time of transfer, including the packetization and de-packetization, is limited.

Many solutions have been proposed, such as VoIP (voice over IP) from the IMTC (International Multimedia Teleconferencing Consortium). In these solutions, a standardized coder first had to be defined. The choice was generally G.723 but other solutions are operational, such as the G.711 coder. The IP packet must be as short as possible and many voice flows must be multiplexed in a same packet, with the goal of reducing the filling time and to limit the network transfer time. If the routers can manage priorities, which is possible when using DiffServ type services, telephonic voice is much more easily transferred in the required lapse of time.

Many standardization committees, are effectively working on this particularly promising subject. As recognized normalization committees, the European standardizing committee ETSI has set up the TIPHON (Telecommunications and Internet Protocol Harmonization Over Networks) group. The project relates to voice and faxes between connected users, in particular on IP networks. The case where a user is working in an IP network and another in a circuit-switching network, whether telephonic, RNIS, GSM or UMTS, is equally in the realm of TIPHON studies. TIPHON activities concern, moreover, the validation of solutions for transporting telephonic voice by means of demonstrators. They are full-scale experiences aiming to demonstrate the efficiency of the solutions. The ETSI is working toward this in collaboration with the ITU-T and the IETF but also with IMTC and VoIP groups.

The ITU-T is actively working on the issue of telephony over IP in three workpackages within the Group 16: WP1 for modems (V series), WP2 for codecs (G series) and WP3 for terminals (H series). The goal of the ITU-T is to develop a complete environment rather than simply a terminal or a protocol.

At the IETF, several workgroups are tackling specific issues, such as:

– AVT (audio video transport) which uses the RTP protocol (RFC 1889 and 1890) to carry out real-time communication;

– Mmusic (multiparty multimedia session control) which uses the SIP protocol and which we will explain later on;

– IPTel (IP telephony) which defines a protocol of bridging localization and a language that enables the communication between circuits and IP flows;

– PINT (PSTN IP Internetworking) which also uses SIP protocol;

– FAX (fax over IP) which stores and sends faxes via electronic messages;

– Megaco (media gateway control) which determines a protocol between a gateway and its controller;

– Sigtran (signal translation) which proposes using the SS7 signaling commands in IP packets;

– ENUM (E.164/IP translations) that manages the translation of E.164 addresses into IP addresses.

Respecting the temporal constraint is a first priority for traditional voice transfer. A second priority concerns the implementation of signaling to connect the two users who want to speak to each other.

Signaling protocols used for the transfer and management voice in IP packets essentially combine H.323 and SIP (session initiation protocol) which we will examine in detail later on. H.323 protocol was defined in a telecommunications environment, contrary to SIP, which derives from data networks and more specifically from the Web. SIP uses the HTTP code as well as the associated security. Moreover, it can accommodate firewalls. SIP implements sessions that are nothing more than telephone calls between a client and a server. Six HTTP primitives are used for this: INVITE, BYE, OPTIONS, ACK, REGISTER and CANCEL.

VoIP has become a traditional application due to the digitization possibilities and the power of PCs that enable the elimination of echoes. The most constraining element is the delay, especially when crossing PC terminals, access networks, gateways, routers, etc.

Consider that a PC requires a transfer time of 100 ms, the modem requires 10 ms, the delay in the gateway also requires 10 ms and the IP network requires 10 ms. The total shows that the 300 ms limit for interaction is quickly reached. Exceeding 150 ms of transit and approaching 300 ms means that the communication quality suffers, as it would during a satellite conversation.

Let us detail the installation of communication. A signaling message is necessary to initiate the session. The first step, the user location, is carried out by a mapping of the destination address (IP address or traditional phone) into an IP address. The DHCP protocol and specialized gateways are solution elements used to determine the receivers' addresses. Establishing a communication involves accepting the destination terminal, whether it is a telephone, a voice mailbox or a Web server. As we have seen, several signaling protocols can be used, such as H.323 from ITU-T, or SIP and SDP from IETF.

18.5. Signaling protocols

Two main signaling protocols have been defined and we will examine them in this section: SIP and H.323.

18.5.1. *SIP*

SIP is a signaling protocol which first version appeared in draft format at the IETF in 1997 in the Mmusic workgroup. Forwarded back to the SIP group in 1999, it was standardized in March 1999 under RFC 2543.

SIP has the goal of establishing, modifying and terminating multimedia sessions between two terminals. It does not define the body of these messages. The body, which contains the media description (video, audio, coders, etc.) is defined by the SDP (session description protocol) protocol. In addition to the description of flow, SIP can transport session media, in particular QoS or security information. Session media is dissociated from SIP exchanges.

By inheriting from the HTTP model, the exchange mode is of client-server type, with client and server being considered equal. SIP messages are requests, also called methods, which generate return messages, the responses. SIP can function on top of several transport protocols. UDP is currently the most used, but the use of TCP is also defined, as well as transport via other protocols, such as SCTP (flow control transmission protocol). With TCP, SIP provides reliability with receipt acknowledgements and timers.

SIP was created to be evolutionary. Only basic functions are mandatory, with the possibility of supporting extensions by the different entities that exchange their capacity.

Two modes of communication are possible and the network chooses the mode:

– direct mode: the two SIP entities which represent the terminals communicate directly;

– indirect mode: intermediate entities, which are part of the network, relay exchanged messages.

18.5.1.1. *SIP entities*

SIP comprises several categories of entities, the most traditional being the user and network entities. These entities exchange messages.

User entities are called UA (user agent). UAs initiate, modify and terminate sessions for the user account in the multimedia application of the user's terminal, i.e. the part of the program that makes it possible to receive and establish sessions. It combines two components: the first which acts as client (UAC) and initiates sessions on the user's request and the second which acts as a server (UAS) and which is responsible for receiving all the sessions bound for the user. The UAs store information on the status of the session and are called *stateful*.

18.5.1.2. *Network entities*

SIP defines three server-type logic entities that are a part of the network and act to extend its functioning:

– *proxy server*. The proxy server has a relay function. It accepts SIP requests or responses from a UA or another proxy server and forwards them. This server can store information on the progress of sessions for which it intervenes. In this case it is called *stateful*. In the opposite case, it is called *stateless*;

– *registrar server*. UAs of its domain register themselves on it and it then informs the location service which is not defined by SIP. The protocol generally used to access this service is LDAP;

– *redirect server*. It responds to requests by giving possible locations of the required UA.

SIP also refers to a fourth entity which does not enter the SIP discussions but offers a location service on which SIP logic entities come to rely.

An example of an exchange between SIP entities is illustrated in Figure 18.2.

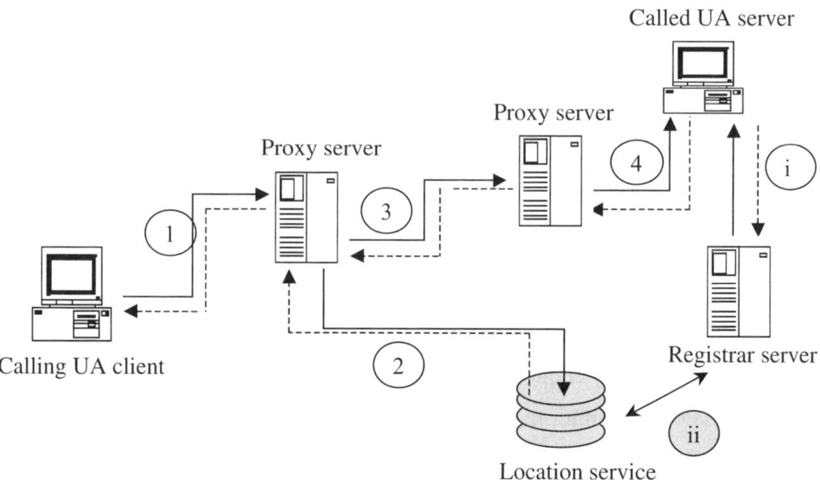

Figure 18.2. *Example of exchanges between SIP entities during the establishment of a session*

18.5.1.3. *SIP messages*

There are two SIP message categories: requests and responses. Messages are coded in textual language. The message is composed of three parts as illustrated in Figure 18.3.

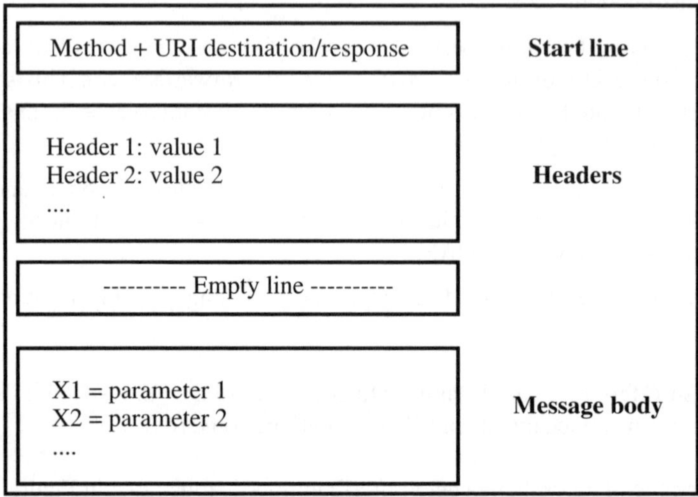

Figure 18.3. *SIP message structure*

Messages initiated by UACs (user agent client) destined for one or more UAS (user agent server) are called requests or methods by analogy with HTTP.

Methods defined in the current version of SIP are the following:

– INVITE: initiation or modification of a session;

– ACK: transmission of a positive response to an INVITE;

– CANCEL: cancellation of a request in progress;

– OPTIONS: capacity request;

– BYE: session termination;

– REGISTER: registering a UA;

– INFO: information relative to the session in progress.

Responses are sent by a UAS or a proxy in response to a request from a UAC. They are grouped in six classes, xx representing detailed codes:

– 1xx: information on the projection of the request;

– 2xx: success of the request;

– 3xx: redirection of the request;

– 4xx: client error;

– 5xx: server error;

– 6xx: global error.

SIP headings are grouped into four categories:

– general: present in requests and responses;

– request: present solely in requests;

– response: present solely in responses;

– entity: related to the body of the message.

The header name is followed by the value of the header.

18.5.1.4. *Session scenarios*

Different session media establishment scenarios are possible. Figure 18.4 illustrates a scenario that highlights a major SIP characteristic:

– during its response, the UAS can start a direct exchange of SIP messages to come (a);

– for their part, proxy servers can oblige the SIP messages to come to go through them (b).

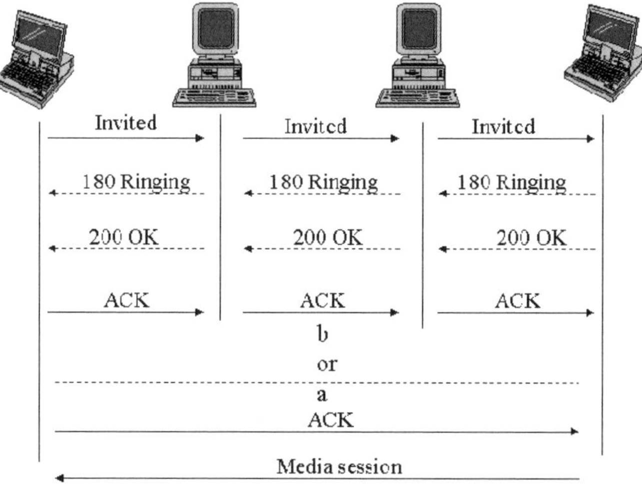

Figure 18.4. *Example of SIP session establishment*

18.5.1.5. *SDP (session description protocol)*

SDP is a media description syntax standardized in April 1998 in the RFC 2327. However, as we will see later on, SDP does not offer the complete possibility to negotiate media capacity. Issued from the IETF Mmusic group preceding SIP, its first application was the description of multicast sessions combined with the SAP (session announcement protocol), which officiates largely in the experimental multicast network Mbone (multicast backbone).

SDP then became the natural protocol of session media descriptions integrated to the establishment of SIP protocol sessions. SDP description syntax follows a textual coding. A session description is in fact an ordered succession of lines, called fields, which are represented by a letter. The number of fields is voluntarily limited to facilitate the coding (via a parser). The only way to extend SDP is to define new attributes.

The general format of a field is x = paramter1 parameter2 … parameterN.

The principal characterizing information for session media are the following:

– IP address (or host name): media flow receiving address;

– port number for flow reception;

– media type: audio, video, etc.;

– encoding diagram (PCM A-LAW, MPEG-2, etc.).

The field format characterizing the session media is m = media port transport format-list, with:

– media: audio, video, application, data, control;

– port: media reception port number;

– transport: RTP/AVP (audio video profiles) or UDP;

– format-list: list of complementary media information, or media payload type.

Many payload types can be listed. If they are listed, it is a proposed choice. To open n audio channels, n media fields must be presented.

18.5.1.6. *Exchange of session media characteristics*

The QoS needs necessary for the session are deduced from the session media characteristics. These characteristics defined by SDP and present in the body of SIP messages are the following:

– extremities of flow exchange: source and destination IP addresses and port numbers;

– type of exchanged media: audio, video, etc.;

– mode of coding used: list of codecs used for each flow, for example, G.711 or G723.1.

During the initialization of an SIP session, only the transmitter's media characteristics are known. They are determined by the UAC according to the request of the user's application, such as telephony over IP. The UAC that initializes the request for a session initiation does not have foreknowledge of information concerning the characteristic part of the UAS media that it is soliciting. It is only after the solicitation has arrived to the UAS that the feasibility of initiating a session is known and, in the event of a positive answer, that the session media characteristics are identified.

It must not be forgotten that a traditional session media results from the exchange of two media flows:

– a media flow from the initiator toward the solicited;

– a media flow from the solicited toward the initiator.

In this version of SIP (RFC 2543), the exchange of session media characteristics gives little scope for negotiation. In fact, the UAS that receives the requested session media characteristics has the type of media and the list of relative codecs proposed by the starting UAC. It is specified that the station that initializes the communication indicates in its codecs list those with which it wants to receive media flow and the fact that it would like to transmit this flow. In its response, the UAS specifies its list of codecs, which can be a subset of the transmitter's list. It also indicates the codecs list with which it wants to receive the media flow. The SDP RFC specifies that when a codecs list is given, it specifies the codecs that can be used during a session in order of preference, the first being considered as the default codec for the session.

In the best case, the solicited UAS accepts the list of proposed coders (solicited flow ⇒ initiating UAC). Inversely, the initiating UAC accepts the list proposed to it (initiating flow ⇒ solicited UAS). In order for a session to take place, each must accept at least the first code specified in the list proposed to it. Otherwise, the session cannot take place.

The exchange of media characteristics is illustrated in Figure 18.5.

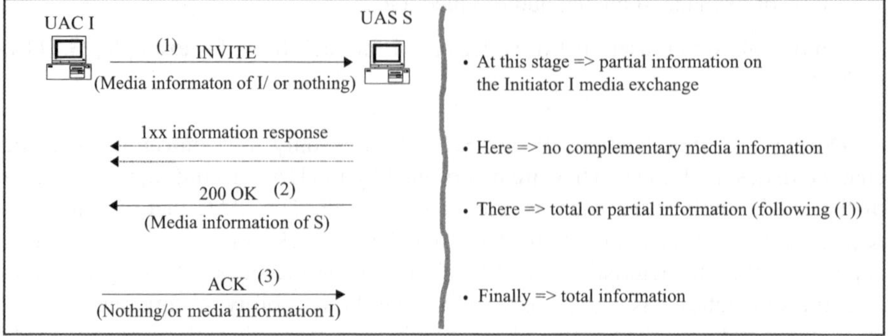

Figure 18.5. *Exchange of media characteristics in the current SIP protocol*

18.5.1.7. *New SIP version*

RFC 3261 describes a new version of SIP. This RFC's goal is to revise RFC 2543, which has now become obsolete, by correcting its errors and by presenting details on the usage scenarios. These modifications are listed at the heart of the RFC.

The following list presents the most significant modifications:

– centering the RFC on the SIP part:

- in the scenarios, the SDP body is no longer presented. SDP exchanges are directly defined in the new RFC 3264 (an offer/answer model with session description protocol),

- SIP server location procedures by DNS are defined in the new RFC 3263 (SIP: locating SIP servers),

- obligation for the UA to support TCP as well as UDP,

- support of TLS (transport layer security) and SCTP (flow control transmission protocol) for the transport level,

- processing of routes and registration of SIP routes that are reworked and largely detailed;

– security modification, which becomes more thorough:

- PGP (pretty good privacy) suppressed and replaced by S/MIME,

- basic authentication suppressed and even forbidden,

- supplemental security functions brought by TLS,

- UA authentication mechanism by server replaced by mutual authentication of the RFC 2617.

Other extensions associated with SIP accompany the new protocol. The following sections present extensions of the RFC that concern the improvement of the negotiation of media characteristics at the heart of the call establishment phase.

18.5.1.8. *The offer/response model*

The objective of the offer/response model is to enable two entities to arrive at a common vision of the session they will have together. This session is described with the help of SDP and is used by SIP. It was defined to specify and complete the exchange of media characteristics during the establishment of the session. In this model, one of the participants offers to the other the description of the session media that it desires and the other responds to this offer. It is specified that when many codecs are listed, the offering participant indicates that it is able to use each of them during the session. The responder can use them in the course of the session without preliminary renegotiation.

The participant making the offer sends the set of media flows and codecs it wants to use, as well as the IP address and the port number on which it is ready to receive. The responder looks at each proposed media flow by indicating if it accepts it or not, with the list of codecs that will be used as well as the IP address and the port number on which it is ready to receive.

In the general case where the session consists of two flows, in other words a sending flow and a receiving flow (sendrecv), the codecs list corresponds to the types of flow encoding which are required at reception and that would be preferred for sending. The first codec listed is the preferred one. It is recommended that the participant being solicited use this codec. In a more general way, the solicited participant attaches in its response the order of the codecs list presented in the offer because this will enable the same codec to be used in both directions. This model also enables the modification of established sessions.

At any time, one participant or the other can launch a new offer to modify the characteristics of the current session. It is possible to modify the parameters of a media flow, to destroy an existing flow or to add a new flow. All the media fields of the first offer are included in the same order and the new medias are added at the end.

18.5.1.9. *The new SIP method*

The goal of this method is to enable a UA to modify the session parameters, such as the media flow and the associated codecs, without impacting the status of the traditional SIP. The method can be started after the establishment of a session or during this phase. This becomes quite useful for modifying the characterizing session information before the session is established, in other words, during the establishment. The example presented is that of the early media welcome message sent to the correspondent. This media flow is sent until the invitation is accepted. At that moment, the session media is modified in order to enable the withdrawal of the welcome message and to open the communication flow. This process is carried out by the UPDATE method.

An extension brought about by the RFC 3262 in July 2002 introduces a mechanism that enables the sending of provisionary responses to be more reliable. These enable the transportation of important data, such as the responses to offers in the offer/response model. This mechanism relies on the introduction of a new header, which is transported in the provisionary response and that specifies the request so that it is made reliable.

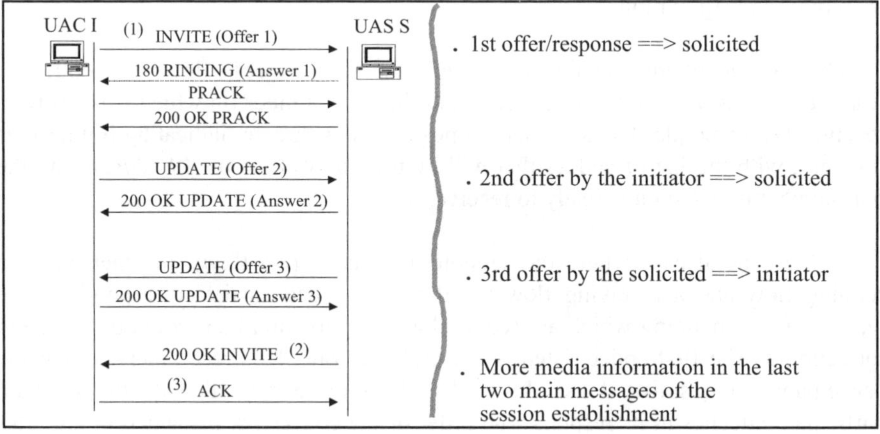

Figure 18.6. *Media characteristic exchange in the new SIP version*

The three extensions introduced in the previous section enrich the exchange of session media characteristics. Figure 18.6 presents an example that highlights all the potential brought by these extensions. These new possibilities have a significant impact. In fact, the information characterizing the media is no longer solely found in the main messages of the establishment of INVITE, 200 OK and ACK sessions but in the provisionary messages, TEL 280 ringing, and the intermediate UPDATE modification messages. This, however, makes the integration more complex.

18.5.2. *H.323*

The H.323 standard, whose standardization was achieved in 1997, describes the network protocols to be implemented in a multimedia terminal, called terminal H.323. Figure 18.7 illustrates the general composition of the protocols that must be implemented in a terminal so that we may talk about multimedia terminal H.323. Instead of using the word protocol, the word coder is sometimes used because protocols can be materially implemented in a coder.

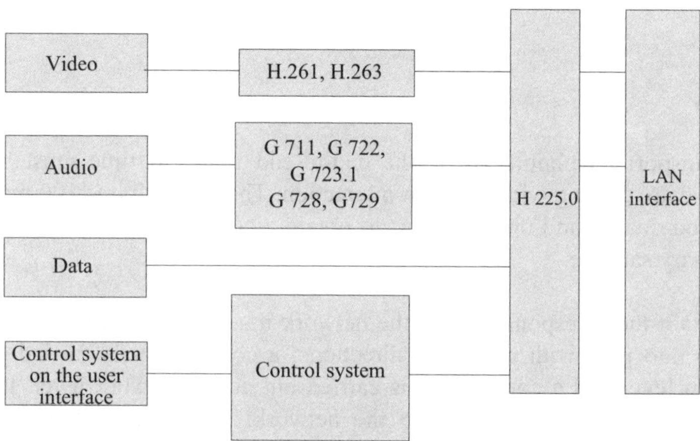

Figure 18.7. *H.323 terminal protocols*

For video equipment, the choice of coders is either H.261, which was previously presented, or H.263.

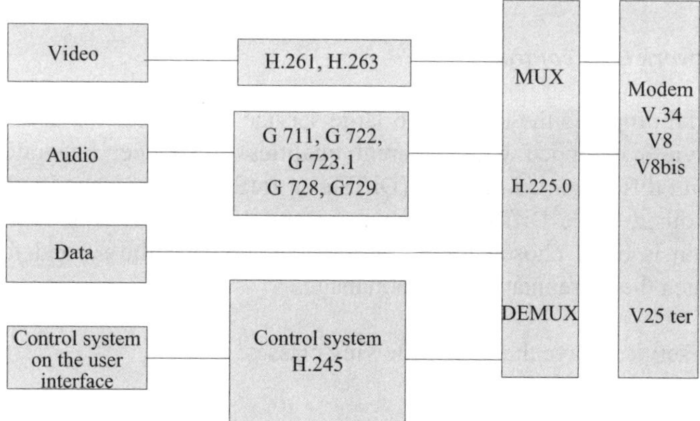

Figure 18.8. *H.324 terminal protocols*

Figure 18.8 illustrates the functioning of H.324 multimedia equipment for traditional phone lines. This includes the V.34 modem that makes it possible to reach transmission speeds of more than 20 kbps but that approaches 50 kbps when associated with compression. It becomes possible to reach similar speeds on analog and digital connections, up to 64 kbps.

Many previously seen coders, found in this multimedia terminal, correspond to the lowest flows in order to respect the transfer capacities of the telephone network interfaces.

18.6. QoS

To transport telephonic voice, the end-to-end transport time must be limited because we are dealing with a human interaction. This limitation is 100 ms to obtain a very good quality and up to 300 ms to obtain a conversation similar to one via a geostationary satellite.

To obtain these response times, the network must offer a QoS. Many solutions can come into play with two main directions: a control that is carried out at the application level and a control that is carried out at the network level. In the first case, the application must adapt to the network: if the network is loaded, the adaptation is carried out on the flow by reducing it; the application adapts. In the second case, it is the network that must adapt to the application. Given a flow and temporal constraints, we need to get an SLA (service level agreement) from the network assuring the quality of the requested service. We will examine in the following several propositions belonging to one category or the other. In the first category we will see RTP/RTCP and in the second IntServ and DiffServ.

18.6.1. *Network level control*

The IETF proposes the use of two large service categories, which are separated by sub-services equipped with different qualities of service: integrated services (IntServ) and differentiated services (DiffServ). IntServ are managed independently from each other, while DiffServ combine several applications simultaneously. The first solution is often chosen for the access network and the second for the core network when there are many flows to manage.

IntServ services have the three following classes:

– guaranteed service;

– controlled load;

– best effort.

DiffServ services have the three following classes:

– expedited forwarding or Premium service

– assured forwarding or Olympic service

– best effort

18.6.1.1. *IntServ*

IntServ service integrates two different service levels with performance guarantees. It is a flow-oriented service, in which each flow can make a specific request for QoS. To obtain a precise guarantee, the IntServ workgroup has considered that only a reservation of resources was undoubtedly capable of bringing the means of guaranteeing the request.

As previously explained, three service sub-types are defined in IntServ: a service with total guarantee, a service with partial guarantee and the best effort service. The first corresponds to rigid services with strong constraints to be respected and the second and third correspond to flexible services with no strong constraints.

When they receive a request *via* the RSVP protocol, routers can accept it or decline it. This request is carried out as it would for the initial receiver's RSVP protocol to the sender. Once the request is accepted, the routers place the corresponding packets in a waiting file of the same class as the service requested.

The IntServ service must have the following components:

– an signaling procedure to alert the traversed nodes. The RSVP protocol is supposed to fulfill this task;

– a method enabling the indication of the user's QoS request in the IP packet so that the nodes can keep track of it;

– a traffic control to maintain the QoS;

– a mechanism to pass the level of quality to the sub-adjacent network, if one exists.

The guaranteed service (GS) affects a limit above the routing time. For this, a reservation protocol such as RSVP is necessary. A reservation request has two parts: a specification of the QoS determined by a FlowSpec and a specification of packets that must be taken into account by a filter, FilterSpec. In other words, not all packets in the flow will necessarily have a QoS. Each flow has its QoS and its filter, which can be fixed (fixed filter), shared with other sources (shared-explicit) or specific (wildcard filter).

The partially guaranteed service called CL (controlled load) must guarantee a QoS approximately equal to the one offered by a minimally loaded network. This class is essentially used for the transport of flexible services. The network transit times of the CL flows must be similar to those of best effort class clients in a minimally loaded network. To achieve this network fluidity, a control technique must be integrated.

The two services must be able to be claimed by the application via the interface. Two possibilities are evoked in the IntServ proposition: the use of the GQoS Winsock2 specification, which enables the transport of point-to-point and multicast applications, and RAPI (RSVP API), which is an applicative interface in Unix.

Packet scheduling in the routers is a second necessary mechanism. One of the most traditionally proposed mechanisms is the WFQ (weighted fair queuing). This algorithm placed in each router requires a queuing of packets in order of priority. The queues are served in an order determined by a scheduling depending on the operator. Generally, the number of packets served at each server depends on the weight parameter of the queue.

Several solutions exist to manage the way the service is affected with queues, generally based on levels of priority. We mention, notably, the Virtual Clock algorithm which uses a virtual clock to determine the emission time and SCFQ (self-clocked fair queuing) which works in minimal time intervals between two packet emissions of the same class, an interval which depends on priority.

The IntServ service poses a scalability problem, which indicates the possibility of behaving properly when the number of flows to be managed becomes very large, as is the case on the Internet. The IntServ control being based on individual flows, the IntServ network routers must in fact store the characteristics of each flow. Another difficulty concerns the processing of different flows in IntServ nodes: which flow should be processed before another when millions of flows arrive simultaneously with different classes and parameters?

In the absence of a solution for all of these problems, the second main control technique, DiffServ, tries to sort the flows in a small well-defined number of classes, by multiplexing flows of a similar nature into larger flows, but always in a limited number. However, IntServ can be applied to small networks such as access networks. Other researches of management processors specialized in QoS have recently appeared on equipment capable of processing several tens or even hundreds of millions of flows.

The ISSLL (integrated services over specific link layers) workgroup of the IETF is seeking to define an IntServ model acting on an integrated level of ATM type, Ethernet type, packet relay type, PPP type, etc. In other words, the objective is to propose mechanisms that enable the passing of a class's priority level to classes that are sometimes not equal, and to choose, in the sub-adjacent network, algorithms which are susceptible of giving a result equal to what would be obtained in the IP world.

18.6.1.2. *DiffServ*

The principal objective of DiffServ is to propose a general configuration which makes it possible to deploy the QoS on a large IP network and to do so rather quickly.

DiffServ separates the architecture in two major components: the transfer technique and the configuration of parameters used during the transfer. This concerns the processing received by the packets during their transfer in a node, as well as queue management and service. The configuration of all the path nodes is carried out by PHB (per-hop behavior). These PHBs determine the different processes corresponding to the flows that have been identified in the network.

DiffServ defines the general semantics of the PHBs and not the specific mechanisms that enable them to be implemented. The PHBs are defined once and for all, whereas the mechanisms can be modified and improved to be different following the type of sub-adjacent network.

The PHBs and the associated mechanisms must be easily deployed in IP networks, which require each node to be capable of managing the flows due to a certain number of mechanisms such as the ordinance, the shaping or the loss of packets crossing a node.

DiffServ aggregates the flows by classes, called aggregates, which offer specific qualities of service. The QoS is assured by processes which are carried out in routers as specified by an indicator located in the IP packet. Aggregation points of incoming traffic are generally placed at the entrance of the network. The routers are configured by the DSCP (differentiated service code point) of the IP packet, which forms the first part of a more general field called DS (differentiated service) and also containing a field called CU (currently unused). In IPv4, this DS field is taken from the ToS (type of service) zone, which is then redefined according to its first use. In IPv6, this field is in the TC (traffic class) zone of the service class.

Figure 18.9 illustrates the DS field in IPv4 and IPv6 packets: the DSCP field takes place in IPv4 ToS fields and in the IPv6 TC field. The DSCP field uses 6 of the 8 bits and is completed by two CU bits. The DSCP determines the PHB service class.

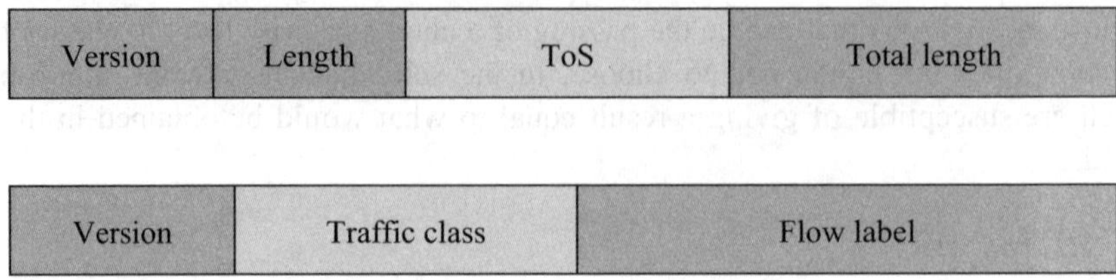

Figure 18.9. *DS field of IPv4 and IPv6 packets*

The 6-bit DSCP field must be interpreted by the node to allot, to the packet, the process corresponding to the indicated PHB class. The two CU bits must be ignored during the process in a standardized DiffServ node. By the intermediary of a table, the DSCP values determine the PHBs acceptable by the node. A default value must always be given when the DSCP field does not correspond to any PHBs.

Telecommunications operators can define their own DSCP values for a given PHB, instead of their IETF recommended values. These operators must, however, provide the standard DSCP value to the exit paths so that the next operator properly interprets this field. In particular, an unrecognized DSCP must always be a default value.

The definition of the DS field structure is incompatible with that of the RFC 791 ToS field that defines IPv4. This ToS field was conceived to indicate the criteria to be privileged in the routing. Among these foreseen criteria there are the delay, the reliability, the cost and the security.

Beyond the BE (best effort) service, two PHBs quite similar to IntServ are defined in DiffServ:

– EF (expedited forwarding), or guaranteed service, which is also called premium service;

– AF (assured forwarding), or assured service, which is also sometimes called Olympic service.

There are four service sub-classes in AF that determine acceptable loss rates for the considered packet flows. These can be classed as platinum, gold, silver or bronze. As this is not standard terminology, it is possible to encounter other names. Within each of these classes, three sub-classes sorted according to their degree of precedence, i.e. their level of priority with regard to each other, are defined. The AF1x class has the highest priority, then comes class AF2x, etc.

There are 12 standardized classes in total, but very few operators use them. As a general rule, operators are satisfied with the three basic AF classes and so with five classes in total, including EF and BE.

The values stored by the DSCP field which are associated with these different classes are illustrated in Figure 18.10. For example, the value of 101110 in the DSCP field means that the packet is of EF type. The BE class's value is 000000.

The 11x000 DSCP is reserved for client classes with a higher priority than those of the EF class. It can, for example, be used for signaling packets.

18.6.1.2.1. EF

The EF PHB is defined as a packet transfer for an aggregate of flows from DiffServ nodes such that the rate of service for packets from this aggregate is greater than a rate determined by the operator.

The EF traffic must be able to receive a rate of service independent from other traffics circulating in the network. More precisely, the rate of the EF traffic must be greater to or equal to the rate determined by the operator as measured in any time interval which is at least equal to the size of one MTU (mean transmission unit). If the EF PHB is implemented due to a priority mechanism in other traffics, the rate of traffic of the EF aggregate must not exceed a limit that would be unacceptable for PHBs of other traffic classes.

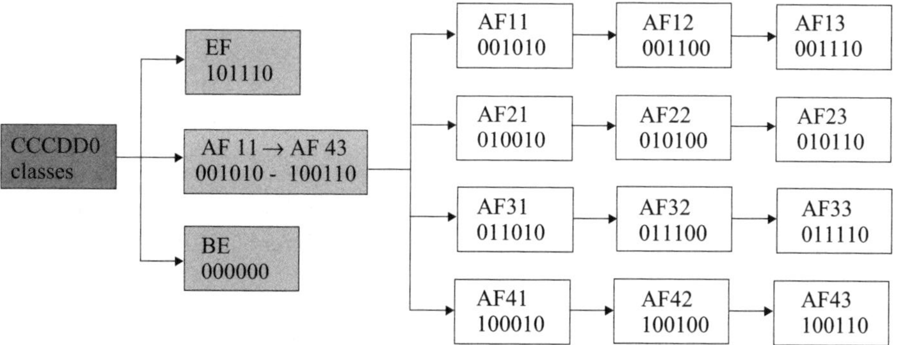

Figure 18.10. *DiffServ service classes and associated DSCP field values*

Many types of ordinance mechanisms can be used to respond to these constraints. A priority queue is the simplest mechanism for the EF service (or PBH) as long as there are no other more highly prioritized queues to preempt the EF packets by more than one packet for an amount of time determined by the rate of service of the EF aggregate packets. It is possible to use a normal queue in a group of queues managed by a weighted round robin mechanism, or to use a sharing of bandwidth of the node's exit queue, which enables the EF queue to reach the rate of service guaranteed by the operator. Another potential mechanism, called the CBQ (class-based queuing) sharing, gives sufficient priority to the EF queue to obtain at least the rate of service guaranteed by the operator.

The EF traffic corresponds to traffic sensitive to delays and to the jitter. It has a high priority in the nodes but must be controlled so that the sum of traffics from different sources traveling along the same path does not exceed the nominal capacity determined by the operator.

Many solutions make it possible to reserve bandwidth for the EF packet flows. An RSVP protocol, for example, can carry out the necessary bandwidth reservations. Another solution consists of using a server specialized in the distribution of bandwidth, called a bandwidth broker. This bandwidth server controls admission by proposing a centralized reservation.

18.6.1.2.2. AF

AF PHBs ensure the transfer of IP packets for which a certain QoS can be guaranteed. AF traffics are subdivided into n distinct AF classes. In each class, an IP packet is assigned a maximum loss rate and a priority of loss, corresponding to precedent classes. An IP packet belonging to class AFi and having a loss rate corresponding to precedence j is marked by an AFij DSCP (see Figure 18.10). As previously explained, there are 12 defined classes for DiffServ, corresponding to four AF classes with packet loss guarantees. The four classes that correspond to the guaranteed loss rate are called platinum, gold, silver and bronze, each of which has three different precedence levels.

AF class packets are transferred independently from those of other AF classes. In other words, a node cannot aggregate flows having different DSCPs into a common class.

A DiffServ node must allocate a series of minimal resources to each AF PHB so that they can fulfill the service for which they were put into place. An AF class must have a minimum of resources in memory and in bandwidth to carry out a minimal rate of service, as determined by the operator, over a long enough period of time. In other words, in a relatively long time frame that can be counted in a second, a guarantee of flow must be given to the AF services.

An AF node must be configured to enable an AF class to receive more transfer resources than the minimum when supplementary resources are available in the network. This supplementary allocation is not necessarily proportionate to the level of the class, but the operator must be able to reallocate the resources freed up by the EF class to the AF PHBs. However, the precedences must be respected: a higher precedence class must not loose more packets than a class with a lower precedence, even if the loss is less than the allowed limit.

A domain that implements AF services must, by the intermediary of border routers, be capable of controlling incoming AF traffic so that the QoS determined for each AF class is satisfied. The border routers must therefore put in place shaper mechanisms, dropper mechanisms, AF packet augmentation or diminution mechanisms and mechanisms for the reassignment of AF traffic into other AF classes. Ordinance actions must not cause a reordering of packets in the same microflow, a microflow being a particular flow within a PHB aggregate.

Implementing an AF strategy must minimize the congestion rate within each class, even if short-term congestions are allowed following the superposition of continuous packet flows (bursts). This requires a dynamic management algorithm in each AF node. RED (random early drop) is an example of such an algorithm. Long-term congestion must also be avoided due to packet losses corresponding to precedence levels and short-term congestion due to queues which make it possible to make certain packets wait. Shaper algorithms must be able to detect the packets that could create long-term congestions.

The basic algorithm that enables the control of AF traffics is WRED, or *weighted RED*. It consists of trying to maintain a fluid state in the network. Packet losses must be proportional to the length of the queues. In other words, surplus packets and then normal packets are eliminated when the traffic is no longer fluid. The time elapsed since the last loss in a single aggregate is taken into account in this algorithm. The procedure attempts to distribute the control to the nodes rather than to the single congested node. Packet destruction algorithms must be independent of the short-term and of the microflows, as well as of the microflows within the aggregates.

The interconnection of AF services can be difficult due to the relative imprecision of the service level of different interconnected operators.

A solution that enables the transfer of an aggregate in an IP network that does not conform to DiffServ consists of creating a tunnel with a higher QoS than that of the PHB. When an aggregate of AF packets uses the tunnel, the assured QoS must enable the basic PHB to be respected at the exit of the tunnel.

A client requesting an AF traffic must negotiate a service agreement called SLA which corresponds to a profile determined by a group of QoS parameters, called the SLS (service level specification). The SLS gives a loss rate and, for EF services, an average response time and a jitter of the response time. The traffic that does not enter the profile is destroyed by priority if a risk of congestion exists that would not enable the conforming traffic to reach its QoS.

18.6.1.2.3. DiffServ node architecture

The architecture of a DiffServ node is illustrated in Figure 18.11. It includes an entrance that contains a classifier, whose role is to determine the right path within the node. The selected junction depends on the class detected by the classifier.

Then there are meters. A meter determines whether the packet can perform as required by its class and then decides on the process. The meter knows all the queues in the node as well as the QoS parameters requested by the aggregate to which the packet belongs. It can decide to eventually destroy the packet, if the class permits it, or to send it to an exit queue. The DiffServ node can also decide to change this packet's class or to multiplex it with other flows, as we will see. The dropper, or suppressor, can decide whether or not to lose the packet, whether or not to destroy the packet, whereas the absolute dropper automatically eliminates the packet.

In other words, the meter can make a decision to destroy and send the packet to an absolute dropper, whereas the meter can only determine the performance parameters and lets the dropper decide whether or not to destroy the packet according to criteria other than the gross measure of performance.

For certain packets, such as the BE packets, it is not necessary to question the performance because there is no guarantee on the aggregate. It is sufficient to know if the packet is to be lost or not. This corresponds to branch D in Figure 18.11. In this same figure, the first branch (A) corresponds to EF or premium clients, the next two (B and C) to AF clients, with gold clients in the higher path and silver or bronze in the other path, and the last branch (D) to BE clients.

The architecture of a DiffServ node is determined by queues destined to make the packets waits before they are sent on the exit path determined by routing. A precedence algorithm is used to process the order of packet emission. The scheduler takes on this task. The simplest algorithm processes the queues in their order of priority and does not allow clients of another queue to get through while there are still clients in a higher priority queue.

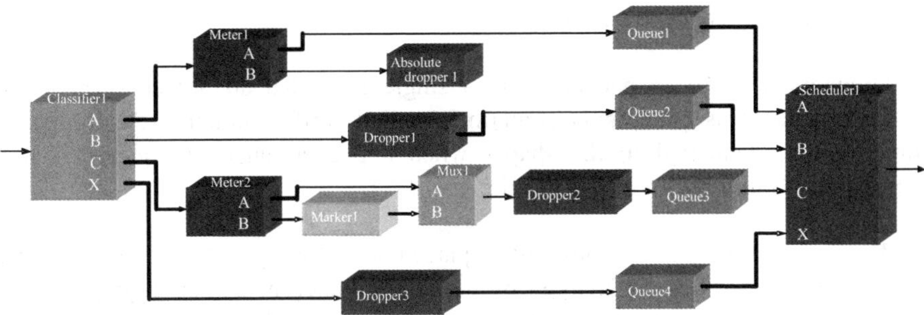

Figure 18.11. *DiffServ node architecture*

Many other algorithms make it possible to give a certain weight to queues such that a non-priority client can be serviced before a priority client. Among these algorithms, we mention WFQ, in which each queue has a weight, for example 70 for the EF queue, 20 for the AF gold queue and 10 for the other AF queue. The scheduler lets EF clients through 70% of the time. If these clients exceed 70%, the schedule lets AF gold clients through for 20% of the remaining time and for 10% of the remaining time it allows AF silver or bronze clients through.

The sum of actions which a packet undergoes in a DiffServ node is carried out by a general body called the conditioner. A traffic conditioner can contain the following elements: meter, marker, shaper and dropper. A flow is selected by the classifier. A meter is used to measure traffic in comparison to the profile. The meter's measurement for a certain packet can be used to determine if the packet should be sent to a marker or to a dropper.

When the packet exits the conditioner, it must have a suitable DSCP value. The meter obtains the temporal properties of the packet flow selected by the classifier according to a profile determined by a TCA (traffic conditioning agreement). The meter sends this information to other parts of the conditioner that implement specific functions adapted to the packets so that they receive the appropriate processing, whether they are in the profile or out of the profile.

Packet markers give the DSCP field a particular value and add the packet to the corresponding aggregate flow. The marker can be configured to mark all the packets with the correct DSCP value or to choose a particular DSCP for a predetermined group of PHBs.

The shapers' goal is to delay the packets within a flow to conform them to a given profile. A shaper generally has a finite memory which makes it possible to delay the packets by putting them on hold. These can be destroyed if there is no available memory space to conform them.

Droppers destroy the packets from a single flow that do not conform to the traffic profile. This process is sometimes called traffic policing. A dropper is sometimes implemented in the shaper when a packet must be rejected if it is impossible to put it back into the profile.

Traffic conditioners are most often placed in entrance and exit nodes of DS domains. Since packets are marked by domain entrance nodes, an aggregate from another operator is supposed to conform to the appropriate TCA.

18.6.2. *Application level control: RTP/RTCP*

The RTP (real-time transport protocol) protocol takes over the transport of the information. Its role is to organize the packets at the network entrance and to control them at the exit to reform the flow with its characteristics (synchronism, loss, etc). This protocol works on the transport level and attempts to correct network weaknesses.

The following are the RTP functions:

– packet sequencing by a classification that makes it possible to detect lost packets, which is essential for reforming the transferred voice. Losing a packet is not a problem in itself as long as not too many others have been lost. On the other hand, recovering a lost packet is imperative as it must be tracked and eventually replaced by a synthesis determined according to the preceding and following packets;

– identifying what has been transferred in the message to enable, for example, a compensation in the case of a loss;

– synchronizing various media, with the assistance of stamps;

– packetization. Audio and video applications are transferred into packets whose dimensions depend on the digitizing codecs. These packets are included in the frames to be transported and must be easily recovered during depacketization so that the application can be simply decoded;

– identifying the source. In multicast applications, the source identity must be determined.

RTP uses the RTCP protocol to transport the following supplementary information for the session management:

– return of the QoS during the session request. The receivers use RTCP to send their QoS reports to the senders. These reports include the number of lost packets, the jitter and the time delay. This data enables the source to adapt itself, in other words to modify the compression rate to maintain the QoS;

– supplementary synchronization between media. Multimedia applications are often transported by distinct flows. For example, voice and image, or even an application digitized on several hierarchical levels, can generate flows that follow different paths;

– identification. RTCP packets contain address information, such as the address of an electronic message, a telephone number or the name of a telephone conference participant;

– session control. RTCP enables the participants to indicate the initiation of a telephone conference (RTCP *bye* packet) or simply to indicate their behavior.

The RTCP protocol requires that the session participants periodically send this information. The periodicity is calculated according to the number of application participants.

Another usable protocol is the RTSP (real-time streaming protocol), whose role is to control a communication between two servers where audio and video multimedia information is stored. The RTSP commands are similar to those of a video tape recorder, such as play, fast forward, rewind, pause, etc. This protocol can be very useful in IP telephony by enabling the recording of a teleconference in order to listen to it again later, the viewing of a video sequence, the recording of a telephone message, etc.

Another important point to enable communication between a sender and a receiver concerns the gateway functions that make it possible to go from a packet transfer network to a circuit switching network, with the inherent addressing, signaling and transcoding problems. These gateways are reduced between ISPs and telecom operators.

To complete the call request, the SIP protocol sends a request to the gateway. The first issue is to determine which gateway is capable of creating the linking circuit to reach the destination. In theory, each gateway can call any telephone number. However, to reduce costs, it is best to choose a local gateway.

18.7. Corporate IP telephony networks

Corporate telephony networks that use the IP protocol have been available on the market since 2000. The goal of these networks is to integrate the data network and the telephonic network into a single network. The IP norm is of course the basis of this integration.

Telephony is digitized and the bytes are put into the shortest IP packets possible to reduce losses in bandwidth. Voice compression using G.729, for example, which is very common, generates 16-bytes packets every 16 ms. For quality purposes, 16 bytes per IP packet should be the maximum. These IP packets are transported with the 150 ms delay constraint to the destination.

In companies, Ethernet packets are used for transport. The IP packet is placed into an Ethernet frame, whose length is 64 octets. A flow rate of 8 kbps in a G.729 codec corresponds to a flow rate of 32 kbps on the Ethernet network. For a Gigabit Ethernet network, the minimum frame size is 512 octets and the flow rate for a single telephonic voice line is 256 kbps.

With respect to the temporal constraints, no time can be lost. The first place where time is lost is potentially in sound processing, which is carried out by a sound card in a PC if the PC is used as a telephone. These sound cards generally have a slow response rate, in the order of 40 ms, which is not acceptable. The use of specific telephones, called IP telephones, becomes important. An IP telephone is a router that directly encapsulates the bytes into an IP packet. This router has Ethernet outlets so that the telephone can be directly connected to the company's network. Figure 18.12 illustrates an IP telephone.

Figure 18.12. *IP telephone*

With the purpose of losing the least amount of time during transport, the network should not have a shared Ethernet network, since the sharing generates a significant loss of time. Therefore, only switched Ethernet networks should be used, if possible with a flow rate of 100 Mbps, to be sure that superimposed voices are not a problem.

Finally, the IP packets or the frame carrying the voice IP packets must have priority in the corporate network. In order to enable this, DiffServ priorities are necessary for the packets, and frame priorities are necessary in the Ethernet. Level 2 priorities correspond to the IEEE 802.1p standard, which defines a 3-bit field to manage up to eight priority classes.

18.8. Implementation of telephony over IP

After having evoked the main characteristics of IP telephony supporting protocols, in this section we describe the process to follow for installing telephony over IP (ToIP) in a company that has several sites. The elements to be considered must conform to a four-level architecture. The lowest level concerns the client, the second level concerns the infrastructure, then the service controller and finally the application level. We will describe these different levels a later on.

A generally standard company network has two sites: one principal and one secondary. The network is made up of a local network in each site. The two local networks are connected by a WAN network, which can be, for example, a private virtual network. This network is made up of Ethernet switches that form a switched local Ethernet network. On this local network there are IP telephones as well as workstations. The workstations can be connected to the IP telephones, or directly to the Ethernet switches. The two local switched Ethernet networks are connected to corporate input-output routers and the two routers are connected to each other by an extended network, for example a private virtual operator network.

By going up in the architecture and starting at the bottom, we first find IP telephones. These are level 3 routers capable of encapsulating telephony bytes. The workstations can also be used as telephones, but the sound card must be of good enough quality so as not to slow down the packetization process of telephonic bytes. These specific workstations are connected by Ethernet, generally at 100 Mbps, with the highest priority available. The DSCP zone of the IP packet is of EF service class and the packet is encapsulated in an Ethernet frame of highest priority, which is indicated by the IEEE 802.1p zone. The other machines that do not produce voice, or more precisely the other applications, position the DSCP to an AF or BE value which is less than that the one used in IP telephony. The company must therefore be equipped with DiffServ routers capable of processing the IP level priority and with Ethernet switches capable of managing IEEE 802.1p priority classes.

If the DSCP value is standardized, the IEEE 802.1p field value is a lot less standardized. Firstly, this field only has 3 bits of priority, which generates 8 priority classes, where DiffServ has 14. Then, the suppliers must conform to the same rules for determining the value of the IEEE 802.1p field. With regard to transport, the maximum transit time in each site must be evaluated in order to deduce the maximum transit time in the WAN network. Once this is known, it is possible to deduce the maximum transit time in the WAN network. The company must negotiate an SLA (service level agreement) with its operator and request that this constraint is respected in the technical aspect, in other words, the SLS (service level specification). The maximum transit time is generally around 50 ms.

18.9. Telephonic IP applications for the general public

A first telephony over IP for the general public is proposed by the operators to offer international communications at the cost of a local call. This consists of gathering a large number of traditional telephonic options and encapsulating them in a single IP packet, which can become quite long. The user connects locally on an IP operator point, which multiplexes all the voice flows on a single IP connection, transatlantic, for example. At the exit of the transatlantic IP connection, the voice recovers its normal composition and is sent in a traditional manner to the destination.

General public applications such as Skype or MNS (Microsoft network service) propose end-to-end telephony over IP. This generally must pass via an ADSL modem on both ends of the communication so that the flow is accepted on the local loop.

Skype uses a P2P technique to remain as simple as possible without having a centralized control. The MNS signal, on the other hand, is managed by a centralized database but can be distributed over several sites.

These telephony over IP protocols use proprietary signals that are not blocked by firewalls or NAT equipment. On the other hand, they do not guarantee as much security for the user as a SIP protocol. However, in these proprietary solutions, voice is encrypted by symmetrical DES keys.

18.10. Telephony-data integration

The integration of telephony and data is not new. Foreseen for a long time, the need for this alliance has been felt since 1995.

This integration, called CTI (computer telephony integration), was created in a CSTA (computer supported telephony applications) workgroup and it was the ECMA that took matters in hand by creating the TGI 1 workgroup.

A first report describes the CSTA's objectives:

– evolved telephony (voice messaging, various network accesses, teleconferencing, etc);

– telemarketing;

– customer service;

– microcomputer as communication centre;

– alarm and service control;

– company data access.

18.11. Conclusion

This chapter has given a brief overview of packet telephony and more particularly, of telephony over IP.

The arrival of telephony over IP is inescapable given the financial profit gained over only a few years by the integration of data and telephony networks. However, telephony is a particularly difficult application to implement due to its real-time and its continuous flow. For the moment, the solutions consist especially of over-dimensioning the equipment and connections by taking into account only the clients carrying out telephony over IP.

18.12. Bibliography

ABRAHAMS J.R., LOLLO M., *CENTREX or PBX: The Impact of Internet Protocol*, Artech House Publishers, 2003.

BLACK U., *Voice Over IP*, Prentice Hall, 2002.

CAMP K., *IP Telephony Demystified*, McGraw-Hill, 2002.

DANG L., JENNINGS C., KELLY D., *Practical VoIP Using VOCAL*, O'Reilly & Associates, 2002.

DURKIN J.F., *Voice-Enabling the Data Network: H.323, MGCP, SIP, QoS, SLAs, and Security*, Pearson Education, 2002.

GHERNAOUTI-HÉLIE S., DUFOUR A., *Ingénierie des réseaux locaux d'entreprise et des PABX*, Masson, 1995.

GRIGONIS R., *Voice Over DSL*, CMP Books, 2002.

HARDY W.C., *VoIP Service Quality: Measuring and Evaluating Packet-Switched Voice*, McGraw-Hill, 2003.

HARTE L., BOWLER D., *Introduction to SIP IP Telephony Systems: Technology Basics, Services, Economics, and Installation*, Althos, 2004.

HÉBUTERNE G., *Ecoulement du trafic dans les autocommutateurs,* Masson, 1985.

JOHNSTON A., *SIP: Understanding the Session Initiation Protocol*, Artech House Publishers, 2003.

MILLER M.A., *Voice Over IP Technologies: Building the Converged Network*, John Wiley & Sons, 2002.

MINOLI D., MINOLI E., *Delivering Voice over IP Networks*, Wiley, 2002.

NELSON M., SMITH A., DEEL D., *Developing Cisco IP Phone Services: A Cisco AVVID Solution*, Cisco Press, 2002.

OHRTMAN F., *Softswitch: Architecture for VoIP*, McGraw-Hill, 2002.

OHRTMAN F., *Voice over 802.11*, Artech House Publishers, 2004.

PULVER J., *The Internet Telephone Toolkit,* Wiley, 1996.

ROB W., *Computer Telephone Integration,* Artech House, 1993.

ROB W., *Computer Mediated Communications: Multimedia Application,* Artech House, 1995.

STAFFORD M., *Signaling and Switching For Packet Telephony*, Artech House, 2004.

SULKIN A., *PBX Systems for IP Telephony*, McGraw-Hill Professional, 2002.

WALTERS R., *Computer Telephony Integration,* Artech House, 1998.

Chapter 19

Wireless Voice over IP

19.1. Introduction

Voice over IP is part of the current climate. More than a technological evolution, it is, today, one of the most promising network applications. Its price is incontestably its major asset. If telephonic communications are generally billed based on call length and distance, Internet communications are most often contractual: there is no distance and no time limit. Billing is fixed for an available flow. As such, all applications whose flows travel on the Internet network are included in the price. We must still find the applications that will most likely attract users. Voice over IP is precisely one of these candidates.

However, the technology of voice over IP has existed for the past 10 years or more. It has undergone multiple international standardizations which, without sheltering it against permanent evolutions inherent in network technologies, have rendered it sufficiently mature to consider a large scale deployment. In order to do this, a number of actors, such as constructors, operators and large companies, proudly present their first solutions of telephony over IP and stimulate the market to follow their example.

Presently, the technological challenge has been placed even higher. Telephony over IP used to offer an alternative to traditional RTC telephony; wireless offers even more advantages. Indeed, wireless networks enable the users to free themselves from cabling constraints on the last meters of connection to their station. The zone of coverage is extended, which provides an increase in utilization flexibility. This

---

Chapter written by Laurent OUAKIL.

gain translates also into an ease of deployment and a network extensibility: this becomes pervasive. Almost naturally, prolonging voice over IP (VoIP) introduces voice over wireless (or VoWLAN for Voice Over Wireless Local Area Network).

However, the constraints of VoWLAN in this new wireless mode of communication are a double innovation. It consists primarily of ensuring wireless portability of elements essential to voice deployment.

This state of the art attempts to review what VoIP represents in such a seductive environment, although difficult to master, as are wireless networks. Initially, it is a question of presenting the characteristics and constraints linked to the transmission of voice and to the use of wireless networks. Since the cost is not the only criteria to consider, the question of the implementation of a complete infrastructure, dedicated to voice and capable of end-to-end managing of audio streams, will be studied secondly. Finally, once the network is mastered, the question of QoS rendered is now considered as being the highest of requirements for users who are used to a guaranteed QoS. A global vision of possibilities will be presented later on.

19.2. Wireless VoIP problems

As we can imagine different supports cohabitating for voice support, VoWLAN technology does not have the goal of dethroning RTC telephony. It rather represents a supplementary step towards the convergence of audio, video and data streams on a single media. However, VoWLAN is not less strong in the competition with RTC telephony as the stakes are often shared between the two technologies. This section puts forward the principal elements that compose VoIP and wireless network technologies and exposes the intrinsic constraints of voice restitution as well as those that characterize wireless networks.

19.2.1. *Definition and principle of wireless VoIP*

Since networks exist in order to manage the exchange of data, the idea of transiting voice is to rely on the backbone IP network protocol by considering that a telephonic conversation is nothing more than an exchange of data streams, with the particularity of being in real-time. VoIP consists of the transportation of voice in an IP data network. It is often confused with telephony over IP, which is nothing but a possible application of VoIP. The principle of VoIP is simple: for the emitter, voice is first digitized and eventually compressed, before being segmented into IP packets that will be transmitted as a data stream through networks, whether digital, Internet or other types of networks. For the receiver, IP packets are reconstituted, decompressed if necessary, and converted into an analog signal transmittable by a regular phone. Voice is therefore treated like any other data stream.

Three types of possible connections are available for the user. The first is from PC to PC. Each of these must be equipped with a sound card, earphones or speakers, a microphone and telephony (or videoconferencing) software that supports the same standards and enables the digitization, compression and transmission of multimedia streams. One of the most popular software is Microsoft's NetMeeting, although its evolution is not ensured. PC to telephone communications is also possible. In this case, voice emitted by the PC user travels within the IP network until it reaches the gateway closest to the correspondent. This ensures the transmission of data between the IP network and the RTC network in both directions. Among the more common software functioning in this mode is Skype. Finally, telephone to telephone communications constitute the third possibility. These require the implementation of a gateway at each end of the network joining the correspondents and as close as possible to them. It is currently a widely used technique for telephony operators to offer an aggressive rate for long-distance destinations. However, audio quality is not always up to par.

The stakes for technology are double: reduce the costs (telephonic as well as cabling) and ensure the convergence of all data on a single support.

19.2.2. *Voice constraints*

Since the telephone constitutes a currently used application, users have important requirements with regards to this service. It is therefore imperative to respect these to be able to offer a service comparable to what is offered in traditional telephony, which is generally called RTC (Réseau Téléphonique Commuté) or PSTN (Public Switched Telephone Network).

In an interactive communication, time is the most fundamental factor. There are essentially two temporal parameters necessary to ensure a reasonable comfort of use: the time of transit and the jitter. The time of transit in the network (or end-to-end travel time) is the lapse of time between the moment when the emitter speaks and the moment when the listener hears the message. It includes a sum of time: the conception time of the emitter's packet (digitization, encoding, compression, packeting), the receiver's packet processing time (decoding, decompression, reassembly, conversion from digital to analog signal), the packet's end-to-end travel time in the network (propagation time, transmission, commutation in the nodes, including the time in commutator queues). So as not to perturb the rhythm of a conversation and to ensure interactivity between the correspondents, this sum of time must be as short as possible. We generally estimate the one-way transit time of a voice packet in the network must not exceed 150 ms so that the communication remains fluid. With an extra 150 ms, the interactivity of the dialog suffers, but communication is still acceptable without being entirely satisfying. Exceeding 300 ms, communication interactivity suffers heavily and the dialog resembles that of an amateur radio announcer. Another essential characteristic is the jitter, representing the variation in time between the predicted receiving time and the actual receiving time of a packet. In fact, it enables the conservation of temporal relations between voice samples emitted by the sender and those retransmitted to the receiver. To obtain a suitable continuity of sound, it is preferable that the jitter is less than 100 ms.

A non-constraining characteristic of PSTN telephony is that it enables the networks which transport voice to establish a communication with a reasonably long length of time. Thus, a tolerance of several hundreds of milliseconds, or of several seconds, is allowed for the establishing time of the communication. This amount of time can be put to good use for configuring the network. However, the percentage of success of an initiated call must be greater than 95% and the conversation must never be interrupted. The availability of the system constitutes in fact a key element in the reliability of a VoIP application.

Moreover, telephony can offer a whole group of supplementary functions to simplify its use and to improve conviviality. In this domain, services can be innovated at lesser costs. Nomadism (the possibility of connecting oneself with identical profile information in distinct locations) becomes possible by attributing the same user identifier to a user, whatever his location, and it is even conceivable to offer mobility (possibility of not terminating a communication during displacement).

19.2.3. *Wireless transmission constraints*

Terminal cabling is often a fastidious operation, especially in new installations. With wireless, there is no need to cable the last meters that link the user to the network. However, in wireless, the rare resource is still the bandwidth: available flows are limited and rarely guaranteed, which constitutes a major constraint in offering QoS indispensable for voice. Moreover, wireless transmission relies on the quality of the radio link. However, this depends on the air support that can degrade with several factors such as interference with other equipment using the same frequency, obstacles between the source and the destination, or the noticeable distancing between the emitter and the receiver.

As significant as they are in IP networks, packet losses are thus particularly common in wireless. However, the loss of a packet means the loss of a bit of voice. To have a good quality, the rate of packet loss must be less than 5%. In addition, confidentiality of communications between users must be protected against clandestine listeners. This precaution is perhaps more indispensable when dealing with wireless transmissions because the air interface is open by nature and consequently accessible to external people in the range of Hertzian waves. Although indispensable, security management involves a supplementary stream, or overhead, which can damage the real-time data and requires a compromise between security and flow rate.

19.3. Voice management indications and designs

With VoIP, a telephone number no longer identifies a telephonic terminal but a specific subscriber. Therefore, no matter the location of the user in the network, his simple connection will suffice to match his name with his geographical location and therefore enable anybody to join him. The principal role of the indication is precisely to exchange messages between the different entities in order to establish a connection between users who want to communicate. The indication makes it possible to locate the services as well as the users and to implement communication parameters, in terms of QoS, reservation, or implementation of flow priorities. It is then possible for IP network users to converse, for example, with users in the commuted telephonic network (as long as they have an address according to the numbering plan E.164 defined by the ITU), or even with internal company numbering plans.

There are two standardized indications for general multimedia flow management and in particular voice flow management: they are protocols H.323 and SIP.

19.3.1. *H.323 protocol*

In order to ensure proper management of multimedia data, the ITU-T (International Telecommunication Union – Telecommunication Standardization Sector) has elaborated a group of standards under the generic name H.32x. Initially conceived for LAN-based (Local Area Network, medium-sized network) multimedia conferences, standard H.323 [H.323] is the standard of reference. It was presented in its first version in October 1996 and has not stopped evolving by increasing its interconnection possibilities with other networks and by presenting a complete architecture for voice management in new versions up to version 6 in June 2006.

The H.323 architecture relies mainly on a group of indication protocols. The RAS (Registration, Admission and Status) protocol is implemented between terminals or gateways and control points in a distinct channel of a voice-element data stream. It makes it possible to authenticate the terminals that are authorized to connect and to control bandwidth use and management. Then, protocol H.225 (*Call Signaling*) opens an end-to-end channel between two H.323 terminals. This same protocol is used to close the channel. Finally, protocol H.245 (*Control Signaling*) establishes a phase of parameter exchange between the communicators to configure the applications used. The channel is then available for communication.

The implementation of H.323 sets up four types of entities:

– terminals: composed of an autonomous system (IP telephone) or simply a PC, terminals are the access support for the user, which makes it possible to transmit and receive multimedia data. They integrate the basic application that establishes the communication and sometimes enrich their functionality to various other applications such as virtual office, white board or file sharing;

– gateways: they are optional, but they are nonetheless essential for ensuring the interconnection of an H.323 network with another telephonic network. They use a group of protocols dedicated to these functions, such as H.320 for the interconnection with an ISDN network, H.321 and H.310 for ATM, H.322 for LANs with QoS management, H.324 for RTC-circuit commutation network. They also ensure cohesion of different media by call control, Q.931 signal correspondence (of the RNIS) to H.225, multiplexing, suppression of echoes and silences, and audio transcoding (compression, decompression);

– gatekeepers: they are also optional and essentially their function is to translate the addressing, thus enabling the routing of flows between the communicators. In other words, they are charged with completing the correspondence of an H.323 identifier (in the form of an email address: name@domain or in a numeric format like a traditional telephone number) into an IP address sent from the terminal. As a concentration point, gatekeepers also enable indication message management, which is useful in the control of calls. They implement AAA (Authorization, Authentication, Accounting) functions for journaling, billing, authentication and access authorization of resources, the counting of calls and, eventually, additional complementary services such as bandwidth limitation, resource allocation, directory consultation and gateway locating;

– multipoint control units (MCUs): these entities are useful for implementing conferences with more than two H.323 terminals. Each user must connect to the MCU, which links the users. These links are unicast when communicating between two communicators and are multicast if there are more than two participants.

The H.323 standard offers a group of associated protocols, which form a complete multimedia communications environment for IP networks. It has imposed itself on the majority of manufacturers as it is still the most commonly used. It was notably chosen for the famous video conferencing software NetMeeting, which now threatens to disappear…

19.3.2. *SIP*

The SIP (Session Initiation Protocol) tends little by little to substitute the use of H.323 to its profile. Work on SIP started in 1995 and it was standardized in February 1999 by the IETF (Internet Engineering Task Force) [RFC 2543 and then RFC 3261]. Its functionalities are quite large because SIP enables communications with more than two users, call forwarding, unique identification of the user independently of his location, identification of the caller, user mobility, physical location masking. Moreover, the SIP is adapted to flow variations, which is a precious asset for Internet communications. And especially, it was considered for the definition and implementation of new services, with the help of XML language.

In practice, the SIP functions according to a client/server model (request/response) similar to the one that uses the HTTP protocol for the Web. Its integration with the Web is much simpler because it only requires a click on an SIP address to initiate a communication.

The format of this address is similar to an email address: it is a URL (Uniform Resource Locators) of type "SIP:name@domain", where the name is a telephone number (or a name with an optional password) and where domain is an IP address (or a domain name with a complementary DNS).

In addition, the signaling is done externally, on a channel dedicated to signaling, separate from data flows that concern voice transport.

The implementation of SIP puts into place four types of entities:

– user agent: this is the terminal on which the application used by he subscriber for telephonic communications is run. Generally, this application is dissociated into two software sub-parts: one client part that emits SIP requests (User Agent Client) and one server part that receives SIP requests (User Agent Server). An SIP communication is established in point-to-point between a user agent server (the called) and a user agent client (the caller). If the caller knows with certainty the location (IP address) of the call receiver, the communication between them is directly possible without going through the other servers;

– proxy server: this is a server that enables logical linking between the caller's terminal and the receiver's terminal, by interpreting and eventually modifying received requests before relaying them to the communicators. All terminals are linked to a proxy server that can change during a single communication if the terminal is mobile. It therefore ensures relay functions for the user agent client and for the user agent server. Contrary to redirection servers, proxy servers do not give the subscriber's location to the requester but relay SIP requests to the subscriber. Of course, it does not forward audio flows, but only indication information;

– redirect server: it sends an address to a subscriber location. It is therefore an alternative to the proxy server because it provides the same locating function. However, it does not enable the initiation by its intermediary, leaving this task to the caller;

– SIP registrar: indispensable for using subscriber location services, the SIP registrar enables registration on the location server;

– location server: stores the correspondence between an SIP address and an IP address by storing the user profile in a database (login, password, rights and especially location). The inventory of this information is carried out by the SIP registrar.

Other equipment can be added to ensure complementary functions such as automatic answer, call logs, billing and access control.

In practice, the architecture is implemented in the following way: initially, all connected terminals must identify themselves to the SIP registrar that contacts the location server to memorize its location information. When a user wants to contact someone, either a redirect server or a proxy server can be used to pinpoint that person's location.

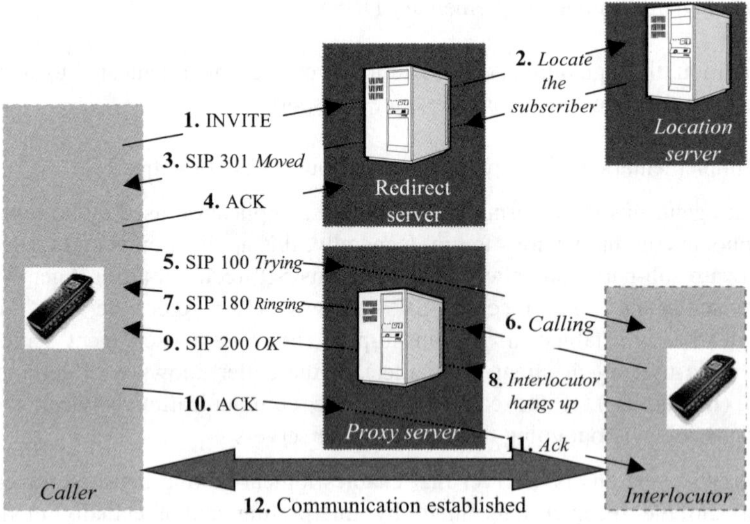

Figure 19.1. *SIP session via a redirect server*

In the first case, the caller must first send an invitation request (INVITE) to the redirect server. The server then sends the request to the location server. Once the response is received, the caller can directly contact his interlocutor's proxy server by sending it a request to establish communication that will be relayed to the interlocutor. This is illustrated in Figure 19.1. The caller here is trying to connect to a user whose location has changed. Message 301 (Moved Permanently) alerted the sender of the change by providing the new valid location.

The other way of establishing communication is by going through a proxy server that interrogates the location server to determine the location of the person being called. Instead of informing the caller, the proxy server contacts the receiver directly. This is shown in Figure 19.2, where the person called has activated a call forward service from his old address to his new address. With a proxy or with a direct connection, the person called decides to accept or decline the call. If the communication is accepted, the terminal agent receives the acceptance for his request via message 200 (OK) that he discharges at once (with an ACK). Communication can then begin.

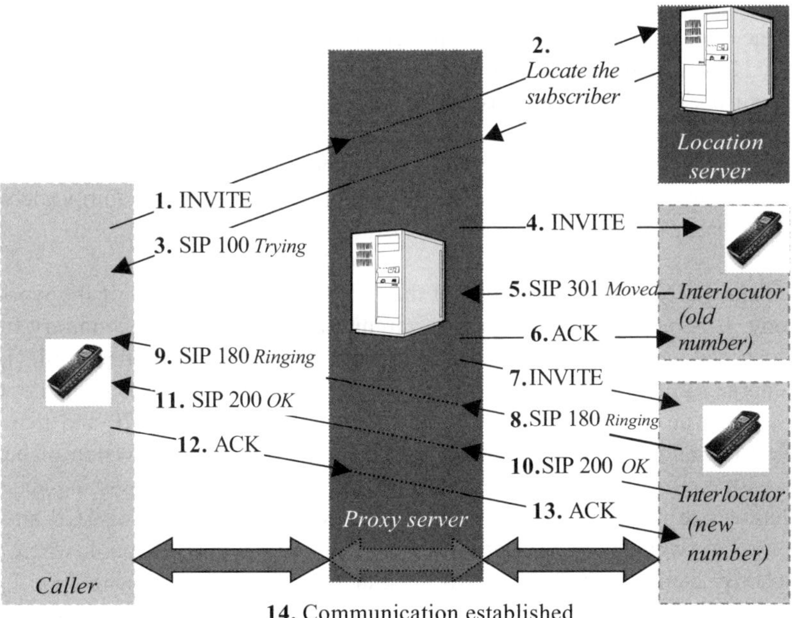

Figure 19.2. *SIP session via a proxy server*

In addition, the SIP message format corresponds to text messages that have a header (which include routing information and session establishment information) and a body (which includes session parameters). The session must be defined using a session language such as SDP (Session Description Protocol) [RFC 2237] which is part of the formulated recommendations in SIP standardization. The SDP protocol makes it possible to describe multimedia session characteristics by defining, notably, the type of exchanged flows (text data, voice, videos, etc.), the coding used, the port and the protocols used, as well as various activity information (session name, date and time of the connection, etc.).

19.3.3. *Comparison between H.323 and SIP*

The reliability of H.323 and SIPs no longer needs to be shown. However, both address the same problem of implementing an infrastructure for managing multimedia streams. Also, the two protocols are concurrent.

The H.323 protocol, being older than the SIP, benefits from its popularity and maturity derived from years of supplementary experimentation. It is also supported by all the operating systems, whereas SIP is only managed by recent operation systems. However, the management of protocol H.323 is complex: as proof, it is written out on 736 pages, whereas the SIP only takes up 153 pages and comprises only six methods. Moreover, H.323 proposes several options without requiring implementation, which risks generating incompatibilities between manufacturers' implementations. Furthermore, H.323 does not enable interoperability between H.323 gateways.

Fundamentally, H.323 is a protocol that follows the philosophy of the world of telephony. It is part of the continuity of circuit switching networks, contrary to the SIP which was initially included in packet switching network problems. SIP takes better charge of the variations in flow and offers capabilities to adapt to the network which are not present in H.323. It is integrated into the world of IP and makes it possible to establish a communication with a simple click. The division of header fields and the message body makes it possible to accelerate the processing speed in the intermediate routers. It is adapted to both the transport protocol TCP and the UDP protocol (whereas H.323 does not work with UDP until version 4, where it is optional). By being more open, it also facilitates the simple implementation of new services. Consequently, SIP is more efficient than H.323 in satisfying constraints in networks such as the Internet and its growing success testifies to that.

19.3.4. *MGCP*

The fruit of common efforts from the IETF and the ITU-T, the MGCP (Media Gateway to Media Controller Protocol) [RFC 2805] is also part of significant developments in network signaling. It is the integration of two protocols: SGCP (Simple Gateway Control Protocol) and IDCP (Internet Device Control Protocol). SGCP is a control protocol between a gateway and an external element called Media Gateway Controller or *Call Agent*. It is on the latter that all the intelligence is deferred: the gateway will execute the requests sent by the *Call Agents*. The IDCP protocol enables the transport of the signalization and the connection control of the Media Gateway. The MGCP is thus used for signaling between a media gateway controller and a group of gateways spread out in an IP network. It enables the VoIP gateways to interface with the SS7 signaling (of the traditional public telephone network) and therefore enables a traditional telephony user to forward his communications on the Internet via gateways that will increase the interoperability of the networks. MEGACO/H.248 may complement SIP: SIP ensures session control, whereas MGCP ensures gateway control. However, MGCP standardization has given way to MEGACO/H.248 (MEdia GAteway COntrol protocol), a standard specified by the IETF and the UIT, but which does not derive directly from MGCP. Consequently, its elaboration rests solely on the functioning model defined by MGCP.

Other IETF workgroups are also working on improving existing architectures, such as IPTEL (IP TELephony) [IPT], which focuses on signaling transport, or SIGRAN (Signal Transport) [SIG] that focuses on signaling transport between the RTC telephony environment and the IP network environment, and several others (MIDCOM, NSIS, etc.).

19.3.5. *Flow transport*

Once the signaling is established, the network is able to manage the audio exchanges in a conversation. The voice's analog signal is converted to a digital signal, which is then encoded and eventually compressed by a codec (coder/decoder). There are, in fact, techniques that compress voice in the order of 8 to 10, without a significant change in audio quality. To this end, a group of recommendations concerning the sampling and compression for pulse modulation and coding of vocal frequencies have been formulated. The G.711 coding or Pulse Code Modulation (PCM) standardized in 1971 is the most used for voice. Proved and performing, it is based on temporal channel multiplexing. Voice is sampled over 8 bits at a frequency of 8 kHz, for a flow of 64 Kbps. Other codes use compression and less frequent sampling to reduce the useful flow, for example, G.726 (ADPCM) at 32 Kbps, G.728 (LD-CELP) at 16 Kbps, G.729 (LD-ACELP) at 8 Kbps, G.723.1

(ACELP) at 5.3 Kbps. The coding choice is made according to transmission time criteria (the shorter the time, the stronger the interactivity) and bandwidth (the larger the bandwidth, the weaker the signal). For comparison, in RTC telephony, a 64 Kbps channel (G.711 coding) is reserved in each direction of the communication. However, the fact that a single user speaks more often and that a conversation involves a number of silences during which the channels remain busy provokes a waste of resources.

The resulting compressed flow is packetized and sent to the network via a transport protocol that must be adapted to real-time flow constraints. Concerning the voice, the use of the UDP (User Datagram Protocol) protocol is preferred to the use of the TCP (Transport Control Protocol) protocol. In fact, retransmission mechanisms implemented in TCP and which are a major asset for the reliability of a communication become a heavy weight in a real-time environment: it is preferable not to retransmit a lost packet because it concerns an old audio signal that is also out of sequence, which risks disturbing the audio message instead of improving it. In fact, the delay generated by its retransmission and its restitution at the receiving end will not only render it obsolete, but also risks compromising the intelligibility of the conversation.

Moreover, audio flow transport is often completed by using both the RTP (Real-time Protocol) [RFC 1899] and the RTCP (Real-time Control Protocol) [RFC 1890]. These two protocols, adapted to unicast and multicast networks whose real-time constraint is present, offer an end-to-end application service. For this, the RTP identifies the content and assigns a sequence number to each packet, making it possible to store packets in memory in order to ensure a continuous sound flow for the interlocutors, which reduces the jitter. As for the RTCP, it is linked to RTP and makes it possible to periodically return control information on the quality and reliability of the transport. With RTCP, the RTP is endowed with a capacity to adapt to network fluctuations.

Although the use of the UDP is not indispensable to the proper functioning of RTP and RTCP, this protocol combination forms a homogenous group adapted to flows whose respect of temporality is essential. However, neither RTP nor RTCP ensure any function of guaranteed QoS or of reliability.

19.4. Adapting wireless QoS for voice

As standards evolve, flows increase. However, at the same time, the most current applications become hungrier for bandwidth, so much so that bandwidth requirements are increasing progressively as flows increase. It is becoming necessary to adapt a management policy appropriate for network transiting flows.

In IPv4 Internet, which is used today, the QoS is of "Best Effort" type. It is an indirect way of saying that QoS is not managed because all flows are treated in the same way. For voice processing, it is an insufficient technique. Flows must be differentiated according to their importance. In fact, it is not enough to implement a mechanism on a single layer, but each process must be optimized to be best adapted to the specificities of a synchronous application in real-time, such as voice. There are two flow management classifications: in end-to-end routing first, where we can reserve bandwidth to guarantee a flow or to differentiate the flows and then in MAC, where we can adapt the support access according to the priority of flows to be transmitted.

19.4.1. *Managing end-to-end QoS*

19.4.1.1. *IntServ/RSVP model*

The IntServ (Integrated Services) model was defined by the IETF [RFC 1633]. Its goal is to offer a guaranteed end-to-end QoS to applications. The type of service and the amount of resources to be allocated is thus broadcast along the path taken by data flows. IntServ is often coupled with the RSVP (Resource reSerVation Protocol) protocol, although this association is not indispensable. The latter takes care of reserving resources useful to an application in the routers. Applications needing bandwidth that is not available in the routers during the request are rejected. However, the management of RSVP is complex. The group of intermediate routers between the emitter and the receiver must in fact make it possible to dynamically allocate resources to the flows. However, currently, few operators use the RSVP protocol.

At the same time, if a certain number of streams are privileged at the expense of other streams, it becomes necessary to dynamically dissociate the streams in order to offer them the expected QoS by the application used. This filtering can be done in a basic manner to the lower layers, in transport, notably port specification. However, the concordance of ports with an application is less and less pertinent, more so as a port can blindly forward any data flow. If no standard protocol of application recognition exists yet, proprietary mechanisms such as NBAR (Network Based Application Recognition) from Cisco can be useful for interpreting application data. The identification of flows essentially addresses a security need because it makes it possible to ensure the legitimacy of assigned priorities. However, it imposes a supplementary processing time on the equipment reserved for this task.

19.4.1.2. *DiffServ model*

Defined by the IETF [RFC 2475], DiffServ (Service Differentiation) brings a more supple solution for managing QoS than IntServ. For each service class, the DiffServ model provides a differentiation of services according to a contract called the SLA (Service Level Agreement) which is specified by the emitter and is applied to a group of nodes that put in place a common management policy. This contract combines parameters that define the QoS required and the types of flows concerned. For each flow, it notably characterizes the bandwidth to guarantee, the time frames to maintain, the size of buffers and the behavior to adopt if the contract cannot be accepted.

Contrary to IntServ, where each node has the same function, DiffServ's architecture has two types of components: border nodes and core nodes. Border nodes ensure the classification and marking of packets according to the traffic class that characterizes them. They exploit ToS fields for IPv4 and traffic class for IPv6, which are renamed in the DiffServ model as DSCP (Differentiated Service Code Point). These nodes can in turn ensure metrology functions in order to verify that flows conform to the specified SLA contract. If flows are in excess according to their SLA specifications, packets can either be rejected or transmitted with a downgraded priority, according to the definition of the SLA. The core nodes ensure a more static management of the QoS: packets are simply stored in queues according to their marking. These queues correspond to the various managed classes and possess their own characteristics (queue processing time, queue size, etc.).

Thus, DiffServ makes it possible to essentially center the implementation complexity in the border nodes, which facilitates its deployment and accelerates processes in the core nodes. Nevertheless, it does not offer the same level of granulation that IntServ does because it aggregates flows by categories. It also imposes the preliminary setting of an SLA in all equipment of the considered domain, which imposes knowing the precise needs of the applications used.

Other protocols are used to favor routing of data in the network. The MPLS (Multi Protocol Label Switching) protocol, which is standardized by the IETF (Internet Engineering Task Force), makes it possible to accelerate the processing of packets in the routers by combining the advanced routing functionalities with the speed of switching. Also, the QOSPF (QoS Path First) protocol proposes a routing algorithm that seeks to maintain stable paths rather than shorter or faster paths. The position taken supposes in fact that a faster or shorter path is inevitably not profitable in the long term and that it is preferable to choose a path that is perhaps longer but whose topology and network state offer a better stability.

19.4.2. *Managing QoS in the MAC layer*

The role of the MAC (Medium Access Control) layer is to enable many stations to access a shared support. In the case of wireless networks, the air interface constitutes the support of Hertzian waves and makes it possible to broadcast the data. The most widespread standard in WLAN networks is without question the IEEE 802.11, which is not very expensive, quite fast and whose evolution is particularly important in recent years, since it has emerged in the private as well as in the business sectors. Although it is specific to wireless networks, the MAC layer defined by the IEEE 802.11 is very close to that of Ethernet networks (802.3) and it is on the latter that we will elaborate.

19.4.2.1. *Basic mechanisms*

19.4.2.1.1. IFS timers

The 802.11 standard defines different timers or IFS (Inter-Frame Space) that enable the establishment of priority mechanisms which are differentiated according to the type of data transmitted. Thus, the shorter the timer, the less the station will wait before being able to emit and therefore the higher priority its data will have with regard to a more significant timer. There are four IFS timers:

– SIFS (Short IFS) that assigns a high priority, essentially for CTS and PCF mode interrogation (by *polling*) discharge messages;

– PIFS (PCF IFS) which is IFS used in PCF mode and that corresponds to a medium priority. It enables access points to have a higher priority access time with regard to stations and is adapted to real-time services;

– DIFS (DFC IFS or Distributed IFS) which is used in DCF mode and corresponds to a low priority of best effort type. It is generally used for the transmission of data without time constraints;

– EIFS (Extended IFS) is the longest of the IFS. It is used in DCF mode when an error is detected.

The 802.11 standard proposes two elementary mechanisms for MAC layer: the DCF (Distributed Coordination Function) method, based on the contention, and the optional PCF (Point Coordination Function) method, without contention (no possible collision).

19.4.2.1.2. DCF method

DFC (Distributed Coordination Function) is the default access method in the 802.11 standard. It was conceived to take care of asynchronous data (without priority). Its functioning relies on the CSMA/CA (Carrier Sense Multiple Access with Collision Avoidance) protocol, which is a multiple access method of carrier and collision avoidance detection. This works according to three procedures: listening to the carrier, the *backoff* algorithm and the use of positive acknowledgments.

Listening to the support is the initial phase before all transmission. It is carried out over two distinct layers of the OSI model. Firstly, on the physical level, the PCS (Physical Carrier Sense) mechanism is used. It makes it possible to detect the activity in other stations by analyzing the broadcast flows and by transiting on the Hertzian support. Then, on the link level, in the MAC sub-layer, it is the VCS (Virtual Carrier Sense) mechanism that determines the support activity. For this, each station manages a NAV (Network Allocation Vector) timer indicating the duration of the support occupation and therefore the time that the stations will have to wait until it is their return to emit. The duration of this timer is fixed by exploiting the contents of special packets called RTS (Ready To Send) and CTS (Clear To Send). These data blocks carry out an explicit reservation of support. Thus, the RTS message is broadcast on the initiative of the station that wishes to emit. It comprises the indication of the source station, of the destination station and of the duration of the anticipated transmission (including the duration for acknowledgment of messages from the receiver). If the recipient accepts the communication, it then returns a CTS message as a response to the RTS request. As all stations that want to emit are listening to the support activity, they also receive the RTS and/or the CTS. The duration field of these blocks of data enables them to update their NAV timer. In addition, the RTS/CTS mechanism makes it possible to mitigate the problem of a hidden station. This problem evokes the case where two stations, A and B, are too far from each other to detect each other and attempt to communicate with the same intermediary station I. In this case, as station B does not hear when station A sends a block of data to station I, it risks emitting a block of data to station I that will collide with the block of data from station A. With the RTS/CTS mechanism, station B will also not hear the RTS request made by A, but it will hear the CTS response sent by I. Thus, the previous request and especially the response of the recipient make it possible to ensure that a single station will then emit.

This VCS mechanism remains optional, however, and can be systematically applied to the request, or never. In general, its use is however recommended when the blocks of data are large. In fact, in case of colliding blocks of data, not only is the support debilitated throughout the duration of its transmission, which constitutes a waste of bandwidth, but its retransmission becomes costly. Inversely, adding RTS/CTS blocks of data can prove less profitable than the loss of data to be transmitted. A compromise is therefore necessary. The loss (by collision, interference, etc.) of a large block of data being costly in terms of bandwidth, a large block is fragmented (in units called MSDU for MAC Service Data Unit) to restrict eventual retransmissions. However, all the fragments will be emitted sequentially (without new support requests) and discharged by the recipient who is reassembling the fragments. The support is therefore not liberated until after the transmission of the entire block of data.

The CSMA/CA protocol starts by listening to the carrier: a station wishing to emit a block of data must verify the availability of support by listening to it for an amount of time equal to DIFS. If this condition is confirmed, the station can then emit. If not, it must continue listening until the support becomes available. In order to reduce the probability of simultaneous emission by several stations, it must then engage a procedure based on the *backoff* algorithm. In this algorithm, the station delays its transmission for a period of time, called *backoff time*, which is chosen randomly within the values 0 and CW. CW is the Contention Window within two predefined limits CWmin and CWmax. If the availability of support is not confirmed during this period of time, the station is authorized to emit. On the contrary, if during this period of time another station begins to emit, the timer is suspended as long as the channel remains busy and a new *backoff time* value is chosen. To minimize the risk of collision even more, the value of the CW variable is incremented (up to the maximum value CWmax) at each attempt, according to the following equation: CWnew = (CWold + 1) x FP – 1, where FP is a persistence factor fixed at 2 in the *backoff* algorithm and which determines the relative growth between two retransmissions.

Finally, the positive acknowledgment makes it possible to confirm the receipt of blocks of data. In Hertzian systems, the transmission of data masks the reception capacity of a terminal, which means that a station cannot emit and receive at the same time. Consequently, collisions cannot be detected during a transmission (following the example of collision detection mechanisms traditionally implemented in telegraphic networks, such as the CSMA/Collision Detection protocol) but only after the complete broadcasting of the block of data. To reduce the number of collisions and increase network performance, the collision prevention mechanism is used. It consists of an explicit acknowledgment by the receiver without which the emitter would be constrained to reemit its data once again, after a fixed waiting time. Thus, upon receipt of a block of data, the destination station confirms its integrity (by the CRC field or Cyclic Redundancy Check Code) and if the data is correct, it emits an acknowledgment (after an SIFS delay). If the emitting station does not receive the acknowledgment of its block of data, it considers that a collision or loss occurs and retransmits its data. After a predefined number of attempts without acknowledgment, the support is considered unstable and the emission is abandoned.

Founded on CSMA/CA, which is a probabilistic method, the DFC protocol is efficient in a light- or medium-loaded network but behaves less efficiently when the network load is heavier. However, especially the QoS with DCF is not managed because by nature the CSMA/CA is a protocol with contention and is equitable, which does not make it possible to guarantee duration of transmission of packets.

19.4.2.1.3. PFC method

PCF (Point Co-ordination Function) is optional and is an access method without contention, proposed by the 802.11 group. It is used solely in a network configuration based on an infrastructure (at the heart of a cell called a BSS or Basic Service Set) and is optimized for regularly spaced transmissions (isochrones), which is particularly useful for real-time streams such as voice.

In this mechanism, the stations cannot emit or receive data unless they have been invited to do so by a special station called the PC (Point Coordinator). In practice, the PC is implemented at an access point. It is the only one to have control on emission rights. Its role is to arbitrate support access by interrogating the terminals to know if they want to emit and to give them access. This is the polling method.

In practice, the time is divided into intervals composed of a CP (Contention Period) and a CFP (Contention-free Period). The CP simply uses the DCF access method. The CFP introduces a PIFS timer, which is smaller than the DIFS. During this CFP period, the PC sends blocks of data (called CF-Down) to the destination stations after a PIFS delay. The stations send their blocks of data (called CF-Up) after an SIFS delay. The contention-free period ends with a particular CF-End (Contention-Free Period End) block of data which is preceded by a PIFS delay.

The PCF method foresees the possibility of introducing QoS in a wireless network, but its constraints rapidly become limited. With each examination of the stations by the coordinator, only one block of data can be sent. Moreover, the examination of the stations is not easily scalable when taking into account the time systematically offered to all the stations, even if several of them have nothing to emit. In addition, in the transmitted blocks of data, nothing indicates their priority according to others (so the coordinator cannot favor one traffic type in the absence of defined profile type). No time delay or jitter is possible. In addition, the PCF method is not efficient in the situation of a hidden station. In fact, a station on the border of a BSS is controlled by the PC of its BSS. Moreover, if its PC authorizes it to emit, nothing indicates that its transmission will not collide with a station from a neighboring BSS. However, the heaviest brake in the use of PCF is without a doubt the fact that this method only functions in infrastructure mode although it is, practically, rarely implemented in access points.

19.4.2.2. *The 802.11e standard*

Neither the DCF nor the PFC method makes it possible to manage the QoS. In a work document (*Draft 4.1*) published in November 2002, the IEEE (Institute of Electrical and Electronic Engineering) group has implemented a new standard, called 802.11e [IEE 11] with the goal of specifying MAC-level QoS management. The ambition of the 802.11e standard is to transmit multimedia data, with not only a

QoS management appropriate to flows but also with security functionalities (by encryption) and mutual authentication (by encryption key distribution).

To manage the QoS, the basic mechanism of these methods relies on the principle of service differentiation in the access control: the stations that manage their flow priorities and implement timers. Thus, a real-time flow will have a shorter emission timer than a flow that is less sensitive to delays, which will favor it for emission. Two access control support mechanisms have been put forward on this principle: the EDCF and the HCF methods.

19.4.2.2.1. The EDCF method

The EDCF (Enhanced Distributed Co-ordination Function) mechanism is an improvement on DCF for managing QoS. It is a contention mechanism because it takes up the CSMA/CA algorithm. It lies on a service differentiation in the flows that are categorized by their priority according to eight classes of traffic, from 7 (the highest priority) to 0 (the lowest priority). The value of this priority is given according to two criteria: the UP (User Priority), which is statically assigned to a user for the duration of the communication and the TSPEC (Traffic SPECification) priority, which is dynamically assigned to the application used. For example, the priority of a flow associated to voice will be higher than that associated to the sending of an email. A TID (Traffic IDentifier) field in the 802.11 header mentions the traffic class to which the block of data belongs. These priorities were indexed in four categories from class zero to three: 0 for best effort services, corresponding to the absence of priority of associated flows, 1 for non-interactive video flows (continuous broadcast type), 2 for interactive video flows (videoconference type) and 3 for audio flows (notably telephony type).

Each of these classes is managed by queues that have a group of parameters defining the way that access to media will be carried out. There are essentially four characterizing parameters for a queue. First, there is the AIFS (Arbitration Interframe Space) variable which defines a timer before emission. It plays a role analog to the DIFS used in the DFS function, but its value here is arbitrarily fixed for each class, which favors some classes over others. Its minimum value is DIFS, as shown in the following diagram. Queues have higher priority when their AIFS value is weak.

The two following variables are defined by CWmin and CWmax, which characterize a contention window's minimum and maximum size. The chosen *backoff time* in the *backoff* algorithm will be between these two thresholds. In case of collision, the contention window will grow according to the *backoff* algorithm by using this same formula: CWnew = (CWold + 1) x FP[i] – 1, but the persistence factor PF is no longer fixed to 2 but varies according to the priority of queue i. The weaker these two threshold values, the shorter the *backoff time* and, consequently, the more privileged the station. Finally, the fourth parameter is the allowed transmission interval called the TXOP (Transmission Opportunity), which defines for each station an ulterior right of emission. This variable defines in fact the precise moment when a station is authorized to emit, as well as the maximum duration allotted for the emission. With this concept, the stations are no longer equivalent, even during their data transmission, some having a longer allocated duration than others. In addition, the contention on wireless media is effective at the access point that distributes the values of the TXOP variables for each station. Each queue disposes of a quadruplet of values, which was introduced by default in each station or imposed by the access point during polling, according to the implementation choices.

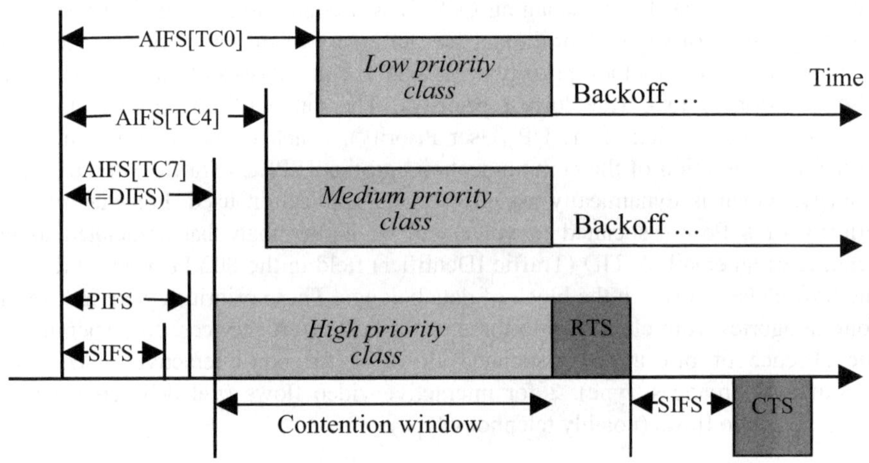

Figure 19.3. *AIFS in the EDCF method*

The EDCF method is only usable in periods with contention, whereas the following method is exploitable in contention periods and in contention-free periods.

19.4.2.2.2. The HFC or EPCF method

The HCF (Hybrid Co-ordination Function) method, also called EPCF (Enhanced PCF), is an improvement of PCF that offers a determinist management of the support access without collision. The mechanism is in fact centralized around a coordinator called the HC (Hybrid Coordinator) which allocates support for each station.

As for PCF, the time is divided into superframes, each composed of a period of contention and a contention-free period. During the contention period, the EDCF method is used. During the contention-free period, the coordinating station interrogates the stations by polling. The transmission of data is carried out by one of two models: either the coordinator emits by transmitting received blocks of data to the stations concerned, or the coordinator lets the stations emit, by distributing to them TXOP intervals associated to the priority they requested during polling. In other words, the coordinator controls support access and allocates the beginning and the duration of the emission of stations according to the priority of their flows. Moreover, the coordinator can access the support in an SIFS time (no matter the type of data it emits), which is less than the PIFS time used in PCF, which confers to it a higher priority than any other station in its BSS. The functional scenario is illustrated in the following figure.

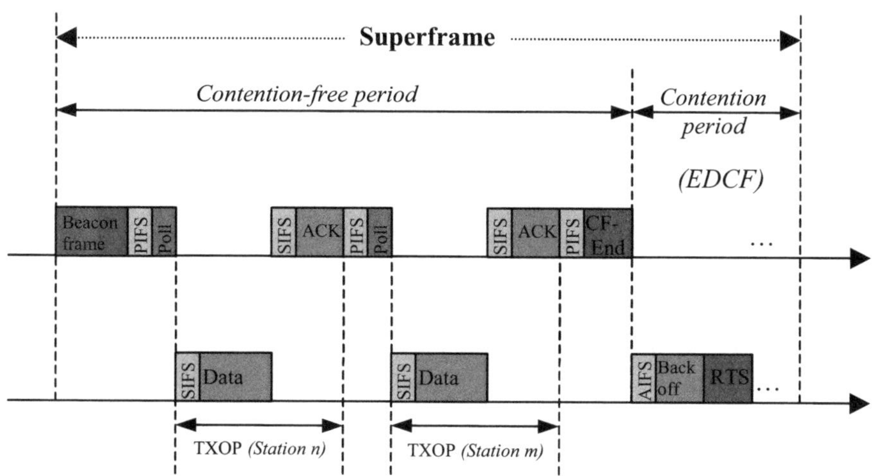

Figure 19.4. *HCF method*

19.4.2.2.3. Complements to 802.11e

Two supplementary mechanisms complement the EDCF and HCF methods. They are the direct link protocol and the burst acknowledgements method.

The direct link protocol (DLP) is a mechanism added to the 802.11e standard in September 2002. It consists of sending blocks of data from one station to another without passing through the access point. It is therefore an additional possibility to connect two interlocutors who are in the same BSS without having to pass in a particular mode (of *ad hoc* type). The transmission time is greatly reduced and the efficiency of use of the bandwidth is optimized in the network. The burst acknowledgements method is an optional technique in the 802.11e standard which improves the efficiency of the channel by aggregating several acknowledgements into a single one. Thus, the emitter is not constrained to be put on standby for acknowledgement receipt at each block of data it sends, but it can send several blocks of data in a row and wait a global acknowledgement of its blocks of data. By reducing the number of signaling messages, this method aims to accelerate communication. However, if the packet losses are frequent in the link between the interlocutors, this mechanism will produce the inverse effect of slowing down, because before notifying the non-receipt of packets to the emitter, the receiver will wait, not only to receive a single packet, but the time it takes to receive a group of packets.

Having had great commercial success with the WPA (Wireless Protected Access) safety requirement, the IEEE 802.11 group is currently developing a standard dedicated to MAC management of QoS in a wireless environment. Called WME (Wireless Multimedia Enhancement), it is destined to commercial availability. It will implement a sub-group, to be determined, of the mechanisms specified in the 802.11e standard.

19.5. Conclusion

Undeniably, for operators as well as for companies that want to develop it, going to VoIP has a cost that is hardly negligible, especially in expanding coverage to wireless networks. However, this entry ticket can prove to be profitable and beneficial if the technology is as advanced as what it is substituting. Simplicity, interoperability, security, mobility and especially QoS must be the basic components for attracting and convincing companies and users to upgrade. These are the current challenges facing VoIP in wireless networks.

Without a doubt, the challenge is also to agree on a single standard, which will be much more easily implemented that the sum of all the proposed standards. The difficulty is keeping the promises put forward in these technologies. Once the standard is found, the interoperability between equipment in a large scale will require that the manufacturers respect the same implementation standards.

However, while VoIP is not sufficiently deployed to be competitive, traditional RTC telephony operators are waging a price war. Also, it is probable in the future that the financial advantages of VoIP solutions will be less noticeable, especially because the entrance ticket for VoIP remains of consequence. Once aligned regarding tariffs, VoIP will undoubtedly be associated with the richness of its added value services. Very often, technologies impose themselves by the intermediary of an additional application, a "*killer application*" like SMS (Small Message Service) for GSM or mail which has been considered emblematic for the Internet for a long time. New services are still to be invented…

By noting the interest in Wi-Fi today, one can imagine that wireless will become the differentiating asset in the emergence of VoIP technologies. And in case of any doubt, the convergence of data to an "all-IP" model for voice, video and data all at once will be proof in favor of future increases in wireless VoIP.

19.6. Bibliography

[H.323] H.323, www.itu.int/itudoc/itu-t/rec/index.html

[IEE 11] Workgroup IEEE 802.11, http://grouper.ieee.org/groups/802/11/index.html

[IPT] IPTEL (IP Telephony), workgroup IETF, http://www.ietf.org/html.charters/iptel-charter.html

[RFC 1633] RFC 1633, IntServ (Intergrated Services), http://www.ietf.org/rfc/rfc1633.txt

[RFC 1890] RFC 1890, RTCP (Real-time Control Protocol), www.ietf.org/rfc/rfc1890.txt

[RFC 1899] RFC 1899, RTP (Real-time Protocol, defined in RFC), www.ietf.org/rfc/rfc1899.txt

[RFC 2237] RFC 2237, SDP (Session Description Protocol), defined by the MMUSIC (Multiparty Multimedia Session Control) group, www.ietf.org/rfc/rfc2237.txt

[RFC 2475] RFC 2475, DiffServ (Differentiation of Service), http://www.ietf.org/rfc/rfc2475.txt

[RFC 2543] RFC 2543, SIP (Session Initiation Protocol), www.ietf.org/rfc/rfc2543.txt

[RFC 3261] RFC 2543, SIP (Session Initiation Protocol), www.ietf.org/rfc/rfc3261.txt

[RFC 2805] RFC 2805, MGCP (MEdia GAteway COntrol protocol), www.ietf.org/rfc/rfc2805.txt

[SIG] SIGTRANS (Signaling Transport), workgroup IETF, http://www.ietf.org/html.charters/sigtran-charter.html

Part 3

The Next Generation of IP Networks

Chapter 20

Pervasive Networks

20.1. Introduction

In the last few years, new terminologies such as pervasive computing, ubiquitous computing, ambient intelligence, wearable computing, context awareness and always best served (ABS) have emerged. We do not know what terms to use when we talk about the famous third age of technology: the age where computing is such an important part of our lives. The goal of this ever present information technology, however, is always the same in its design which is to make our environment easier by creating, in a manner which is transparent to the user, a physical relationship with the information technology and telecommunications world. The environment becomes more conscious, adaptive and attentive to the user.

This chapter introduces ambient intelligence and its problems, with the emphasis on a very important aspect communication networks. Ambient networks are actually an environment which enables users to easily access different rich and varied services throughout one or more access networks that might be wireless. These networks offer a global connectivity. They are completely transparent and only show the necessary functions to their users.

The chapter is organized as follows: section 20.2 introduces ambient intelligence and its specific problems; section 20.3 presents ambient networks and their different components such as wireless networks (local networks – WLAN, personal networks – WPAN, etc.), ad hoc and sensor networks and the integration of these different

---

Chapter written by Sidi-Mohammed SENOUCI.

environments, as well as an explanation of the available services within these networks; finally section 20.4 concludes the chapter.

20.2. Ambient intelligence

Ambient intelligence refers to a new paradigm in information technology in which the processing of information is spread between all the objects within the network and the environment of customers who can then be assisted by an ambient intelligence of that environment. Ambient intelligence is a new paradigm whose goal is to create an intelligent daily work area that is user-friendly and that is integrated within the walls of our homes, in our offices and just about anywhere else. An example of an ambient living room and of a non-ambient living room is shown in Figure 20.1.

a)

b)

Figure 20.1. *(a) Non-ambient living room (b) ambient living room (source: HomeLab Philips)*

In order to offer this functionality to users, a study has to be made in the different information technology areas. As shown in Figure 20.2, ambient intelligence is based on three technologies: ubiquitous computing[1], which consists of integrating microprocessors in everyday objects; ubiquitous communication, which enables these same objects to communicate between themselves and with the user through a wireless network and man-machine intelligent interfaces that enable users to control and interact with these objects in the most natural way possible.

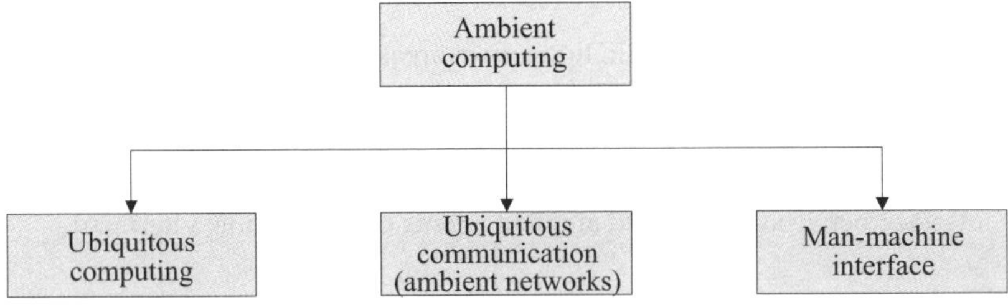

Figure 20.2. *Ambient computing components*

Ambient intelligence promises a world filled with small electronic elements, inexpensive, interconnected, autonomous, sensitive to context and with a certain degree of intelligence, all this in our own everyday environment be it personal or professional (in our cars, buildings, trees, the street, etc.). Their use would be multiple; from prevention (fires, accidents) to assistance (guidance, remote control) and comfort (see the example in Figure 20.1). One of their main qualities would be their total transparency, they would be present but completely invisible and interaction with them should also be transparent.

ISTAG (IST Advisory Group) [IST] published a document in 2001 entitled *Scenarios for Ambient Intelligence in 2010*. It groups together four scenarios, showing what an ambient world could be: the first scenario *Maria, road warrior* talks about ease of travel with the help of ambient intelligence. It includes avoiding customs formalities, handling customs without us, using a rental car without the need for keys, which also possesses an automatic guidance system and the hotel room which is personalized (temperature, music and lighting). The other three scenarios are entitled *Dimitrios and the digital me (D-Me)*, *Annette and Solomon in the ambient for social learning* and *Carmen*: *traffic, sustainability and commerce*.

---

1 *Ubiquitous computing* by Marquez Weiser: it is the exact opposite of virtual reality. Whereas virtual reality immerses people inside a world generated by the computer, ubiquitous computing forces the computing devices to integrate in people's everyday lives. Virtual reality relies mostly on computing power. Ubiquitous computing is the result of a difficult integration of human factors, information technology and social sciences.

20.2.1. *Problems related to research in ambient intelligence*

Ambient intelligence is not really a subject for research per se. It is in the middle of multiple other research areas, which are not usually cooperative. Since ambient intelligence is related to so many different areas, large organizations are the only ones able to launch research studies on the subject. Therefore, consortiums have recently been put in place between laboratories and large companies in order to promote research in this domain (ITEA Ambience project [AMB], for example).

These domains are organized in three main areas [BAS 03]:

– internal ambient agents (ubiquitous computing);

– interaction between digital ambient agents (ambient networks);

– interface between human and ambient agents (man-machine interface).

20.2.1.1. *Internal ambient agents*

An ambient system is usually composed of small multiple interconnected systems. They must be small, inexpensive and as simple as possible. These aspects which are particular to ambient systems are mostly processed at hardware level and concern the following research areas: machine architecture (electronics, energetics, nanotechnologies, etc.) and software agents in general (AI, logic, etc.). Intelligent and communicating objects must integrate perfectly and adapt to people's daily lives. A good example is that of the camera in the movie *Minority Report* which, by analyzing the character's retina, identifies him and sends him a targeted ad according to his consumer habits.

20.2.1.2. *Interaction between digital ambient agents (ambient networks)*

This axis concerns the ubiquitous communication concept and therefore the all-to-all connection and makes all ambient agents communicating. These agents can transmit information with the help of wireless communication support. The goal for ubiquitous computing is to be grouped with ubiquitous communication, processing and data transmission thus becoming an integral part of all objects of everyday life. Let us remember that protocol IPv6 has been designed with these potential applications in mind.

One of the first objectives is therefore to spread wireless communication networks and to make the technology behind the scenes transparent to the user. A second objective is to give the user access to information no matter the location or context.

In order to attain these goals, an important part of the research specifically deals with local and wireless personal networks (WLAN, WPAN), ad hoc and sensor networks. Another part deals with distributed software systems architecture which enables the access to information and services from any terminal, anywhere. These studies focus on with middleware, which requires standards definition (UpnP, HAVi and JINI, for example) enabling interoperability between components of an ambient network.

This axis is the backbone of this chapter and it is detailed in section 20.3.

20.2.1.3. *Interface between human and digital ambient agents*

Finally, the third axis deals with areas studying interaction between humans and ambient agents, in particular, the IHM design area, the automatic processing of the language as well as the sociological aspects of relations between the human and the digital societies. The goal is for this interaction to be more user-friendly and thus more natural.

20.3. Ambient networks

The ambient network, as shown in Figure 20.3, is a spontaneous environment that enables users (terminals of any type) to access rich and personalized services through one or more fixed or wireless access networks (integration of infrastructures, wired or wireless). This type of network must provide a global connectivity and free itself from its computing nature in order to show only the accessible functions in an intuitive manner.

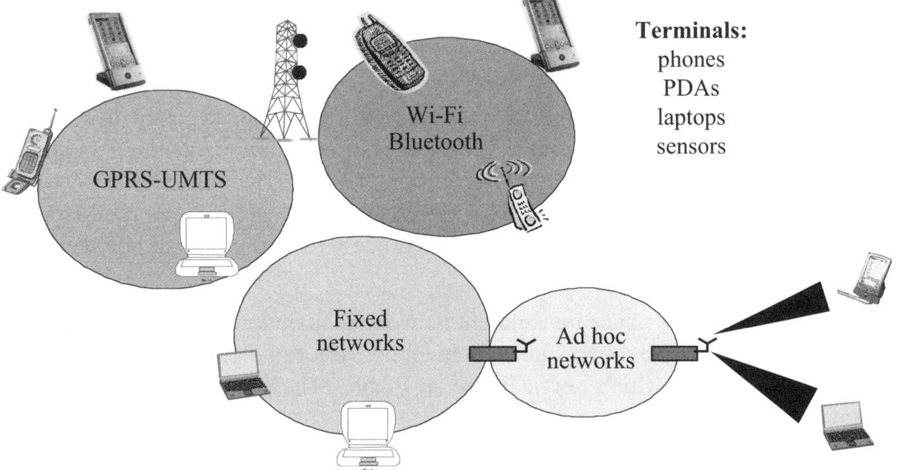

Figure 20.3. *Ambient networks*

The fact that this network is ambient implies that it must also guide the user with the help of a group of functions that adapt to his preferences, his current location, the actual context and the capabilities of his terminal. In order to fulfill these objectives, different aspects must be taken into account [AND 03]:

– wireless terminals: there are more and more devices used in our daily lives (phone, PDA – Personal Digital Assistant, sensor, etc.) that are linked through different types of networks, often wireless. Different studies are being done in order to make these terminals less expensive, easier to use, smaller and less power hungry. In section 20.3.3, we will see a category of these terminals, the sensors;

– links: ambient networks use different types of links and different characteristics (bandwidth, delay, coverage). Currently, different networks coexist without real integration. Terminals will have a choice between different technologies: wired (DSL, for example) or wireless such as WLAN (802.11b, 802.11a, 802.11g), WPAN (Bluetooth, 802.15) or UMTS/GPRS. We will present these wireless technologies in section 20.3.1;

– ambient network integration: the user takes advantage of a connectivity that is always present offered by ambient networks and demands adequate application performance in spite of the fact that he may move from one location to another or when he changes terminals. The network must negotiate communications in the name of the user and configure the services found while taking into account the application's quality of service (QoS) constraints. The complete process should be designed with minimum user intervention. However, the integration of these ambient networks, the mobility management (vertical handover between these different technologies), QoS management and security remain open concerns. These problems are explained in detail in section 20.3.4;

– ad hoc networks: most wireless networks are based on a fixed infrastructure. In order to widen the coverage, new wireless communication solutions are emerging. Among the solutions that eliminate the need for fixed infrastructure in order to communicate are ad hoc networks. The major goal of ad hoc networks is to spread mobility notions to enable access to information and to "anywhere anytime" communication. One of the biggest challenges with those networks is routing since each station can be used by other stations to process data routing. Ad hoc networks are studied in section 20.3.2;

– service discovery: once the ambient network is operational, it should then offer a group of services. The devices in an ambient network can offer their services while others can discover which services are available. These aspects are studied in section 20.3.5.

20.3.1. *Wireless links*

In an ambient environment, the terminals can choose between different wireless technologies. In this section, we will talk about cellular networks GPRS/UMTS, micro networks, known as wireless LANs – WLANs and nanonetworks, known as wireless personal area network (WPAN).

20.3.1.1. *GPRS/UMTS*

20.3.1.1.1. GPRS

GPRS (general packet radio service) network is a step up from GSM in packet mode. It is stacked over the GSM network and it benefits from the same radio interface but possesses a new core network adapted to packet transfer. The purpose is to use available slots in the TDMA frame within the GSM to transport packets. The theoretical throughput varies between 9.6 and 171.2 Kb/s based on the number of slots dedicated to packet mode. However, for many reasons, in practice, the useful throughput is of approximately 10 Kb/s output and 30 to 40 Kb/s input. GPRS service enables point-point connections in connected or disconnected mode and multipoint in broadcast or multicast mode. It offers a standardized access to the Internet as well as data exchange volume pricing.

GSM and GPRS can be associated with EDGE, which is a new more powerful radio interface modulation system. Raw throughput is then three times higher, but the deployment of EDGE is more expensive, the hardware modifications are more significant and it can reduce the reach of BTSs (base transceiver station). E-GPRS is EDGE's main application which respects the specifications of the third generation at a lower cost than those of the UMTS. It therefore becomes the transition solution toward the third generation.

20.3.1.1.2. UMTS

The third generation of mobile networks has not been completely defined. The ITU (International Telecommunication Union) standards organization has launched an international program in the form of a request for proposal, the IMT 2000, to define an international standard. The UMTS (Universal Mobile Telecommunications System) is one of the propositions retained and is backed by Europe and Japan (3GPP), among others. The USA supports the CDMA 2000.

In its current version, the UMTS uses the core of the GSM/GPRS network which corresponds to a compatibility and profitability concern. Globally, the architecture remains the same. The user domain is still made up of a terminal with a chip (U-SIM). The main differences are found in the radio area; the antennae or base stations are called Node B (instead of BTS) and Node B hubs are RNCs (BSC – base station controller).

Theoretical throughput has obviously increased: from 144 Kb/s in rural zones to up to 2 Mb/s in the case of low mobility with good radio conditions (they are inferior in practice). However, this version of UMTS, release 99, is considered too heavy and the following ones will be designed to reduce this complexity. In time, the third generation should no longer be based on the second generation architecture but it should work progressively on an all-IP architecture. The B Node, the gateway to RTC as well as a certain number of servers, will be directly interconnected on one IP network.

20.3.1.2. *Wireless micro networks (WLAN)*

A wireless micro network (WLAN) is a network where the stations are no longer linked physically but instead they are linked through wireless support. Although there is no longer a physical link between the different stations of a WLAN, it still retains the same functions as LANs: the interconnection of machines capable of sharing data, services or applications. The WLAN is not an alternative to LAN and it is not designed to replace it. Generally speaking, WLAN is used as an extension to an existing LAN. There are currently two WLAN standards derived from two different standard organizations that are incompatible between themselves: Hiperlan and IEEE 802.11.

20.3.1.2.1. 802.11 networks

IEEE 802.11 [BRI 97, LAN 99] is the first WLAN standard since 2001. It is currently very successful due to its ease of installation, its performance and its competitive price. It relies on IEEE 802's MAC technique, with a different access technique to the CSMA/CA[2] channel. Within the unlicensed 2.4 GHz band, this wireless local network has a cellular architecture described in Figure 20.4. A group of stations, usually one or more computers (laptops or desktops) equipped with an 802.11 network interface card, get together to establish communication between themselves and form a set of basic services or BSS (basic set service). The zone occupied by stations within a BSS is a Basic Set Area (BSA) or a cell. The different APs are linked together through any wireless or wired network.

2 CSMA/CA: Carrier Sense Multiple Access with Collision Avoidance.

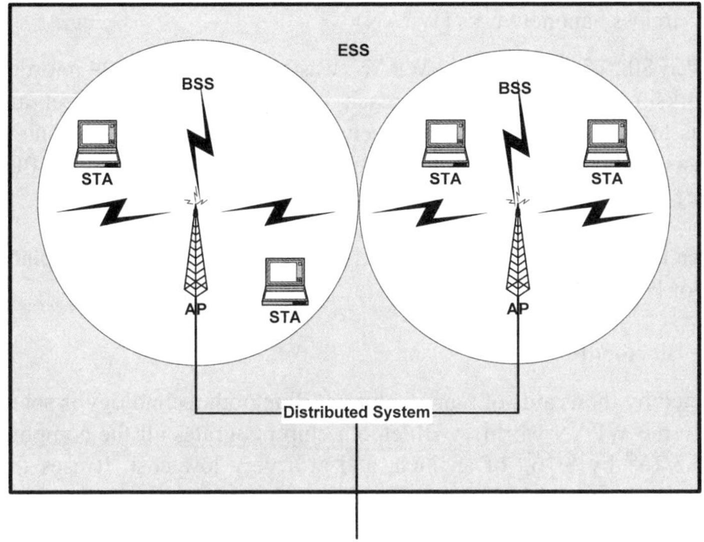

Figure 20.4. *802.11 network architecture*

The most widely used version is 802.11b, also called Wi-Fi (wireless fidelity). This version offers a throughput of 11 Mb/s and does not propose any real handover mechanism. Its security is also weak because it is possible to listen for carriers and thus intercept network traffic (security mechanisms exist but they are insufficient). However, this technology is still young and it evolves rapidly toward versions offering better performance. 802.11a offers communication speed of up to 54 Mb/s within the 5 GHz band. 802.11e and 802.11i propose the introduction of QoS and better security.

20.3.1.2.2. Hiperlan

Originally from Europe (ETSI – European Telecommunication Standards Institute), the Hiperlan (high performance radio LAN) standard is often cited as a direct competitor to IEEE 802.11, which has retained some aspects of it. Standards developers have favored ATM interfaces, which may explain the competitive advantage of 802.11, based on Ethernet.

The current standard defines two types of Hiperlan, both using the 5 GHz band ratified during the European Conference of Postal and Telecommunications Administrations (ECPT): Hiperlan 1 and Hiperlan 2. Hiperlan 1 offers speeds between 10 and 20 Mb/s, whereas Hiperlan 2 should reach throughputs of 54 Mb/s. This standard is designed better than 802.11, but it is pricey and slow in coming.

20.3.1.3. Wireless nanonetworks (WPAN)

The IEEE 802.15 group, called WPAN, was established in 1999 in order to study nanonetworks or networks with a reach of approximately 35 feet and whose objective is to create connections between the different terminals of one user or of multiple users. The network can interconnect a laptop, a cell phone, a PDA or any other device of this type.

In order to reach those goals, industrial groups have been put in place such as Bluetooth or HomeRF.

20.3.1.3.1. Bluetooth

Supported by thousands of manufacturers, Bluetooth technology is set to become the norm in the WPAN world. A Bluetooth chip integrates all the components on a surface of 5/16th by 5/16th of an inch, and at a very low cost. It uses frequencies within the unlicensed 2.4 GHz band. The terminals interconnect to become a piconet (maximum of 8 terminals), which can also be interconnected to form a scatternet (see Figure 20.5).

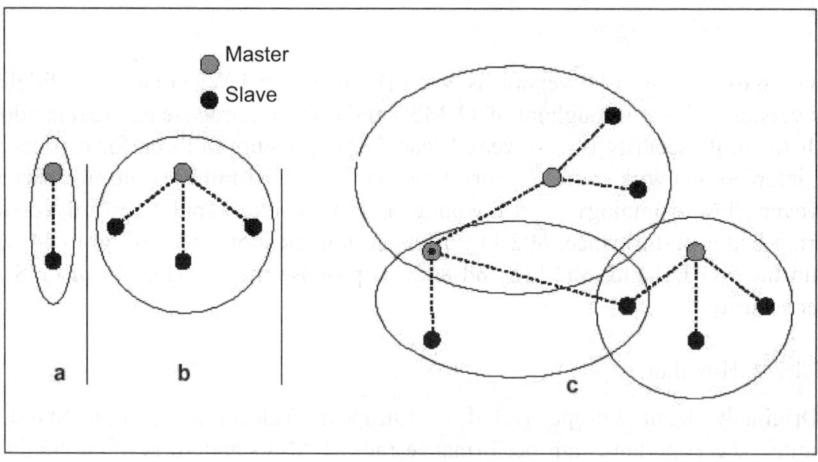

Figure 20.5. *Bluetooth network architectures: (a) piconet to only one slave, (b) piconet with many slaves and (c) scatternet*

Version 1.0 of Bluetooth enables a communication between two terminals with a maximum throughput of 1 Mb/s which may not be enough for some applications. The goal of the 2.0 workgroup is to increase this throughput and to put in place enhancements (handover, routing optimization, coexistence with the other networks within the 2.4 MHz band).

20.3.1.3.2. HomeRF

Another technology, which emerged from home automation, competes with Bluetooth: HomeRF. A base station communicates with different units in a range of 165 feet within the 2.4 GHz band. Basic throughput is at 1.6 Mb/s but should increase. Its characteristics are comparable to those of Bluetooth but it is not endorsed by as many manufacturers as Bluetooth.

20.3.2. *Ad hoc networks*

Due to the commercial success of wireless local networks, mobile equipment development has been steadily growing. The user is becoming more and more mobile with the help of PDAs and laptops. In this context, new, more powerful wireless communication solutions are starting to come out. Among these main solutions, which eliminate the need for a fixed communication infrastructure, are ad hoc networks.

Ad hoc networks are particular wireless local network architectures based on technologies such as Wi-Fi. However, while each Wi-Fi network user is connected via a radio base, in an ad hoc network the terminals can also communicate between each other without outside help and thus without any infrastructure. They can even serve as relays for each other. It is a sort of "soft architecture", progressive and automatic. The key is flexibility and autonomy, since the network evolves according to accesses and current users. It can also be completely independent of any infrastructure. Besides, the reach of the signal is scaled according to the number of users and the throughputs can be preserved, but they must be shared with a Wi-Fi radio base.

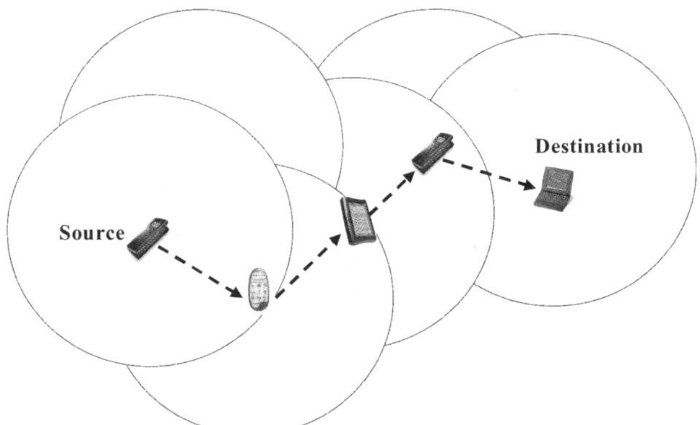

Figure 20.6. *Example of an ad hoc network (the circles represent the radio reach of each ad hoc station)*

Each station can be called by other stations to route data. In this way, when a transmitting station is out of reach of the destination station, network connectivity is maintained by intermediary stations. An example of an ad hoc network is illustrated in Figure 20.6 in which a transmitting station communicates with a destination station by using intermediary nodes. Globally, there are two families of ad hoc routing protocols: "proactive" routing protocols, which anticipate the request for packet transmission and "reactive" routing protocols that react to the demand. Between these two families, another approach is starting to emerge, "hybrid" protocols that are based on both proactive and reactive protocols. You will find below a general overview of these protocols and their main characteristics.

20.3.2.1. *Proactive protocols*

Proactive routing protocols in mobile ad hoc networks are based on the same principle as the routing protocols used in wired networks. The two main methods used are "link state" and "distance vector". These two methods require a periodic routing data update that must be transmitted by the different network routing nodes. The most important protocols in this class are LSR (link state routing) [HUI 95], OLSR (optimized link state routing) [CLA 03] and DSDV (dynamic destination-sequenced distance vector) [PER 94].

20.3.2.2. *Reactive protocols*

Routing protocols that belong to this category create and maintain the routes as needed. When the network needs a route, a route search procedure is launched. The most important protocols in this class are DSR (dynamic source routing) [JOH 03b] and AODV (Ad hoc on demand distance vector) [PER 03].

The main goal of ad hoc networks is to extend mobility notions in order to enable access to information and to communicate "anywhere anytime". IETF[3]'s MANET Group [MANET] research shows that development of these wireless and infrastructureless networks is growing rapidly. Developers are already imagining all sorts of applications: military, obviously, to create mobile tactical networks, but also civil for emergency procedures, communications with vehicles, reconfiguration of wireless networks in companies or even the creation of temporary networks surrounding special events. Without a doubt, the major advantages of this new generation of mobile networking are flexibility and low cost.

The absence of a fixed infrastructure, however, poses a number of non-trivial problems such as security, adoption of global network management policies (invoicing, QoS offering) and the autonomy of batteries [SEN 03].

3 IETF: Internet Engineering Task Force.

20.3.3. *Sensor networks*

Sensor network technology is still very young, miniaturization is insufficient, but the trend is there to stay: from flower pots to vehicles, from bridge structures to production chains. We will see more and more sensors in all aspects of our lives, designed to measure a large number of signals and to communicate them to control centers via wireless networks. We can find these network applications anywhere permanent data collection can be useful; health, war, climatology, process control and seismology. In fact, sensors can function within large machines, at the bottom of an ocean, in biologically or chemically contaminated areas, in a battlefield far into enemy lines, inside a building or a home.

In order to get there, though, the challenges remain significant. However, laboratories and companies are starting to offer innovative products in that sector.

But what is a sensor node? As shown in Figure 20.8, a sensor node is made up of four basic components: a detection unit, a processing unit, a communication unit and a battery. There might be other components such a tracking system, a power generator and a motor. The detection unit is often composed of two sub-units: the detector and the analog-digital converter (ADC). Analog signals transmitted by the sensor are converted into digital signals through an ADC and are then introduced into the processing unit. The processing unit, which usually has a small storage unit associated with it, manages the necessary procedures in order to collaborate with the other nodes in the network in order to complete assigned tasks. The communication unit links the node to the network. One of the most important components of a sensor node is the battery. Batteries can be fed by power generators such as solar batteries. Most of the routing techniques in sensor networks require an exact location of the sensors. It is therefore common practice that a sensor node has its own tracking system. The motor can be necessary at times in order to move a sensor node.

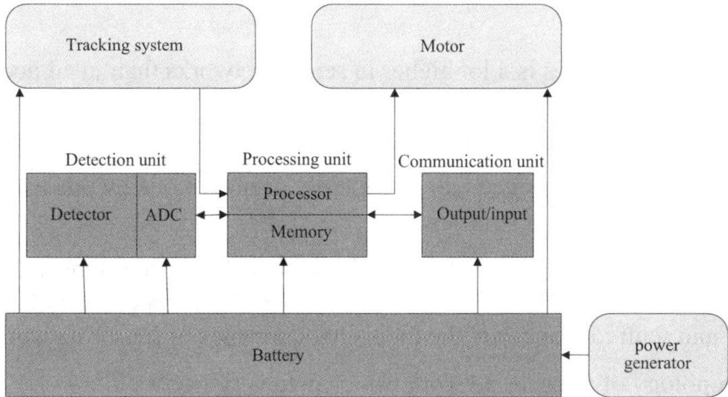

Figure 20.7. *Sensor node architecture*

A sensor network is composed of a large number of sensor nodes deployed either one by one or in bulk and in an ad hoc manner inside or very close to the observed phenomenon. The position of the sensors does not need to be predetermined. This aspect is important for areas that are inaccessible, for example (biologically or chemically contaminated areas). However, and at any moment, other sensor nodes can be redeployed in order to replace failing nodes or if there is a change in the task to accomplish.

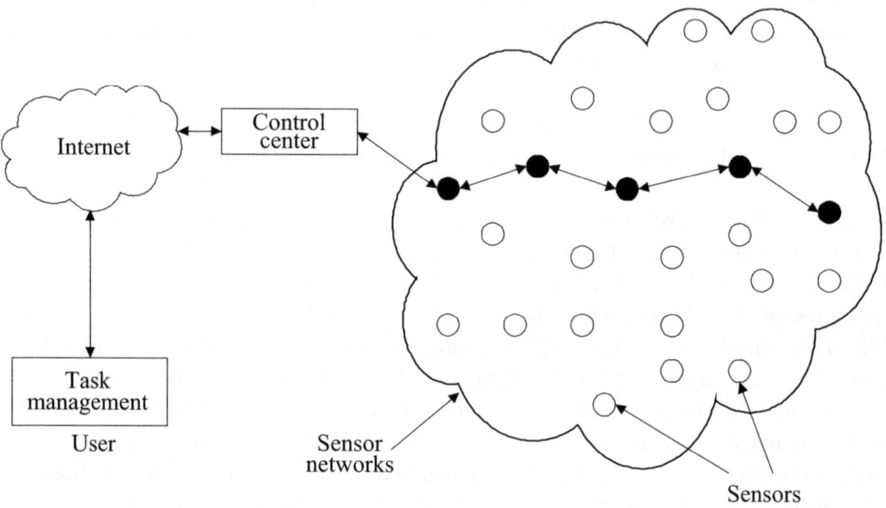

Figure 20.8. *Sensor network architecture*

The differences between sensor and ad hoc networks presented previously are summarized below:

– the number of nodes is a lot higher in sensor networks than in ad hoc networks. In a sensor network the distance between the nodes is often less than 33 feet and the density can be higher than 6 nodes/f^3;

– sensor nodes are deployed either randomly or placed one by one by a robot or a human;

– sensors can fail (loss of power, physical damage, interference, etc.). This breakdown should not affect the global task of the sensor network. Therefore, reliability and fault tolerance are the main characteristics of sensor networks;

– the topology of a sensor network can change very frequently;

– sensor nodes mainly use broadcast communication, whereas most ad hoc networks are based on point-to-point communication;

– sensors are nodes limited in power, computing capabilities and memory.

So, contrary to ad hoc networks, the design of a sensor network is influenced greatly by different factors such as fault tolerance, scaling, cost of production that must be low, the environment, the network's topology, hardware constraints, the transmission medium as well as energy consumption.

The authors of [AKY 02] explain the layered architecture of a sensor network which comprises a physical layer, a data link layer, a network layer, a transport layer and an application layer. For the authors, energy consumption is the main factor to take into consideration for each of these layers. For example, in the case of routing, routes are often traced according to energy metrics. They list different algorithms; here are the most important:

– routing based on residual power: in this type of routing, the route chosen is the one with the maximum of the sums of residual energy (E_r) of the nodes in the route;

– routing based on transmission power: in this type of routing, the route chosen is the one with the minimum of the sums of required transmission power (E_t) of the links in the route.

The authors also mention a group of research studies that are needed for each of these protocol layers.

Without a doubt, the major advantages of these networks are their flexibility, fault tolerance and deployment speed. These advantages will make the networks an integral part of our daily lives. However, new technologies are needed due to the specific constraints of this type of network.

20.3.4. *Interoperability between ambient networks*

The different ambient networks form a cell hierarchy of different sizes and throughputs (Wi-Fi, UMTS, Bluetooth, etc.). At one geographic point, one or more of these networks will be available. This concept of hierarchy has been explained clearly for the first time in the article *The Case for Wireless Overlay Network* [KAT 96]. The promise consists of ending competition between the different networks and to make them coexist in order to offer the best technology for each type of application. A user wanting to use a service can not only choose one available access network, but the most adequate and that which will satisfy his needs the most. He may also change networks if another network is better than the one he is using. This process should be designed for a minimal user involvement. This philosophy has produced what we call ABC for "always best connected".

In order to illustrate this concept, we can use the classic example that we see in the majority of articles about this problem and which is shown in Figure 20.9 [GUS 03].

The ABC concept enables a better connectivity for the user by using terminals and access technologies that are most adapted to the needs and context. However, an ABC scenario, where we allow the user to choose the best access networks and terminals available, can create significant complexity not only technical but also in terms of policy between the different access operators and service providers. Therefore, different aspects must be considered [GUS 03]: user's experience, agreements between these access operators and service providers and technical solutions for the terminal and the network.

20.3.4.1. *User experience*

User experience concerns the manner in which to supply ABC service to the users. Many factors have been identified [GUS 03]: subscription or payment for ABC service, information delivery, mobility management and user interaction:

– subscription: the simplest scenario is that the user has only one subscription to one access operator, which makes invoicing much more simple. That way, a unique identifier is associated with the user for all visited ambient networks. This is similar to the notion of roaming in cellular networks. On the other hand, this implies that the access networks which will communicate authentification and authorization information through an AAA infrastructure (Authentication, Authorization and Accounting) are adequate.

EXAMPLE.– Laetitia wakes up in the morning and links her laptop to the Internet using a DSL connection (supplied by her Wanadoo service provider). She wants to check her email and connects to her company network via a VPN (Virtual Private Network) connection. She can then read and send email. Laetitia starts a download on her laptop and realizes that she is late for a meeting. She unplugs her computer from the DSL modem and leaves her house. When the laptop is disconnected from the DSL, it commutes automatically to the access network that is available, in her case GPRS (provided by Orange). The VPN connection is maintained and the download continues. In the subway in Paris, she starts a conversation on NetMeeting with colleagues in Holland by using her PDA. The PDA is linked by using GPRS. Due to the limited text/voice capabilities of her PDA and the GPRS network, NetMeeting adapts and grants access only to voice/text services. When Laetitia arrives at her office, she transfers her NetMeeting session from her PDA to her laptop, which is now linked to the WLAN of the company. Because the capabilities of the laptop and the WLAN are much better, NetMeeting adapts by adding a video service.

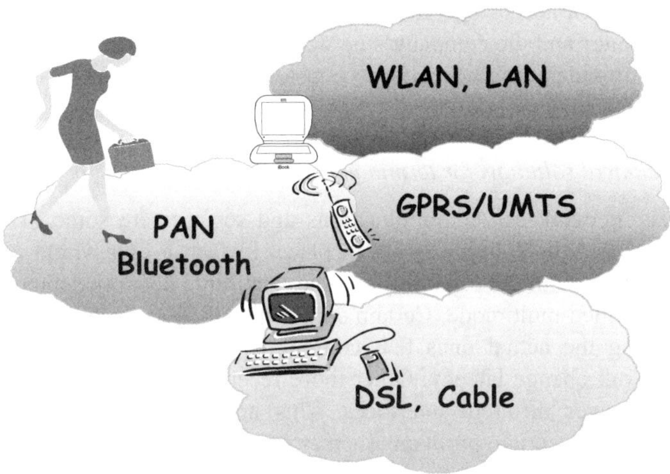

Figure 20.9. *ABC concept's application scenario*

– information delivery: a major part of user experience is the behavior of the application. Applications must adapt to environmental changes in order for the delivery of information to be the fastest and adjust to the terminal's capabilities and the access network's technology while optimizing the presentation of information. The applications may, autonomously or by using the user profile, require a QoS (QoS) in order to provide an optimal execution;

– mobility management: in its most simple form, ABC service provides the possibility for the user to access different network technologies without any mobility management behind. Therefore, another service to consider consists of offering the possibility for a terminal to move between the different technologies, while maintaining current connections. That is what is called a vertical handover and is the object of section 20.3.4.3. Other services could also enable a user to transfer a session from one terminal to another, without any loss of data or having to restart the application (see the example of Laetitia above);

– user interaction: the user should have the tools that provide necessary ABC information while giving him the possibility of changing the access manually, for example. These tools must be simple and easy to use by an average user.

20.3.4.2. *Agreements between ABC players*

The user can go from one network to another. This implies that there are agreements between the access operators in order to manage the roaming, agreements between access operator and the service provider, agreements between the access operator and the company's network and between the operator and/or the ABC service provider. Another point to consider is the subscriptions, user profiles and AAAs management.

20.3.4.3. *Technical solutions for terminals and networks*

Terminals, network servers, applications and services are some of the entities affected when an ABC solution is put in place. In fact, a user would need a more intelligent terminal with the capability to access multiple networks: a multi-interface terminal, also called multimode. Certain studies [KAT 98] propose only having one interface among the actual ones that communicate at a given time, in order to optimize network change latency. Other more recent studies describe simultaneous data reception over different interfaces. This approach is therefore technically possible, but it poses certain problems such as energy consumption.

There are different physical architectures involved in the realization of multi-interface terminals [ARE 01, KON 00]. The most simple consists of setting different radio interfaces in parallel, as shown in Figure 20.10(a). However, in this way, the terminal is not progressive: it cannot adapt to the arrival of a new network. On the other hand, as presented in Figure 20.10(c), a single programmable interface can connect to any frequency with any received signal process. This solution is obviously the most flexible. An intermediate solution exists and is presented in Figure 20.10(b): a fixed number of radio-frequency receivers are controlled by a unique signal process unit, which is configured by necessary parameters. These different approaches are still being discussed based on economic, performance and adaptability criteria. A major criterion is energy consumption, which is essential in portable terminals [STE 96].

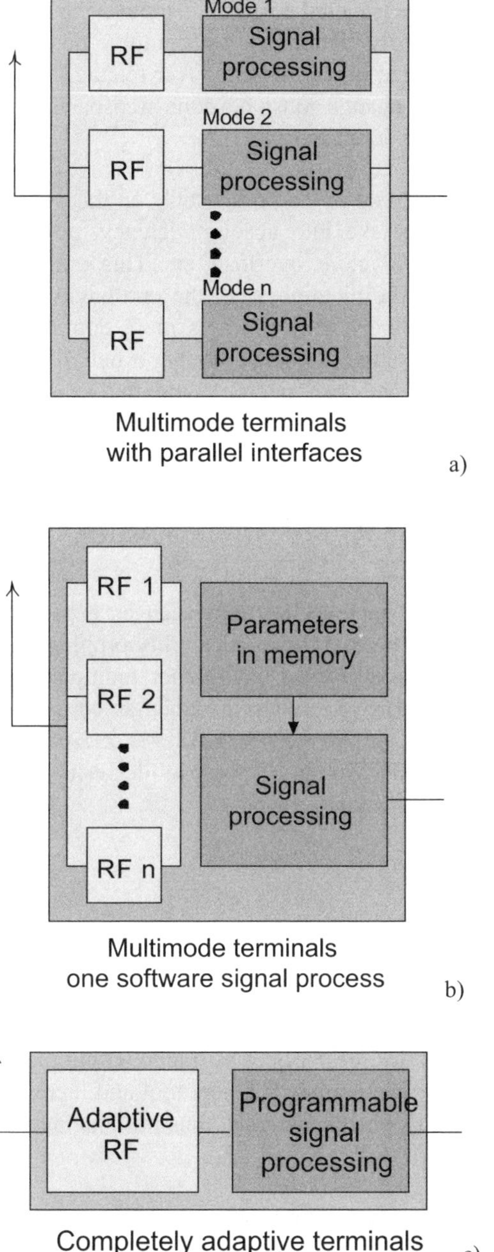

Figure 20.10. *The different architectures of multi-interface terminals*

The change of interface is called a vertical handover: the terminal changes access networks without necessarily changing its geographic position as is the case with classic handover, also called horizontal handover (change of base station in a same access technology). This change must be done transparently to the user, without interfering with his current services.

Many studies have been done. The first article on this subject [KAT 98] exposes the different problems to take into account: latency, power supply problem for interfaces that must remain active, overload, etc. This study considers that the best network is the lowest in the hierarchy, i.e. the one having the largest throughput. However, the choice for the best network is more complex and must take other criteria into account. This is the subject of another article [WAN 99] which proposes separating the choice of the best network and the handover mechanism. After picking up parameters on the different available networks and the user's preferences, the terminal executes a cost function for each network and retains the best. The user has very little input: it takes into account only certain aspects such as cost or performance. However, application requirements such as QoS parameters are not taken into account.

Multiple interface management is often seen as a particularity of Mobile IP [PER 96, JOH 03a] or more generally as a mobility problem. Instead of having only one temporary multicast address, we have either multiple, or a unique temporary multicast address. Therefore, an interface change corresponds to a temporary address change. Each article [WAN 99, KAT 98, ZHA 98, CHE 96] defines an extension of the Cellular IP, Mobile IP or a Mobile IP-like protocol which enables multi-interface and mobility management.

20.3.5. *Service discovery*

One of the objectives of ambient networks is to enable mobile user to access and provide services independently from time and place. From there, a new requirement (function) has come out, that of discovering automatically and without external interference the group of desired services available within the current domain of the mobile terminal. A new function needs a new tool and therefore several studies are in place in industrial and research contexts, thus introducing service discovery (SD) protocols.

A service discovery protocol is a process supplying automatic and spontaneous mechanisms for the location of services provided in a network. It enables servers to publish, or notify their services and clients to discover these services in some form of necessary autoconfiguration.

In order to illustrate the requirement for service discovery protocols, we can use the classic example found during each protocol specification. Let us imagine a mobile terminal that enters a network for the first time and that wishes to print a document. It must locate the network printers, compare them according to their capabilities (printing postscript documents, printing in color, printing as fast as possible, etc.) and then choose one for its document. Service discovery protocols enable the execution of this location task, spontaneously, without operator intervention. Figure 20.11 illustrates an example of a meeting room where a new lecturer wants to project his presentation and print it in order to give copies to the different participants.

Service discovery protocols eliminate the need to know the name or address of the equipment within the service. The user only has to name the service, by entering its type, and to supply the list of attributes needed. The discovery protocol will take care of matching the user's description with the address of the appropriate service, thus providing an automatic configuration mechanism for the applications within local networks.

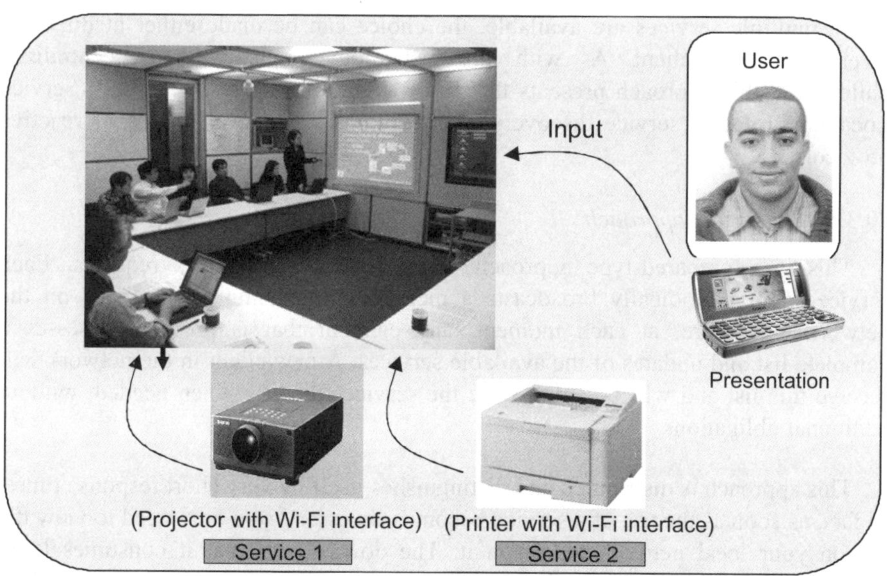

Figure 20.11. *Discovery of a printer and of a video-projector*

Different approaches exist concerning the design of service discovery protocols: the reactive approach and the proactive approach.

20.3.5.1. *Reactive approach*

This is an on-demand concept, where the discovery process is only executed when there is a user request for a specific service. There are two variables: a distributed approach, and a centralized approach.

Within the distributed reactive approach, a request is sent over the network (for example, in multicast) specifying the wishes of the user requiring a specific service. On the service side, a service agent, always monitoring the network, hears this request and responds to it (unicast) if it corresponds to the service that implements it. The client then has to choose between all the responses obtained. This method is better suited to small networks with limited spread. It presents the disadvantages linked to transmission in wireless networks.

The centralized reactive approach requires that a services directory is present. Each available service is registered in this directory by the delivery of a registration request at launch. As soon as the client has a request for a specific service, he just needs to notify the central node of his request in order to receive the required information and therefore communicate with the appropriate service. In the case where multiple services are available, the choice can be made either at directory level or by the client. As with any technology implementing a centralized philosophy, this approach presents the disadvantage of failures. The SLP (Service Location Protocol) service discovery protocols and Jini are examples of reactive protocols.

20.3.5.2. *Proactive approach*

This is a prepared-type approach which anticipates clients' requests. Each service agent periodically broadcasts a message announcing its presence on the network. Therefore, at each moment, any element that is present possesses a complete list and updates of the available services. A new client in the network will receive this list and will be able to use the services directly when needed, without additional obligations.

This approach is distributed and distinguishes itself by very short response times. In fact, as soon as the use of a service becomes necessary, you only need to view the list on your local network to obtain it. The downside is that it consumes large amounts of bandwidth, thus decreasing the network's performance. The UPnP (Universal Plug and Play) service discovery protocol is an example of proactive protocols.

In conclusion, if our objective is to shorten response times, this approach seems the most logical one. On the other hand, if our major concern is to conserve bandwidth, it is best to use a reactive method.

20.4. Conclusion

Ambient intelligence is a nice name, representing a new paradigm in information technology and grouping a multitude of technologies. People, in an ambient world, are assisted by a digital environment that is sensitive, adaptable and responsible for their needs, habits, movements and emotions.

Ambient intelligence is located at the junction of many domains which are not usually cooperative (IA, IHM, telecoms, etc.) and that we have introduced in this chapter. This new paradigm has set innovative communication technologies, often wireless, which enable the formation of ambient networks that will contribute to people's quality of life. The future communication networks will supply a global connectivity and therefore will assist, in a transparent manner, the user with very advanced services. They will make the underlying technology invisible to the user. Access to information and control of the environment will become more transparent to the location, terminal and context (new user interfaces, rapid response time, control of different types of wireless terminals, availability and security, etc.).

This chapter describes different aspects of ambient networks where user satisfaction is the dominant factor: wireless communication networks and their integration, mobility management, ad hoc networks, sensor networks as well as the discovery of available services. One of the major qualities of these networks would be their total transparency: they would be present, but completely invisible to us, interaction with them should be just as transparent.

In order to achieve these objectives somewhat futuristic and promised by ambient networks, several questions are still open. We expect them to offer global and varied connectivity at minimum cost, a larger capacity and QoS guarantees. The terminals must also be more sophisticated, more user-friendly and equipped with more natural interfaces, multimodal and auto-configurable. Accumulating these multiple functions obviously generates a large number of challenges and a considerable complexity.

20.5. Bibliography

[AKY 02] AKYILDIZ I. F., SU W., SANKARASUBRAMANIAM Y., CAYIRCI E., "A Survey on Sensor Networks", *IEEE Communications Magazine*, vol. 40, no. 8, p. 102-114, August 2002.

[AMB] ITEA Ambience project, http://www.extra.research.philips.com/euprojects/ambience.

[AND 03] ANDRZEJ D., "Ambient Networking", *Smart Object Conference – SOC'2003*, Grenoble, France, May 2003.

[ARE 01] ARETZ K., HAARDT M., KONHÄUSER W., MOHR W., "The Future of Wireless Communications Beyond the Third Generation", *Computer Networks*, vol. 37, no. 1, p. 83-92, September 2001.

[BAS 03] BASKIOTIS N., FEREY N., *Cours en introduction à l'intelligence ambiente*, 2003.

[BRI 97] BRIAN P. CROW, WIDJAJA I., KIM J. G., SAKAI P. T., "IEEE 802.11 Wireless Local Area Networks", *IEEE Communication magazine*, p. 116-126, September 1997.

[CHE 96] CHESHIRE S., BAKER M., "Internet Mobility 4x4", *ACM SIGCOMM*, p. 318-329, Stanford, August 1996.

[CLA 03] CLAUSEN T. (ED.), JACQUET P. (ED.), "Optimized Link State Routing Protocol", MANET working group, IETF, http://www.ietf.org/rfc/rfc3626.txt, October 2003.

[GUS 03] GUSTAFSSON E., JONSSON A., "Always Best Connected", *IEEE Wireless Communications*, p. 49-55, February 2003.

[HUI 95] HUITEMA C., *Routing in the Internet*, Prentice Hall, 1995.

[IST] ISTAG – IST Advisory Group, http://www.cordis.lu/ist/istag.html.

[JOH 03a] JOHNSON D., PERKINS C., "Mobility Support in IPv6", Mobile IP Working Group, IETF Internet-Draft, draft-ietf-mobileip-ipv6-24.txt, June 2003.

[JOH 03b] JOHNSON D. B., MALTZ D. A., HU Y.-C., "The Dynamic Source Routing Protocol for Mobile Ad-Hoc Networks (DSR)", MANET working group, IETF Internet Draft, draft-ietf-manet-dsr-09.txt, April 2003.

[KAT 96] KATZ R. H., BREWER E. A., "The Case for Wireless Overlay Networks", *Proc. SPIEE Multimedia and Networking Conference (MMNC'96)*, p. 77-88, San Jose, CA, January 1996.

[KAT 98] KATZ R. H., STEMM M., "Vertical Handoffs in Wireless Overlay Networks", *Mobile Networks and Applications*, vol. 3, no. 4, p. 335-350, 1998.

[KON 00] KONHÄUSER W., MOHR W., "Access Network Evolution Beyond Third Generation Mobile Communications", *IEEE Communication Magazine*, December 2000.

[LAN 99] LAN/MAN Standards Committee, *ANSI/IEEE Std. 802.11: Wireless LAN Medium Access Control (MAC) and Physical Layer (PHY) Specifications*, IEEE Computer Society, 1999.

[MANET] IETF MANET Working Group (Mobile Ad-Hoc NETworks), www.ietf.ora/html.charters/manet-charter.html.

[PER 94] PERKINS C. E., BHAGWAT P., "Highly Dynamic Destination-Sequenced Distance-Vector Routing for Mobile Computers", *ACM SIGCOMM'94*, p. 234-244, October 1994.

[PER 96] PERKINS C., "IP Mobility Support", *IETF RFC 2002*, October.

[PER 03] PERKINS C. E., BELDING-ROYER E. M., DAS S. R., "Ad-Hoc On-Demand Distance Vector (AODV) Routing", MANET working group, IETF, http://www.ietf.org/rfc/rfc3561.txt, July 2003.

[SEN 03] SENOUCI S., Application de techniques d'apprentissage dans les réseaux mobiles, PhD Thesis, Pierre and Marie Curie University, Paris, October 2003.

[STE 96] STEMM M., GAUTHIER P., HARADA D., KATZ R. H., "Reducing Power Consumption of Network Interfaces in Hand-Held Devices", *Proc. 3rd Workshop on Mobile Multimedia Communications (MoMuC-3)*, September 1996.

[WAN 99] WANG H. J., KATZ R. H., GIESE J., "Policy-Enabled Handoffs Across Heterogeneous Wireless Networks", *Second IEEE Workshop on Mobile Computing Systems and Applications (WMCSA '99)*, New Orleans, LA, February 1999.

[ZHA 98] ZHAO X. *et al.*, "Flexible network support for mobility", *MOBICOM'98*, p. 145-156, October 1998.

Chapter 21

Wi-Fi/IEEE 802.11

21.1. Introduction

Ambient networks, and more importantly wireless networks, are the main networks that surround us. In time, the user will have the possibility of connecting to one of these networks and of changing it during his communication. The ambient Internet is starting to emerge with the first wireless networks and more specifically with the explosion of Wi-Fi networks. The objective of this chapter is to give an overview of this first vision of the ambient Internet.

After an introduction of the technology used in these Wi-Fi networks, we will explain in detail its functionality and its improvement throughout the changes before concluding on its prospects.

21.1.1. *The origin*

After the first fundamental experiment on radio conduction executed by the French physicist Edouard Branly in 1890 in his laboratory at the Institut Catholique de Paris[1] and the transmission of the first wireless telegram between Wimreux and Douvre by the Italian physicist Gulglielmo Marconi to Edouard Branly in 1899, wireless communication, i.e. using radio waves, has continuously progressed.

Chapter written by Denis BEAUTIER.

1 http://museebranly.isep.fr, website of Branly from the Institut Supérieur d'Electronique de Paris museum (ISEP, http://www.isep.fr), engineering school founded in 1955 in order to transmit the knowledge acquired in Branly's research laboratory.

21.1.2. *Context*

With the emergence of laptop computers, the notion of mobility has grown rapidly. The need to connect without wires in order to attain real wireless communication has quickly become a necessity.

At the beginning, many solutions have been proposed by companies, but due to their diversity, they were not interoperable. Defining a common norm has quickly become necessary in order to develop this market.

Two organizations intervened in this approach:

– IEEE (Institute of Electrical and Electronics Engineers) is responsible for the technological certification that is the 802.11 standard;

– WECA (Wireless Ethernet Compatibility Alliance) has created the Wi-Fi (wireless fidelity) label in order for the manufacturers to validate their products' interoperability. WECA has since been replaced by the Wi-Fi Alliance.

The IEEE was already responsible for the standardization of wired networks for the cabling, the physical layers and linking, under the 802 (LAN, Local Area Network) label. IEEE 802.11 has then become the name for this standard which guarantees interoperability and Wi-Fi is its commercial name.

There are different needs for wireless connections:

– the interconnection of peripherals to a computer or between different pieces of equipment together within a close proximity, WPANs (wireless personal area network);

– local networks of computers to cover an office or a building, within 300 feet, WLANs (wireless local area network);

– metropolitan networks covering an area or a city of several miles, WMANs (wireless metropolitan area network);

– extended networks to cover a territory of hundreds of miles, WWANs (wireless wide area network).

Wi-Fi/IEEE 802.11 is a part of WLANs. It connects desktop computers, laptops or PDAs (personal digital assistant) equipment types through radio waves. The current performance can reach 54 Mb/s covering about three hundred feet. The main service offered is the Internet access without bothersome cables where the network is ensured through bridges.

21.1.3. *Topology*

Topology is based on the cell. A cell is called a BSS (Basic Set Service). It is used in two modes:

– the infrastructure mode (see Figure 21.1) is a topology where the cell is centered around a base station that is called Access Point (AP) and a BSS is therefore controlled by an AP. The APs can be linked together by a Distribution System (DS) that plays the role of backbone between BSSs in order to form an ESS (Extended Set Service);

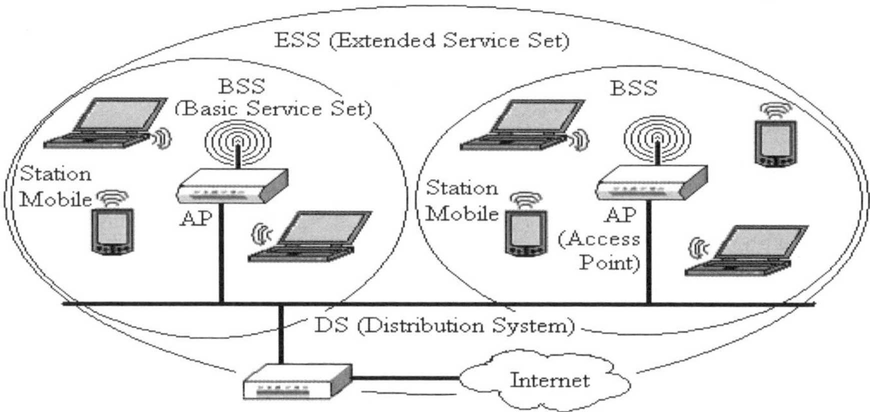

Figure 21.1. *Topology in infrastructure mode*

– the ad hoc[2] mode (see Figure 21.2) is a topology without infrastructure made up of IBSSs (Independent BSS) and there are no APs in this case. It connects mobile stations which are in the same reciprocal coverage zone and therefore they "see" each other.

2 It is not possible in the ad hoc mode to route information from one mobile station to another through intermediate mobile stations as is the case with ad hoc networks. In order to have this functionality, you must install an appropriate protocol such as OLSR (optimized link state routing protocol) proactive protocol (i.e. by updating possible routes before activation) or AODV (ad hoc on demand distance vector) reactive protocol (i.e. by flooding during route activation).

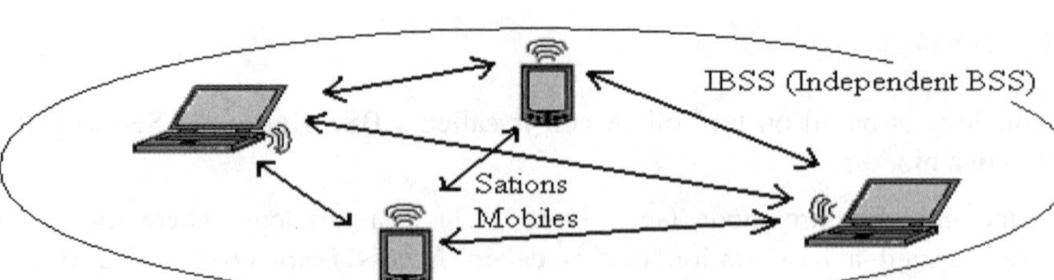

Figure 21.2. *Topology in ad hoc mode*

21.1.4. *Hardware*

There are two kinds of hardware in Wi-Fi/IEEE 802.11 networks:

– wireless stations which are laptop computers or PDAs equipped with a WNIC (wireless network interface card);

– access points which are boxes that control the exchanges and ensure bridge function with the cable network.

21.2. Technology

Wireless technology must resolve numerous problems that cable networks do not face. Among others, there is the necessity to find an available and adequate global bandwidth to support it (see Appendix B for the concerns that this problem causes with the diversity of amendments d, h and j), controlling signal degradation[3], user mobility, preserving communication confidentiality, taking into account limited battery life, being concerned for human health[4], etc.

21.2.1. *List of protocols containing IEEE 802.11*

This list is compatible with all those with the norm IEEE 802, which is close to the ones defined within OSI (open system interconnection) (see Figure 21.3). There

---

3 The attenuation, reflection, diffraction, diffusion, fading (produced by the echo's multiple paths), the Doppler effect and the electromagnetic interference (such as microwave ovens, garage door remote controls, etc.) can provide degradation of signal and affect wireless transmissions.

4 ¾ of the human body is made up of water. Water resonance frequency is at around 2.5 GHz and this frequency is used in Wi-Fi. We can naturally wonder if Wi-Fi use might have consequences on the human body.

is a physical layer compatible to that of OSI and a linking layer made up of two sub-layers contrary to that of OSI:

– an MAC (medium access control) sub-layer which determines the way that the radio channel is allocated, i.e. which is the next station authorized to transmit;

– an LLC (logical link control) sub-layer which presents to the higher network layer a homogenous interface whatever the type of IEEE 802 network used.

OSI model	IEEE model	Standards				
Linking	LLC	802.2				
	MAC	802.11				
Physical	PHY	IR	FHSS	DSSS	OFDM	HR-DSSS

Figure 21.3. *The list of IEEE 802.11 model protocols*

21.2.2. *Radio transmission techniques of the physical layer*

The goal of this layer is to transmit a MAC frame from one station to another. Electric engineering is the competency required in order to achieve its goal. The major differences concern the properties of the frequencies used (sensitivity, directivity, reach) and the modulations providing the appropriate throughputs and responses based on the requirements.

It is divided in two sub-layers:

– underneath, PMD (physical medium dependent) controls coding and modulation;

– above, PLCP (physical layer convergence protocol) listens to the support and provides a CCA (clear channel assessment) to the MAC layer to indicate that the support is available.

Several types exist.

21.2.2.1. *Infrared*

This transmission which broadcasts frequencies at 0.85 μ or 0.95 μ offers two throughputs of 1 and 2 Mb/s according to the encoding system used.

It does not go through the bulkheads and therefore it enables good area isolation. However, it is sensitive to sun rays. It has not been successful.

21.2.2.2. *FHSS/DSSS*

Two transmission methods by radio wave within the low reach 2.4 GHz[5] band use spread spectrum techniques.

21.2.2.2.1. FHSS (frequency hopping spread spectrum)

The band is divided in 79 channels[6] of 1 MHz each with a GFSK (Gaussian frequency phase keying) modulation. A random number generator produces the sequence for the frequencies to follow. Two stations which feed their generator with the same starting value (seed) at the same time as they are synchronized can communicate. Dwell time for each frequency is parameterizable but must be lower than 400 ms.

The advantages are: a fair allocation within the total allocated spectrum, an important improvement in security (listening-in is impossible without knowing the sequence and the utilization period of each frequency), good resistance to multipath propagation over long distances (because the utilization period over a frequency is short), good immunity to radio interference, a 26 network colocation in one location.

The disadvantage is its low throughput, which stays limited to 2 Mb/s.

21.2.2.2.2. DSSS (direct sequence spread spectrum)

The band is divided in 14 channels[7] of 20 MHz with 1 channel per BSS.

Throughputs are limited to 1 Mb/s with Binary Phase Shift Keying (BPSK) modulation of 1 bit per baud and 2Mb/s and a Quadrature Phase Shift Keying (QPSK) modulation of 2 bits per baud.

In order to limit noise effects, a technique (chipping) enables the extension of 1 data bit into 11 bits used by a Barker sequence. A theoretical maximum colocation of 3 BSS is possible.

---

5 We are referring here to the 2.4000-2.4835 GHz band also called ISM (Industrial, Scientific, Medical) band. The transmission power is controlled in France by ART (Autorité de Régulation des Télécommunications), in Europe by ETSI (European Telecommunications Standard Institute) and in the USA by the FCC (Federal Communication Commission).

6 35 or 79 channels for France according to the band, 79 for Europe and the USA.

7 13 channels for France and Europe, 11 for the USA.

21.2.2.3. *OFDM (Orthogonal Frequency Division Multiplexing)*

This technique offers a maximum theoretical throughput of 54 Mb/s. It is used in the IEEE 802.11a 5 GHz[8] band. It uses 52 frequencies of which 48 are attributed to data and 4 to synchronization.

It is a complex coding technique which is based on a phase leading modulation to offer high throughputs even with QAM (Quadrature Amplitude Modulation). At 54 Mb/s, 216 data bits are coded in 288 bits[9] symbols.

This offers the advantage of an efficient use of the spectrum in number of bits per hertz and a good immunity against multipath propagation interferences.

This technique is also used in the IEEE 802.11g 2.4 GHz band.

21.2.2.4. *HR-DSSS (High Rate DSSS)*

This technique, which works in the 2.4 GHz band, offers a maximum theoretical throughput of 11 Mb/s. It is used in the IEEE 802.11b norm.

A CCK (complementary code keying) coding technique is added to QPSK modulation.

Wi-Fi proposes the adjustment of throughputs according to the quality of transmission which is related to the environment and the distance. This mechanism (variable rate shifting) enables the automatic variation of the maximum throughput for each station from 11 to 5.5, then 2 and finally 1 Mb/s in the worst of cases. In the case of IEEE 802.11a products, the different throughputs go from 54, 48, 36, 24, 18, 12, 9 to 6 Mb/s.

The advantage is mandatory communication in the worst case scenario. On the other hand, the disadvantages can be numerous: high throughput stations must wait for slower exchanges from low throughput stations and real-time flows (video flows, for example) would suffer from this variation.

21.2.3. *Sub-layer MAC protocols*

The goal of this sub-layer is to determine the way in which the channel is allocated, i.e. which station will be the next one able to transmit. Computer engineering is the competency required in order to achieve this goal. It also controls

8 It is the U-NII (unlicensed national information infrastructure) band.

9 In order to ensure compatibility with the European HiperLAN/2 system (see footnote 13).

frame addressing and formatting, error control from CRC, fragmentation and re-assembly, Quality of Service (QoS), energy, mobility, security.

The characteristics of the radio medium are completely different from those of wired medium.

Indeed, contrary to Ethernet, it is impossible to transmit and listen at the same time (in full-duplex) to detect if a collision is happening during the transmission. The CSMA/CD (carrier sense multiple access with collision detection) technique is therefore ineffective. Furthermore, there are inherent problems with the radio medium.

In fact, let us consider 3 stations (see Figure 21.4a) on the same frequency A, B and C where A and B see each other, B and C see each other, but A and C do not see each other. A communicates with B. C also wants to communicate with B, but cannot detect that A is communicating with B. If C transmits, then there is a collision in B. This situation is called "hidden station problem". Here, A is hidden from C and vice versa.

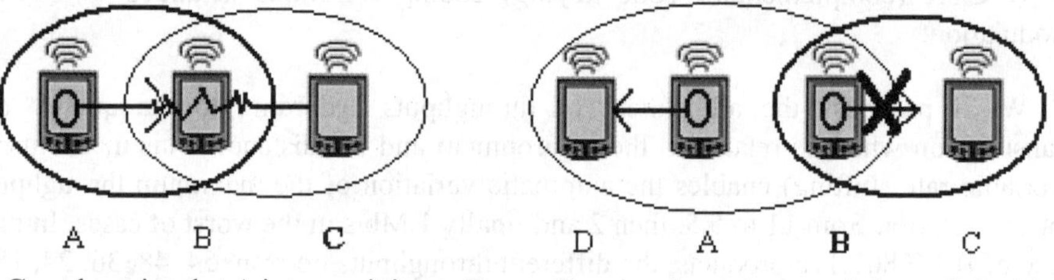
A
B
C
D
A
B
C
C not knowing that A is transmitting to B can decide to transmit and therefore interfere with the reception in B
B seeing that A is transmitting to D decides not to transmit to C although it is possible to do so

Figure 21.4. *(a) Hidden station and (b) shown station problems*

This situation brings up another problem (see Figure 21.4b). B sees A and C, but A and C still do not see each other. A transmits to a D station which is not within B's reach. B chooses not to communicate with C due to the perceived activity in the cell shared with A. This situation is called "shown station problem". In this case B is shown by A.

In order to resolve these problems the IEEE 802.11 norm accepts two modes of operation.

21.2.3.1. *DCF (distributed coordinated function)*

In this mode, the coordination function is distributed. It resembles Ethernet where it does not call on any centralized entity. It is distributed evenly between all stations in the cell (AP included in infrastructure mode). All transmitters therefore have an equal opportunity to access support (it is a sort of best effort). It is viable for asynchronous applications. This is a mandatory mode in IEEE 802.11. It is the only usable one in ad hoc mode.

The recommended technique is CSMA/CA (carrier sense multiple access with collision avoidance). It stems from the traditional CSMA. Two listening methods are possible.

Each frame is separated by an appropriate inter frame space (IFS) as we will see later (see Figure 21.8).

21.2.3.1.1. Wait for collision stochastic algorithm method (Backoff)

The transmitting station listens to the channel (see Figure 21.5) and initializes a transmission counter N to 0.

IF the channel is available

THEN

IF it remains available for a time higher than a DIFS (DCF Inter Frame Spacing)

THEN

(1) The station can transmit the data. It informs the group of listening stations of the theoretical maximum exchange time. This time is placed in a NAV field within the data frame (see Figure 21.10).

The transmission is done in its entirety no matter what (collision, interference, etc.).

The receiving station verifies if the CRC (Cyclic Redundancy Check) of the received message is valid, then sends or not an ACK (acknowledgment) after a SIFS (Short IFS) such as SIFS<DIFS in order to ensure that the current exchange is not interrupted by a new exchange from other stations.

IF the receiving station does not receive an ACK,

THEN

It considers that the message has not been received and will attempt to resend it (N is incremented) a certain number of times and go to (2).

OTHERWISE

Go to (2).

OTHERWISE

(2) The transmitting station listens to the channel and tries to extract the NAV field in order to find out the theoretical maximum time for the current exchange to avoid disruption. Once the channel is available again.

IF it stays available for a time higher than a DIFS

THEN

IF no random delay has been drawn (Timer Backoff)

THEN

Draw a random delay (Timer Backoff) between 0 and ($CWmax=2^{K+N}-1$)*time_slot; where CWmax is the Contention Window, N is the retransmission number and K is the minimum size of the contention window

Waiting for an event:

EITHER the Timer Backoff has expired

THEN to transmit go to (1).

OR the channel is busy once again (by another station whose Timer Backoff has expired previously)

THEN the current Timer Backoff is reduced according to the elapsed time (entire time_slot number because all stations are synchronized) then go to (2)

OTHERWISE

Go to (2).

By multiplying by two the contention window size for each collision (because N is incremented in this case), the stochastic wait algorithm in the case of exponential collision (Backoff) enables a greater reduction of the probability that a collision will happen for all the stations that have drawn the same Timer Backoff.

The only time when this algorithm is not used is when a transmitting station notices that the channel is available when needed and that it remains that way during a DIFS (the case where the first two algorithm tests are true).

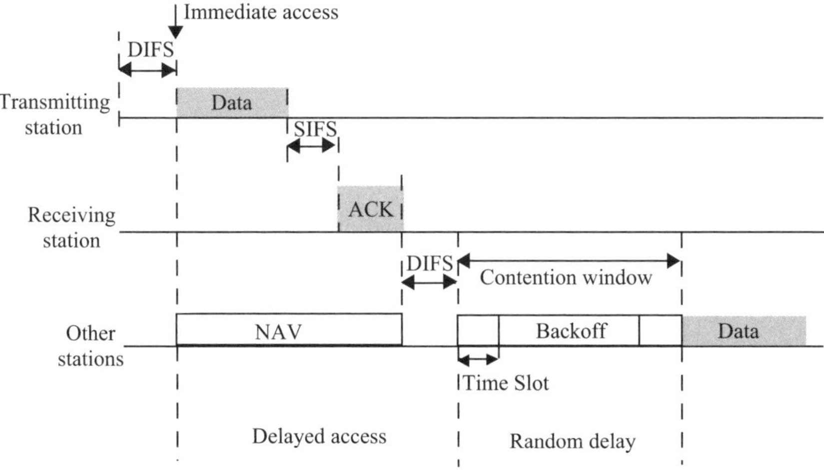

Figure 21.5. *Access to support by the collision stochastic algorithm the case of collision*

21.2.3.1.2. VCS (Virtual Carrier Sense) method

This is derived from MACAW (Multiple Access With Collision Avoidance For Wireless LAN) which starts a preliminary exchange between 2 parties to send a signal to the environment that an exchange will take place.

This enables the resolution of the "hidden station" problem and thus reduces collision risks.

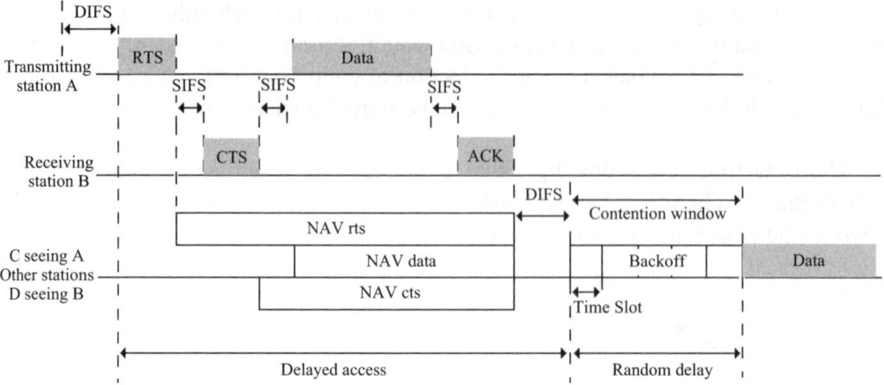

Figure 21.6. *Virtual channel method*

In this example A and B see each other and A wishes to transmit to B (see Figure 21.6):

– A listens to the virtual channel (VC);

– if the channel is available, then A requires permission to transmit by sending an RTS (request to send) which is a little 30 byte frame;

– B receives the RTS and decides whether to reject or accept the request;

– if B accepts the request then B transmits an invitation to transmit CTS (clear to send) after an SIFS;

– A receives the CTS, also waits for an SIFS then transmits the data;

– after the transmission and the wait time for an SIFS, A receives an ACK transmitted by B. If the acknowledgement is not received then the complete transmission has to be restarted.

After the RTS/CTS exchange, A and B are sure that the channel has been reserved for the data transmission without collision. Indeed:

– let us presume that C sees A but not B:

- seeing the RTS, C declares the virtual channel busy and will therefore not transmit until the timer expires (network allocation signal) indicated in a NAV (network allocation vector) field (see Figure 21.10) which is located in the RTS frame;

- therefore C does not collide with the current exchange with A,

- A can then receive the CTS and the ACK without collision;

– or D sees B but not A:

- seeing the CTS, D declares the virtual channel busy and therefore does not transmit until the timer expiration indicated in a NAV field which is located in the CTS frame,

- then D will not collide with the current exchange with B,

- B can then receive data without collision.

This support allocation mechanism does not decrease the probability of collision over RTS/CTS control frames. In fact:

– the RTS frame can collide with B with a transmission from D; in this case the channel is not globally reserved because no CTS has been sent;

– the CTS frame can collide within the coverage zone for B; in this case the channel is not globally reserved in this zone and therefore data reception from B can be disrupted by an activity within this same zone.

On the other hand, the cost of retransmission of these little 30 byte frames is much lower than the cost of a data frame that would be a maximum of 2,500 bytes.

21.2.3.1.3. Fragmentation

Since radio transmissions are noisy and unreliable and the probability that a transmission is successful is inversely proportional to the size of the frame, the IEEE 802.11 protocol accepts fragmentation of long frames in order to limit disruptions.

Fragmentation is done for frames whose size is larger than a certain threshold (fragmentation threshold).

Each fragment is numbered ($Frag_i$) and acknowledged (ACK_i) through a "stop-and-wait" source protocol. The virtual channel is reserved by RTS/CTS which contains the NAV of the first fragment. Then, each fragment contains the NAV of the next fragment in the manner of RTS/CTS in order to extend channel reservation until the last fragment. ACK, RTS, CTS and Fragment are separated by SIFSs in order to guarantee continuity of exchange. This enables transmission by bursts (see Figure 21.7).

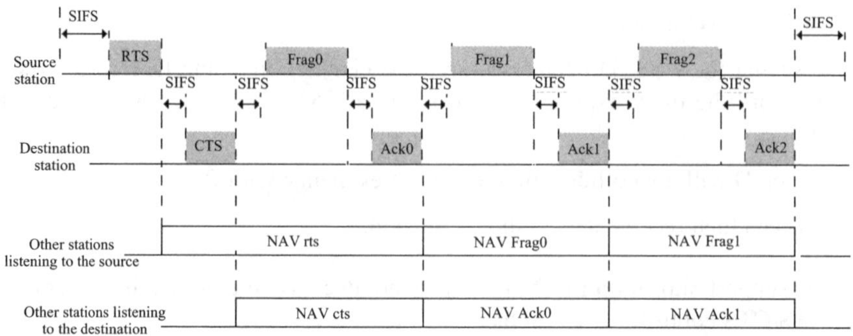

Figure 21.7. *A fragmented transmission*

An error will only produce the retransmission of a fragment and not of the entire frame. This increases effective throughput. The size of the fragment is not set by the norm, it is different for each cell and can be adjusted by the base station.

21.2.3.2. *PCF (Point Coordination Function)*

In this mode, the coordination function is centralized on the base station (AP) to control the cell's activity. It sends an offer to transmit (poll) to each station that needs this service. Data exchange can be either increasing or decreasing. Stations are all polled one at a time. Collisions are therefore impossible since the channel is allocated successively for each station. The standard only proposes the mechanism, but it indicates nothing for frequency, order and polling evenness between the stations.

This mode is optional. It can be used alternately with the DCF mode. It is recommended for real-time applications such as voice or video. Its centralized character makes it useless in ad hoc mode.

The AP can take control of the support because it has a PIFS (PCF IFS) timer that is smaller than the one in the DCF mode (PIFS < DIFS). Therefore, it has the most priority than all the other types of exchange. PCF mode frames are also separated by the smallest SIFS timer (SIFS < PIFS < DIFS) in order to guarantee continuity.

Even if this mode enables a certain bandwidth guarantee, it is still not implemented in current cards.

The exchanges that we have shown are done with the help of 4 different time intervals (IFS = Inter Frame Space) (see Figure 21.8):

– the short inter-frame interval (SIFS, Short ISF) enables a current exchange to continue without interruption. It precedes: a CTS after an RTS, an ACK or ACK_i after data or a $fragment_{i+1}$ after an ACK_i;

– the PCF (PIFS, PCF IFS) inter-frame interval enables a base station to send a beacon frame or an invitation to transmit (poll);

– the DCF (DIFS, DCF IFS) inter-frame interval enables any station to transmit a frame within the traditional contention rules controlled by the backoff algorithm;

– the extended inter-frame interval (EIFS, Extended IFS) enables a station to signal that it has received an erroneous or unknown frame without interfering with the other exchanges (that is why it has the lowest transmission priority) and therefore it has the longest timer.

This interval delay order makes it possible to define priorities and to complete more urgent exchanges that are current before the less urgent exchanges.

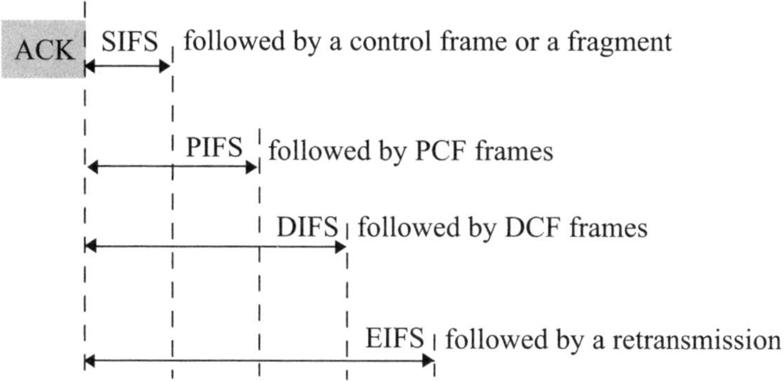

Figure 21.8. *Inter-frame spacing within the IEEE 802.11 standard*

21.2.3.3. *Other capabilities*

In infrastructure mode, the base station broadcasts a beacon frame at regular intervals (10 to 100 times a second) which enables all the stations to communicate in the BSS and which contains such system parameters as:

– the sequence of the frequencies to use and the maintenance times for these frequencies (for FHSS);

– clock synchronization;

– the reminder for new stations to subscribe to the list of stations called to transmit for a certain throughput in order to guarantee corresponding bandwidth and services that can use it in PCF mode;

IEEE 802.11 controls power management in order to ensure optimized energy usage for mobile stations that only have limited power capacity. Two modes are proposed:

– the continuous aware mode by default;

– the power save polling mode where mobile stations decide to go in sleep mode. They subscribe to this service through AP. They only become active when AP sends a transmission of its beacon frame which contains information in a TIM (Traffic Information Map) table on data that might have been received during sleep mode. If that is the case, stations request to receive the data from AP and they return to sleep mode. Once that is done, the AP will buffer the new received data until the new beacon frame. This mechanism decreases power consumption but also decreases transmission throughput.

21.2.4. *Frame structure*

At physical level, a frame is made up of 4 fields (see Figure 21.9):

– a preamble for synchronization and beginning of frame marking;

– a PLCP (Physical Layer Convergence Protocol) header to define the length and throughput of data;

– encapsulated MAC data;

– a CRC (Cyclic Redundancy Check);

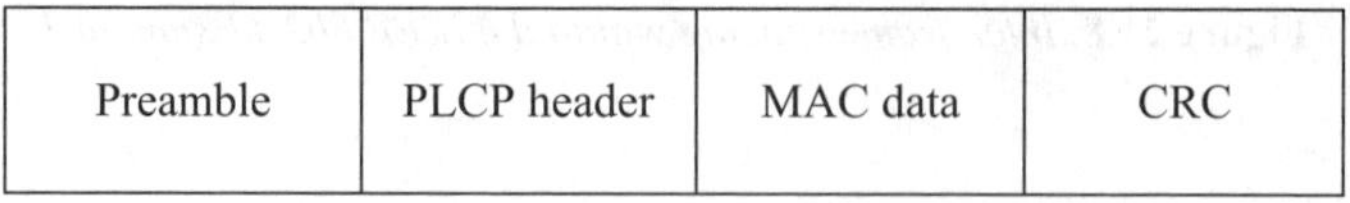

Figure 21.9. *Physical frame structure*

At MAC level, IEEE 802.11 defines 3 types of frames.

21.2.4.1. *Data frames*

This frame encapsulates IP packets. It is made up of the following fields (see Figure 21.10):

– a control frame (2 bytes) with the 11 following sub-fields (see Figure 21.11):

- current protocol version in the cell (2 bits),

- the type (data, control, management) (2 bits),

- the sub-type (e.g., RTS, CTS) (4 bits),

- to DS: the AP sends the frame toward the intercellular Distribution System (1 bit),

- from DS: the frame comes from the intercellular Distribution System (1 bit),

- a fragment will follow (more fragments) (1 bit),

- indicates that it is the retransmission of a frame (retry) (1 bit),

- sleep/active switch for power management (1 bit),

- indicates whether there are more frames for the receiver (more data) (1 bit),

- indicates whether the core of the frame is encrypted with WEP (Wireless Equivalent Privacy) (1 bit),

- indicates to the receiver to process according to the order of received frames (1 bit);

– duration in milliseconds (1 CTS + 1 Data or Control + 1 ACK + 3 SIFS) corresponding to the time the data frame will occupy and its acknowledgement in order for the other stations to manage the NAV reservation mechanism (2 bytes);

– 4 address sub-fields in the IEEE 802 format (24 bytes):

- destination (@1) (6 bytes),

- source (@2) (6 bytes),

- destination base station if transmitting to another cell (@3) (6 bytes),

- source base station if receiving from another cell (@4) (6 bytes);

– Sequence number (2 bytes):

- fragmented frame number (12 bits),

- fragment number in this frame (4 bits);

– data (up to 2,312 bytes);

– an FCS (Frame Check Sequence) control (4 bytes).

Control	Duration	@1	@2	@3	Sequence	@4	Data	FCS
6 bytes	2 bytes	6 bytes	6 bytes	6 bytes	2 bytes	6 bytes	0 to 2,312 bytes	4 bytes

Figure 21.10. *MAC data zone structure*

Protocol version	Type	Sub-type	To DS	From DS	More fragments	Retry	Power managt.	More data	WEP	Order
2 bits	2 bits	4 bits	1 bit	1 bit	1 bit	1 bit	1 bit	1 bit	1 bit	1 bit

Figure 21.11. *MAC header control sub-field structure*

21.2.4.2. *Management frames*

These frames manage the connection via association and authentification mechanisms. Their format is essentially the same as the data frame format. The difference is indicated in the frame control type field.

21.2.4.3. *Control frames*

These frames manage access control to support via RTS, CTS and ACK frames.

They are made up of 4 fields (see Figure 21.12):

– a frame control (2 bytes);

– NAV duration (2 bytes);

– address sub-fields in the IEEE 802 format (6 or 12 bytes):

  - destination one for CTS, ACK and RTS,

  - source only for RTS.

– an FCS (Frame Check Sequence) control (4 bytes).

Control	Duration	@dest	@source	FCS
2 bytes	2 bytes	6 bytes	6 bytes	4 bytes

Figure 21.12. *RTS control frame structure*

21.2.5. *Proposed services*

The IEEE 802.11 standard lists 9 services of which 5 are for distribution and 4 are for stations.

21.2.5.1. *Distribution services*

They manage the members of a cell and the interaction with the other cells in the case of mobility. They are controlled by the base stations.

– the association that takes place after authentification enables a mobile station to connect to a base station in infrastructure mode or to other mobile stations in ad hoc mode. This can be done at the beginning, in active mode (after sleep mode). Two modes are possible according to the power of the received signal and the power consumption necessary for the exchange:

- passive listening where AP regularly sends a beacon frame with synchronization information,

- active listening where the mobile station transmits a request frame (probe request frame) to the AP which indicates its identity and its capabilities (throughput, whether PCF is required or not, power management) and waits for a response;

– dissociation of the base station's initiative (for maintenance reasons, for example) or of the mobile station (for start or extinction reasons, for example);

– reassociation enables the implementation of the move from one mobile station to another without communication loss[10] (see section 21.3.3). In order to do this, the mobile station regularly tests the strength of received signals on every channel. It can decide to associate with another channel and this enables the creation of WLANs by recovery of cells whose channel frequencies are disjointed in order to avoid interference;

10 It could also be that a fixed station notices that another access point is better in terms of signal or load quality than the one on which it is connected. Wi-Fi integrates a load balancing mechanism to avoid this kind of situation.

– the distribution indicates how to route frames to destination from a base station. If the destination is local, then the base station sends it directly. If not, then the base stations routes it to the distribution system;

– integration makes it possible to adapt the addressing strategy if the frames are on a different network.

21.2.5.2. *Service stations*

These are used for intracellular activity:

– authentification happens either after having accepted the new mobile station in the cell by reassociation mechanism or after the listening mechanism for the authorization to exchange; there are two mechanisms for this function:

- by default, where all stations are accepted (open system authentification),

- if the WEP is activated (shared key authentification) then the AP sends a challenge to the mobile station for the password (or the secret key) that was assigned to it. The mobile station enters its key and then returns to the AP. If the result is correct, the mobile station is registered in the cell;

– authentification cancellation will terminate a session;

– confidentiality can be guaranteed by data encryption (WEP; see section 21.3.1);

– data delivery, which is the main objective for a communication system, is not optimal in the case of wireless networks and therefore higher layers must also manage the detection and correction of errors which are not controlled by the IEEE 802.11 level.

21.3. Amendments, progress and characteristics

The first IEEE 802.11 norm provided a maximum theoretical throughput of 1 Mb/s. It quickly became clear that this value was considered a limit that would impede its commercial development. The first enhancement was the first normalized IEEE 802.11b amendment offering a maximum theoretical throughput of 11 Mb/s.

It was followed by the second normalized IEEE 802.11a amendment offering a maximum theoretical throughput of 54 Mb/s. The way was then paved for other amendments in order to improve the norm and respond to market demand and take advantage of the solutions provided by the IEEE with the help of engineers and researchers. The list of amendments can be found in Appendix B.

21.3.1. *Security (amendment i)*

No need for alligator clips to connect to a wireless network. An antenna is sufficient to hear what can be received wherever it is[11]. It is therefore normal to protect ourselves from unwanted threats.

Wi-Fi/IEEE 802.11 exposes 3 security mechanisms:

– the first one is mandatory. It controls the access to the network. It starts at the network's name or SSID (Service Set ID). A station that does not have it cannot access the network;

– the second is optional. It also controls access to the network from the MAC address list (ACL, Access Control List) of the cards authorized to connect. A station without the right MAC address cannot connect to the network;

– the third is optional. It authenticates, encrypts and verifies the integrity of the data. It is a secure data linking protocol (at MAC level) called WEP (Wired Equivalent Privacy) which enables a confidentiality similar to cable systems. Since by default there are, this objective seems easy to reach. It belongs to the family of symmetrical algorithms with a shared secret key of 40 or 104 bits.

It is based on the RC4 (Ron's Code 4) algorithm made up of two parts:

– production of a key (key scheduling algorithm) by concatenation of the shared secret key and of an IV (Initialization Vector) to generate a state table;

– which in turn makes it possible to generate a pseudo random sequence through a PRNG (Pseudo Random Number Generator) derived from the RC4.

At transmission (see Figure 21.13):

– clear IV is transmitted in the frame;

– the secret key and the IV are concatenated (key scheduling algorithm);

– the result is injected in the PRNG/RC4;

– clear data is mixed by an operator or exclusive to the sequence generated by the PRNG/RC4 and then transmitted in the frame;

– an ICV (Integrity Check Value) is executed from the data and then mixed by the operator or exclusive to the sequence generated by the PRNG/RC4 and then transmitted in the frame;

– the header and the CRC of the MAC frame are not encrypted in order to let the other stations listen to the information useful for proper execution of the access protocol.

---

11 WarDriving is a technique which consists of recording what is happening in a given area with the objective of penetration or even mapping via a GPS (Global Positioning System).

At reception (see Figure 21.14):

– the IV is extracted from the frame;

– then concatenated with the secret key (key scheduling algorithm);

– the result is injected in the PRNG/RC4;

– the encrypted data are mixed by an operator or exclusive to the sequence generated by the PRNG/RC4;

– an ICV' integrity control is executed from the encrypted data;

– the ICV integrity control is extracted from the frame;

– the ICV and the ICV' are compared in order to verify data integrity, i.e. to ensure that they were not modified.

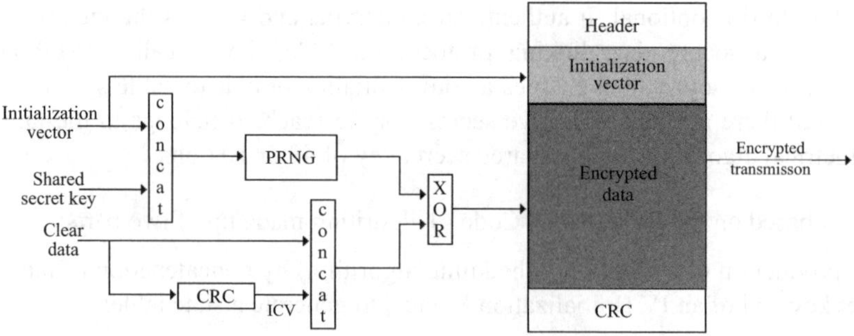

Figure 21.13. *Encryption at transmission*

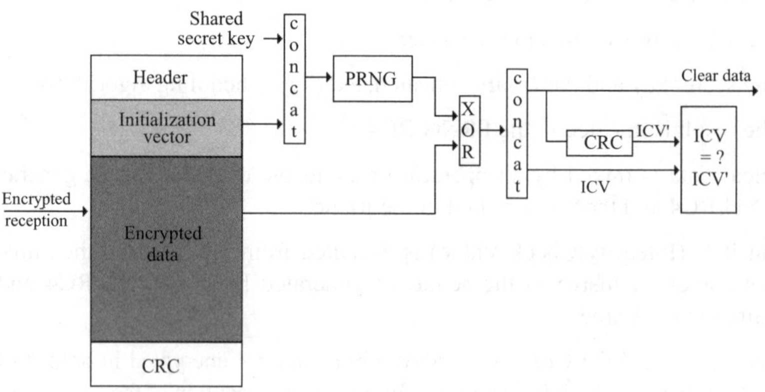

Figure 21.14. *Decryption at reception*

The concern is that there are generally significant problems with these security measures. Indeed, the SSID circulates clearly on the network and can therefore be exploited again (spoofing), a MAC address can also be processed again and make the ACL useless and the RC4 key is breakable if we have a certain number of encrypted frames.

IEEE 802.11i recommends using the AES (Advanced Encryption Standard) to improve encryption and authentification. It is a fast and reliable algorithm but a heavy consumer of resources.

While we are waiting for this solution integrated to Wi-Fi, there are other ways to secure such a network, such as:

– regularly changing the key;

– IEEE 802.1x (port-based network access control) for authentification and key management;

– EAP (Extensible Authentification Protocol) for authentification;

– RADIUS (Remote Authentification Dial-In User Service) for authorization and authentification;

– AAA (Authentification, Authorization, Accounting) for accounting, authorization and authentification;

– VPN (Virtual Private Network) for offering a secure end-to-end tunnel via IPSec;

– a smart card for authentification and other services.

In practice, we must not hesitate to multiply the protection in order to make a system that is less vulnerable. In the case of Wi-Fi, in addition to the solutions that were just explained, we must opt for the infrastructure mode rather than the ad hoc mode which offers no security, not use DHCP (Dynamic Host Configuration Protocol) but instead attribute fixed IP addresses and set the transmission power to the minimum desired coverage zone.

In the meantime, the Wi-Fi alliance proposes a new encrypting protocol with the name WPA (Wi-Fi protected Access).

21.3.2. *Quality of service (amendment e)*

Quality of Service (QoS) is critical for real-time applications (such as voice, video, multimedia, etc.). In order to achieve this we must define priorities between data in order to best guarantee its delivery when it is required. PCF mode is a good start in offering QoS but it is not or barely implemented in current cards. And it is still pretty much elementary.

In amendment e, this is done at frame level (MAC), which is the lowest and therefore the fastest. Priority management is executed by reducing IFSs. in this way priority flows (at short IFS) will always go before the flows with less priority (at longer IFS).

For this, amendment e proposes 2 methods:

– the EDCF (Extended DCF) is an improvement from DCF which implements 8 priority levels that represent 8 traffic categories (TC) through 8 queues which have their own AIFS (Arbitration IFS), contention window (CW) persistent factor (PF) such as DIFS < AIFS < EIFS and the CW is doubled in the case of collision only if PF = 2, as in the case of the traditional backoff algorithm. If two traffic classes wish to transmit at the same time, the class with the most priority, determined by a TxOP (Transmission Opportunities) timer is selected;

– the HCF (Hybrid Co-ordination Function) combines the EDCF and the PCF in such a way that it is possible to transmit bursts of PCF even within the core of the EDCF. This is done in order to guarantee access to the support of the periodic traffics.

In the meantime, priority management can be assured by delaying acknowledgements of low priority packets at IP level. We can only hope that this amendment will not have the same outcome as PCF.

21.3.3. *Mobility (handover), amendment f*

The goal is to maintain communication during handover[12].

Juxtaposing cells (BSS) one next to the other by linking them through a backbone forms a coverage surface (ESS) as large as needed (see Figure 21.15).

This juxtaposition is done by recovering neighboring cells that must have different sequences. In this way, a mobile station can see multiple APs in a given area.

---

12 This concerns horizontal handover between cells of a same network. It should not be confused with vertical handover which operates between cells of different networks such as between Wi-Fi, GPRS (General Packet Radio Service), UMTS (Universal Mobile Telecommunications System) or CDMA 2000 (Code Division Multiple Access 2000) without interruption of communication. Roaming enables the access of its usual network through another one via a disconnection of its home network and a reconnection to the new one.

In the case of mobility it can choose the AP which corresponds best to it in each area. The handover describes the mechanism which makes it possible to go from one cell (BSS) to another without interrupting the current communication.

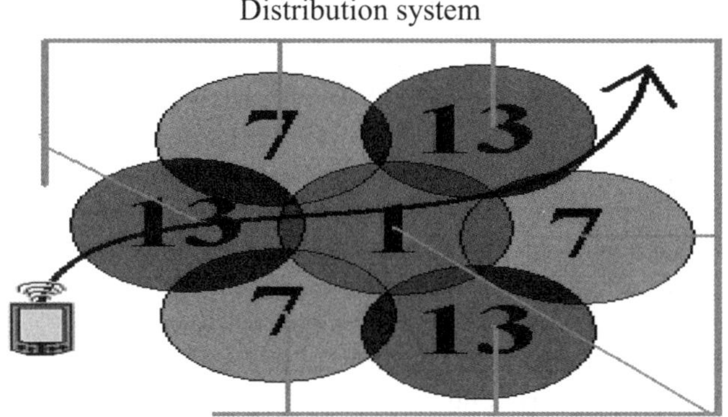

PDA in mobility situation between multiple cells. Cell APs are linked by the DS. Cell frequencies correspond to channels 1, 7 and 13.

Figure 21.15. *Example of intercellular mobility*

This mobility relies upon the proprietary Inter-Access Point Protocol (IAPP). Security can be assured by the RADIUS protocol. Protocols such as DHCP or Mobile IP are recommended to ensure distribution of correct IP addresses. While we are waiting for f (standardized mechanism), the association/reassociation/ dissociation mechanisms are used via cellular IP/Mobile IP.

The handover recommends the development of ambient networks which enable a connection to the Internet from anywhere, at any time and at reduced cost. They are at the junction between wireless networks and mobility.

21.3.4. *Throughputs (amendments g and n)*

IEEE 802.11g, which has been finalized since 2003 and considered as an extension of a and b, uses the ISM band of 2.4 GHz as well as CCK coding and OFDM transmission techniques in order to reach a throughput of 54 Mb/s.

g is compatible with b in the sense that it operates in the same frequencies and when all the correspondents are in g, then the exchange is done in g. However, when a station connects in b then all exchanges are done in b.

IEEE 802.11b proposes a maximum theoretical throughput of 11 Mb/s. By taking out the overhead caused by headers, signaling, IFS timers and backoff and by retransmissions and possibly by encryptions, it is practically decreased by half to 6 Mb/s. It is the same with amendments a and g which propose a maximum theoretical throughput of 54 Mb/s for a practical throughput which almost never exceed 22 Mb/s.

The future standard 802.11n will be officialized at two times. A first version, envisaged in 2007, will ensure the compatibility of the materials of different marks. It will be necessary to wait until 2008 for the final standard. It is greatly inspired by the HiperLAN[13] norm. It must offer throughputs of 108 and even 320 Mb/s, with the characteristic that practical throughputs should be close to these theoretical throughputs with a desirable efficient range of 100 feet. It integrates new compression algorithms, better error and interference management as well as a new antenna technology. It must enable the control of power, the control of frequencies currently communicating, use both 2.4 and 5 GHz bands and integrate i, e and f amendments. This new standard would constitute a technological break that would require a change of all the cards. Considering these requirements, ensuring compatibility between cards of different manufacturers will remain a challenge.

21.4. Conclusion

21.4.1. *Uses*

Wi-Fi/IEEE 802.11 is really a revolution in information technology in general and in Internet access in particular. It rapidly deploys WLANs in the form of HotSpots[14]. It brings to the Internet what the laptop computer has brought to information technology: mobility.

The attraction for Wi-Fi/IEEE 802.11 is explained by its ease of deployment, its integration with existing infrastructure, the low cost of material and the wide utilization spectrum.

---

13 HiperLAN is a WLAN technology standardized by the ETSI (European Telecommunications Standard Institute) which does not benefit from international support although it offers excellent security, QoS, coverage and throughput services, among others.

14 Public access points to the Internet via a wireless connection in high traffic areas such as train stations, airports, universities, shopping centers, hotels, specific events, user groups, etc.

The liberalization of Wi-Fi usage outside[15] brings new operators, the WISPs (Wireless Internet Service Providers).

Apart from HotSpots there are other uses for Wi-Fi, such as interoffice connection for linking together two sites with a distance of 300 feet with two directional antennae, management of production centers, warehouses, workshops with hardened material, real-time follow up of patient files in hospitals, applications linked to mobility for companies where the personnel need their computers and can go from offline to online and work wherever they need to.

Uses linked to transport are also expanding rapidly. The transport mode obviously being the car, it can now communicate. Concept cars already exist with Wi-Fi and TabletPCs. The vehicle must be linked outside via GPRS/UMTS and even Wi-Fi. Services offered can be: downloading files (e-fuel) in gas stations and parking areas, help with navigation, access to the Internet, vehicle follow-up by the dealer (communicating black box), vehicle follow-up in the case of theft, user mobility services offered in urban or rural areas.

21.4.2. *Currently*

The IEEE 802.11 produces 6 norms (see Table 21.1): 3 for low throughput (1 InfraRed, 2 FHSS and DSSS spread spectrum radio waves) and 3 for high throughput (with a, b and g amendments). This may seem like a lot and contradictory with the objective of a norm that is supposed to define a common technical base to continue the economic expansion. The latter enables manufacturers to make products according to requirements, but also to develop new services and therefore generate new needs. In fact, this objective is the source for the diversity of the solutions. Furthermore, the economic challenges generate significant pressures to create a new standard that would help define a common base for interoperability. An important dynamic exists around WLAN and especially Wi-Fi/IEEE 802.11 definitions. This process seems to frantically auto-fill.

15 CraieFiti (war-chalking) indicates the presence of access points and their characteristics with easily recognizable symbols.

Version		802.11	a	b	g	e	f	i	n
Throughput (Mbit/s)		1-2	6-54	2-11	2-54				320
Band (GHz)		2.4	5	2.4	2.4				2.4 & 5
PHY layer		FHSS DSSS	OFDM	HR-DSSS	OFDM				
MAC layer	QoS					Yes	No	No	Yes
	Handover					No	Yes	No	Yes
	Security					RC4	RC4	AES	AES

Table 21.1. *Wi-Fi/IEEE 802.11 and amendments, toward a complete protocol*

21.4.3. *Currently being studied*

Reservation mechanisms by virtual channel (VCS) and PCF are rarely implemented. The WEP is considered somewhat weak. There is a good probability that theoretical throughputs will be compared to processor frequencies in the next few years. Progress is therefore necessary. It will mainly be:

– physical aspects through other types of coding and modulation:

- to improve throughputs in amendment IEEE 802.11n,

- to improve range, coverage;

– software aspects with:

- QoS in amendment IEEE 802.11e,

- a better handover control in amendment IEEE 802.11f,

- better security in amendment IEEE 802.11i.

21.4.4. *The future*

Wi-Fi/IEEE 802.11 is currently the main technology used with WLANs and the ambient Internet. Products exist, but do not technically respond to all requirements.

It may very well become a master player of future 4G systems. There is still significant development potential with it. At the very least, it will have a place as a complement to other products and techniques. It could even become a serious competitor due to its rapid growth. In order to achieve this, multiple amendments are being finalized. Can we presume that these amendments might be assembled in one standard that will put together all the solutions to the known problems? Will amendment n finally be the standard that we have been waiting for?

Wi-Fi/IEEE 802.11 plays its part in this trio: miniaturization of electronic components, energy autonomy and mobility. The future is promising.

Wi-Fi/IEEE 802.11 and wireless networks in general[16] constitute an alternative and a complement to cable networks. There is still a lot of work to be done to reach maturity. Reinforcing security, making installation and management easier, facilitating coexistence, developing interoperability among them and mobile telephony are among the improvements yet to happen[17]. These rapidly deploying networks are expected to grow quickly. There is no doubt that wireless networks and Wi-Fi/IEEE 802.11 in particular constitute the future of networks in general.

21.5. Appendices

21.5.1. *Appendix A: Web references*

Normalization organism:

– IEEE: http://www.ieee.org;

– IEEE for the 802.11norm: http://IEEE 802.org/11/;

Standardization organism:

– WECA: http://www.weca.net.

21.5.2. *Appendix B: amendments to IEEE 802.11 standard*

The first basic IEEE 802.11 norm offers 3 types of products: 1 in IR and 2 in radio in the ISM band at 2.4 GHz with DSSS and FHSS modulations for a maximum throughput of 2 Mb/s (finalized in 1997, corrected in 1999).

---

16 The explosion of wireless networks does not only concern Wi-Fi but also HiperLAN (WLAN), BlueTooth (WPAN, IEEE 802.15.1), ZigBee Alliance (micronetworks – mesh network), UWB (UltraWide Band), WiMAX (IEEE 802.16), etc.

17 We can already see the merging of PDAs and cell phones.

The list of known amendments:

– a: 54 Mb/s in the 5 GHz band over 8 modulation channels OFDM for ranges of approximately 300 feet outside and 30 inside with a maximum throughput at 15 feet; incompatible with the basic IEEE 802.11 norm and amendment b (finalized in 2001);

– b: 11 Mb/s in the 2.4 GHz band over a maximum of 14 channels with HR-DSSS modulation for ranges of approximately 900 feet outside and 90 feet inside with a maximum throughput at 30 feet; compatible with basic IEEE 802.11 norm (finalized in 1999);

– c: enables the definition of a bridge with 802.11 frame at linking level (finalized in 1998);

– d: equivalent to b but adapted to the regulations of countries in terms of frequencies and transmission power (finalized in 2001);

– e: for QoS at MAC level (applies to a, b and g); project approved in 2000;

– f: for handover management between APs of one DS via the IAPP protocol in order to enable interoperability of APs and DS of different manufacturers (project approved in 2000);

– g: 54 Mb/s in the 2.4 GHz band with CCK/OFDM modulation; compatible with basic IEEE 802.11 norm and amendment b (finalized in 2003);

– h: equivalent to a but adapted to regulations of European countries in terms of frequencies and transmission power (finalized in 2003);

– i: to improve security and authentification at MAC level (applies to a, b and g), replacing WEP and based on IEEE 802.1x (project approved in 2000);

– j: equivalent to h but adapted to Japanese regulations (project approved in 2002);

– k: must supply information to optimize network resources by adding other information such as position determination technology in order to offer new services (project approved in 2002);

– l: not used for risk of confusing it with i;

– m: workgroup that must correct interpretation errors of the norm (project approved in 2003);

– n: increase of throughputs to 108, maybe even 320 Mb/s, enables power management, management of current communication frequencies, uses both 2.4 and 5GHz bands, integrates amendments i, e and f (project approved in September 2003, expected for 2005/2006);

– o: not used for risk of confusing it with 0.

21.6. Bibliography

[MAL 02] MALES D., PUJOLLE G., *Wi-Fi par la pratique*, Eyrolles, April 2004.

[MUL 02] MÜHLETHALER P., *802.11 et les réseaux sans fil*, Eyrolles, 2002.

[STD 97] IEEE Std 802.11-1997. Part 11 "Wireless LAN Medium Access Control (MAC) and Physical Layer (PHY) Specifications", IEEE, 1997.

[STA 99] IEEE Std 802.11a-1999. Part 11 "Wireless LAN Medium Access Control (MAC) and Physical Layer (PHY) Specifications", IEEE, 1999.

[STB 99] IEEE Std 802.11b-1999. Part 11 "Wireless LAN Medium Access Control (MAC) and Physical Layer (PHY) Specifications", IEEE, 1999.

[STX 01] IEEE Std 802.1x "Standard for Local and Metropolitan Area Networks: Port Based Access Control", IEEE, 2001.

[AES 01] FIPS PUB 197AES "Advanced Encryption Standard (AES), FIPS, 2001.

Chapter 22

Mobility and Quality of Service

22.1. Introduction

Internet protocols as defined by the IETF[1] have been designed for fixed networks. Used in mobile or wireless environments, their performance is greatly impacted.

Users expect a flexible access to Internet services which is not only limited to data services, but to multimedia services as well. This brings new problems for Quality of Service (QoS), since it must include terminal mobility. Different QoS architectures have been defined in the mobile networks context but none are considered as the complete solution for offering QoS in a mobile context.

The first efforts for an Internet access system through wireless networks have been developed by the telecommunications community where different systems had been proposed such as the GPRS (General Packet Radio Service) [ETSI 98] and the UMTS (Universal Telecommunication System) [KAA 00] systems. These systems are able to transport IP packets by using a packet switching network in parallel with the circuit switching phone network. These architectures use proprietary protocols and are subjected to strict licensing rules in terms of usage and price.

This chapter will identify interaction requirements between QoS and mobility architectures. To achieve this, a summary of QoS and mobility solutions will be presented, followed by a description of the problems linked with interaction between

Chapter written by Hakima CHAOUCHI.

1 Internet Engineering Task Force. www.ietf.org.

QoS and mobility. Finally, a few architectures supporting QoS and mobility will be presented.

22.2. Summary of QoS and mobility architectures

In order to understand the interaction between QoS and mobility, a summary of these architectures is provided in this section.

22.2.1. *QoS architectures*

since the Internet was not designed to support applications with QoS constraints such as telephony or video, the objective of manufacturers and thus of the IETF must be to attain a global architecture guaranteeing end-to-end QoS. IETF's studies have arrived mainly at three basic architectures: the IntServ (Integrated Services) architecture, the DiffServ (Differentiated Services) architecture and the RTP (Real-Time Protocol) architecture for real-time applications. Other options have been defined around these basic architectures, such as ISSLL (Integrated Service over Specific Link Layer) which defines IntServ architecture over DiffServ. More recently, MPLS (Multipath Label Switching) technology has opened a new approach to QoS control. Below is a brief presentation of the two basic architectures which are IntServ and DiffServ, as they are the ones with the most dependable interaction and mobility offerings.

22.2.2. *QoS in brief*

QoS defines non-functional parameters of a system which affects QoS perceived by results [CHA 99]. In multimedia, it could mean the quality of an image or the speed of a response. Tables 22.1 and 22.2 below summarize the parameters linked to QoS technology and user respectively. The QoS requirements of the user are transformed in QoS technology parameters by QoS management tools. This is defined as a group of functions and mechanisms of supervision and control to ensure that the desired QoS level is in place and especially that it is maintained [CHA 99]. Two types of QoS management functions are identified: static and dynamic, respectively summarized in Tables 22.3 and 22.4 below. Static QoS management functions are applied at system start-up, whereas dynamic functions are applied, as much as possible, during system execution [CHA 99].

22.2.3. *IntServ/RSVP architecture*

In IntServ [BRA 94] architecture, the objective is to provide circuit switching advantages on a packet switching network. In order to do this, the routers reserve resources for data flow. A route with the necessary resources is established between the transmitter and the receiver. These routes are refreshed periodically in order not to block resources indefinitely. New network elements are defined at router level as presented in Figure 22.1 to support the control for this architecture. Three levels of service are defined by IntServ architecture: Guaranteed Service (GS), Controlled Load (CL) and Best Effort (BE). The GS service guarantees that users will benefit from fixed bandwidth, fixed end-to-end delay and no packet loss. The CL service ensures that users will have a level of service close to Best Effort in a network that might be overloaded. And finally, the BE service is characterized by the absence of QoS: the network provides only the level of service that it can.

Category	Parameter	Description/Example
Time	Delay	Necessary time for a packet to be transmitted
	Response time	Round Trip Time (RTT), which is the necessary time for sending a packet and receiving confirmation
	Jitter	Variation in the delay or response time
Bandwidth	System bandwidth	Required or available bandwidth, which is measured in bit/s
	Application bandwidth	Required or available bandwidth, which is measured in units specific to the application, e.g.: video sampling bandwidth
	Transaction throughput	Number of operations required or executed per second
Reliability	Mean failure time (MFT)	Time between two failures
	Mean repair time (MRT)	Minimum time from failure to normal function
	Mean time between failures (MTBF)	MTBF = MFT + MRT
	Percentage of available time	MFT/ (MRT +MFT)
	Loss or error ratio	Total proportion of data not received properly, e.g.: network error loss

Table 22.1. *Technological parameters of QoS [CHA 99]*

Category	Parameter	Description/Example
Importance	Priority	An importance factor can be allocated to users, to multimedia traffic, etc.
Perceived QoS	Detail of an image	Resolution in pixels
	Color accuracy	Color information in pixels
	Video ratio	Sampling ratio
	Video quality	Jitter of samples
	Audio quality	Ratio of audio sampling, number of bits
	Video/audio synchronization	Video and audio flow synchronization; e.g.: synchronization of mouth movements
Cost	Cost by usage	Cost of setting up a connection or of access to a resource
	Cost by unit	Cost by unit of time or by unit of data, e.g.: Cost of a connection on a period basis or by rate of downloaded data
Security	Confidentiality	Prevent access to information
	Integrity	No modification of data sent
	Non-repudiation	Ensure that the sender cannot deny having sent data. Use of signatures
	Authentification	Prove identity of users

Table 22.2. *User parameters of QoS [CHA 99]*

Function	Definition	Technique examples
Specification	QoS requirements and capability definition	Requirements at different levels, user, environment, technology, application, etc.
Negotiation	Process to achieve a specification that will be accepted by all parties	Modification of parameters after failure; these modifications must consider relations established between the parameters and user preferences
Admission control	Comparison of QoS requirements and capacity to reach these needs	Available resources can be estimated according to reservation information and to performance models
Resource reservation	Allocation of resources to connections	Techniques of network resource reservation (e.g.: RSVP)

Table 22.3. *Static functions of QoS management [CHA 99]*

Function	Definition	Technique examples
Supervision	Measure of actual QoS provided	Monitoring of the parameter in relation with the specification
Control	Ensure that all parties adhere to the QoS agreement	Control of parameters which are in relation with the agreement
Maintenance	Modification of parameters by the system in order to maintain QoS	Use of filter, queue in order to maintain the delay or throughput parameter
Renegotiation	Renegotiation of the agreement	Necessary when the maintenance functions of the QoS can no longer respect the agreement
Adaptation	The application adapts to the system's QoS changes; probably after a renegotiation	Adaptation techniques at application level such as reduction of data output throughput if bandwidth decreases
Synchronization	Combination of multiple QoS multimedia flows with time constraints requires synchronization	It is important to link time information of both multimedia flows that we wish to synchronize

Table 22.4. *Dynamic functions of QoS management [CHA 99]*

In order to establish this reservation of resources, signaling has been defined, which is based on the RSVP (ReSerVation Protocol) protocol [BRA 97]. Signaling can be used in other architectures apart from IntServ [WRO 93]; it proposes resource reservation by the receiver and it enables a dynamic control of the bandwidth. This protocol sits on top of the IP protocol and it uses unicast or multicast routing protocols to route the messages. Seven types of RSVP messages have been planned: Path, Resv, PathErr, PathTear, ResvTear, ResvConf (Optional). To support this architecture, the routers have been equipped with new mechanisms such as traffic classifier, admission controller and scheduler. The RSVP module in the router processes the RSVP request based on the QoS policy which is configured by the operator and, based on the availability of resources, accepts or rejects this request. Traffic is identified and classified in the corresponding service class. Finally, the scheduler provides the process corresponding to the service class by applying the corresponding scheduling algorithm.

In the resource reservation procedure, the transmitter first sends a PATH message to the receiver, specifying traffic characteristics. The message is updated for each router along the way. The receiver responds with a RESV message and indicates the necessary resources to support the appropriate traffic. A router on the route can reject the reservation request if it does not have enough resources to support the traffic. Reception of the RESV message by the transmitter confirms that the resources are available and have been reserved throughout the route.

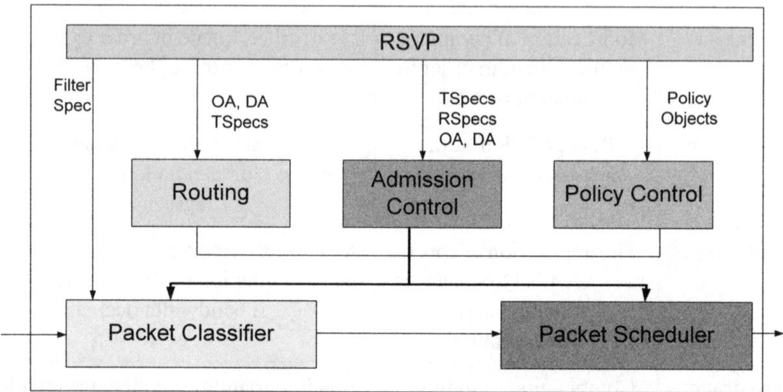

Figure 22.1. *RSVP router [CHI 98]*

The biggest problem with this architecture is scaling. Indeed, it is now important to maintain a data flow state, which can become very significant according to the increase of the number of flows in the network.

22.2.4. *DiffServ architecture*

The DiffServ [BLA 98] architecture has been designed to deal with the IntServ architecture's weakness in terms of scaling. The DiffServ architecture is based on flow aggregation and provides QoS by flow class. In this architecture, network routers are classified in two categories: edge routers and core routers. Input and output edge routers define a DiffServ Domain (DS). Processing complexity is pushed back to these edge routers. Three service classes are defined: Expedited Forwarding (EF), which provides a packet delivery guarantee in the assigned time, Assured Forwarding (AF) which defines four sub-classes and which provides a guaranteed service in terms of no loss of packets guarantee and Best Effort (BE) which offers no QoS guarantee. Each IP packet is marked by a DSCP (DiffServ Code Point) field; the packets are conditioned at DS domain input and are processed according to their level of service from core routers. The new functions of DiffServ edge routers are classification and conditioning of traffic as illustrated in Figure 22.2. The internal DiffServ routers to DS domain execute a particular behavior for each service class called Per Hop Behavior (PHB). The DSCP field of each packet determines at internal router level which PHB to apply. In IPv4, the DSCP field is provided by the unused ToS (Type of service) field. In IPv6, it is supplied by a new field in the header called FL (Flow Label). A new SLA (Service Level Agreement) concept has been introduced to describe the service contract between two DS domains or between a client and a DS domain. This SLA contains technical specification (Service Level Specification, SLS) which serves to configure DS routers in order to provide the level of service promised in the contract. The SLA also contains the administrative and legal contract specifications. In order to do this, the IETF has introduced the PDB (Per Domain Behavior) which only contains technical specifications, the PHB and the different necessary configurations to satisfy the SLA.

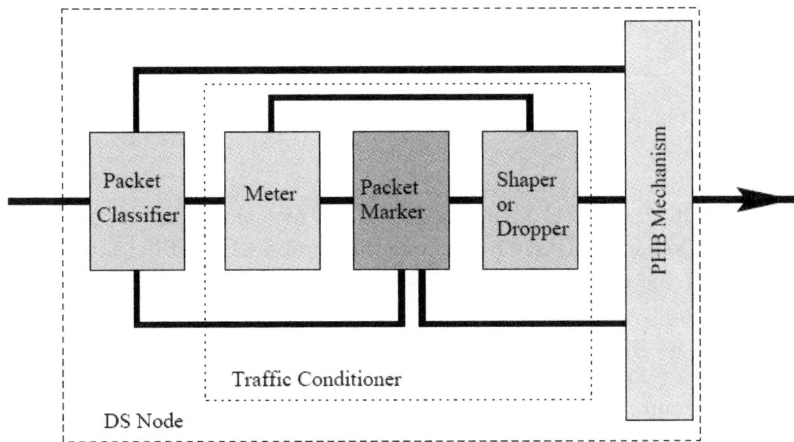

Figure 22.2. *DiffServ router [CHI 98]*

The high point of this architecture compared to the IntServ architecture is scaling, however, this architecture can only supply approximate QoS. Furthermore, the absence of signaling cannot provide a dynamic QoS.

22.2.5. *Other QoS architectures*

In order to obtain end-to-end QoS guarantee as well as an architecture that is scalable, an architecture combining IntServ and DiffServ has been proposed by the ISSLL workgroup [BER 00]. The workgroup has proposed the use of IntServ at access network level and DiffServ at core network level. Since the access network can shrink to one node, the goal here is to combine the service guarantee offered by IntServ and scaling enabled by DiffServ in order to provide end-to-end QoS.

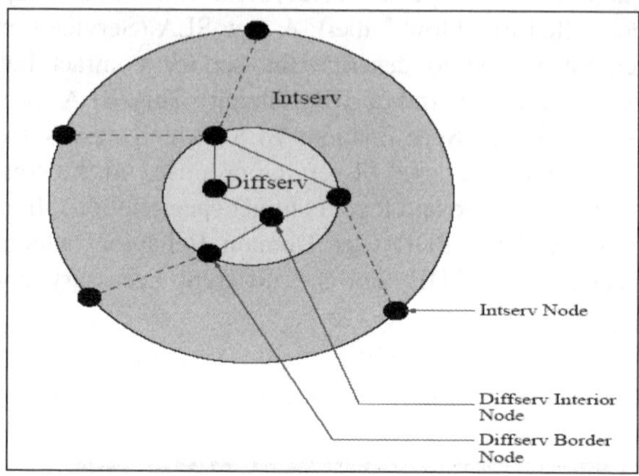

Figure 22.3. *End-to-end QoS architecture: IntServ/DiffServ*

MPLS (Multi Path Label Switching) [PUJ 03] technology has brought a breath of fresh air to QoS architectures by enabling the reproduction of the advantages of a circuit switching network over a packet switching network like IP. This technology has been used alone or combined with other architectures such as DiffServ to ensure end-to-end service guarantee. Packets are marked during MPLS domain entry by a label and they are then switched instead of routed due to this labeling plan put in place by an IP control packet, which is itself routed by the IP routing plan.

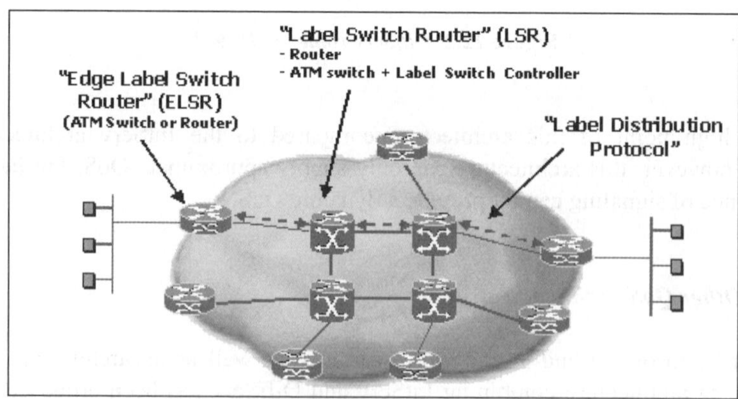

Figure 22.4. *MPLS architecture*

22.3. Mobility architectures

In this section, we will present the foundation of mobility management in IP networks, as well as a description of architectures and protocols proposed by standardization organizations such as IETF or by academic institutions.

22.3.1. *Mobility overview*

Studies that have been done to resolve the mobility problem in IP networks have resulted in two major directions. Macro mobility support and micro mobility support. Macro mobility is concerned with a shift between two different domains. Macro mobility can happen during an active mobile user session, or when a new session is launched by a domain user through a visited network, which is know as nomadism. Micro mobility means shifting between two points in the same domain. Basic functionalities of mobility management are mainly localization management and handover management. For this the following functions are all, or in part, necessary in the mobility management protocol [MAN 02]:

– authentification and authorization;

– packet transfer;

– route updates;

– handover management;

– inactive mobile device support;

– address management;

– security support.

The criteria of choice for mobility management protocol are mainly duration of handover and the rate of lost packets during handover procedure. We talk of smooth handover if a handover was handled with a minimum packet loss of fast handover if a handover was handled with a minimum delay and finally of seamless handover if a handover was handled with a minimum delay and minimum packet loss. The major problem with handover management in IP networks is that level 3 handover (network layer) happens only at the end of level 2 handover (link layer). This means that the mobile node has disconnected from its previous access router and has connected to a new access router but only at the physical level and it must wait for level 3 handover to happen in order to use the network's resources. The objective is this time that separates the execution from level 2 to level 3. In order to do this, we must improve movement detection techniques at level 3 and attempt maximum synchronization the start of level 2 and level 3 handovers.

22.3.2. *Macro mobility support*

Proposed by the IETF, Mobile IP is the macro mobility support standard in IP networks [PER 97, PER 02] and its goal is to maintain the connection of the mobile node to the network while changing the attachment point. This is done in a transparent way to the higher TCP/IP layers pile as well as to the IP routing plan. In order to do this, it proposes a localization management mechanism and a mechanism for rerouting or packet transfer to the mobile destination.

Mobile IP defines a mobile source domain where there is a mobility agent HA (Home Agent) and a visited domain where this is a mobile agent FA (Foreign Agent).

The HA is responsible for maintaining location information of the mobile node which it updates regularly. It is also responsible for transferring the packets destined for the mobile node through an IP tunnel to the FA or the mobile node if the mobile node has a routable address (co-located CoA, CCoA which can be allocated by DHCP) or non-routable (Care of Address; CoA allocated by the FA). When the mobile node cannot obtain a new address from the HA or the DHCP, the FA will provide a new IP address in the visited domain and will transfer to the HA the registration request of the location sent by the mobile node. The FA is also responsible for the recovery of packets sent to the mobile node in the IP tunnel established by the HA and for transferring them to the mobile node.

The basic operation of Mobile IP is the same as for Mobile IPv4 and Mobile IPv6. However, due to IPv6's new functionality, Mobile IPv6 has brought solutions to certain problems found in Mobile IPv4, such as triangular routing.

The mobile node, after obtaining its transient address (CoA or CCoA), will send a request to HA to register its new location. The HA keeps a correspondence between the permanent address of the mobile node which is linked to its HA (Home Address) and its CoA. The correspondent nodes of the mobile node send their traffic using its permanent address. The HA intercepts these packets and can redirect them toward the mobile node by constructing an IP tunnel [PER 96]. Encapsulated and redirected packets have an additional IP header which contains the mobile node's CoA address as a destination address. The IP tunnel is established with the FA if the mobile node contains a CoA. In Mobile IPv4, we call this redirection a triangular routing. An enhancement of this routing has been proposed and called "route optimization" [PER 01], so that the mobile node can inform the corresponding node as well as the HA of its new location. This enables the sending of packets directly to the node corresponding to the mobile node. Unfortunately, this solution has had security problems that have been resolved in Mobile IPv6. Thus, in Mobile IPv6, triangular routing does not exist anymore. Furthermore, the FA is not needed since the mobile node can construct its own address which can be routable with the IPv6 addressing plan.

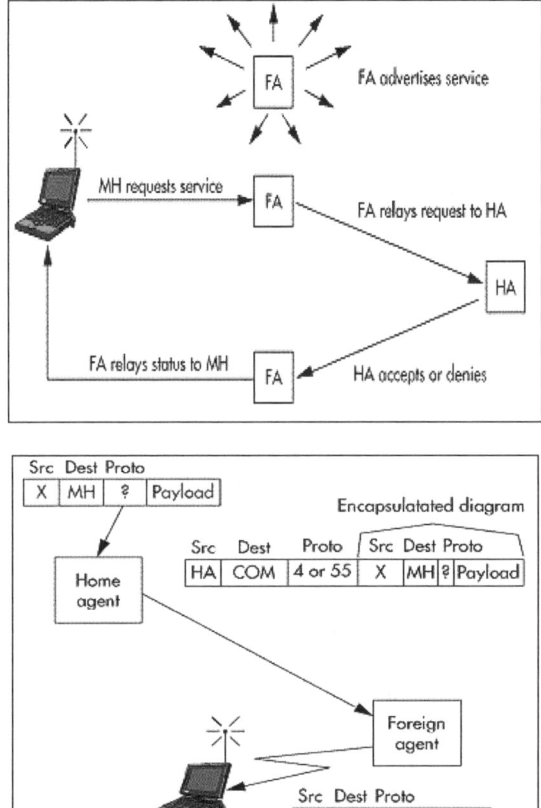

Figure 22.5. *(a) Registering with Mobile IP, (b) IP tunneling*

22.3.3. *Micro mobility support*

Mobile IP performance studies concerning moves between two base stations of one domain have revealed its inability to support this type of mobility called micro mobility [CAM 02, REI 03]. For this, other approaches have emerged from different studies such as hierarchic approaches (PAA: Proxy Agents Architectures), or localized routing modification-based approaches (LERS: Localized Enhanced Routing Schemes) [MAN 02]. In the first category we find hierarchical Mobile IP

(HMIP) [CAS 00, MAL 00b] and Fast handball (FMIP) [MAL 00a and c] as well as other Mobile IP enhancements [MAL 01]. In the second category we will find Cellular IP [CAM 00b], HAWAII [RAM 00], Multicast-based approaches as well as ad hoc routing methods (MANET) [MAN 02].

22.3.4. *Proxy-based architectures*

This type of architecture introduces the hierarchy notion of mobile agents (FA and/or HA) to locate registration messages and to minimize the necessary time for the handover procedure.

One of the Mobile IP modifications is hierarchical Mobile IP [CAS 00] which proposes the improvement of Mobile IP performance in micro mobility. An FA is installed at the gateway of the visited network thus forming a GFA (Gateway Foreign Agent). This GFA is responsible for the regional registration procedure [GUS 01, MAL 00b] by hiding all movements within the visited network to the HA. The mobile node has a permanent address (home address) and a temporary address named CoA (Care of Address). The CoA is provided by the FA of the visited network. In this way, the HA keeps the correspondence between the permanent address (home address) and the CoA (GFA), the GFA keeps the correspondence between the local CoA and the CoA (GFA). The registration procedure is similar to the one for Mobile IP, the only difference is that the registration with HA is done only if the mobile node changes GFAs, if not, registration within the visited network is done after the GFA, which plays the role of a localized HA. Packets destined for the mobile node are redirected by the HA toward the GFA which then transmits them to the mobile node.

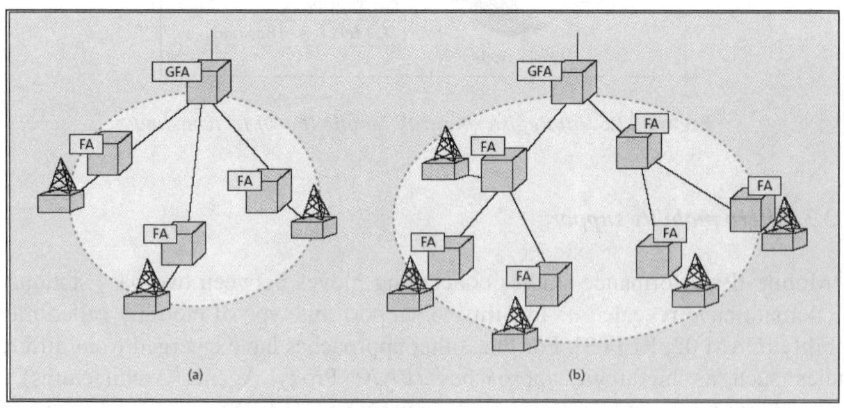

Figure 22.6. *FA hierarchy [REI 03]*

Another Mobile IP improvement is Fast Handoff [MAL 00a and c] which is an attempt to improve the HMIP handover delay even more in micro mobility. To achieve this, it proposes the improvement of movement detection procedure by using information from the link layer concerning the start of handover at level 2 (link layer). This information is used to anticipate level 3 handover and thus utilizes the network resources in the new cell as soon as possible after level 2 handover. For this, the mobile node will start its registration procedure with the new FA through the old FA even before level 2 handover is finished. Furthermore, a tunnel will be established between the old FA and the new FA in order to transmit the packets which continue to arrive to the old FA during handover. This is possible due to the "route optimization" mechanism. This cancels the triangular routing problem. In fact, the mobile node sends its new location to the HA at the same time as the corresponding nodes and in this way the corresponding nodes directly send their packets to the mobile node.

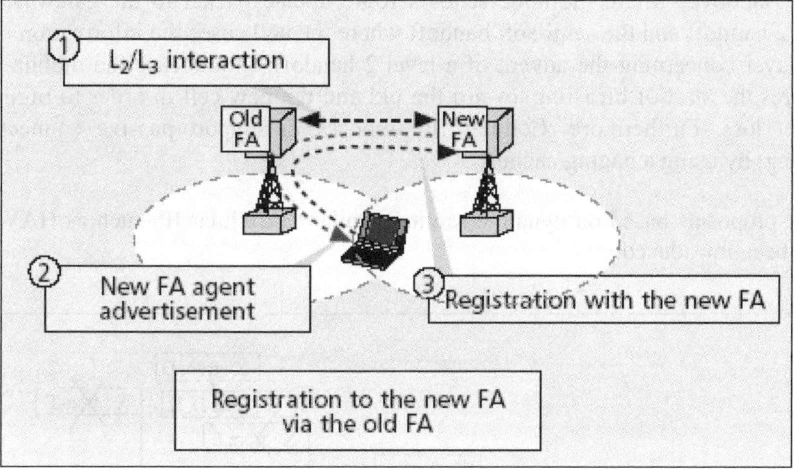

Figure 22.7. *Mobile IP Fast Handoff [REI 03]*

Other proposals have been introduced to improve the quality of handover in Mobile IPs such as Proactive Handover or Telemip [REI 03, TAN 99].

22.3.5. *Localized routing modification-based architecture*

This micro mobility management approach introduces a dynamic routing in certain areas of the network. Three solution categories are identified [MAN 02]. Per host forwarding architectures, such as Cellular IP or HAWAII, Multicast and MANET (Mobile Ad hoc NETwork) based architectures.

Cellular IP [CAM 00b] was designed to replace IP in an access network. A Cellular IP domain is made up of MAs (Mobility Agents) where one is a gateway to the Internet and acts as an FA and executes Mobile IP. Each MA contains a routing cache which contains the next node toward the mobile node and an entrance to reach the gateway. This cache is used by the MA to transfer packets from the gateway to the mobile node or from the mobile node to the gateway. Routes are established and maintained with the transmission of two control packets. A beacon message periodically floods the network by the gateway in order to create the route to the gateway for all MAs. The route update packet is sent by the mobile node to its first connection to the network when it changes adherence point but usually periodically. These packets are transferred hop by hop to the gateway creating or updating the entries in each MA's routing cache. Cellular IP proposes two types of handover: Hard Handover where the node sends a route update packet to the gateway after level 2 handoff and the semi-soft handoff where the node uses the information of the link layer concerning the advent of a level 2 handoff. In this case the mobile node requires the start of bicasting toward the old and the new cell in order to minimize packet loss. Furthermore, Cellular IP proposes to support passive connectivity (paging) by using a paging cache.

Other proposals based on dynamic routing similar to Cellular IP, such as HAWAII, have been introduced.

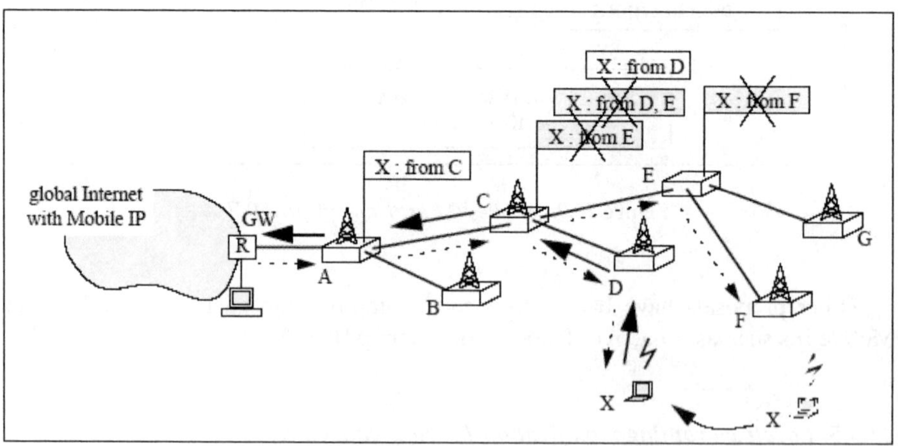

Figure 22.8. *Cellular IP*

22.3.6. *Multicast architectures*

Multicast architectures have been designed to support point-to-multipoint connections independently of the location, addressing and routing. This type of architecture has proven to be adequate for mobility support. In the multicast-based architecture proposal for mobility support, the mobile node will have a multicast-CoA address. The mobile node can request to join its multicast group either before or after handover from its neighboring access networks. Dense mode multicast and sparse mode multicast are examples of multicast proposals for mobility support [MAN 02, MYS 97, MIH 00].

Finally, the MANET architectures for mobility management such as MER-TORA [MAN 02] have been proposed. MANET protocols have been designed for ad hoc network support where the mobile node and the routers are mobile. In the case of mobility management, it is considered that the access network part is the ad hoc part that is fixed and only the terminals are mobile. Therefore, the ad hoc approach applies in this case as well.

Figure 22.9 below summarizes the different mobility support proposals in IP networks.

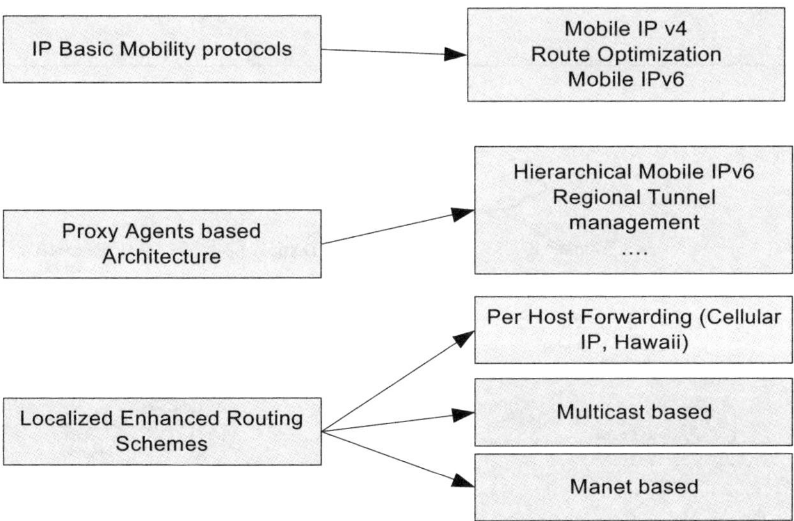

Figure 22.9. *Classification of mobility support proposals [MAN 02]*

22.4. Impact of mobility on QoS

The interaction between QoS management and IP networks mobility is still difficult as these two controls have evolved separately from each other. Indeed, studies done in the maintenance and control of QoS domains have been more involved with cable networks than mobile networks. Major problems that can be encountered by the QoS management process are mainly due to the network topology, to user macro mobility and micro mobility.

22.4.1. *Network topology impact*

Depending on the access points between which the mobile node moves, the type of handover is different and consequently signaling rate in the access network will be different. The deeper the handover in the access network topology, the longer the reestablishment of the end-to-end QoS in the mobile node's new location. Different handover types are defined whether they are limited to an access router (Intra Access Router), to two access routers in one gateway (Access Network Gateway), to a same domain (Inter Access Router), to different gateways but from a single domain (Inter Access Network Gateway), or still to different domains (Inter Access Domain). Figure 22.10 below illustrates these different handovers.

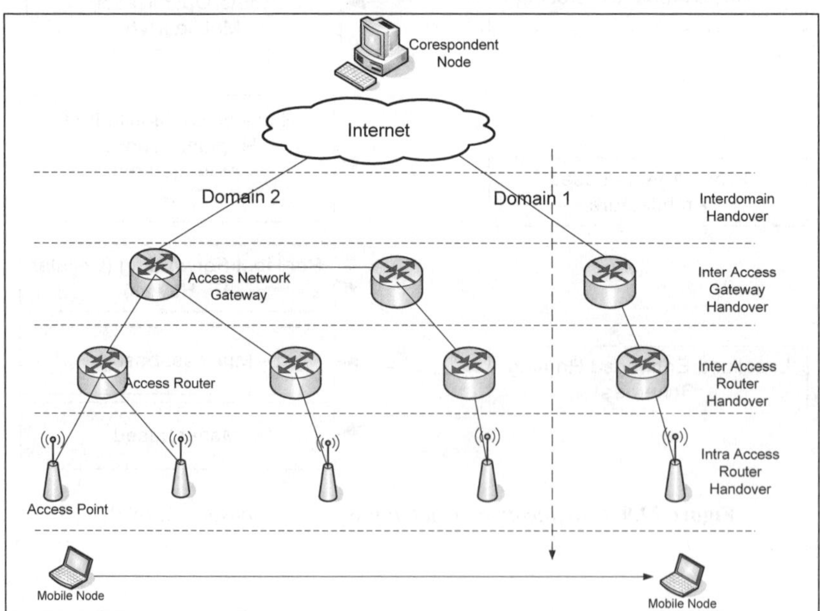

Figure 22.10. *Depth of handover [MAN 02]*

In the case of IntServ architecture, RSVP has a problem guaranteeing reservation since the reservation, routing and data transmission functions are independent [MAN 02]. During handover, the route changes and the packets can only receive a Best Effort service level until the route is updated by periodic PATH and RESV messages from the RSVP protocol. This problem is much more significant if the handover is deep and requires more time to update QoS, which causes a bigger degradation of the level of service.

In the case of DiffServ architecture, as there is no signaling mechanism, mobile node traffic disrupts the sharing of resources for other mobile devices which are already connected in the cell. It is necessary to have a resource controller (bandwidth broker) in order to regulate resource sharing within DiffServ.

In the case of Intra Access Router handover, the route to ANG does not change since this type of handover, called level 2 or radio handover, only affects radio resource; it is transparent to superior layers. The access control is therefore not necessary. In the case of Inter ANG handover, the ARs change during handover. In this case, the handover affects availability of radio resources as well as those of the access network. Moreover, the new access router must use access control. In the case of Inter ANG handover, mobility has a significant impact on resource coordination. First, the mobile device needs a new IP address. Then, QoS must be affected until route refresh by the RSVP and finally, the most complex handover is the one that requires the change of the administrative domain. In addition to the basic operations linked to mobility (new IP address, packet re-routing), the mobile device needs to authenticate again to be authorized to use the resources of the new domain. This causes signaling traffic overload which can affect the delay of recovery of mobile device resources.

22.4.2. *Macro mobility impact*

Macro mobility requires a more significant delay of route recovery than micro mobility. In addition to the control operations for localization or rerouting, other operations such as reauthentification or authorization to use resources can be necessary. Recovery and reconfiguration of QoS on the new route also require some time to set up. This is in addition to the necessary time for macro mobility. Other factors such as mobile device speed or the number of active mobile devices in the network cause an overload of the network during control packet traffic. Another problem concerns address management. Indeed, as the mobile device changes IP addresses, end-to-end QoS architectures such as IntServ require that the whole reservation operation over end-to-end route must be redone. This will result in a significant waiting time before QoS recovery, which will affect QoS guarantee.

In such architectures as Mobile IPv4, certain procedures are particularly problematic for QoS recovery. For example, triangular routing causes a problem in most QoS architectures because the arriving packet route (downstream) to the mobile device (by the IP tunnel) is different from the route of packets exiting the mobile device toward the network (upstream). A solution has been proposed, Reverse Tunneling [MON 01], for using the same route for packets entering and exiting the mobile node. In this case, triangular routing is avoided. However, the routers in the IP tunnel are incapable of knowing the QoS parameters specified in the IP header. Indeed, the only information available through the IP tunnel is the IP address. For example, in the case of an encapsulated RSVP message in the IP tunnel, with the Router Alert field used to indicate to the routers that the traffic requires specific QoS, this field unfortunately cannot be picked up by the routers due to the IP tunnel. Multiple solutions have been suggested to solve this problem; however, they increase the complexity of QoS management in the mobile world.

In the case of Mobile IPv6, problems such as triangular routing or Tunneling are resolved with Route Optimization [PER 01] for the former and with Routing Header for the latter [JOH 01]. This last one reduces network overload due to IP Tunneling elimination. In addition, a movement detection mechanism is integrated in order to enable a better performance during handover. In fact, the detection of a movement of mobile device promptly launches the Mobile IP procedure more quickly and minimizes handover time. Despite Mobile IPv6's willingness to reach transparency and handover support objectives, this protocol remains non-optimized to offer a Seamless Handover control (minimum delay and minimum loss) in a cellular network supporting real-time applications. This is caused by the large number of localization registrations and of the route recovery latency after handover.

22.4.3. *Micro mobility impact*

Movement of nodes within a same domain (micro mobility) generally affects QoS architectures according to the IP encapsulating, tunneling, multicast or dynamic routing used. In the IntServ architecture, each router keeps a state by flow. In this way, the movement of the mobile device launches routing localized repairs as well as resource reservations in the network.

In a DiffServ architecture that has no signaling mechanism (no state to update), the service level provided varies according to the number of mobile devices arriving or leaving a cell.

The interaction between QoS and micro mobility must consider the following points [MAN 02]:

– the use of IP tunneling hides the header information of the packet which affects the classification of this packet;

– the change of CoA address during the session affects the recovery time of the route and the necessary resources;

– packet multicast over multiple routers consumes a lot of network resources;

– the use of fixed routes using the same gateway to outside networks is not scalable;

– adaptability to route changes is necessary;

– finding an optimal route from gateway to the access routers;

– routing support based on QoS.

Improvements have been proposed in order to better support IP address changes and also to support IP Tunneling in IntServ [TER 00] and in DiffServ [BLA 00] but these proposals do not essentially resolve the problems described earlier.

22.5. Interaction architectures between QoS and mobility

In this section, we present some solutions recommended by the IETF and by the academic community for the problems regarding the interaction between QoS and mobility. The final goal is to combine management of macro mobility, micro mobility and QoS for end-to-end guarantee of service in mobile device networks. Two methods are possible based on the transport mode of signaling, which can be either separately transported by data packets (out-band signaling) or transported in data packets (in-band signaling).

22.6. Band interactions

Interaction between QoS and micro mobility can be handled in different ways depending on whether resource reservation is done at the end of handover, during handover, at the same time as handover or before handover.

22.6.1. *Admission and priority control*

In an architecture such as DiffServ, it is possible to put an admission control mechanism in place that will limit Best Effort traffic so that not all leftover resources will be used and also to reserve a part of the resources to support traffic for arriving mobile nodes. In order to do this, it is not necessary to install new devices, admission control rules just need to be applied for network access routers in order to limit traffic in each service class. Furthermore, we must use a priority mechanism to take into account mobile device traffic during handover as well as mobile device traffic which starts a new session in the cell. This technique is applied to cellular networks within telecommunication networks [MAN 02]. The problem is to know the proportion of the bandwidth in order to reserve for priority and non-priority traffic [MAN 02].

22.6.2. *Resource reservation at the end of the handover*

The simplest approach for combining QoS and mobility management is to execute them separately. Indeed, during handover, mobility management is executed and after this execution, the QoS management procedure can start to restore resources on the new route established by the mobility management procedure. In the case of RSVP, route changes are taken over by the soft-state character of the reservation. RSVP contains a periodic refresh of the route mechanism which enables the repair of a route or a portion of the route depending on the depth of the handover.

The problem with this approach is that the service may see a degradation of service during the length of time RSVP needs to restore QoS on the new route. If the refresh message is generated before the complete restoration of the new mobile device route, there may be resource reservations on a route that does not correspond to the one for the mobile device. This route unfortunately will not be corrected before the next refresh message. In certain cases, the reservation can be rejected for lack of resources on one of the routers on the new route for the mobile device. Finally, this resource reservation disruption may happen as often as the number of times the mobile node changes access routers, even during a same RSVP session. This solution provides a very poor global QoS in the network.

22.6.3. *Resource reservation during handover*

In this approach, events generated by the mobility management procedure launch the resource reservation process of the QoS management procedure. The RSVP reservation mechanism is launched as soon as the new route is established. It is called Local Path Repair mechanism and is defined in the RSVP specifications; this option minimizes service disruption time between the start of handover until restoration of resources on the new route. It also avoids the execution of resource reservations in the network before information concerning the new mobile device route is available. Restoration delay for the new route can also be minimized by localizing reservation repair where the route has changed, i.e. between the Crossover router shown in Figure 22.11 and the new access router [MAN 02]. However, in the case of a new QoS negotiation, end-to-end signaling process is inevitable. In addition, the old route must be explicitly freed.

One of the problems of this approach is the complexity introduced in the network to intercept the RSVP message in order to limit repair in the area between the Crossover router and the mobile device. Furthermore, the fact that the RSVP mechanism is launched by events linked to mobility generates more significant signaling than what would happen in the case of a reservation independent from mobility where the RSVP procedure is launched only by refresh messages.

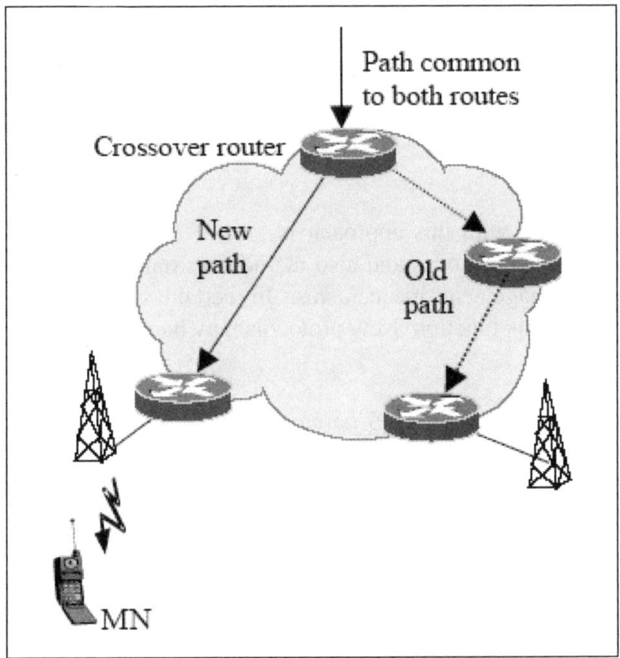

Figure 22.11. *Crossover router [MAN 02]*

In the case of proxy agent-based mobility management, the event that will launch the RSVP procedure is receiving of a new location registration ACK. In the case of MANET-based mobility management, receiving the Route Update message is what will launch the RSVP procedure. Finally, in management based on localized routing modification, as soon as the routing information is distributed in the network, the RSVP procedure is launched.

22.6.4. *Resource reservation at the same time as handover*

This approach proposes the use of unified QoS and mobility signaling. This is possible either by extension of existing QoS or mobility signaling or by QoS routing. The goal of this approach is to minimize service disruption time by ensuring that the QoS will be in place as soon as possible since QoS parameters are propagated at the same time as the start of handover. The idea of this approach is to install resource reservations at the same time as route update. Thus, it avoids the execution of reservations before the end of propagation of the new route. It also avoids executing them after the end of the route update procedure as in the approach presented earlier. It also enables the installation of multiple reservations by using the same signaling message. Generated signaling traffic can then be decreased as in previously discussed approaches. Reservation repair may also be located, except in the case where a new end-to-end reservation is required. Resources of the old route should be explicitly freed.

The main problem with this approach is first of all processing the complexity introduced in the network nodes and also the need to transport QoS information by micro mobility management mechanisms. In certain cases, new messages are necessary to ensure this function. New protocols may be necessary to unify QoS and mobility signaling.

In proxy agent approaches, QoS parameters are transported in the localization registration message. An integration attempt of QoS with Mobile IPv6 is proposed in [CHA 01]. A new QoS Object Option is introduced in IPv6 for use in Mobile IPv6 localization registration messages. These are sent to corresponding nodes or to mobility agents. In the approach proposed by the MANET Group, QoS information is transported in the route update messages. In the routing modification approach, QoS information is transported by the path set-up and refresh messages. Finally, knowing QoS needs during handover makes it possible to choose the next cell that responds to these needs.

22.6.5. *Resource reservation in advance*

In order to reach a guarantee of service independent from mobility, the techniques presented previously, such as access control, priority, integration of QoS signaling into mobility signaling are not sufficient. The mobile node needs to execute resource reservation of cells that it can visit during its route. Two approaches are presented next, one applies to the IntServ architecture (MRSVP) and the other applies to the DiffServ approach (ITSUMO).

22.6.5.1. *MRSVP*

Mobile RSVP [TAL 98] is a reservation in advance protocol in the IntServ architecture in a network that supports mobile nodes. It proposes three levels of service for mobile users: Mobility Independent Guarantees (MIG), Mobility Independent Predictive (MIP) and Mobility Dependent Predictive (MDP). The MIG services offer guaranteed service to the mobile device, MIP offers a predictive service and MDP is a predictive service with high probability but which can be deteriorated in certain circumstances. Two reservation types are possible: passive and active reservations. Active reservations are launched by the mobile node in the cell where it is located, whereas passive reservations are executed at locations where the mobile node could visit. These are described in the MSPEC (Mobility Specification) object. This localization list can be approximately determined by the network during movement of the mobile device or simply expressed by the mobile device that knows its route. New localizations can be added to MSPEC and passive reservations are also executed. Resources passively reserved by mobile devices can be used by other lower quality traffic such as Best Effort. As soon as the mobile devices become active, resources are recovered which can affect the QoS of the flows using them. A unicast packet is sent by using Mobile IP routing. Reservation is all done through the IP tunnel route established by the Mobile IP.

22.6.5.2. *ITSUMO*

ITSUMO (Internet Technologies Supporting Universal Mobile Operation) architecture [ITS] is founded on DiffServ. It defines a group of domains. Each domain has a QoS Global Server (QGS) and local QoS servers (QLN, QoS Local Node). Resource reservation is not done by the mobile device but by the QGS server. When it arrives in a domain, the mobile device interacts with QGS by negotiating a Service Level Specification (SLS) in a dynamic manner, by specifying its reservation requirement and by signaling its mobility profile. Based on negotiated service level information (SLS), the QGS calculates the mobility model and the bandwidth to reserve for all cells of the mobile device's domain. It then sends this configuration to local QoS servers (QLN). The QGS regularly updates the QLNs in order to include mobile device reservations that may be visited. However, this approach could be more efficient if there were passive reservations or a handover guard band mechanism [MAN 02]. The protocol used to put in place this SLS negotiation signaling and reservation request is DSNP (Dynamic Service Negotiation Protocol) [CHE 01]. The difference with the MRSVP approach is that the reservation must not be signaled by the mobile device itself for each access router that is in the mobility profile; it is the responsibility of the QGS to execute this reservation instead of the mobile node.

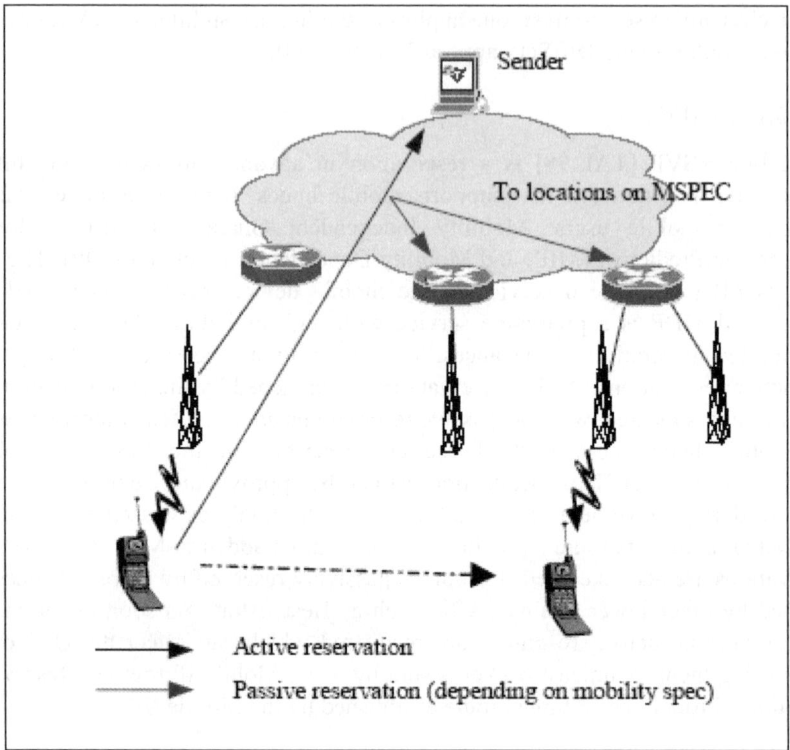

Figure 22.12. *MRSVP architecture [MAN 02]*

22.6.6. *Context transfer*

The goal of resource reservation is to guarantee a level of service to mobile users; however, it is not an optimized use of network resources. In order to solve this problem, a new approach proposes asking neighboring cells for the availability of their resources and informing them of QoS requirements before handover. To do this, a specific architecture is defined, called Context Transfer Framework [SEA]. In this architecture, each access router maintains a context of the mobile node. Within the context, characteristics are defined such as QoS or security. These parameters are linked to the processing of mobile device traffic at access router level, where the mobile device is located. During movement of mobile device from one cell to another, this context is transferred to the new cell during handover in order to prepare QoS in the new cell. No other signaling is necessary for QoS since the process corresponding to mobile device traffic is already in place in the new cell, due to context transfer from the mobile device. Initially, context transfer was designed for the communication between access routers to recover QoS in the new cell. The goal for the extensions of this architecture is the communication between access routers and gateways to recover QoS over the route by sending the context of the mobile device to all routers up to the gateway [MAN 02].

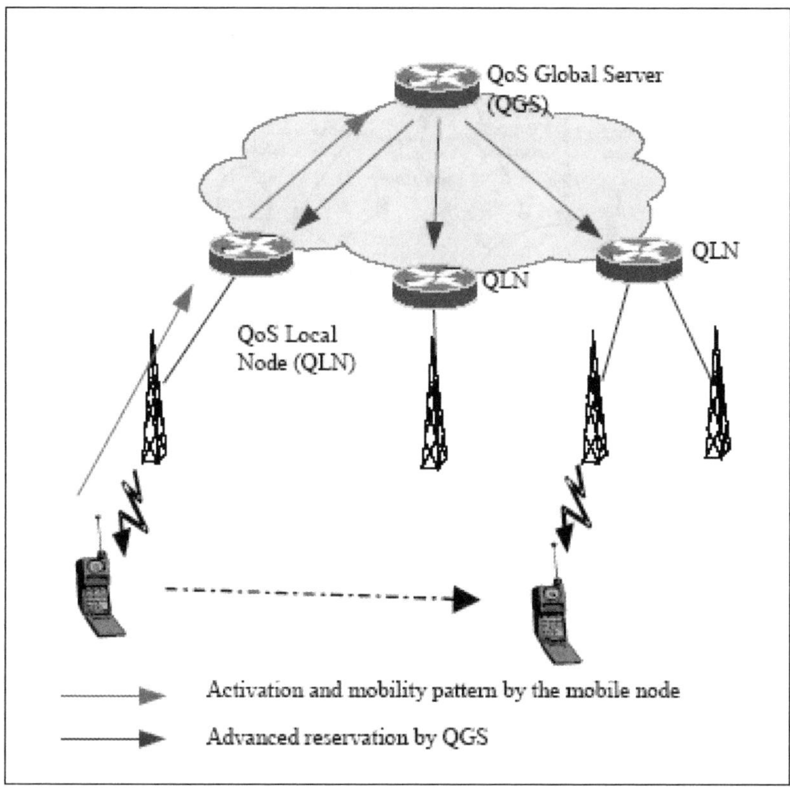

Figure 22.13. *ITSUMO architecture [MAN 02]*

22.7. Interaction with band signaling: INSIGNIA

Contrary to out-band signaling which uses specific control packets, the INSIGNIA architecture (INband SIGnaliNg support for QoS In Ad hoc networks) uses data packets to transport QoS signaling, from which we get the name in-band signaling [LEE 02]. In-band signaling is well adapted to support end-to-end QoS and mobility in dynamic environments where network topology and node connectivity are very dynamic. INSIGNA supports reservation, rapid restoration and end-to-end adaptation based on IP robustness and flexibility.

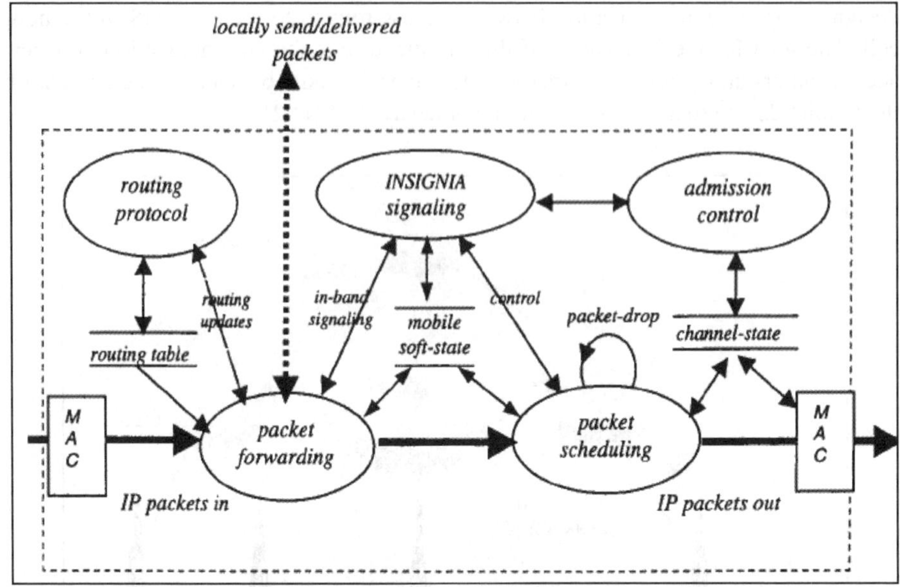

Figure 22.14. *QoS INSIGNA architecture*

22.8. Other communities

The third generation of mobile systems (3G Mobile Telecommunication Systems) represents the evolution of the mobile telephony system to provide multimedia services over cellular networks [3GPP]. A packet switching network is added as a core network in 3G networks to transport application IP packets. Mobile nodes are connected at the core network by radio interfaces. From the core network, packets can reach the Internet network. Concerning mobility and QoS management, 3G systems propose three types of mobility: terminal mobility, which corresponds to the one defined in IP networks, personal mobility whose goal is to deliver services to users independently from their network or access terminal and finally services mobility. Terminal mobility support, at IP backbone level, Mobile IP is considered an important candidate for mobility management. However, at access network level, proprietary solutions are now recommended. Concerning QoS, four service classes have been defined: conversational, streaming, interactive and background [3GPP].

ISO (International Standard Organization) and ITU (International Telecommunication Union) have a common workgroup whose goal is to define a reference model for ODP (Open Distributed Processing) systems [BLA 97]. This group is working on the design of an architecture for QoS support. However, this architecture is not considering the mobility aspect. The ODP group has developed

CORBA (Common Object Request Broker Architecture) architecture which can serve as support for mobility management. Functionalities necessary for QoS have been introduced in version 3.0; the mobility aspect is more difficult to integrate [CHA 99].

Other works deal with adaptability support, which is concerned with application layers. Indeed, adaptability support is critical in wireless and mobile environments. An example for adaptability is proposed by the TCP protocol which uses an adaptability mechanism, slow start, which can react to the state of the network. Another example comes from the RTP (Real-time Protocol) protocol used to support real-time applications. This protocol notifies the application of the state of the network so that the application can adapt to the constraints of the network by reducing generated traffic, for example [SCH 96].

22.9. Conclusion

The road toward a definitive and operational end-to-end QoS guarantee architecture in mobile and wireless environments is still long and riddled with challenges. In fact, current studies only partially resolve the problems. Some propose solutions for mobility problems, others resolve QoS problems and more recently some have combined QoS and mobility. Concerning mobility, the biggest challenge is to resolve such problems as localization management or optimal control of handover quickly in order to avoid disruption of the current service. Concerning QoS, the focus has to be on offering a better service to users with a minimum of resources in the network in order to increase profit for network and service operators. In the mobile context, maintaining QoS means recovering its configuration in the new mobile device location as fast as possible. Finally, the goal of the interaction between QoS and mobility is to launch a QoS configuration procedure by events linked to mobility. This can be done in different ways as presented in this chapter. The main idea is to execute resource reservations in advance of the handover, during handover, after handover, or at the same time as handover. Each technique has its advantages and its disadvantages. It is necessary to combine these different techniques and to put a dynamic mechanism in place which can choose the execution of one of them for each type of traffic. The underlying idea is to provide a dynamic QoS control in the network. This should be accompanied by adaptability of services and critical functionality in dynamic areas as is the case in mobile and wireless networks. Finally, good network provisioning, good traffic engineering and dynamic QoS management are necessary for the realization of QoS based mobile networks.

22.10. Bibliography

[3GPP] 3GPP Third Generation Partnership Project, http://www.3gpp.org.

[ALW 93] ELWALID A. I., MITRA D. "Effective Bandwidth of General Markovian Traffic Sources and Admisssion Control of High-Speed Networks", *IEEE/ACM Transactions on Networking*, vol. 1, no. 3, p. 329-343, June 1993.

[AWD 97] AWDUCHE D. O., AGU E., "Mobile Extensions to RSVP", *proceedings of ICCCN'97*, Las Vegas, NV, p.132-136, September 1997.

[BER 00] BERNET Y. *et al.*, "A Framework for Integrated Services Operation over DiffServ Networks", *Internet Engineering Task Force*, Request for Comments (RFC) 2998, November 2000.

[BLA 97] BLAIR G., STEFANI J.-B., *Open Distributed Processing and Multimedia*, Addison-Wesley, 1997.

[BLA 98] BLAKE S., BLACK D., CARLSON M., DAVIES E., WANG Z., WEISS W., "An Architecture for Differentiated Services", *Internet Engineering Task Force*, Request for Comments (RFC) RFC 2475, December 1998.

[BLA 00] BLACK D., "Differentiated Services and Tunnels", *Internet Engineering Task Force*, Request for Comments (RFC) 2983, October 2000.

[BRA 94] BRADEN R., CLARK D., SHENKER S., "Integrated Services in the Internet Architecture: an Overview", *Internet Engineering Task Force*, Request for Comments (RFC) 1633, June 1994.

[BRA 97] BRADEN R., ZHANG L., BERSON S., HERZOG S., JAMIN S., "Resource reSerVation Protocol (RSVP) – Version 1, Functional Specification", *Internet Engineering Task Force*, Request for Comments (RFC) 2205, September 1997.

[CAM 00a] CAMPBELL A. T., GOMEZ J., "IP Micromobility Protocols", *ACM Mobile Comp. and Commun. Review*, vol. 4, no. 4, October 2000.

[CAM 00b] CAMPBELL A. T. *et al.*, "Cellular IP", *Internet draft*, draft-ietfmobileip-cellularip-00.txt, January 2000.

[CAM 02] CAMPBELL A. T. *et al.*, "Comparison of IP Micromobility Protocols", *IEEE Wireless Commun.*, vol. 9, no. 1, February 2002.

[CAS 00] CASTELLUCCIA C., BELLIER L., "Hierarchical Mobile IPv6", *Internet draft*, draft-castelluccia-mobileip-hmipv6-00.txt, July 2000.

[CHA 99] CHALMERS D., SLOMAN M., "A Survey of Quality of Service in Mobile Computing Environment", *IEEE Communication Surveys,* second quarter 1999.

[CHA 01] CHASKAR H., KOODLI R., "A Framework for QoS Support in Mobile IPv6: Another In-Band Approach Introducing a QoS Hop-by-Hop Option in Mobile IPv6", *Internet draft*, March 2001.

[CHE 01] CHEN J. C. *et al.*, "Dynamic Service Negotiation Protocol (DSNP)", *Internet draft*, July 2001.

[CHI 98] CHIMENTO P., "Tutorial on QoS Support for IP", *technical report 23*, http://ing.ctit.utwente.nl, 1998.

[ETSI 98] European Telecommunication Standard Institute, "General Packet Radio Service (GPRS), Service description, Stage 2", *Technical Report ETSI*, digital cellular telecommunications system (Phase 2+), TS 101 344 v7.7.0, ETSI, June 2001, 3GPP TS 03.60 version 7.7.0 released 1998.

[GUS 01] GUSTAFSSON E., JONSSON A., PERKINS C., "Mobile IP Regional Registration", *Internet draft*, draft-ietf-mobileip-reg-tunnel-04.txt, March 2001.

[ITS] ITSUMO Group, "A reference Architecture for All IP Wireless Networks", 3GPP2, http://www.3gpp2.org.

[JOH 01] JOHNSON D., PERKINS C., "Mobility Support in IPv6", *Internet draft*, 2001.

[KAA 00] KAARANEN *et al.*, *UMTS Network: Architecture, Mobility and Services*, John Wiley, Chichester, UK, 2000.

[LEE 00] LEE *et al.*, "INSIGNIA: An IP-Based Quality of Service Framework for Mobile Ad-Hoc Networks", *Journal of Parallel and Distributed Computing*, vol 60, no. 4, p. 374-406, April 2000.

[MAL 00a] EL MALKI K., SOLIMAN H., "Fast Handoffs in Mobile IPv4", *Internet draft*, , draft-elmalki-mobileip-fasthandoffs-03.txt, September 2000.

[MAL 00b] MALINEN J., PERKINS C., "Mobile IPv6 Regional Registrations", *Internet draft*, draft-malinen-mobileip-regreg6-00.txt, July 2000.

[MAL 00c] EL MALKI K., SOLIMAN H., "Hierarchical Mobile IPv4/v6 and Fast Handoffs", *Internet draft*, draft-elmalki-solimanhmipv4v6- 00.txt, March 2000.

[MAL 01] EL MALKI K. (Ed.) *et al.*, "Low Latency Handoff in Mobile IPv4", *Internet Draft*, draft-ietf-mobileip-lowlatency-handoffsv4-01.txt, May 2001.

[MAN 02] MANNER J., LÓPEZ A., MIHAILOVIC A. *et al.*, "Evaluation of Mobility and QoS Interaction", *Computer Networks*, vol. 38, no. 2, p. 137-163, February 2002.

[MIH 00] MIHAILOVIC A., SHABEER M., AGHVAMI A. H., "Multicast for Mobility Protocol (MMP) for Emerging Internet Networks", in *Proceedings of PIMRC2000*, London, September 2000.

[MON 01] MONTENEGRO G., "Reverse Tunneling for Mobile IP", revised, *Internet Engineering Task Force*, Request for Comments (RFC) 3024, January 2001.

[MYS 97] MYSORE J., BHARGHAVAN V., "A New Multicasting-based Architecture for Internet Host Mobility", *Proceeding of ACM Mobicom*, September 1997.

[PER 96] PERKINS C., "IP Encapsulation within IP", *Internet RFC*, RFC 2003, October 1996.

[PER 97] PERKINS C. E., "Mobile IP", *IEEE Commun. Mag.*, vol. 35, no. 5, p. 84-99, May 1997.

[PER 01] PERKINS C., JOHNSON D., "Route Optimization in Mobile IP", *Internet draft*, draft-ietf-mobileip-optim-11.txt, September 2001.

[PER 02] PERKINS C. ED., "IP Mobility Support for IPv4", *Internet RFC*, RFC 3220, January 2002.

[PUJ 03] PUJOLLE G., *Les réseaux*, Eyroles, 2003.

[RAM 00] RAMJEE R. *et al.*, "IP Micro-Mobility Support using HAWAII", *Internet draft*, draft-ietf-mobileip-hawaii-01.txt, July 2000.

[REI 03] REINBOLD P., BONAVENTURE O., "IP Micro-Mobility Protocols", *IEEE Communication*, Surveys, vol. 5, no. 1, 2003.

[SCH 96] SCHULZRINNE H., CASNER S., FREDERICK R., JACOBSON V., "RTP: A Transport Protocol for Real-Time Applications", *Internet Engineering Task Force*, Request for Comments (RFC) 1889, January 1996.

[SEA] IETF SEAMOBY WORKING GROUP, http://www.ietf.org/html.charters/seamoby-charter.html.

[TAL 98] TALUKDAR A., BADRINATH B., ACHARYA A., "MRSVP: A Resource Reservation Protocol for an Integrated Services Packet Network with Mobile Hosts", in *Proceedings of ACTS Mobile Summit'98*, June 1998.

[TAN 99] TAN C., PINK S., LYE K., "A Fast Handoff Scheme for Wireless Networks", *Proceedings of the 2nd ACM International Workshop on Wireless Mobile Multimedia*, ACM, August 1999.

[TER 00] TERZIS A., KRAWCZYCK J., WROCLAWSKI J., ZHANG L., "RSVP Operation Over IP Tunnels", *Internet Engineering Task Force*, Request for Comments (RFC) 2746, January 2000.

[WRO 97] WROCLAWSKI J., "The Use of RSVP with IETF Integrated Services", *Internet Engineering Task Force*, Request for Comments (RFC) 2210, September 1997.

Chapter 23

Sensor Networks

23.1. Introduction

The sensor network domain has been in existence for a long time but for a number of years, research efforts on this domain were only concerned with improvements in transducers, analog-to-digital conversion of information captured by sensors, cable-based sensor network deployment and in the design of sensor-specific communication protocols. In addition, the applications for these networks have been confined to the military and industrial sectors. Today this scenario is changing: advances in electronic systems, micro-electromechanical systems (MEMS) and in wireless communications have made sensors evolve toward a set of sophisticated miniaturized electronic circuits, which are associated with wireless communication interfaces. These new devices are capable of processing their own data, of making decisions on alerts and of communicating together. As a consequence of these advances, the sensor network design has been highly disrupted these last few years.

Given this novel sensor networking technology, it is enough, for example, to deploy sensors on a field for them to auto-organize. A network is created and the sensors cooperate to execute given tasks. We are then in the presence of sensor networks with automatic layout. The "intelligence" of sensors is not limited to communication intelligence; local processing can be done. A sensor is not passive anymore; it can, for example, locally do a soil analysis, with very little time between two analyses. Thus, the central server finds itself discharged of a part of the processing. For example, it only needs to inquire the sensor when needed in order to

---

Chapter written by Paulo GONÇALVES.

receive analysis results. A sensor is no longer just an information generator; it can receive configuration data, operation system updates and operation parameters issued from a central server or a local controller (PDA, etc.). The controller can, for instance, transmit the limit values between which a physical variable monitored by the sensor must normally evolve. The controller can also indicate the operation mode of a sensor: sensing data transmitted at each state change, at each variation of a monitored value (e.g., the temperature), systematically at given frequency, or for a request. The sensor behavior is therefore defined according to the application and the process that it monitors.

There are many advantages to using sensor networks. Due to their capability to offer a virtual presence in isolated and/or risky regions, numerous applications can be developed for the surveillance of resources or of the environment. We can find applications based on the use of thousands of highly integrated and small electronic sensors. In industry, wireless sensor networks are interesting when it is time to replace or even avoid expensive cabling.

A sensor network is a particular kind of ad hoc network in which the protocol that organizes the communication must take into account energy and scaling constraints. However, it is important to note that sensor networks have unique characteristics compared to traditional ad hoc networks. In an ad hoc network there is no other specific task to execute except for communication. In a sensor network there is always a goal or a task to execute. The nodes that form sensor networks collaborate toward a specific goal. On the other hand, in an ad hoc network, each node can have its own goal. Sensor networks can be formed by thousands of sensors that have no identifier. However, ad hoc networks are made up of a small number of nodes with a unique identifier. For sensor networks, the most important metric is energy, whereas in the case of ad hoc networks, throughput is the fundamental element. Finally, ad hoc networks rely on point-to-point communication, whereas sensor networks generally use flooding to communicate.

A new research trend focuses on the implementation and the organization of communications between a large number of sensors, while taking into account severe energy constraints of sensors. Other problems that also need to be addressed concern the development of low-power radio interfaces, the development of miniaturized devices capable of extracting energy from the environment (energy-scavenging), the development of very small operating systems, the research of new protocols, the development of algorithms for signal processing and the miniaturization of power sources. This chapter offers an overview of the different capabilities marking the new generation of sensor networks.

23.2. Definitions

Some definitions are necessary before we present an overview of the different elements of sensor networks.

23.2.1. *Sensor*

The word sensor has different definitions. The most common definition is of a device which transmits an output signal (often electric) in response to a physical input stimulus. In other words, a sensor is a transducer: a device capable converting a signal such as sound, temperature, pressure and light, for example, into an electric signal.

The sensor that is discussed in this chapter mainly corresponds to the integration of multiple modules: sensing, processing, communication and power. The sensing module is made up of one of more traditional sensors (e.g., temperature, pressure, humidity, vibration, movement), of components for signal conditioning and of an analog-to-digital converter. The processing module is composed of a low-power and low consumption microprocessor and of memory.

The communication module uses a low-power wireless communication interface which enables the sensors to communicate. In many platforms this interface is a radio interface.

The power module is responsible for providing the necessary power to the sensor's operation. Normally, this module uses a battery. The different modules which make up the current sensor are illustrated in Figure 23.1.

We currently define a sensor as an organism capable of picking up information from a physical process, of processing data and of communicating.

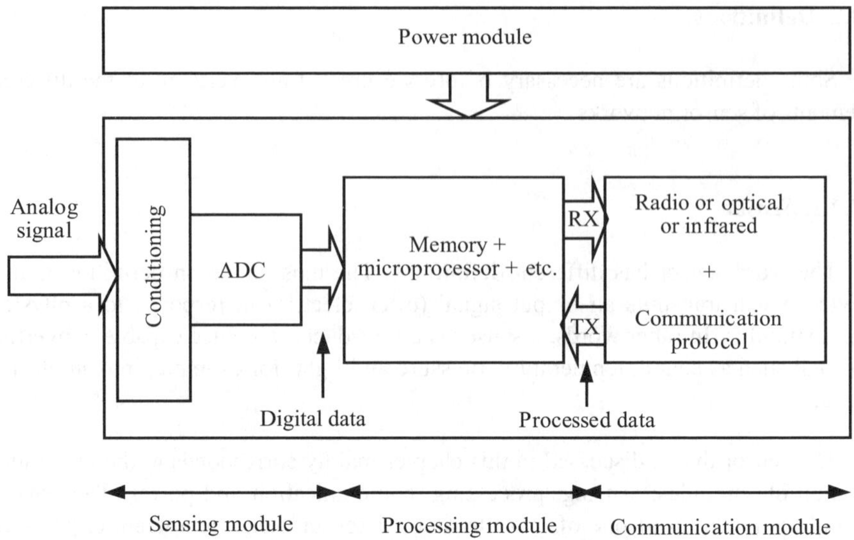

Figure 23.1. *Different sensor modules*

23.2.2. *Observer*

The observer is simply a user interested in the data that is collected by sensors. The observer can transmit requests in order to receive required data. He can also notify the sensor of the operation mode: sensing data transmitted at each state change, at each variation of a monitored value or systematically at a given frequency. Multiple observers can be present in the network.

23.2.3. *Phenomenon*

It is the physical process that is of interest to an observer. The data generated from monitoring this process can be analyzed and even filtered by the network. Multiple events can happen at the same time and at the same place.

23.2.4. *Sensor network*

A network of distributed sensors is composed of sensors geographically spread in a generally large zone according to the transmission range of the sensors. This network can have hundreds and even thousands of sensors depending on the application. In applications using radio communications, the sensors' arrangement can ensure coverage of a large area even if the communication range of each sensor is relatively small. In other words, it is not the sophistication of the network but the number of sensors within the network that will make the difference. Each sensor that composes the network also plays the role of a relay station. In this way, information is transmitted to destination by going from one sensor to the other by radio links. The sensors within the network collaborate and look for the best path to route data to the observer. The installation of such a network is done automatically, without human intervention. The network is able to configure itself; if certain sensors are out of service, the number and meshing of sensors will compensate the loss. In other words, the active sensors in the network will search for an alternative route to transmit data.

The integration of new sensors in the network is also possible due to the data processing capability of each sensor. In this way, it is possible to get precise and diverse information on the environment they cover.

23.2.5. *Sensor network models*

According to the behavior of sensors, of the observer and of the phenomenon to detect, sensor networks can be classified into two categories: static and dynamic networks. In the case of static networks, the sensors, the observer and the phenomenon to detect do not move. The dynamic sensor network model extends the mobility notions to the sensors themselves, to the observer and to the physical process of interest (e.g., oil spills). Due to this, mobility management within sensor networks is different from mobility management in traditional ad hoc networks where only the nodes can be mobile.

23.3. Transmission medium

Sensors communicate together through a wireless transmission medium. This transmission can be done by radio, infrared or optics. In order to enable global use of such networks, the transmission media must be available in the whole world. A first choice for wireless communication is the use of radio links in one of the ISM bands, the available bands that are used by wireless networks. The main advantage of using ISM bands is the wide spectrum of frequencies and the fact that they are available everywhere. Today most sensor platforms use radio support in one of the ISM bands.

Another wireless communication possibility between sensors is infrared or IrDA (Infrared Data Association) [INF]. IrDA is an association started in 1993 and groups together approximately 150 members. The speed of infrared transmission is not uniform all over the world and there is no norm. It can reach 16 Mb/s according to the new infrared IrDA-V VFIR specifications. The IrDA technology is normally designed for particular applications such as synchronization between cell phones, personal digital assistants and computers. However, with these new specifications, other applications can be found. However, the communication would necessitate a line of sight (i.e. obstacle-free path) between the transmitter and the receiver and must be done at short range (3 to 10 feet). In addition, infrared does not work well outside due to luminous rays which can cause interference. This all makes the use of infrared technology difficult for communication between sensors.

Optical transmission is also a solution for communication between sensors in some applications. In [WAR 01], the smart dust mote is an autonomous sensor able to communicate with other sensors via an optical communication support. Two transmission scenarios have been studied: a passive transmission scenario which uses a CCR (Corner-Cube Retro-reflector) and a second communication scenario that uses a laser diode with mobile mirrors.

23.4. Platforms

There is a lot of industrial and academic activity surrounding wireless sensor networks. Multiple projects have been started in the last few years. Some of them focus on the development of platforms. In this section, we will present some of these platforms. In most of the platforms available today, the size of sensors is practically determined by the power supply, which is most often two little AA batteries. Miniaturization of the different sensor components mostly comes from advances in micro-electromechanical systems (MEMS). MEMS are miniature components (the size of a millimeter) whose functioning associates very different techniques such as microelectronics, optics, electromagnetism, fluid mechanics, thermodynamics, microsurgery, etc. to semiconductor substrates. Battery miniaturization and the improvement of its autonomy remain the major challenge. Indeed, it will be important that the volume of the power supply is about the same size (a few square millimeters or less) as the size of the sensor and that the power supply has a comparable lifetime, if not better, to the batteries available today. For now, there are no commercially available mini-batteries with long autonomy that fill this need. In order to reach the demands of sensor networks, researchers are studying innovative energy sources using new concepts and material. We will continue on this subject in section 23.6. Figure 23.2 illustrates the size of a sensor from the Mica Motes family which we are presenting in the next section.

Figure 23.2. *Mica Mote: a new generation of wireless sensors. (Source: Intel Berkeley Research Laboratory)*

23.4.1. *Mica Motes*

Mica Motes [SMA] represent a sensor family developed by University of California, Berkeley. The Mote sensor is a small system, currently about the size of a box of matches and constantly in miniaturization stage. The Motes hold an Atmel Atmega 103 (4 MHz; 8 bits; 128 Kb of instruction memory; 4 Kb of RAM memory) microcontroller, a 512 Kb flash memory card, a analog-to-digital converter and peripheral interfaces. Each device is equipped with sensors to measure different environment parameters: magnetic fields, temperature and atmospheric pressure. Support to miniaturized microphones and cameras is currently being developed. The core of the system is a very small operating system (TinyOS [HIL 04]) that can operate on 8 bits and whose role is to control input-output, communications and energy resources, in order to conserve energy as much as possible. The wireless communication interface uses a TR1000 [RFM] module that operates on the 916.5 MHz frequency and offers a throughput of 2.4, 19.2 or 115.2 Kb/s.

23.4.2. *MIT μAMPS*

The μAMPS (μ-Adaptive Multidomain Power-aware Sensors) [MIN 01] project is being developed at MIT (Massachusetts Institute of Technology). The μAMPS-1 node measures 55 mm × 55 mm. It is equipped with a sensing system, a microcontroller and a radio interface. The sensing system is made up of an acoustic sensor and a low-power analog-to-digital converter. Data processing and network functions are supported by a StrongARM SA-1110 microcontroller. The radio interface uses an LMX3162 module and operates in the 2.4 GHz frequency band. The node is able to transmit and receive at 1 Mb/s in half-duplex mode. CSMA (Carrier Sense Multiple Access) is used as the media access protocol. The small operating system called μOS controls input-output, communications and energy resources. The sensing module consumes 40 mW and the electronic components 110 mW. The analog-to-digital converter consumes 1.75 μJ/bit. The radio interface consumes 175 mW in transmission or receiving mode.

23.4.3. *PicoNodes*

Sensors that are smaller than 1 cm^2, lighter than 100 grams, cheaper than $1 and that consume less than 100 mW of energy are the idea behind studies performed by Berkeley researchers for the PicoRadio [RAB 00] project. These sensors, called PicoNodes, will be powered by energy extracted from the environment. PicoNodes will need to be able to organize themselves in order to form a network infrastructure. Researchers estimate that networks containing between 100 and 1,000 nodes can be established. Data transmission throughput must be inferior to 100 Kb/s.

23.4.4. *Rockwell WINS*

The Rockwell WINS [ROC] node measures 3.5 × 3.5 × 3 inches. It uses a StrongARM SA-1100 microcontroller at 133 MHz. This microcontroller offers 16 Kb of instruction cache and 8 Kb of data cache. Data and software storage is ensured by 1 Mb of SRAM and 4 Mb of flash memory. The radio interface uses a RDSSS9M DCT chip manufactured by Conexant Systems. The radio interface operates on the 900 MHz frequency band and uses the DSSS (Direct Signal Spread Spectrum) technique. TDMA is used as media access protocol and the throughput is 100 Kb/s. The sensing module is composed of a magnetometer, of an accelerometer, of an acoustic sensor and a seismic sensor. The node consumes 1 watt maximum. The microcontroller consumes 300 mW, the radio interface 600 mW in transmission mode and 300 mW in receiving mode. The sensing module consumes less than 100 mW.

23.5. Energy consumption

The energy consumption of a sensor is based on its three activities: data sensing, data processing and communication. Of all of these activities, communication is the one that consumes the most energy. There is much less energy used for data processing than for communicating. On the Mica Motes platform, for example, transmission of 1 byte consumes the energy equivalent to 11,000 of computation. This implies that data processing (i.e. aggregation and elimination of redundancy) plays an important role in increasing the network's lifetime. The energy consumed for data sensing depends on different factors, particularly the number and type of microsensors installed on the sensing module and the sampling frequency. However, in many platforms the energy consumed by the sensing module is low compared to the energy necessary for data processing and communication.

Radio operation mode	Consumption (mW)
Transmission	14.88
Receiving	12.50
Idle listening	12.36
Sleeping	0.016

Table 23.1. *Energy consumption example of a radio interface*

Efficient control of the radio interface is also important to ensure a long lifetime for the network. Radio interface control must consider the energy necessary to initialize the interface, to transmit and receive packets and for idle listening mode. Table 23.1 illustrates energy consumption of the radio interface on the platform presented in [SCH 02]. The energy consumed by the radio interface in idle listening mode is 12.36 mW, which compares to the energy consumed in transmission or receiving mode. On the other hand, the energy consumed by the radio interface in sleeping mode is only of 0.016 mW. The strategy to minimize energy consumption of the radio interface consists of using periodic sleep-wakeup cycles. However, going from sleeping to listening cannot be done too often because the energy needed to start the radio interface can be significant compared to the energy used during the interface's active period (i.e. receiving, transmission or idle listening). All this is demonstrated by Shih *et al.* [SHI 01]. Furthermore, there exists a compromise between the periodic sleep-wakeup cycles and the detection delay and the transmission of information.

23.6. Power supply

Today, electrochemical energy supplied by batteries is the main power supply for portable devices. However, other forms of power supply will be useful to sensors. Independently from the power supply type, the lifetime of a sensor obviously depends on the quantity of energy that its reserves can store. The first interesting metric for energy storage is the energy per unit of volume (J/cm^3). An important characteristic concerning energy storage is that the instant power supplied depends on the size of the reservoirs. So in some cases, as with microbatteries, the maximum power density (μW/cm^3) must be taken into account.

23.6.1. *Energy reservoirs*

Primary batteries (non-rechargeable) are the most versatile and the smallest source of energy on the market today. Table 23.2 shows the energy density of some primary batteries. Zinc-air batteries are the most dense, but their lifespan is very low. Primary batteries show a very stable voltage. This enables the electronic devices to be directly powered by the battery without any help from additional energy regulating circuits which would consume much more energy.

Chemistry	Zinc-air	Lithium	Alkaline
Energy (J/cm^3)	3,780	2,880	1,200

Table 23.2. *Energy density of some primary batteries*

Secondary batteries (rechargeable) are usually used by electronic products such as cell phones, personal digital assistants and laptop computers. The energy density of some secondary batteries is presented in Table 23.3.

Chemistry	Lithium	NiMHd	NiCd
Energy (J/cm^3)	1,080	860	650

Table 23.3. *Energy density of some secondary batteries*

23.6.2. *Microbatteries*

As the size of electronic circuits has decreased by an order of magnitude, the size of batteries has decreased very slightly. Miniaturization of batteries remains a challenge because the output power of a battery depends on its active surface. In fact, the maximum output current of the battery depends on the surface of its electrodes. Since microbatteries are very small in size, the electrodes also have a reduced surface and therefore the output current of microbatteries is very low.

Maintaining or even increasing the performance of a sensor by reducing the size of the batteries is a very real problem and many research studies are focusing on this subject. Thin film batteries have large energy density, long lifetime, smooth discharge profile, wide temperature range, stable conservation and short recharge time. When these batteries reach a long lifetime, they can be incorporated in chips. Bates *et al.* (Oak Ridge National Laboratory) [BAT] have created a thin film battery by alternating between layers of Lithium Cobalt Oxide, Lithium Phosphate Oxynitride and Lithium metal. The maximum potential is 4.2 V and the maximum direct current is 1 mA/cm^2 and 5 mA/cm^2 for Lithium Cobalt Oxide batteries. Thickness of the batteries is 10 µm. However, the studies have been conducted on batteries with a surface of a few cm^2. Thin Film Lithium Ion Cells from Cymbet [CYM] are already incorporated in semi-conductors, RFID labels, medical electronics, military or space applications or in communication equipment. These batteries have an energy density of 900 watts-hour per liter, voltage of 3.6 V/cell and up to 70,000 charge cycles. Different dimensions are proposed, from 0.1 µm^2 to 10 cm^2, and thickness from 5 to 25 µm.

Infinite Power Solutions [RFI 03] produces thin film batteries (< 5 mm) which supply 4 V with a capacity of 200 µA/cm^2. Different methods have been developed to enable battery charging, in particular by using renewable energy (difference in temperature, movement or vibration) or RF energy. The field of application covers RFID labels, microelectronics and medical equipment. The cost for microbatteries is relatively high; it varies from $1 to $10. Economic production is therefore critical for commercial success.

3D microbatteries use a new space concept, based on 3D microstructures. Hart *et al.* [HAR 03] have developed a theory on a 3D battery that is made up of a series of cathode and anode bars suspended in a solid electrolyte matrix. The theoretical output power of a 3D microbattery has been shown to be much higher than the one for a 2D battery of the same size. This performance has been obtained due to a higher electrode/volume surface ratio as well as to low ohmic losses due to short ionic diffusion distances. Researchers estimate that it will take another five years to develop a light 3D commercial battery.

23.6.3. *Micro fuel cells*

Fuels using hydrocarbons have a higher energy density than batteries. The density of methanol is 17.6 KJ/cm^3, i.e. six times higher density than a lithium battery. Micro fuel cells could be integrated into sensors. However, the technology is not quite ready. The maximum efficiency that can be obtained by a micro fuel cell is not yet known, but will certainly not be high. Today, the efficiency of micro fuel cells is less than 1%.

Hahn *et al.* [FRA] and Lee *et al.* [LEE 02] have done research on micro fuel cells based on hydrogen. Hahn *et al.*'s system produces a power of 100 mW/cm^2 from a 0.54 cm^2 device. Lee *et al.*'s system provides 40 mW/cm^2. In collaboration with the University of Freiburg, Imtek has developed a micro fuel cell [MIK] which is 2.5 mm high with an active surface of 1 cm^2. This cell can produce electric densities of 1 W/cm^3 when it is hydrogen or air powered. This innovation is based on microchannels in bipolar plates made from metal sheeting.

Other research on methanol micro-cells is also being done. Holloday *et al.* [HOL] has developed a methanol fuel processor of a few mm^3. This processor, combined with a 2 cm^2 micro-cell, has provided from 25 mA to 1 V. The efficiency obtained is 0.5%. However, target efficiency is 5%.

23.6.4. *Atomic micro-energy*

Radioactive material contains high energy densities. This type of energy has been used for about 10 years. However, studies for the development of atomic energy micro-supply have started only recently. Researchers at Cornell University extract the necessary energy from a radioisotope. The energy from the radioactive material is converted directly in movement that drags an MEMS. A prototype uses an isolated brass bar that is 2 cm long over the thin film of a nickel-63 radioisotope. With a radioactivity period of 100 years, this isotope offers a very long lifetime. Future versions produced with the use of nanotechnology can be smaller than 1 mm^3. The mobile bar can move a small linear device, a cam or a small wheel by energy transfer or can produce electricity with the use of magnetic winding fixed to a rotating bar.

23.6.5. *Energy extracted from the environment*

The energy extracted from the environment will also be able to serve as auxiliary supply for sensors. However, in order for that to happen, we still need to come up with devices capable of extracting the energy from the environment and efficiently converting it into electric energy. Another factor that is just as important is the miniaturization of these devices which must be compatible with the sensors miniaturization. Table 23.4 shows that a solar cell can supply up to 15 mW per square centimeter in full sun at noon and 0.15 mW when cloudy. However, power density in full sun at noon is of approximately 100 mW per square centimeter [RAN 03]. This shows that the conversion efficiency of a solar cell peaks at 15% in strong light conditions.

23.7. Evaluation metrics

The main metrics to evaluate sensor network protocols is as follows: energy and lifespan of the network, latency, precision, robustness against outages, scalability.

23.7.1. *Energy and lifetime of the network*

In many scenarios, the non-rechargeable battery is the only power supply of a sensor. It is therefore important to put in place optimized protocols in order to maximize the lifetime of the network. This lifetime can be measured in different ways depending on the needs of the applications. For example, the lifetime of a network can be the time when only one sensor in the network fails or the time when a determined number of sensors in the network breaks down. The moment when the network stops sending important data concerned with an application may also be the lifetime of the network. A uniform use of the energy of the sensors is the key to delay the arrival of partitions in the network for as long as possible.

Energy source	Power density	Energy density
Primary battery	–	1,050-1,560 mWh/cm^3
Secondary battery	–	180.5-300 mWh/cm^3
Solar (outside)	15 mW/cm^2 (full sun at noon) 0.15 mW/cm^2 (cloudy skies)	–
Solar (inside)	0.006 mW/ cm^2 (office) 0.57 mW/ cm^2 (< 60W lamp)	–
Vibrations	0.01 – 0.1 mW/cm^3	–
Acoustic noise	3E-6 mW/cm^2 at 75 dB 9.6E-4 mW/cm^2 at 100 dB	–
Micro fuel cell	–	972 mWh/cm^3
Radioactive (Ni-63)	0.52 μW/cm^3	–
Temperature	40 μW/cm^3	–

Table 23.4. *Energy supply comparison. The values provided are estimations found in other works. Vibrations' power densities have a high dependence on the frequency and the amplitude of vibrations*

23.7.2. *Delay and precision*

The observer is interested in obtaining information on a phenomenon in a short time. The acceptable limit for this delay depends on the application. Obtaining precise information is the goal of the observer. Just as with delay, precision is a specific parameter of the application. There has to be a compromise between precision, delay and efficient use of the energy. The infrastructure of a sensor network must be adaptive in order for the precision and delay requirements to be met with a minimal use of energy.

23.7.3. *Robustness against outages*

Sensors in a network can be down due to the dynamic conditions of the environment in which they are placed. A sensor will also be down after a complete discharge of its battery. In a network with a large number of sensors, the replacement of broken down sensors may not be possible. It is therefore important to develop mechanisms enabling a quick adaptation of the network and robustness with regard to the state of the sensors. It would be important that the outages are hidden from the application.

23.7.4. *Scalability*

Scaling sensor networks is an important, even critical factor. The use of a hierarchical infrastructure and of aggregations is increasingly important for scalability.

23.8. Network protocols

Routing protocols developed for ad hoc networks such as AODV, DSR and DSDV are not quite adapted to the particular characteristics of sensor networks. It is therefore important to put new routing protocols in place which are able to organize the communication while taking into account the energy and scalability constraints.

Direct diffusion [INT 00] is a new routing protocol in which the observer broadcasts his requirements to indicate to the sensors which tasks are to be executed. The observer broadcasts his requirements by flooding the network. Each sensor stores the requests in its buffer. Each request has a time flag and one or more gradients. The time flag indicates for how long the request is valid and the gradients identify (local identifier) the neighboring nodes that have broadcast the request. In this way, during the propagation of the request in the network, the gradients establish a return path toward the observer. This path is used when a sensor sends data corresponding to the observer's request. The use of one of the neighbors among many possibilities must be determined according to the constraints. The data aggregation is determined by local policies. The observer must regenerate and sometimes detail his requests as soon as he starts to receive data from the responding sensors. Geographic routing techniques can be used in order to execute a partial flooding of the network.

Gossiping [HED 88] is a probability flooding scheme. The nodes broadcast messages only to a few neighbors instead of all of them. The majority of the nodes in the network will still receive the broadcast message anyway. Link redundancy within the network is the key with this method. The Gossiping scheme can be used to broadcast requests and data collected by the sensors. However, this protocol must be adapted to the energy constraints of the sensors. Furthermore, propagation time for messages to all nodes can be long, which limits the use of this protocol.

SPIN (Sensor Protocols for Information via Negotiation) [HEI 99] is a family of protocols whose goal is a more efficient spreading of data in the network. The traditional data dissemination approaches such as Gossiping and flooding enable the diffusion of redundant messages in the network. This leads to a significant waste of energy. SPIN proposes a negotiation protocol used by sensors before sending their data. If a sensor has data to share, it announces its data to the other nodes in the network. If the node is interested in receiving the data, it sends a request to the sensor. Negotiation for data transmission uses three messages: advertize (ADV), request (REQ) and data (DATA). ADV and REQ messages only contain a description of the data. When a sensor detects a phenomenon, it sends an ADV message to the neighboring nodes. If a neighboring node is interested by the data, it sends a REQ message to the sensor which will then send the data. Then, the node that has received the data repeats the same process. The result of this approach is the spreading of data in the network. The SPIN-1 protocol has the above described functionality. The SPIN-2 protocol also uses a heuristic method for energy conservation. When the node's energy is lower than a certain threshold, it reduces its participation in the message diffusion process.

ACQUIRE (Active Query Forwarding in Sensor Networks) [SAD 03] treats the network as a virtual database. An active request to the database corresponds to a request for problem resolution by considering the information collected by the sensors. The observer sends a request on the network in order for the sensors to resolve the specified problem. The request is sent over a series of nodes. At each intermediate hop, the node that processes the request uses the data from the neighboring nodes that are up to *d* hops of it. In this way, it receives a partial response to the request. Then, it chooses the next node to which it will send the request. In this way, the request keeps on circulating within the network until it arrives at a node which has all the partial results necessary to resolve the specified problem. At that time, the response to the problem is routed back to the observer by using the same path in reverse or the shortest path.

LEACH (Low-Energy Adaptive Hierarchy) [HEI 00] is a hierarchical routing protocol which minimizes energy consumption of a sensor network. The observer and sensors are considered as fixed. The nodes are grouped in clusters. Each cluster is represented by its cluster-head (i.e. the leader within a cluster). LEACH distributes to each sensor in the network the responsibility to be a cluster-head. The cluster-head election is accomplished periodically. This prevents that a sensor will consume more energy than others due to it always being leader. LEACH is executed in phases. In the first phase, the sensor calculates the value of a parameter which enables it to determine if it is cluster-head or not. The sensors that are leaders notify all the sensors in the network that they are the new cluster-heads. When it receives the signal announcing the leaders, the sensor determines which cluster it depends on. This choice is based on the signal quality of cluster-heads at the moment it receives the signal. Then, the sensor will inform its cluster-head that it is a member of the cluster. Then, the leader organizes the group in order for the communication to be TDMA-based. In the second phase of LEACH, the sensors are ready to collect data and to transmit them to their cluster-head. The data received by the cluster-heads can be aggregated before being transmitted to the observer. After a determined time, the network restarts in the first phase.

23.9. Auto-organization

The large number of expected sensors in applications makes manual configuration of each device impossible. Therefore, it is important that the sensors are capable of creating the network's topology themselves. In addition, due to each sensor's dynamic (e.g., down sensors in the network), sensors must be able to update the topology.

ASCENT (Adaptive Self-Configuring sensor Network Topologies) [CER 02] is an adaptive topology control mechanism for wireless sensor networks. ASCENT uses multiple phases to organize the sensors. During initialization, the sensor starts with a "listening phase" in order to estimate the number of neighboring sensors that are active. Then, the sensor initializes the phase called "joining phase", where it will decide whether to join the network or not. Such a decision is based on the measured packet loss ratio and the estimated number of active neighboring sensors. During joining phase, the sensor can join the network for a short period of time in order to determine if it contributes to the improvement of the network's connectivity. If it joins the network for a long period of time, it enters a new phase called "active phase", where routing messages and data exchanges can be executed. If the sensor decides not to join the network, it initializes the "adaptive phase", which extends for a determined time period, where the communication range is reduced. ASCENT is situated between the MAC layer and the routing protocol. Thus, it does not route and does not depend on a particular routing protocol. Results show that ASCENT increases the efficiency in energy use and reduces message loss when it controls the number of active neighbors.

The auto-organization of sensors can also be executed by the MAC layer. SMACS (Self-Organizing Medium Access Control for Sensor Networks) [YE 04] is a control mechanism that constructs a non-hierarchical topology. The protocol enables the nodes to discover their neighbors and to establish a planning for synchronized packet transmission and receiving. SMAC periodically puts nodes in sleep mode in order to reduce energy consumption. In addition, to save even more energy, the node is put in sleep mode during transmission in other nodes.

23.10. Applications

It does not take a whole lot of imagination to understand that wireless sensor networks have a great future ahead of them. We will progressively find sensors everywhere, measuring a large number of signals and communicating them to control centers. Sensor networks will be useful to applications in many sectors such as military, environment, health, industrial and home automation just to name a few.

23.10.1. *Military*

Ease of deployment, automatic arrangement and robustness against failures make sensor networks a promising information acquirement technology for military applications. Sensors can be deployed on battlefields to detect enemy troop movements. We can imagine systems sensitive to heartbeats making it possible to locate precisely each soldier or sniper on a battlefield. The sensors could also indicate the presence of explosives, chemical or bacteriological weapons.

23.10.2. *Environment*

Sensor networks will soon participate in the surveillance of the environment. Multiple catastrophes such as widespread fires and pollution can then be avoided. You will simply need to spread from an airplane a cloud of sensors over a forest. As soon as a temperature spike is detected near one of them, the information is transmitted from sensor to sensor to a centralized observation system. Presuming that the sensors are capable of providing location information even approximately, any fire is detected and located during the first few seconds.

Ambient air quality monitoring is a part of the many ways in which we can protect the environment in which we live. In fact, human activity (industries, transportation and use of solvents) releases substances in the ambient air in the form of gas and dust. The installation of sensor networks will enable to measure air and water quality. The data from the sensors will be transmitted to control centers in order to evaluate risks linked to pollution or follow (and anticipate) air and water quality.

Sensor networks will be very useful in plantations. Sensors dispersed within a plantation will not only collect traditional information such as the temperature and humidity. They will communicate together and will generate a soil analysis. The irrigation of the plantation will be completely automatic. It will also be possible to connect to one of them to recover the data and act accordingly.

23.10.3. *Surveillance and rescue*

Sensor networks will also be used for surveillance of goods and people. The Loccatec [LOC] European project foresees the installation of video microsensors within the walls of every building located in a seismic risk zone. If there was a collapse, rescue services could have an instant estimation of the number of potential victims in each room and locate them.

23.10.4. *Artificial conscience*

Creators of smart dust foresee that sensor networks will help in the development of artificial intelligence systems composed of thousands and tens of thousands of information technology agents. In such systems, sensors such as Mica Motes will be useful as input-output and local processing systems whose role will be similar to neurons in the brain. They hope to put multiple artificial conscience systems in place on a whole territory.

23.10.5. *Home automation*

Sensor networks are also useful to the management of a home. Lighting and air conditioning will be controlled on the basis of sensor measures. Lamps, for example, will communicate with nearby sensors and will be automatically commanded to increase intensity if necessary.

23.11. IEEE 802.15.4 standard or ZigBee™

Slow inexpensive wireless networks with low throughput are incorporated in the IEEE 802.15.4 standard, called ZigBee™, in reference to the bees that individually are not worth much but together can accomplish major achievements. ZigBee functions at a maximum throughput of 250 Kb/s (2.4 GHz frequency band). The communication range of Zigbee devices varies from 30 to 245 feet. With regard to power consumption, wireless Zigbee modules must be able to function for 6 to 24 months with a pair of AA batteries. In order to do this, the radio interface must be in sleeping mode as often as possible. The price for Zigbee devices is approximately $5.

23.12. Production cost

Since networks can contain hundreds and even thousands of sensors, cost reduction is of course a critical issue. Cost reduction means components that are more compact, consume little energy, have more functionality and are mass produced. Unfortunately, we are very far from this reality today; Mica Motes, for example, cost between 50 and $100 per unit. Researchers estimate that a sensor will cost less than $1 in 5 years.

23.13. Conclusion

There is a lot of research activity surrounding the still very young wireless sensor network world. Besides the communications themselves, interesting challenges are before us: the development of low-power miniaturized radio interfaces, the design of devices able to extract energy from the environment, the development of small operating systems, the development of new protocols, the development of signal processing algorithms and the miniaturization of power sources. These sensor networks will be useful for applications in many sectors such as military, environment, health, industrial control and home automation just to name a few.

23.14. Bibliography

[BAT] BATES J., DUDNEY N., NEUDECKER B., UEDA A., EVANS C.D., "Thin-film Lithium and Lithium-ion Batteries", *Solid State Ionics*, 135: 33-45.

[CER 02] CERPA A. and ESTRIN D., "ASCENT: Adaptive Self-Configuring Sensor Network Topologies", in *Proceedings of Infocom*, New York, June 2002.

[CYM] CYMBET CORPORATION – Thin-Film Energy Systems, URL: http://www.cymbet.com.

[FRA] FRAUNHOFER INTITUT ZUVERLÃSSIGKEIT UND MIKROINTEGRATION, "Micro (MEMS-based Fuel Cells)", URL: http://pb.izm.fhg.de.

[HAR 03] HART R. W., WHITE H. S., DUNN B. and ROLISON D. R., "3-D Microbatteries", *Electrochemistry Communications*, 5:120-123, 2003.

[HED 88] HEDETNIEMI S. and LIESTMAN A., "A Survey of Gossiping and Broadcasting in Communication Networks", *Networks 18* (4), p. 319-349, 1988.

[HEI 00] HEINZELMAN W. R. *et al.*, "Energy-efficient Communication Protocol for Wireless Microsensor Networks", in *IEEE Proceedings of the Hawaii International Conference on System Sciences*, p. 1-10, January 2000.

[HEI 99] HEIZELMAN W. R. *et al.*, "Adaptive Protocols for Information Dissemination in Wireless Sensor Networks", *Proceedings of the ACM Mobicom*, p. 174-185, Seattle, Washington, 1999.

[HIL 00] HILL J. *et al.*, "System Architecture Directions for Networked Sensors", in *Proceedings of the 9th International Conference on architectural Support for Programming Languages and Operating Systems*, p. 93-104, ACM, Cambridge, MA, USA, November 2000.

[HOL] HOLLODAY J.-D. *et al.*, "Microfuel Processor for use in a Miniature Power Supply", *Journal of Power Sources*, 108:21-27.

[INF] INFRARED DATA ASSOCIATION, URL: http://www.irda.org.

[INT 00] INTANAGONWIWAT C., GOVINDAN R. and ESTRIN D., "Directed Diffusion: a Scalable and Robust Communication Paradigm for Sensor Networks", p. 56-67, *Proceedings of ACM Mobicom*, MA, 2000.

[LEE 02] LEE S. J. *et al.*, "Design and Fabrication of a Micro Fuel Cell Array With Flip-Flop Interconnection", *Journal of Power Sources*, 112:410-418, 2002.

[LOC] LOCCATEC PROJECT, http://www.loccatec.org.

[MIK] Mikrobrennstoffzelle, http://www.ise.fhg.de/german/fields/field5/mb3/index.html.

[MIN 01] MIN R. *et al.*, "Low-Power Wireless Sensor Networks", *Proc. 14th Int'l Conf. on VLSI Design*, p. 205-210, Bangalore, India, January 2001.

[RAB 00] RABAEY J., AMMER J., DA SILVA JR. J. L., PATEL D. and ROUNDY S., "PicoRadio Supports Ad-Hoc Ultra-Low Power Wireless Networking", *IEEE Computer Magazine*, vol. 33, no. 7, p. 42-48, July 2000.

[RAN 03] RANDALL J.-F., "On Ambient Energy Sources for Powering Indoor Electronic Devices", PhD Thesis, Ecole Polytechnique Federale de Lausanne, Switzerland, May 2003.

[RFI 03] THIN-FILM BATTERY FOR RFID SENSORS, *RFID Journal*, May 7 2003, URL: http://www.rfidjournal.com/article/articleview/411.

[RFM] RF Monolithics Inc., ASH Transceiver TR1000 Data Sheet, URL: http://www.rfm.com.

[ROC] ROCKWELL WINS, URL: http://wins.rsc.rockwell.com.

[SAD 03] SADAGOPAN N., KRISHNAMACHARI B. and HELMY A., "Active Query Forwarding in Sensor Networks (ACQUIRE)", *SNPA '03*, May 2003.

[SCH 02] SCHURGERS C., TSIATSIS V., GANERIWAL S. and SRIVASTAVA M., "Optimizing Sensor Networks in the Energy-Latency-Density Design Space", *IEEE Transactions on Mobile Computing*, vol. 1, no. 1, January-March 2002.

[SHI 01] SHIH E. *et al.*, "Physical Layer Driven Protocol and Algorithm Design for Energy-Efficient Wireless Sensor Networks", *Proceedings of ACM Mobicom*, p. 272-286, Rome, Italy, July 2001.

[SMA] SMART DUST, URL: http://www.cs.berkeley.edu/~awoo/smartdust.

[WAR 01] WARNEKE B., LIEBOWITZ B., PISTER K. S. J., "Smart Dust: Communicating with Cubic-Millimeter Computer", *IEEE Computer*, p. 2-9, January 2001.

[WIL 95] WILLIAMS C. B. and YATES R. B., "Analysis of a Micro-Electric Generator for Microsystems", *Transducers 95/Eurosensors IX*, p. 369-372, 1995.

[YE 04] YE W., HEIDEMANN J. and ESTRIN D., "Medium Access Control with Coordinated, Adaptive Sleeping for Wireless Sensor Networks", in *IEEE/ACM Transactions on Networking*, 2004.

Chapter 24

Mobile Ad Hoc Networks: Inter-vehicle Geocast

24.1. Introduction

For many years now, new technologies regarding Intelligent Transport systems [ITS 04a] constantly flood the automobile market. Their main goal is the improvement of security, efficiency and comfort of transportation. Recent research in this domain is of particular interest to many countries in Europe, Japan and North America. We are seeing more and more investments. Indeed, according to a report published in 1997, the USDT (US Department of Transportation) and the ITSA (Intelligent Transportation Society of America) have estimated that the intelligent transport market will reach 420 billion dollars over a period ending in 2015 [BUR 00].

In Japan, progress in IVHS (Intelligent Vehicle/Highway System) is more advanced than in Europe and the USA. In fact, due to its nature, i.e. small area and big population, it has seen an acceleration of transportation problems. Since 1973, the Japanese government has encouraged different projects for the development and the evaluation of IVHS technologies. These different projects have produced the VERTIS (Vehicle, Road and Traffic Intelligence Society) association in 1994.

In the USA, the first research efforts were done individually by some states and cities. In 1987, the Mobility 2000 project, created in the IVHS context, has

---

Chapter written by Abderrahim BENSLIMANE.

encouraged the creation of national IVHS programs. From this, ITSA [ITSA 04b] has emerged.

In Europe, the two main programs are Drive [DRI 99] and Prometheus [MAR 99]. In 1986, Prometheus was created within the Eureka platform in order to improve the quality of information given to the driver and to help him with direct intervention mechanisms. It also anticipates cooperative driving and traffic management mechanisms for better traveling efficiency.

Driven by the European Commission, Drive's intent is to bring Europe toward a transportation environment based on the communication between the vehicle and the road. It plans the installation of infrastructures throughout Europe's roads. In this way, vehicles circulating on these roads will benefit from the many services offered by this infrastructure. For example, the vehicles can be notified of traffic flow, accidents, get navigational information and even use the Internet.

The success of these different projects has emphasized the importance of communication in the transportation environment. Indeed, the emergence of solutions based on communication helps to overcome the traditional systems limitations (radar, cameras and video) and makes the availability of advanced new services possible.

Traditional guidance services based on bulletin boards and FM station broadcasting must become more efficient and more precise with the introduction of communication. Information on the state of traffic is transmitted and received faster when the vehicles communicate directly between each other.

Cooperative driving services, based on the exchange of information between the members of a group of vehicles (convoy), guarantee more comfort and efficiency. Indeed, exchanged data: speed, acceleration and positioning of vehicles help to maintain a secure distance between vehicles in all kinds of situations (rapid speed change, etc.) and can also help to detect bottlenecks [BRE 01]. Exchanged data can be simply presented to the driver in a convenient way (voice), or processed by automatic auto-acceleration and auto-brake agents in order to optimize traffic flow.

Security services are the first systems to establish; they are based on detection of anomalies (obstacles, fog, etc.) and the broadcasting of alert messages to inform road users of the problems. The detection of anomalies is done by image processing and the broadcasting of alert messages is based on mobile communication technologies. In this context, two different approaches are used: wireless networks with infrastructure or mobile ad hoc networks.

Solutions based on the use of wireless networks with infrastructure have been widely studied and evaluated in multiple projects [DRI 99]. However, due to the infrastructure, the deployment of these solutions takes more time and is more expensive than ad hoc-based solutions. In addition, solutions based on ad hoc networks are more portable. Indeed, you only need to equip vehicles with a means of communication (radio equipment and adequate protocols).

The rest of the chapter focuses on inter-vehicle communication using ad hoc networks, in particular, the diffusion of alert messages. Section 24.2 briefly presents routing and broadcasting techniques in ad hoc networks. Studies done in the transportation environment are presented and analyzed in section 24.3. The solution that we propose, Inter-vehicle Geocast, is explained in section 24.4. Section 24.5 presents the simulation results obtained and evaluates the performance of the IVG method compared to methods described in section 24.3. Finally, a conclusion is given in section 24.6.

24.2. Mobile ad hoc networks

Mobile ad hoc networks (or MANET: Mobile Ad hoc NETwork [MAN 03]) are distinguished from the other types of wireless networks, such as cellular networks, by a total absence of fixed infrastructure. A certain number of mobile devices equipped with radio interface cards and adequate protocols are sufficient to make up an ad hoc network. This includes cell phones, vehicles, satellites, laptops and even desktop computers.

Mobile ad hoc networks are expected to play an important role in mobile communication in the future. Not only because their deployment is less expensive, but also because it is the only viable solution when wired networks are either inefficient or simply impossible. In fact, because they can be organized spontaneously and in a completely distributed way, they offer more flexibility and robustness. Actually, research surrounding wireless ad hoc networks is of great interest to the military, sales and marketing, and universities. Applications of this type of network vary widely: conferences, meetings, emergency situations (natural catastrophes, etc.), military operations and cooperative driving between vehicles.

Contrary to networks with infrastructures, routing in ad hoc networks remains difficult. This problem mainly resides in the lack of fixed routers, from the moment where all the elements of an ad hoc network can be mobile. Consequently, vehicles in an ad hoc network must cooperate together in order to ensure routing and thus each vehicle is considered as a potential router.

In order to ensure routing in ad hoc networks, different protocols have been proposed. Multiple strategies have been used to classify routing protocols. In [MAU 01], the authors give a classification according to topology and position. A finer classification of protocols based on topology in two categories, flat and hierarchical, has been introduced in [XIA 02].

Topology-based protocols use information on existing links between nodes in the network in order to ensure routing. They can be classified in three categories. Among the proactive protocols (periodic table broadcast), we can mention DSDV [PER 94], OLSR [CLA 01], FSR [PEI 00a] and TBRPF [OGI 01]. Among reactive protocols (on demand routing calculation), we can quote AODV [PER 99] and DSR, [JOH 01]. As for hybrid protocols (reactive and proactive), we will mention ZRP [HAA 02] and LANMAR [PEI 00b].

The very large number of nodes and high mobility (high speed, approximately 80 mph) make these approaches not feasible for inter-vehicle communication.

Position-based protocols [MAU 01] use additional information: the geographic position of nodes for routing, thus enabling them to overcome the limitations of topology-based protocols. A localization service is used by the transmitter to determine the receiver's position. This position, included in the packet's destination address, will ensure routing based on the position. The routing decision only depends on the position of the destination and the position of the direct neighbors. Thus, the routing algorithms based on position do not need to establish or maintain routes. The nodes do not need to store routing tables or to exchange control messages. In addition, routing protocols based on position enable the broadcasting of messages toward a given geographical zone in a natural way (geocast). In this category of protocols, we can quote LAR [KOY 98], which is an on demand protocol, DREAM [BAS 98] which is a proactive protocol and GPSR [KAR 00] which uses location information of neighbors, periodically received, to transmit packets.

In our case, a vehicle needs to communicate with all neighboring unknown vehicles instead of sending a message to a specific identified vehicle.

The broadcast is planned to be frequently executed in ad hoc networks. Indeed, it is used by routing protocols for the construction and maintenance of their tables. It is also used for multicasting in highly mobile networks as well as to send alert messages in emergency situations.

Due to the difficulty of synchronization and the lack of information on global network topology, broadcast scheduling is a complicated task. A simple solution is network flooding [PER 01]. Each of the vehicles receiving the broadcast message it rebroadcasts immediately toward all its neighbors. However, this solution presents problems such as contentions, collisions and excessive redundancy of broadcast messages. These problems are known as a broadcast storm [NIS 99].

To overcome the contention problem, the received message is immediately rebroadcast, but its rebroadcast is differed for a time called defertime. With different defertime values, simultaneous rebroadcast or collisions are reduced. Multiple metrics are used for the defertime value: random generation, distance function separating the transmitter from the receiver, etc. [NIS 99, BAC 03]

To reduce redundancy, a mechanism based on heuristics is planned to classify useless rebroadcasts. Several heuristics have been proposed: random classification, approach based on the use of a counter, approach based on localization and approach based on the notion of cluster [NIS 99]. The authors in [SUN 01] also use coverage tree structures.

24.3. Communication in intelligent transport

24.3.1. *Role-based multicast (RBM)*

The solution proposed in [BRE 00a], called RBM, is designed to reduce the redundancy of transmitted messages. The disadvantage with this solution is that it uses more bandwidth due to the exchanged signals for the detection and maintenance of the neighbors. As shown in [BRE 01], RBM is not reliable since it fails in certain situations to transmit the alert message to destination at the right moment and it fails to prevent the fragmentation problem.

In RBM, each vehicle must maintain two lists: S and N. The N list represents its neighbors and S is the list of vehicles from which it has already received alert messages. A vehicle V receiving an alert message must first verify S and N values. If the N and S lists are equal, V does not have uninformed new neighbors. In this case, V must not rebroadcast the message. It goes from its initial state to another state where it waits for new neighbors to broadcast the message and as soon as it detects a new neighbor, it broadcasts the message and goes into final state.

On the other hand, if the N and S lists are not equal, V must rebroadcast the message after a given time. V first goes from its initial state to another state where it waits some time to decide about the rebroadcast. Once this time has expired, V verifies the N and S lists again. If it finds the N and S lists to be equal, it goes to the state where it waits for new neighbors. If not, it rebroadcasts the message and goes into final state.

The problem with this solution is that there is no return from final state to initial state and due to this, the solution does not make it possible to overcome the network's fragmentation in the case where there are a small number of vehicles on the road.

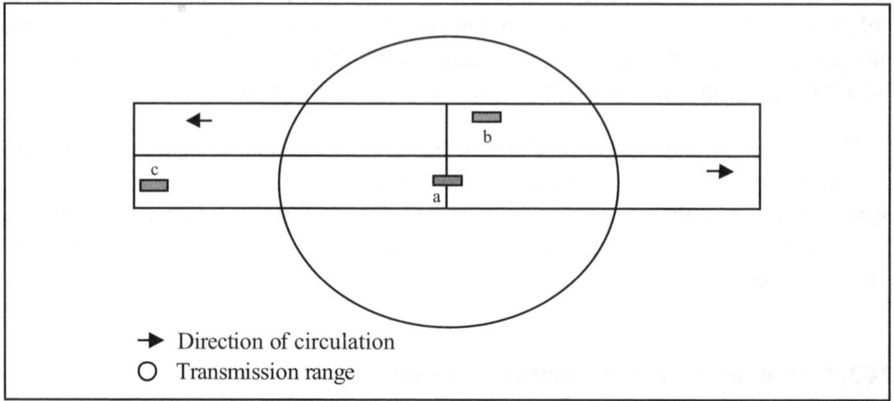

Figure 24.1. *Example of RBM failure to overcome fragmentation*

In the situation shown in Figure 24.1, vehicle (a) detects a new neighbor (b) and then it sends an alert message and goes into final state. Vehicle (b) circulating in the opposite direction to (a) will detect a new neighbor (c), it then rebroadcasts the message and goes into final state. Then, (c) will detect (a) as a new neighbor, it will therefore rebroadcast the message and go into final state. When (a) receives the message from (c), it will not react and it will ignore it because (a) is in final state. Finally, we will only have notified three vehicles (a, b and c) in this scenario and therefore RBM is not adapted to all situations.

24.3.2. *Detection of lanes (Trade) and Distance Defer Time (DDT)*

In [SUN 00], the authors propose two other scenarios, Trade (TRAck DEtection) and DDT (Distance Defer Time), in order to improve the performance of broadcast with regard to reliability (facing fragmented networks) and scalability (reducing the number of messages exchanged, particularly those used for the maintenance of neighbors).

In Trade, the objective is to guarantee reliability with a minimum of alert messages. Each vehicle wanting to broadcast an alert message must designate a certain number of vehicles among its neighbors to resume rebroadcast after it. This reduces to a minimum the number of alert messages sent over the network. However, Trade does not reduce all the exchanged messages on the network as it is based on the calculation of neighbors by all the vehicles and the explicit exchange of location information. Even if the messages are short, they consume radio bandwidth.

While using location by GPS, DDT was designed to reduce the number of messages exchanged on the network, thus to broadcast without the calculation of neighbors. DDT is based on the defertime method for broadcast scheduling. Each vehicle receiving an alert message waits for some time before deciding to rebroadcast. If during this time it receives the same alert message from another vehicle, then it will not rebroadcast the message. The problem with DDT is that the vehicle that broadcasts only does it once, which makes DDT unreliable in a network that is not very loaded or with remote fragments. Results published in [SUN 00] show that the reliability ratio of DDT increases with the increase of transmission range; however, no perfect reliability was obtained even with a range of 2,000 m (6,000 feet).

In this chapter, we propose a new method called IVG, which generalizes the previous methods and helps to overcome the problems presented above: fragmentation, reliability and neighbor calculation. In order to do this, we propose periodic rebroadcasts of alert messages by introducing dynamic relays. These relays are designed according to the distance to the transmitter. The analytical study of the complexity of IVG shows a significant improvement compared to the other methods, RBM, Trade and DDT. In addition, results obtained with simulation show the efficiency of the IVG method whatever traffic conditions are prevalent (rural or urban zones).

24.4. Inter-vehicle geocast

In this section, we describe the context and description of IVG protocol operations.

24.4.1. *Context*

Each vehicle that executes IVG must be equipped with a device which enables it to obtain its geographic position in real-time. In reality, GPS gives the position in 3D coordinates but for simplicity reasons we only use (x, y) coordinates in the plan. Vehicles are equipped with an R range omnidirectional radio antenna and circulate on an L large road.

Communication between vehicles is assumed to be bidirectional and is based on message broadcast. Each vehicle possesses a node_id identifier to be uniquely identified it in the network. Each broadcast message on the network must have a unique identifier in order to distinguish the different messages exchanged. The (node_id, msg_id) pair is used for the identification of the message, where node_id is the vehicle identifier that broadcasts the message and msg_id is the message's sequence number.

24.4.2. *Protocol description*

When a vehicle detects a problem (accident, bottleneck, etc.), it broadcasts an alert message informing the other vehicles of the danger. Several methods have been used to detect a crash. For example, during an accident, the airbag trigger can initiate broadcasting of the alert message by the vehicle involved. Among all the neighbors of this vehicle, only those who are in the risk zones (see Figure 24.2) will take it into consideration. Among all the neighbors in the same zone, only one vehicle, called the relay must react to ensure the rebroadcast to notify the vehicles that have not yet received the message. The relay is chosen in a totally distributed way. Each vehicle can know, with only the information in the alert message received, if it will become the relay or not. In addition, the relay must be chosen to ensure coverage of the largest transmission zone which is not covered by the transmitter. Consequently, the relay must be the neighbor that is the farthest away from the transmitter.

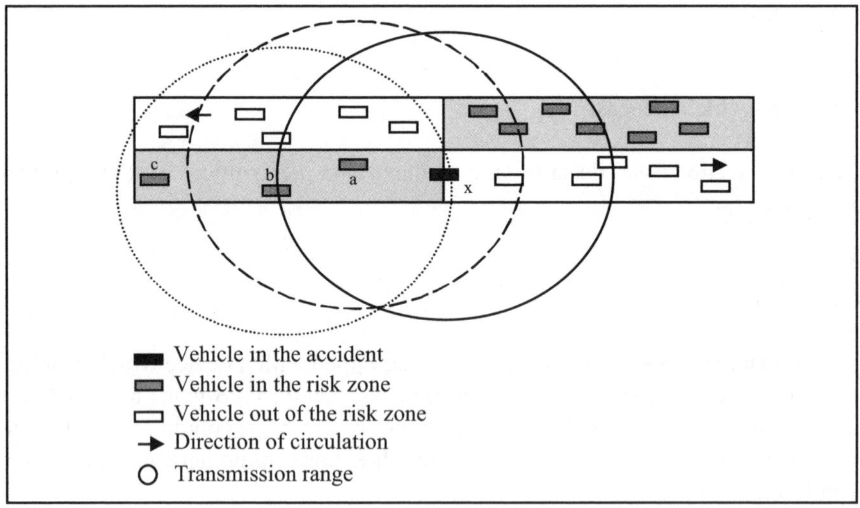

Figure 24.2. *Risk zones and relay selection*

In Figure 24.2, the vehicle involved in the accident x broadcasts an alert message. We can see that if vehicle (a) is chosen as the relay, vehicle (c) is not informed because it is out of the transmission range of vehicle (a). However, if vehicle (b) is chosen as the relay, vehicle (c) is informed. The way in which a vehicle is designated as a relay is based on the defertime method. In defertime, the vehicle receiving an alert message should not rebroadcast directly but must defer this rebroadcast for a time, called defertime. If after this given time it does not receive another broadcast of the same message from another vehicle, then it rebroadcasts the message.

To allow the vehicle that is the farthest away from the transmitter to become the relay, the defertime of each vehicle must be inversely proportional to the distance separating it from the transmitter. The more distance between the two vehicles, the less defertime. The defertime (x), which is calculated by a vehicle (x) receiving a message and retransmitting it, is explained in the following formula:

$$defertime(x) = \max\_defer\_time \cdot \frac{\left(R^{\varepsilon} - D_{sx}^{\varepsilon}\right)}{R^{\varepsilon}} \qquad [24.1]$$

where ε is a positive integer.

Supposing that the distribution of vehicles is uniform, the choice of ε = 2 will give a uniform distribution of the different defertime values in [0,max_defer_time]. D_{sx} is the distance between the transmitter (s) and the receiver (x). The value of max_defer_time is equal to twice the average transmission delay.

The behavior of vehicles executing IVG is represented by a finite state machine. Initially, all vehicles are in state E_0. When a vehicle is involved in an accident it goes into state E_1, where it periodically broadcasts an alert message. All the vehicles receiving this message go into state E_2 where they wait according to a timer. The first vehicle whose timer expires goes into state E_3 (becomes the relay) after which it periodically rebroadcasts the alert message. The other vehicles that were in state E_2, go into state E_4 since it is useless that they rebroadcast as there is a relay behind them to ensure rebroadcast. When the relay receives another message from another vehicle behind it, it goes into state E_4. Vehicles in state E_4 then go into state E_2 in the case where they receive an alert message from a relay that is in front of them (in an overtaking situation).

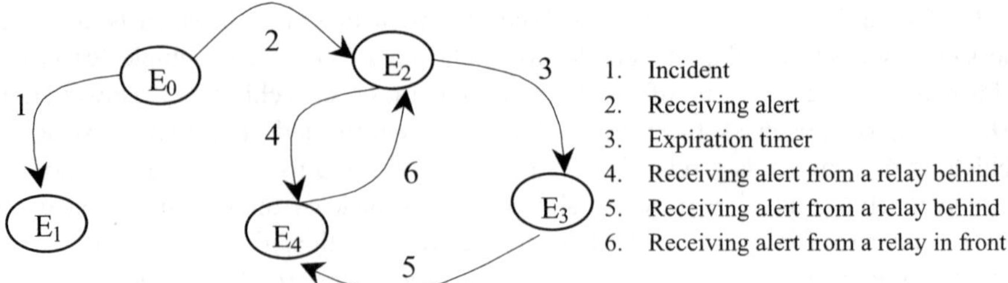

Figure 24.3. *Modeling of IVG algorithm with finite state machine*

Broadcast of alert messages is limited to risk zones, i.e. all vehicles receiving alert messages outside of this zone must ignore them in order to avoid infinite broadcasts.

A finite state machine is associated to each new alert message. For example, for an m_1 message the vehicle can be in state E_1 and for another m_2 message it can be in state E_2. Then, each vehicle maintains a t_state table where each entry is constituted of the (msg_cle, state) pair such that msg_cle is the (node_id, msg_id) pair and state is the state of this vehicle for this message.

State E_0 is initial and final at the same time. It means that there is no entry in table t_state for this message. When a vehicle exits the risk zone with regard to an accident, it releases the corresponding entry in table t_state.

The case of multiple relays caused by the fact that two (or more) vehicles, of equal distance to the broadcast source are designated as relays at the same time is not a problem. It can be resolved with the help of vehicle identifiers. The vehicle that will remain a relay is the one that has the smallest node_id identifier and the others must go into state E_4.

Determining the position and direction of the circulation can be easily calculated according to the vectors of direction of both vehicles and of the angles between them. For example, let us suppose that the displacement vector of the relay vehicle is V_1 and the displacement vector of the vehicle receiving the alert message is V_2. If the absolute value of the angle between V_1 and V_2 $|(V_1,V_2)|$ is lower than a given δ threshold, then both vehicles are going in the same direction. Or, if $|(V_1,V_2)-\pi| < \delta$, then the two vehicles are going in opposing directions. On the other hand, if none of these conditions is verified then the vehicles are driving on different roads. Value δ can tolerate lane changes and traveling effects. In general, δ is considered equal to $\pi/4$.

To generalize this method to take into account turns, we propose the use of the average vector of elementary displacements.

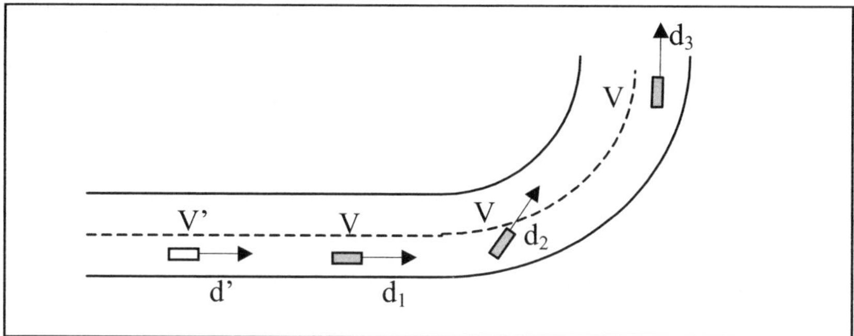

Figure 24.4. *Circulation direction with turns*

In Figure 24.4, we show the example of how the generalized (or average) displacement vector approach can resolve the problem of the direction of circulation during turns. In this example, the displacement vector for vehicle V is $d = (d_1 + d_2 + d_3)/3$ and the displacement vector for vehicle V' is d' (d' is also the average of three elementary moves for V'). Finally, the angle between vectors d and d' has more chance to be lower than δ, so both vehicles travel in the same direction.

In general, the generalized displacement vector d is given by the formula:

$$d = \frac{\sum_{i=1}^{n} d_i}{n} \qquad [24.2]$$

where d_i represents the elementary displacement vectors. The choice of the correct n value for the right number of elementary vectors is based on the different angle values used for the turns.

24.4.3. *Rebroadcast period for a relay*

In order to calculate the value of rebroadcast period Δθ, we use the diagram of Figure 24.5. Let us suppose that vehicle (A) is designated as a relay and vehicle (B) travels in the direction indicated in Figure 24.5.

Vehicle (B) was in position B_0 on the date θ_0 and will be in position B_1 on the date θ_1. The period $\Delta\theta$ between rebroadcasts must be lower than period $\Delta\theta_{max}$ such that $\Delta\theta_{max} = \theta_1 - \theta_0$. In fact, vehicle (B) must be informed of the situation before it enters the zone "too late". In this zone, braking is useless because the distance it takes for the vehicle to stop, after braking, is higher than the distance that separates it from the relay. In the worst of cases, vehicle (A) is stopped and vehicle (B) is going at a speed of $V = V_{max}$. In this case, the value of $\Delta\theta_{max}$ is equal to:

$$\Delta\theta_{max} = \theta_1 - \theta_0 = \frac{R}{V} - \frac{D_{braking}(V)}{V} \qquad [24.3]$$

where $D_{braking}(V)$ is the braking distance (see [24.4]) and R is the transmission range 9. In this formula it is important to note that the transmission range must be higher than the braking distance.

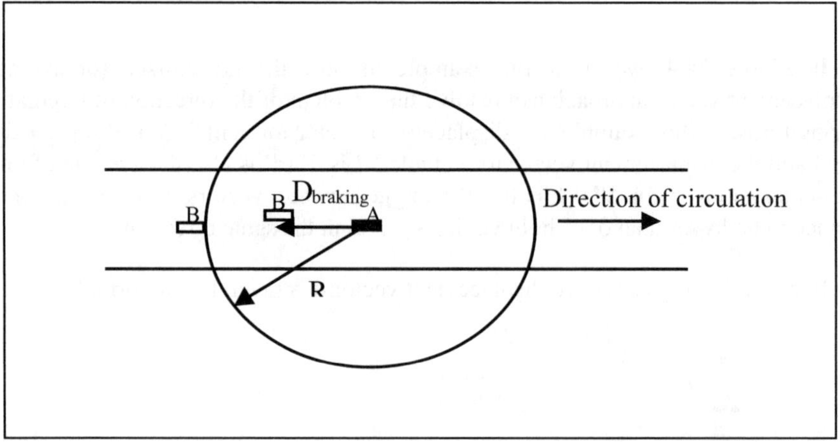

Figure 24.5. *Calculation of retransmission period $\Delta\theta$*

The braking distance used is the distance covered by the vehicle from the time of the driver's reaction until complete stop and is given in the following formula:

$$D_{braking}(V) = V \cdot \Delta t_{reaction} + \frac{V^2}{2 \cdot b_{max}}(m) \qquad [24.4]$$

where:

– $\Delta t_{reaction}$ represents the reaction time of a normal driver, which is generally equal to 1 ms;

– b_{max} represents the maximum deceleration, which is generally equal to 4.4 m/s^2;

– V is the speed of the vehicle in m/s.

24.5. Performance evaluation

In order to evaluate IVG performances we have created a model of mobility to simulate the behavior of vehicles on the road. Vehicles are uniformly distributed on roads at N vehicles per kilometer per lane and travel at constant speed along their lane. The speed of each vehicle is chosen randomly in the interval $[V_{avg} - \varepsilon, V_{avg} + \varepsilon]$, where V_{avg} represents the average circulation speed and ε its variation. The road has 2C lanes: C lanes each way.

In section 24.5.1, we compare the complexity of the IVG algorithm with Trade, which is the only algorithm providing a reliable broadcast. We show that the load of the network, in terms of bandwidth usage, is much more significant with Trade. This shows the scalability advantage of IVG.

In section 24.5.2, we show by simulation the gaps in terms of reliability of the RBM and DDT algorithms in light traffic situations. We emphasize the high level of reliability of IVG whatever the traffic conditions.

24.5.1. *Network load*

To determine the usage ratio of the bandwidth, we calculate O_{IVG} which is the number of messages exchanged on the network to get the alert message to all vehicles located in the risk zones (multicast group).

The number of relays used for IVG is:

$$number\_relays = \frac{M}{C \cdot N \cdot R} \qquad [24.5]$$

where M represents the multicast group size.

Consequently, the maximum number of messages exchanged is:

$$O_{GIV} = \left\lfloor \psi / \Delta\theta \right\rfloor \cdot \frac{M}{C \cdot N \cdot R} \qquad [24.6]$$

where ψ represents total time of alert and $\Delta\theta$ transmission period. The symbol $\lfloor \ \rfloor$ represents the upper integer part.

On the other hand, the number of messages exchanged in Trade is equal to the number of messages used for neighbor calculation, plus the number of alert messages sent through the relays.

$$O_{TRADE} = 2 \cdot M \left\lfloor \psi / \Delta\theta \right\rfloor \cdot \frac{M}{C \cdot N \cdot R} \qquad [24.7]$$

where $\Delta\theta$ represents the refresh period of calculating the neighbors.

24.5.2. *Simulation*

We have implemented the IVG, RBM and DDT methods in ns2 [ISI 95] in order to compare their performances in different situations. Numerical values used for C, L, V_{avg} and ε are 4, 40m/s, 25m/s and 5m/s respectively. The capacity of the channel used is 36 Kbps in the 2.4 GHz ISM frequency band. The protocol used for the MAC layer is IEEE 802.11. The size of messages used is 64 bytes (see the appendix).

Figure 24.6 shows the ratio of vehicles alerted at the right time (before entering the risk zone) based on different Δθ values. The results observed for Δθ correspond to those obtained by the formula [24.3]. Indeed, with a transmission range of 200 m and a braking distance of 150 m, we have $\Delta\theta_{avg} = (200 - 150)/30 = 1.66$ s. Although the braking distance (corresponding to maximum speed $V_{avg} + \varepsilon = 30$ m/s) is approximately 102 m, we have used 150 m in order to take into account the deteriorated driving conditions (rain, etc.). The range of 200 m used is not a problem in reality as long as there are wireless cards whose external range exceeds 400 m [XUK 02].

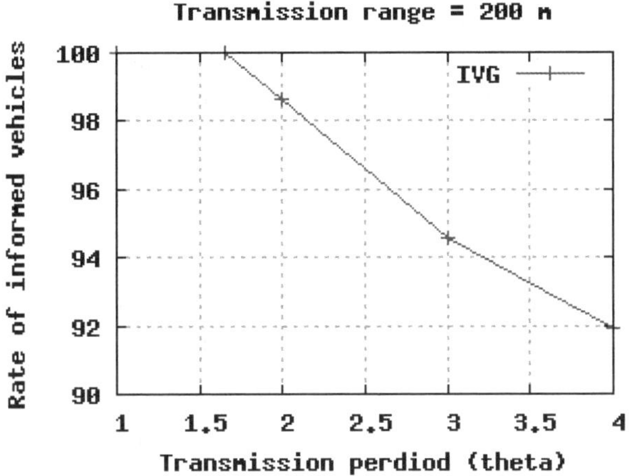

Figure 24.6. *Number of uninformed vehicles according to $\Delta\theta$*

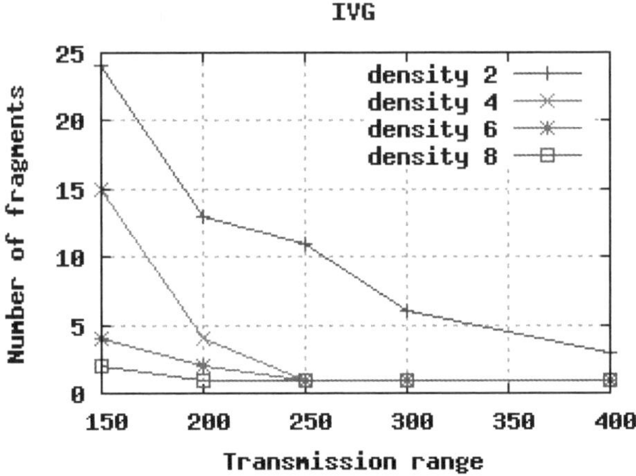

Figure 24.7. *Fragmentation of network according to R and N*

Several simulation scenarios with different traffic densities (N = 2, 4, 6 and 8 vehicles/km/lane) have been executed in order to evaluate the ratio of reliability of IVG, RBM and DDT. Variations in traffic density and in the transmission range enable us to give different values to the fragmentation ratio. A fragmented ad hoc network is a network containing at least two vehicles that cannot communicate together at a given time. All the values in the transmission range used in the simulation enable the complete coverage of the network (the highway). Indeed, minimum R range of the radio equipment used is 150 m (higher than the braking distance) and ensures the complete coverage of a highway whose width is equal to $L_{max} = \sqrt{3/2}\ R = \sqrt{3/2} * 150 = 129.90$ m. In practice, this value is largely superior to the width of existing highways.

Figure 24.7 illustrates the number of fragments in the multicast group (vehicles located in risk zones) for the different executed scenarios. We notice that the multicast group remains fragmented, even with a transmission range of 400 m, in situations of low traffic (N = 2). This shows that the use of algorithms such as RBM and DDT is dangerous in such situations (for example, rural roads). In fact, Figure 24.8.a shows that the ratio of informed vehicles at the right time (before entering the "too late" zone) does not exceed 70%, even with a transmission range of 400 m, for DDT. We notice that DDT enables a perfect reliability only when the network is not fragmented: from a transmission range of 250 m for densities of 4 and 6, and from 200 m for a density of 8 vehicles/km/lane.

Figure 24.8a shows that the reliability of IVG is equal to 70% with a transmission range of 150 m. This is mainly due to the short transmission range which is equal only to the braking distance. Indeed, all vehicles receiving the first alert message from the accident are already in the "too late" zone. In addition, in certain situations the relays can leave risk zones without notifying all the members of the multicast group. In this case, the vehicle with the accident takes the responsibility to inform the other uninformed vehicles. Consequently, all new vehicles notified by rebroadcast are already in the "too late" zone.

Figure 24.8b shows a case where reliability of RBM is better than IVG. Indeed, a scenario with a traffic density of 4 vehicles/km/lane and a transmission range of 150 m has shown reliability at 100% for RBM and at 93% for IVG. The reason for having certain vehicles uninformed at the right moment for IVG is mainly caused by the equality between the transmission range and the braking distance. Reliability of RBM in this situation is due to the fact that rebroadcasts are not limited to multicast members but to all vehicles. This means that certain vehicles may be informed by rebroadcasts of other vehicles located outside risk zones. These vehicles that are not part of the multicast group rebroadcast the alert messages to overcome the lack of relays in multicast zones. On the other hand, these messages overload the network and can then compromise reliability in case of collisions, particularly in dense traffic situations or when network use is too high from other applications, which causes scalability problems.

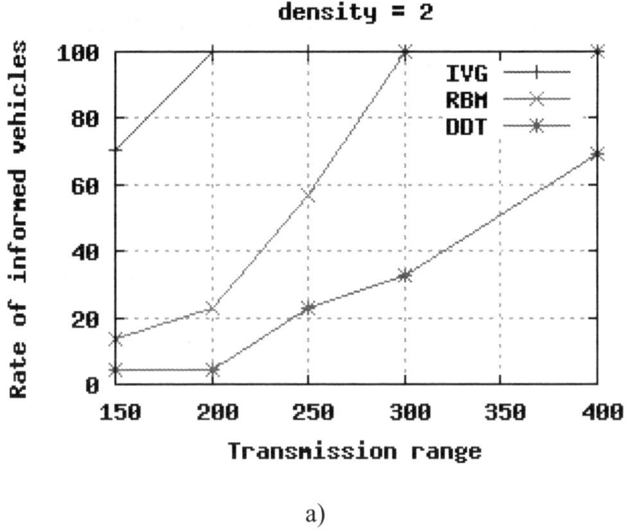

a)

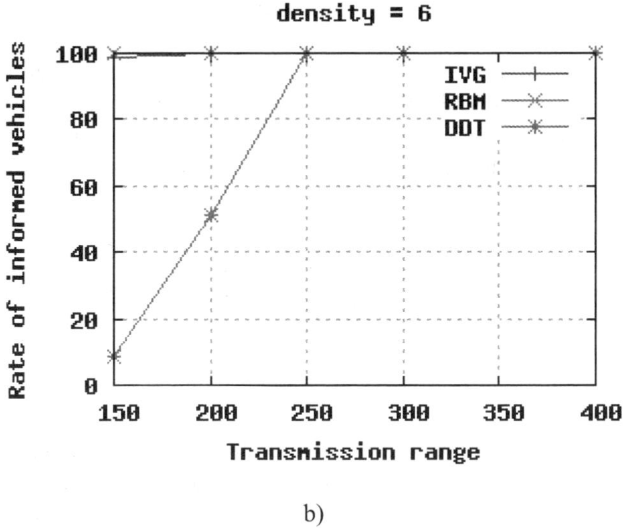

b)

Figure 24.8. *Comparison between IVG, RBM and DDT*

Figure 24.8 shows that IVG becomes reliable in all situations (N = 2, 6) with a transmission range higher than or equal to 200 m. The results obtained with DDT show the inability of this algorithm to overcome network fragmentation. Results obtained with RBM correspond to the analysis of section 24.3.1 which shows that RBM does not overcome fragmentation in all situations, particularly in the case of light traffic.

24.6. Conclusion

In this chapter, we have presented methods used for disseminating alert messages in the case of inter-vehicle communication. Some of these methods are based on neighbor calculations and/or on location by GPS. We have presented a new method called IVG, based on GPS and independent of neighbor calculations, which broadcasts the alert message in a reliable way to all the vehicles in risk zones for a given accident, while minimizing bandwidth use. Saving bandwidth is very important for the scalability of wireless applications. Indeed, several applications such as cooperative driving, Internet browsing, etc. are already planned for tomorrow's vehicles.

The main objective for the design of the IVG method is to improve the existing solutions in the sector. In fact, none of the RBM, Trade and DDT solutions could ensure reliability and scalability at the same time.

On the one hand, RBM and Trade are based on neighbor calculations to provide reliability, which causes scalability problems especially in busy networks or in networks with multiple communication sources. In addition, we have shown by analysis and simulation that the fragmentation problem is not overcome by RBM in all situations.

On the other hand, DDT, which is designed for scalability, does not make it possible to overcome the network fragmentation and therefore does not ensure the reliability required in emergency situations.

Simulation results with ns2 have shown the efficiency and reliability of the IVG protocol. However, certain problems can be found during actual use, such as transmission reliability, precision and availability of GPS localizations. In fact, reliability and transmission range are linked to radio equipment. Technological progress in the sector leads us to believe that there will be major improvements in the near future.

Concerning GPS localization reliability, we have proposed an extension to IVG, which can also be used with the other methods in order to enable its use in hybrid environments where some vehicles may not be equipped with GPS [BEN 03]. The results obtained show that the IVG method is efficient even in an environment where only 40% of vehicles are equipped with GPS.

24.7. Bibliography

[BAC 03] BACHIR A., BENSLIMANE A., "Multicast in Ad-Hoc networks: Inter-Vehicle Geocast", *Proceedings of 57th IEEE VTC,* Jeju, Korea, April 2003.

[BAS 98] BASAGNI S. *et al.*, "A Distance Routing Effect Algorithm for Mobility (DREAM)", *ACM/IEEE Inter. Conf. Mobile Comp. Net.*, p. 76-84, 1998.

[BEN 03] BENSLIMANE A., BACHIR A., "Inter-Vehicle Geocast Protocol Supporting Non-Equipped GPS vehicles", *Proceedings from LNCS 2865 ADHOC-NOW'03 Int. Conference on Ad-Hoc, Mobile and Wireless Networks*, Montreal, Canada, Springer Publisher, p. 281-286, October 8-10, 2003.

[BRE 00a] BRIESEMEISTER L., HOMMEL G., "Overcoming Fragmentation in Mobile Ad-Hoc Networks", *Journal of Communications and Networks,* vol. 3, no. 2, September 2000, p. 182-187.

[BRE 01] BRIESEMEISTER L., "Group Membership and Communication in Highly Mobile Ad-Hoc Networks", PhD Thesis, School of Electrical Engineering and Computer Science, Technical University of Berlin, Germany, November 2001.

[BUR 00] BURSA M., "Big Names Invest Heavily in Advanced ITS Technology", *ITASA Magazine*, no. 8, December/January 2000.

[CLA 03] CLAUSEN T., JACQUET., "Optimised Link State Routing Protocol", *IETF Internet draft*, draft-ietf-manet-olsr-10.txt, May 2003.

[DRI 99] http://www.ist-drive.org, 1999.

[HAA 02] HAAS Z., PEARLMAN M., SAMAR M., "*The Zone Routing Protocol (ZRP) for Ad-Hoc Networks*", IETF Internet draft, draft-ietf-manet-zone-zrp-04.txt, July 2002.

[ISI 95] http://www.isi.edu/nsnam/ns, 1995.

[ITS 04a] http://www.its.dot.gov, 2004.

[ITS 04b] http://www.itsa.org, 2004.

[JOH 01] JOHNSON D., MALTZ D., BROCH J., *Ad-Hoc Networking*, chapter "DSR: The Dynamic Source Routing Protocol for Multihop Wireless Ad hoc Networks", p. 139-172, Addison Wesley, 2001.

[KAR 00] KARP B. and KUNG H. T., "GPSR: Gready Perimeter Stateless Routing for Wireless Networks", *Proceedings of 6th Annual Int. Conf. Mobile Comp. and Networking (MobiCom 2000)*, p. 43-54, Boston, MA, USA.

[KOY 98] KO Y. B. and VAIDYA, "Location-aided Routing (LAR) in Mobile Ad-Hoc Networks", p. 66-75, *ACM/IEEE Inter. Conf. Mobile Comp. Net.*, 1998.

[LIA 03] http://www.lia.univ-avignon.fr/equipes/RAM/index.html, 2003.

[MAN 03] http://www.ietf.org/html.charters/manet-charter.html, 2003.

[MAR 99] MARTIN A., MARINI H., TOSUNOGLU S., "Intelligent Vehicle/Highway System: A Survey – Part 1", *Florida Conference on Recent Advances in Robotics,* Gainesville, Florida, April 1999.

[MAU 01] MAUVE M., WIDMER J., HARTENSTEIN H., "A Survey on Position-Based Routing in Mobile ad hoc Networks", *IEEE Network Magazine*, vol. 6, no. 15, p. 30-39, November 2001.

[NIS 99] NI S. *et al.*, "The Broadcast Storm Problem in Mobile Ad-Hoc Network", *Proceedings of Mobicom 1999*, p. 151-162, Seattle, Washington, USA 1999.

[OGI 02] OGIER R., TEMPLIN F., BELLUR B., LEWIS M., "Topology Broadcast Based on Reverse-Path Forwarding (TBRPF)", draft-ietf-manet-tbrpf-03.txt, November 2001.

[PEI 00a] PEI G., GERLA M. and TCHEN W., "Fisheye State Routing: a Routing Scheme for Ad-Hoc Wireless Networks", *Proceedings of ICC 2000*, New Orleans, LA, June 2000.

[PEI 00b] PEI G., GERLA M., and HONG X., "LANMAR: Landmark Routing for Large Scale Wireless Ad-Hoc Networks with Group Mobility", *Proceedings of IEEE/ACM Mobihoc 2000*, p. 11-18, Boston, MA, August 2000.

[PER 94] PERKINS C. E. and BHAGWAT P., "Highly Dynamic Destination-Sequenced Distance-Vector Routing (DSDV) for Mobile Computers", *Computer Communications Review*, vol. 24, no. 4, p. 234-244, October 1994.

[PER 99] PERKINS C., ROYER E., "Ad-Hoc On-Demand Distance Vector Routing", *Proceedings of the 2nd IEEE Workshop on Mobile Computing Systems and Applications,* p. 90-100, New Orleans, LA, February 1999.

[PER 01] PERKINS C., ROYER E., DAS S., "IP Flooding in Ad-Hoc Mobile Networks", IETF Internet Draft, draft-ietf-manet-bcast-00.txt, November 2001.

[SUN 00] SUN M., *et al.*, "GPS-based Message Broadcast for Adaptive Inter-vehicle Communications", *Proceedings of IEEE VTC Fall 2000*, no. 6, p. 2685-2692, Boston, MA, September 2000.

[SUN 01] SUN M., FENG W., LAI T., "Location-Aided Broadcast in Wireless Ad-Hoc Networks", p. 2842-2846, *Proceedings of IEEE GLOBECOM 2001,* San Antonio, Texas, November 2001.

[XUK 02] XU K., GERLA M., BAE S., "How Effective is the IEEE802.11 RTS/CTS Handshake in Ad-Hoc Networks?", *Proceedings of IEEE Globecom*, Taipei, Taiwan, November 2002.

24.8. Appendix

Message format

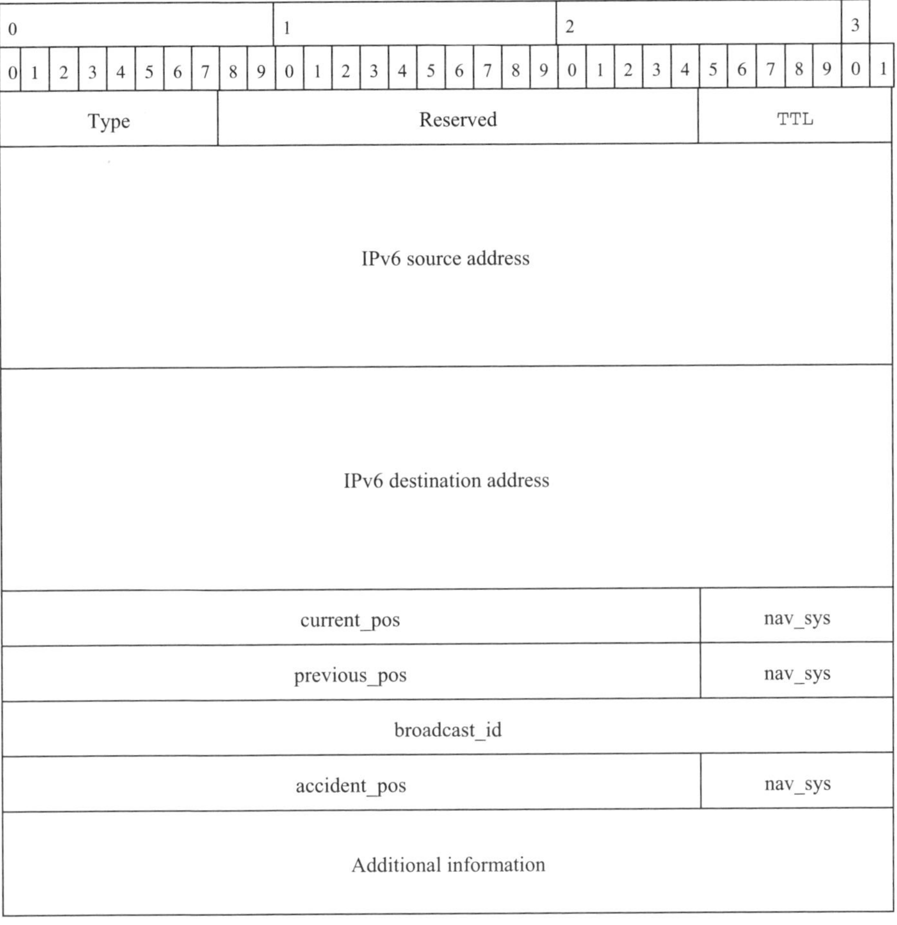

Figure 24.9. *Alert messages format*

– Type: type of information sent. For example, 00000001 means there is an accident so the receiver can read the position of the accident in the accident_pos field.

– Reserved: field reserved for future use.

– TTL: lifespan of message. If this value is equal to 0 then the message is ignored.

– IPv6 source address: address of the transmitter. We have planned IPv6 addresses (128 bits) to enable the protocol to operate with the Internet. IPv4 may also be used but due to the large number of vehicles, the addressing space must be very large.

– IPv6 destination address: address of destination. In our case it is a broadcast address.

– current_pos: current position of the transmitter of the message.

– previous_pos: previous position of the transmitter.

– nav_sys: the navigation system used (GPS, Galileo, Glonass, etc.).

– broadcast_id: the sequence number of the broadcast.

– accident_pos: geographic position of the accident.

– additional information: field reserved for future use. It can be used for cooperative driving or to introduce security functions.

Chapter 25

Pervasive Networks: Today and Tomorrow

25.1. Introduction

Pervasive networks are still in their infancy. They will continue to grow until a point where, from anywhere on the globe, we will have access to several networks. In this chapter, we will start with an overview of networks that can be part of this pervasive world. Among these networks four main categories will be included: personal networks of small diameter in the order of a few feet, local networks of a maximum of a few hundred feet, metropolitan networks of a few miles and finally extended networks covering entire countries. These different categories are represented in Figure 25.1.

In addition, in these categories we must differentiate wireless networks from mobile networks. Wireless networks require that their clients do not exit the cell in which they are. In mobile networks, the client can move and change cells while still being connected. We will examine these two main solutions knowing that mobile networks are classified in the extended network category since they can develop over large areas. We will compare them all in the context of pervasive networks.

Then, after this presentation of today's and tomorrow's networks, we will examine their properties in terms of control, security and mobility management complementing what was presented in the preceding chapters. Control is absolutely necessary in order to execute multimedia applications with time and throughput constraints. Security has in part been illustrated by the use of chips which can be a solution for these networks that must be scalable. Finally, mobility management

---

Chapter written by Guy PUJOLLE.

benefits from the help of handovers (i.e. intercellular changes) between cells of different technologies such as going from a Wi-Fi network to a UMTS.

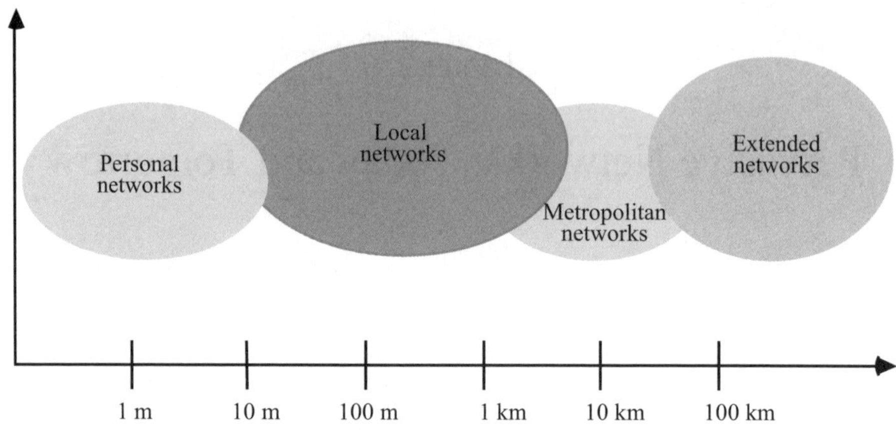

Figure 25.1. *Different wireless network categories*

Finally, we will examine applications for these pervasive networks and their arrival on the market. One of the biggest problems is continuity of service in the pervasive environment where throughputs will vary after a handover from one network to another and will not give the same quality of service (QoS). In addition, to actually arrive at a pervasive network, there needs to be cooperation between a large number of operators in order to cover the world.

25.2. Networks of the pervasive Internet

The pervasive Internet is made up of a large number of heterogenous networks for total coverage and maximum throughputs. Obviously, extended networks are the ones that will enable basic coverage with large enough cells so that there are no holes in the coverage. In this context, today mobile networks are essentially the ones participating but unfortunately throughput is relatively low even if it will increase significantly.

In this category we find the major solutions retained by 3G, i.e. the third generation of mobile devices. These solutions mostly include UMTS, from Europe and Japan, and CDMA 2000 supported by the USA. Two major groups have been put in place to commercially develop underlying standards; 3GPP and 3GPP2, respectively.

These third generation mobile systems are presented as competitors to second generation infrastructures already deployed. However, it was important to think about the transition between both generations, that could not be immediate, and the possibility of creating a maximum number of common services. The third generation is an improvement over the previous one with a QoS that is now at least comparable to the one provided by wired networks. Furthermore, 3G networks, such as UMTS, strive to provide new and significant technological advances including worldwide roaming, a large range of services, with or without high throughput, audiovisual services and the use of a single terminal in different radio environments. The range of telecommunications services must adapt in a flexible manner to the users' needs and enable them to communicate independently of their location and their access method.

Figure 25.2 illustrates the different environments defined for 3G networks.

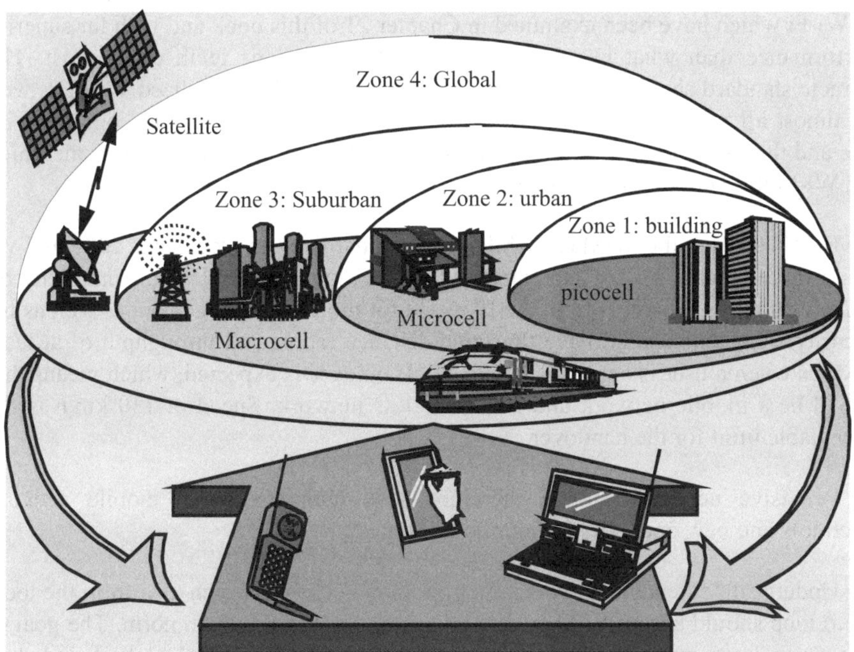

Figure 25.2. *The different 3G network cells*

Each of the four zones represented in the figure characterizes a specific environment. The first zone, of a few feet, corresponds to a high density environment, in a busy downtown, for example, with picocells in which traffic is very dense. In this type of environment, the maximum throughput can reach 2 to

10 Mbps, thus equaling and even exceeding the ADSL offered throughput. In the next two zones, from a few dozen to a few hundred feet, mobility increases, whereas throughput decreases. In the last zone, of a few dozens to a few hundred miles, which is characterized by the satellite component, mobility is global, with a throughput that can reach 2 Mbps. Land, maritime and aeronautical situations are included in the objectives of these umbrella cells. In this way, the user can be in a vehicle, on a boat or on an airplane and take advantage of a continuous availability of services.

3G networks simultaneously use cellular and wireless techniques as well as the satellite component, complement of fixed and mobile networks, which provides a global coverage, inside and outside of buildings, which does not enable conventional land deployment.

The wireless network world wants to compete with 3G with networks as simple as Wi-Fi which have been examined in Chapter 21 of this book and with far superior performance than what is achieved in 3G, all this at one tenth of the cost. The miracle standard should come from IEEE that has already normalized Wi-Fi as well as almost all wireless network standards. The standards that should compete with 3G, and that we sometimes call 4G, are IEEE 802.16e and IEEE 802.20 from which the Wi-Mobile product will emerge.

IEEE 802.16e, the WiMax mobile, was standardized in December 2005 and we can estimate that it is very late compared to 3G. However, it plays on simplicity by borrowing all its protocols from the IP world for mobility management as well as for security. The cells should have the same distance radius and throughput of at least 1 Mbps when a user is accepted in the cell. Handover is expected, which means that it will be a mobile network and not a wireless network. Speed of 130 km/h is the acceptable limit for the handover.

Pervasive networks should therefore have both traditional mobile network operators and new operators for mobile ISP types.

Underneath extended networks, metropolitan networks which also form the local radio loop should accept WiMax networks from the IEEE 802.16 norm. The goal of this norm is to propose xDSL wireless modems that are fixed at first and then mobile. The basic IEEE 802.16 standard corresponds to the fixed xDSL links but in a radio channel in frequencies lower than 11 GHz and higher than 10 GHz with aggregation of technologies between 10 and 11 GHz. The arrival of IEEE 802.16e in December 2005 made available the use of wireless mobile xDSL links.

From a performance standpoint, WiMax networks reach a total throughput of 60 Mbps on which users are multiplexed.

In the case of local networks, the dominance of the IEEE 802.11 standard is undeniable. The network category has been presented in Chapter 21.

One of the major modifications in the wireless world comes from small personal networks whose goal is to interconnect all devices that the users have in their pockets. In fact, personal wireless networks may not be part of the pervasive Internet but they will be connected to it. From any personal network terminal, it must be possible to find an entryway to the pervasive Internet by going through the personal network equipment that possesses it.

The IEEE 802.15 standard is the basic structure of personal network products. It contains three main sub-standards: IEEE 802.15.1 or Bluetooth, IEEE 802.16.3 or UWB (Ultra-Wide Band) and IEEE 802.15.4 or ZigBee which corresponds to the very low throughput networks that we have seen in sensor networks in Chapter 7: ZigBee. Let us describe a few of the major elements in these different standards.

Bluetooth has not in fact been normalized by the IEEE but recovered. There is therefore nothing new to this standard that is not already part of Ericsson's studies. The piconet cell corresponds to basic networks containing a maximum of 8 machines that can interconnect together to form scatternets. These architectures have been taken over by the other members of the IEEE 802.15 standard family.

The performance expected by UWB is of 1 Gbps but because of the technology which consists of transmitting over the whole spectrum with power under noise, the norm has chosen a speed of 480 Mbps as the speed to achieve. It is interesting to note that developers such as Intel have already chosen this norm for tomorrow's input/output interfaces with the WUSB (Wireless USB) proposition.

The ZigBee norm is much more focused toward signaling between all pieces of equipment instead of the transmission itself. The goal is then low throughputs at a particularly low cost with a battery's lifespan of several years, more precisely for the lifespan of the device. We should see the emergence of ZigBee chips on most commercialized equipment whether they are appliances, electronic as well as health and entertainment, etc. Energy consumption of ZigBee chips must be minimized by a switch to sleep mode that can be very long if there is no client needing to communicate.

We have grouped in Table 25.1 the main wireless techniques in order to compare them in terms of throughput and of maximum distance to access point. In this table we have added solutions that could be used in pervasive Internet such as DAB and DVB systems for sound and image. DVB is used, for example, in terrestrial digital TV and could eventually enable video broadcast over the pervasive Internet.

Network	*Throughput*	*Maximum distance*
Bluetooth	1 Mbps in total	10 m
Wi-Fi	11 or 54 Mbps in total	100 m
2G	10 Kbps by user	30 km
2.5G	30 Kbps by user	30 km
3G	384 Kbps – 2 Mbps by user	10 km
802.16 – WiMax	60 Mbps in total	30 km
DAB	128 Kbps by user	50 km
DVB	2 Mbps by user	50 km
GPS	insignificant	100 km cell
Satellite	Several hundreds of Mbps in total	2,000 km cell
ZigBee	20 or 250 Kb/s in total	10 m
WiMedia	480 Mbps in total	10 m

Table 25.1. *Comparison of different networks in the pervasive Internet*

To conclude this first section, we must mention the solutions that increase the network's coverage. In particular, to increase the surface covered by operators, clients can themselves become antennae. This idea has enabled ad hoc networks to emerge, i.e. networks in which each mobile terminal is actually a router and has a routing duty to transport packets to their destination.

Before going into too much detail about ad hoc networks, let us explain that meshed networks can be a first solution to increase coverage without it being too complex. A meshed network is a group of wireless networks directly interconnected through their radio wave. Several solutions are already on the market: first of all, access points that are directly linked together by the frequency used to connect the clients. In other words, this means that an access point becomes one of the clients of another access point in the same way as a mobile terminal. It is of course important that access points are not too far from one another and it is also important to control the particular flows that go from access point to access point and that can have a heavier throughput than the throughputs of each client individually. Another solution is to reserve a specific radio channel to make the connection between access points. In this case, access points must possess two Wi-Fi cards, one to connect the mobile terminals and another for communications between access points. Finally, a last solution is the use of another frequency band, for example, by using IEEE 802.11b or g for mobile terminals and an IEEE 802.11a card for the connection of access points between each other. We must determine a routing algorithm in the linking network between access points. In fact, there can be several possible routes to go from one user to another user or from one user to the Internet gateway. The routing problem in a meshed network can become very complex when the same wavelength is used for the connection of mobile terminals and antennae. Indeed, it is important to try to pass through cells with the least amount of traffic.

A second possibility for extending the pervasive Internet can be found in ad hoc networks. In ad hoc networks, the infrastructure only contains mobile terminals: there is no fixed station anymore. Mobile devices accept to act as routers to enable the transmission of information from one terminal to another. An ad hoc network is illustrated in Figure 25.3.

Contrary to what we might think, ad hoc networks are dozens of years old. They are designed to implement a communication environment deployed with no other infrastructure than the mobile devices themselves. In other words, mobile devices can act as gateways to enable communication from one device to another. Two mobile devices too far apart to communicate directly together can find an intermediate mobile device able to be a relay.

The major drawback of this type of network comes from the network topology's definition, i.e. how to determine which are the neighboring nodes and how to go from one node to another. Two extreme solutions can be compared: an ad hoc network in which all nodes can communicate with all the others implying a long range from the transmitters. The second solution, on the other hand, is the one where the radio range is the shortest one possible: for communicating between two nodes, it is necessary to pass through several intermediate machines. The advantage of the first solution is high transmission security, since we can go directly from the transmitter to the receiver, without depending on intermediate equipment. The throughput of the network is at a minimum as the frequencies cannot be reused. In the second case, if a terminal breaks or is shut off, the network can divide itself in two distinct sub-networks, without communicating from one to the other. Obviously, in this case, the global throughput is optimized as there is a high level of frequency reuse.

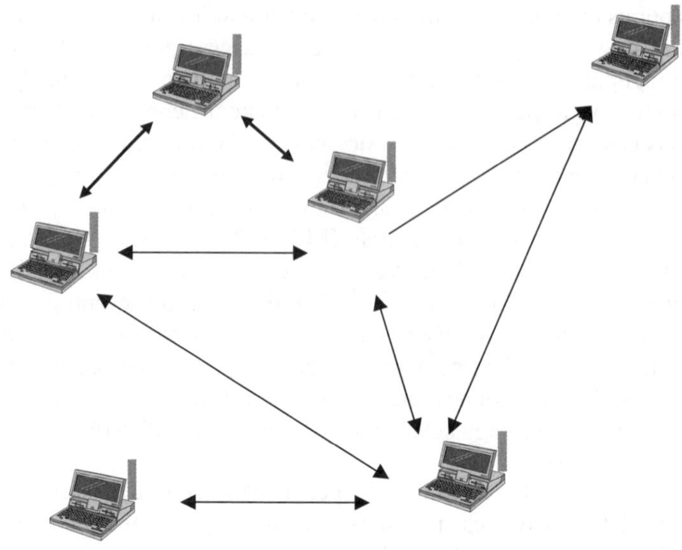

Figure 25.3. *Ad hoc network*

The access techniques are the same type as for the mobile networks. However, since all portable devices act as BSS and since they are themselves mobile, new properties must be added to the management of user addresses and to routing control.

The solution developed for ad hoc networks is based on the IP environment. Mobile devices that act as gateways – most often all mobile devices – implement a router within their circuits, meaning that most of the problems come down to routing problems within Internet since mobility is controlled by the Mobile IP protocol.

The advantages of ad hoc networks are their very simple extensions, their physical coverage and their cost. However, in order to benefit completely from these advantages, a certain number of obstacles need to be overcome, such as QoS and security because of the mobility of nodes.

MANET (Mobile Ad hoc NETwork) is the IETF workgroup that focuses on standardizing ad hoc protocols working on IP. This group has started with traditional Internet protocols and has perfected them so that they can function with mobile routers.

Two main protocol families have been defined:

– *reactive protocols*. The terminals do not maintain a routing table but do take it into account when a transmission is waiting. In this case, flooding techniques are mainly used to index the mobile devices that can take part in the transmission;

– *proactive protocols*. Mobile devices strive to maintain a coherent routing table, even during the absence of communication.

Ad hoc networks are useful in several cases. They enable the establishment of networks in a short time, for example, in the case of an earthquake or for a meeting with a large number of participants. Another possibility is to extend access to the cell of a wireless network like Wi-Fi. As illustrated in Figure 25.4, a terminal located outside of a cell can be connected to the machine of another user in the cell's coverage zone. The latter acts as an intermediate router to access the antenna of the cell.

Ad hoc networks have many problems because of the mobility of all their equipment. The main problem is the routing necessary to transfer packets from one point to another point in the network. One of the goals of the MANET group is the proposal of a solution to this problem. At this time, four main proposals are on the table, two of reactive type and two of proactive type. Among the other problems, we can mention security, QoS and mobility management during communication.

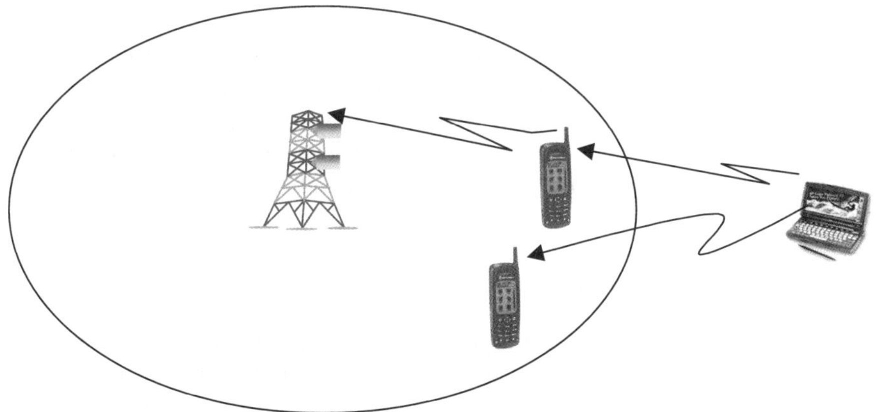

Figure 25.4. *Coverage extension by an ad hoc network*

Routing is the main element in an ad hoc network. There needs to be a routing application in each node of the network to control the transfer of IP packets. The simplest solution is obviously having direct routing, as illustrated in Figure 25.5, in which each station in the network can directly reach another station, without the need of an intermediary. The simplest case is of a small cell, of a diameter smaller than 35 feet, as in an 802.11 network in ad hoc mode.

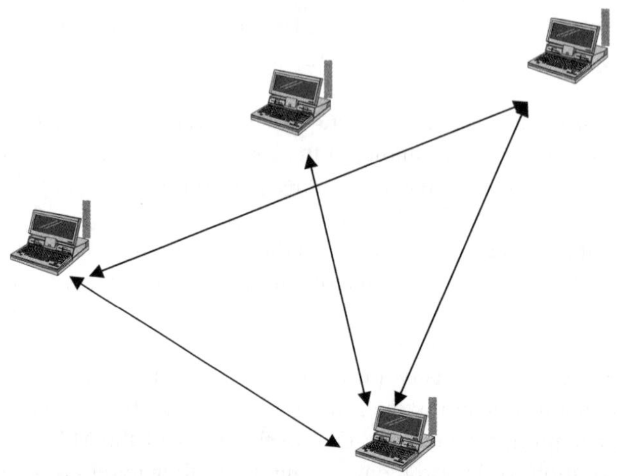

Figure 25.5. *Direct communication between machines in an ad hoc network*

The traditional case of routing in an ad hoc network consists of passing through intermediate nodes. These must contain a routing table able to direct the packet to the receiver. The strategy of an ad hoc network is to optimize routing tables by generally regular updates. If the updates are too regular, there is a risk of network overload. However, this solution has the advantage of maintaining updated tables and thus to enable rapid packet routing. Updating only during a new arrival flow restricts the load circulating in the network, but takes away several supervision flows from the network. It is then important to put routing tables in place that can execute the transmission in acceptable delays.

Figure 25.6 illustrates the case of an ad hoc network in which, to be able to go from one node to the other, it may be necessary to go through intermediary nodes. Several obstacles can be found along the way while constructing a routing table. For example, in terms of signal transmission, it is possible that the link is not symmetrical: communicating one way is acceptable but not the other way. The routing table must take that into account. Radio signals are sensitive to interference; the asymmetry of the links can then become complicated by the possible fading of links.

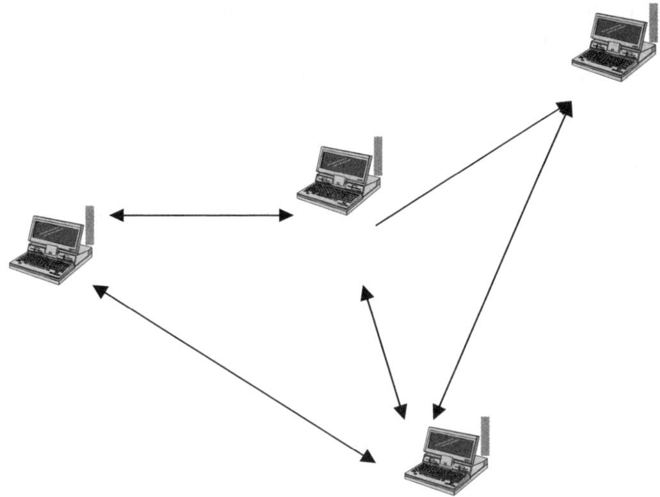

Figure 25.6. *Routing by intermediate nodes*

For all these reasons, the network's routes must always be modified, which brings us to the question always asked at IETF: should the routing tables be maintained in the mobile nodes of the ad hoc network? In other words, is it worth it to update routing tables that continually change or would it be not better to determine the routing table at the last moment?

As explained previously, reactive protocols work by flooding to determine the best route when a packet flow is ready to be transmitted. There is no exchange of control packets outside of the supervision to determine the flow's route. The supervision packet that is broadcast to all neighboring nodes is broadcast again by neighboring nodes until it reaches the receiver. Depending on the chosen technique, we can use the route determined by the first supervision packet that arrives to the receiver or plan several routes in case of a problem on the main route.

Proactive protocols behave in a totally different way. Supervision packets are continuously transmitted in order to maintain the routing table updated by adding new lines and deleting others. Routing tables are therefore dynamic and vary according to the supervision packets reaching the different nodes. A problem in this case consists of calculating a routing table that is compatible with the routing tables of the different nodes so that there are no loops.

Another possibility consists of finding a compromise between both systems. That is, to regularly calculate routing tables as long as the network has a small load. In this way, the performances of user flows in transit are not modified too much. When traffic increases, updates slow down. This method simplifies the creation of a reactive routing table when a request gets to the network.

The protocols that have been discussed at the IETF in the MANET group are summarized in Table 25.2. Different metrics can be used to calculate the best route:

– distance vectors assign a weight to each link and add the weights to determine the best route, which corresponds to the one that has the lowest weight;

– source routing enables to determine the best route as the one that allows the supervision packet to arrive first at destination;

– link states indicate the links that are more interesting to take compared to those that are not.

Metric	**Reactive**	**Proactive**
Distance vector	AODV (Ad hoc On demand Distance Vector)	DSDV (Destination Sequence Distance Vector)
Source routing	DSR (Dynamic Source Routing)	
Link state		OLSR (Optimized Link State Routing Protocol)

Table 25.2. *Ad hoc protocols*

Another environment that should flourish in future pervasive networks comes from sensor and actuator networks. In fact, sensors should be positioned everywhere and should assist in information transmission. Actuator networks are also very important since everything can remotely be controlled: heating, lighting, appliances, etc.

25.3. QoS and security

Once all the networks are installed, pervasive Internet must be completed by enabling a mobile terminal to go from one cell to another, by ensuring communication security, by giving a QoS to flows that need it and by controlling users' nomadism.

Let us start with intercellular changes or handovers. There are three types:

– Horizontal handovers enable passing from one cell to another within the same technology, vertical handovers enable passing from one cell to another with different technologies but only at the MAC layer; over the MAC layer, the LLC layer is common and the transported packets are IP packets. Finally, diagonal handovers enable passing from one cell to another with different technologies like going from a Wi-Fi network to a UMTS network.

– Horizontal handovers are the most traditional and are directly controlled by the material itself when the network is a mobile network. The horizontal handover can be added to a wireless network as is the case with Wi-Fi where the IEEE 802.11f standard proposes an acceptable solution for all manufacturers. In this wireless network context, passing from one technology to another is studied by the IEEE 802.21 workgroup, between the 802.11, 802.15, 801.16 and 802.20 standards.

– Diagonal handovers have been widely studied in the last few years and most of the larger operators and manufacturers are capable of managing them. Terminals able to execute diagonal handovers are already poised to be marketed between the Wi-Fi and GPRS technologies.

The part that is the most delicate at this point is the application handover. In fact, if on the hardware level we are able to pass a communication, for example, from Wi-Fi to GPRS, the application may have problems keeping up because it has to go from a throughput of several megabits per second to a few kilobits per second. In order to do this, we must be able to adjust throughout according to the user's SLA (Service Level Agreement) and to determine if a high degradation is acceptable by the user or not. We have discussed these problems in Chapter 22.

With regard to telephony which could become the number one application in the pervasive Internet world, we have examined the solutions in Chapter 19. QoS in Wi-Fi networks has been examined in Chapter 5. Finally, security which can be recommended in a pervasive environment with the help of a smart card has been discussed in Chapter 16.

25.4. Services

We will end this chapter with services that could be available on a large scale in the pervasive Internet.

In principle, pervasive Internet services must be compatible with those of fixed telecommunication networks when it comes to functionality, user interface, cost and quality. Pervasive Internet networks must take responsibility for multipart multimedia services, like those provided by fixed networks. This point is important because then a user equipped with a "mobile device" can benefit more freely from the services to which he is subscribed, without experiencing any QoS loss.

The services provided go from audio to video, including data and multimedia. The user of one single terminal must establish and maintain several simultaneous connections. He must also receive offers for applications requiring different QoS parameters. The services proposed depend on the properties, or capabilities, of the terminal as well as on the offer from the operator concerned. Services requiring high transmission throughputs are concentrated in high density zones, such as business centers, instead of urban zones. Pervasive Internet users will not be aware that a radio link connects their terminal to worldwide telecommunication networks.

The strategy deployed to integrate new services in the pervasive Internet is similar to what has been designed in the intelligent networks context. It consists of defining service capabilities and not the services themselves. Service capabilities focus on transport techniques needed by QoS parameters as well as the mechanisms required for their execution. These mechanisms include functionalities provided by the different network elements, communication between each other and storage of associated data. These normalized capabilities provide a platform supporting voice, video, multimedia, messages, data and other teleservices, user applications and additional services. They enable the creation of a market for services determined by the users and service providers.

Global mobility is impossible to execute today because of the multiplicity of systems and networks. It is almost impossible for a user to benefit from the same services in the same conditions in his subscribed network and in the visited networks. When it becomes reality, service mobility will give users the same services in the same conditions, wherever they are. Roaming, or terminal mobility, will enable the construction of a universal network. Future networks will in addition support the VHE (Virtual Home Environment) concept which will offer a complete set of services with the same appearance, whether the user is located in his subscription network or in another network.

Personal mobility will only be possible if the user has an identifier independent from the network and service provider. Network operators and service providers are often two separate entities and it would therefore be complicated to provide services, particularly call set-up, if the user had a specific number with each of these entities. This is why the user identifier must be unique. With this unique personal number, which is the only one known by the callers, the user can be contacted anywhere in the world.

Pervasive Internet networks, 3G and 4G, must support number portability at the level of service providers, location and services. Number portability in 3G mobile networks, in the case of service providers, implies that an identifier, called IMUN (International Mobile User Number), is allocated to each new 3G subscriber. A subscriber can change service providers while still keeping his IMUN, on the condition that the new service provider offers the given service in the same geographic zone. Similarly, a service provider can change network operators while still keeping its IMUN. Location portability means that each subscriber can be called independently from his location and therefore his mobility. Number portability means that the number to dial to reach a user is independent of the required service.

Radio spectrum must be used in an efficient way and eventually be shared between different operators in order to provide global radio coverage, which is transparent to the user, with the satellite portion. It must be possible for a pervasive Internet terminal to adapt to the radio interface provided in a specific region and to determine the capabilities of the services provided in this region. Furthermore, as more than one radio interface is available in a given region, the standard must plan for a mechanism which enables a pervasive Internet terminal to select the radio interfaces capable of providing the appropriate service capabilities.

Individual clients have a profile of particular services, with a definition of the types of subscribed services including, for example, the time of day when these services are used. The service provider is responsible for controlling all aspects of the subscriber and user service. When the subscriber and the user are separate entities, the subscriber controls the services profile of the user within the limits of the subscription. The subscriber can easily modify this profile and decide if the user controls the activation-deactivation of the services defined in the user services profile. Any change to the services profile must be done in a secure environment. Information concerning the user, which is required to identify him without ambiguity and enabling his registration for a service, is registered on his integrated circuit card which contains an equivalent to USIM (User Service Identity Module).

The multiple profile concept must be introduced in pervasive Internet networks. A multiple profile enables a user to obtain certain services from a service provider and other services from another provider with a single card. In the case where multiple subscriptions are possible on the same card and when they involve several service providers, a different number is allocated for each one of them. For outgoing calls, the user must be able to select the service provider that he prefers for each call or according to his subscription. Several multiple subscriptions can be active at the same time. The standard enables the support of multiple registrations on a single terminal by inserting multiple chips.

Security of the future pervasive Internet will be a critical success factor. As explained in Chapter 13, we must plan for a mutual security: the network can authenticate the user in order to verify that he is the person authorized to use the services, and the user can request network authentication at the time of registration before initializing a service. Similarly, networks can authenticate each other. Other authentication relations are defined in the networks and have been detailed in Chapter 14. These new services will be flexible in order to be as close as possible to user demands.

25.5. Bibliography

[ALA 04] AL AGHA K., PUJOLLE G., VIVIER G., *Réseaux de mobiles et réseaux sans fil*, Eyrolles, 2004.

[ALE 01] ALESSO H. P., SMITH C. F., *The Intelligent Wireless Web*, Addison Wesley, 2001.

[DAV 01] DAVIS H., MANSFIELD R., *The Wi-Fi Experience: Everyone's Guide to 802.11b Wireless Networking*, Que, 2001.

[DOR 02] DORMAN A., *The Essential Guide to Wireless Communications Applications*, Prentice Hall, 2002

[FLI 01] FLICKENGER R., *Building Wireless Community Networks*, O'Reilly, 2001

[GOL 95] L. GOLDBERG, "Wireless LANs: Mobile Computing's Second Wave", *Elect. Design*, vol. 43, 6, pp. 55-72, June 1995.

[MAX 02] MAXIM M., POLLINO D., *Wireless Security*, Osborne McGraw-Hill, 2002.

[NIC 01] NICHOLS R. K., LEKKAS P. C., *Wireless Security: Models, Threats, and Solutions*, McGraw-Hill, 2001.

[OHA 99] O'HARA B., PETRICK A., *The IEEE 802.11 Handbook: A Designer's Companion*, IEEE Press, 1999.

[PAH 01] PAHLAVAN K., KRISHNAMURTHY P., *Principles of Wireless Networks: A Unified Approach*, Prentice Hall, 2001.

[PER 00] PERKINS C., *Ad-Hoc Networking*, Addison-Wesley, 2000.

[PRA 98] PRASAD R., *Universal Wireless Personal Communications,* Artech House, 1998.

[SAY 01] SAYRE C. W., *Complete Wireless Design*, McGraw-Hill, 2001.

[TOH 01] TOH C. K., *Ad-Hoc Mobile Wireless Networks: Protocols and Systems*, Prentice Hall, 2001.

[WEB 00] WEBB W., *Introduction to Wireless Local Loop,* Artech House, 2000.

Chapter 26

Optical Networks

26.1. Introduction

In the early years of this new millennium, we see great changes in telecommunications networks. These changes are tightly linked to the evolution of lifestyles and to the behavior of users. They are mostly due to the increasing users' demand in terms of bandwidth and to their requirements in terms of quality of service (QoS). In fact, the unprecedented success of the Internet and of the World Wide Web, whether it is the number of users or the time each user spends during a connection, has contributed to the evolution of networks. For example, the average telephone communication time is three minutes, whereas the average Internet connection time is 20 minutes. In addition, Internet traffic doubles each four to six months and this trend will very probably continue in the years to come.

This has led to a great disruption in the nature and the characteristics of traffic exchanged in the network. The dominating traffic was voice and now the trend is the opposite with data traffic that is starting to take over. These changes have pushed service providers to reexamine their networks and to adapt them to face these new challenges.

All these factors have contributed to the development of high throughput optical networks and to their rapid transition from a research product into a commercial product which is largely deployed in current networks. In this chapter, we will introduce optical technology. The goal of this chapter is to present the fundamental characteristics of optical networks.

Chapter written by Nizar BOUABDALLAH.

26.2. History

An optical fiber is a dielectric guide which makes it possible to conduct the light over a large surface. In 1966 the idea was launched to transport optical signals on a fiber over long distances, but it took years to be able to control manufacturing processes and to control material composition which has a decisive influence on losses. We could then obtain low enough attenuations to make the transmission of signals over sufficiently long distances in order to create enough practical interest and to make the optical technique competitive. Started in 1960 at 1,000 dB/km, the attenuation has reached 20 dB/km in 1975, then 0.2 dB/km in 1984.

Compared to other existing transmission supports, fiber represents an almost constant attenuation on a huge frequency range (several thousand gigahertz) and thus offers the advantage of gigantic bandwidths, enabling us today to consider the transmission of very large digital throughputs (several terabits per second) which are required by the multiplication of services and the higher requirements for image transmissions. It also became apparent very early on that optical systems brought a notable gain on the distance between repeaters that went from a few kilometers to dozens of kilometers, compared to coaxial cable systems of similar capacity. Starting in 1978, systems working at an optical wavelength of 0.8 μm were installed, transmitting a throughput between 50 and 100 Mb/s, with space between repeaters of 10 km, i.e. approximately three times those of coaxial cable systems of similar capacity. This was the first generation of optical fiber transmission systems.

The second generation of optical fiber transmission systems appeared in the 1980s and is directly descending from the development of the monomodal fiber and of the semi-conductor lazer at 1.3 μm, a wavelength for which intramodal distortion (i.e. distortion induced over the signals by propagation) is minimal. Throughputs higher than 1 Gb/s, with a space between repeaters of several dozens of kilometers are then reached. The reach of these systems is limited by fiber loss, 0.5 dB/km at the best, and so the idea was born to develop sources emitting at a wavelength of 1.55 μm for which attenuation is minimal. Figure 26.1 presents the variation of attenuation according to the optical fiber spectrum.

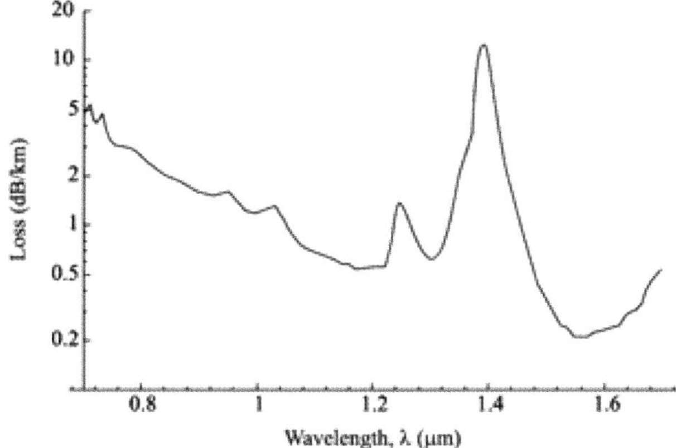

Figure 26.1. *Attenuation loss in silica as a function of wavelength*

However, the fiber does not decrease to a perfect attenuator: the variation of the refraction index according to the wavelength is the main cause of intramodal distortion, which will lead to a distortion of transmitted signals. This linear effect manifests itself all the more when the distance is high and the bandwidth of transmitted signals is significant. With loss reduction and the appearance of very high capacity systems, intramodal distortion has become a fundamental effect.

In this context, the gain obtained by using sources transmitting at a wavelength of 1.55 µm is destroyed by the effect of the intramodal distortion, which is the direct result of the fact that all the wavelengths do not spread at the same speed. This intramodal distortion of the fiber material is much stronger than at 1.3 µm and from it comes the bandwidth limitation and thus the throughput. Simultaneous progress on lazers transmitting on a single mode as well as in the transmission environment (shifted dispersion fibers) will bring solutions to these problems and the first systems working at 1.55 µm appeared at the end of the 1980s, with throughputs higher than 2 Gb/s.

Appearing at the end of the 1980s and quickly becoming developed products, fiber amplifiers have brought considerable disruption in the fiber optic communications sector: inserted in the transmission line, they make it possible to compensate fiber attenuation and therefore to increase the reach of transmission systems, at the cost of higher noise [DES 96]. The compensation is therefore the occurrence of non-linear effects, which are also a source of signal degradation, but can be used in a positive way in certain situations to compensate the influence of intramodal distortion.

Used as head amplifiers, the amplifiers also increase the sensitivity of optical receivers. Finally, their huge bandwidth (30 nm and even much more today) enables us to consider the simultaneous amplification of several juxtaposed optical carriers in the spectrum, constituting what we call a multiplex. Hence, we have the emergence of the wavelength division multiplexing (WDM) concept; each fiber transporting a multiplex to N channels is then equivalent in capacity to N fibers, each transporting one channel, and it is easily conceivable that this approach will potentially increase the capacity of a network in a significant way without modifying its physical infrastructure.

Systems using this technique, for the most part with a throughput of 10 Gb/s per channel, are today being installed by all the large international operators in their transport network to be ready for the traffic increase expected in the next few years. Systems with $N \times 40$ Gb/s have already been proposed and installed by manufacturers and the evolution toward multiplex with a large number of channels and/or with high capacity per channel will in all likelihood continue in the future in order to be ready for the need in capacity growth that transport networks such as metropolitan networks experience.

Finally, today the optical transmission makes it possible to reach a quality (expressed in terms of error ratio) far superior to previous systems, in particular wireless beams. The optic fiber is also used in video transmission networks to transmit a multiplex of electric sub-carriers which modulate an optical carrier in intensity. Each of these sub-carriers, which corresponds to a TV channel, is modulated itself in an analog way (frequency modulation, single side band amplitude modulation) or digital (phase modulation, amplitude modulation over two carriers with quadrature, etc.).

26.3. Evolution of optical networks

In this context, optical networks present a solution to the problems linked to the growing demand for bandwidth from users, by offering high transmission capacity in the network. In addition to their capacity to deliver bandwidth in a flexible manner, optical networks provide a common infrastructure capable of supporting different types of services.

The optical fibers offer a much larger bandwidth than coaxial cables and, at the same time, they are less susceptible to electromagnetic interference effects. Due to this, the optical fiber has become the preferred support for data transmission with a throughput of more than a terabit per second and a range of thousands of kilometers. The last statistics from the Federal Communications Commission in the USA show that optical networks are deployed on a large scale. Optical networks are deployed everywhere in the telecommunications network, except in the last kilometers, i.e. in residential access networks. Indeed, even if the optical fiber reaches several cities today, it has not gone into users' homes due to the high cost of cabling homes and the uncertain rate of return on the investment by service providers.

For more than a century, connecting to the telephone network was based on the use of twisted pair. The mid-1960s coincided with the invention of the optical fiber but it is only with the development of lazer diodes and the constant improvement of manufacturing processes in the mid-1980s, that the optical fiber started to create a practical interest for the transmission of signals over long distances. At the end of the 1980s, the optical fiber seemed to constitute the ideal transmission media for the simultaneous transport of voice, data and video to the end subscriber. However, significant progress has been made since then in signal processing techniques as well as in video compression algorithms enabling to constantly push back the limits of twisted pair transmissions.

The 1990s coincided with the emergence of the Internet which has even surpassed voice in volume of data exchanged. The explosion of the number of users connected to the worldwide network clearly shows the limitations of the current networks, which are mostly based on the analog telephone network and on digital access (RNIS). The network of networks has developed new services (e-commerce, video on demand, teleconference, ultrafast Internet, distance learning, etc.) requiring high transmission throughputs that traditional twisted pair will not be able to provide.

When we talk about optical networks, we are actually talking about two generations of networks. In the first generation, the optical fiber is used only as high capacity transmission support. In this phase, the optical fibers have replaced copper cables because optical fiber provides a relatively low error ratio per bit and a higher transmission capacity. On the other hand, all the intelligent functions whether it is switching/routing or controlling are still executed from the electronic area. The best illustration of this first generation is shown in SONET (synchronous optical network) networks and their equivalent SDH (synchronous digital hierarchy) which form the basis of the core American, European and Asian network architectures.

We are today seeing the deployment of the second generation of optical networks where routing, switching and controlling functions are progressively migrating to the optical area. However, before going deeper in this new generation, we will start by presenting the structure of the optical transmission system. Then, we will talk about multiplexing techniques adopted in optical networks that provide the required capacity to develop these networks.

26.4. Structure of an optical transmission system

A digital transmission system over optical fiber contains the following devices and components (see Figure 26.2):

– a source (diode lazer) and a modulator printing the information to transmit over the optical carrier. The modulation used is an intensity modulation. In the ideal case, it is an all or nothing modulation, where one of the binary data modes is associated with the transmission of a certain intensity and the other with the absence of signal. In practice, the transmitter is characterized by a rate of extinction, the difference (in decibels) between the powers transmitted in each of the two modes that is not infinite. In the case of a WDM system, we have as many sources as channels. Besides, the use of WDM prevents the use of the frequency modulation;

– in the case of a WDM system, a multiplexer is put in place which makes it possible to juxtapose the different channels in the band;

– in general, a power amplifier that enables the injection in the fiber of sufficient power is used to increase the signal range;

– online amplifiers inserted along the fiber to compensate for the attenuation. The distance between amplifiers, called no amplification, is a critical characteristic of the link;

– in general a receiving head amplifier (optical) is used just before receiving;

– in the case of a WDM system, a demultiplexer that makes it possible to separate the different channels is put in place. In general, a receiver can use either a demultiplexer in order to separate the channels, or a frequency filter to extract the channel addressed to it;

– for each channel, a photoreceptor which converts the optical signal into an electronic signal that, after filtering and sampling, will restore the transmitted information.

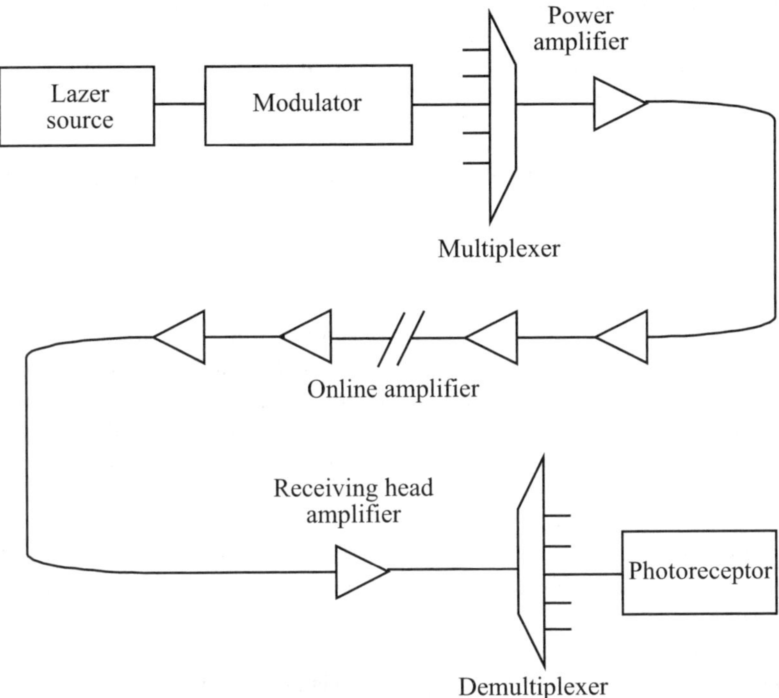

Figure 26.2. *General structure of a transmission system over optical fiber*

The transmitted signal can degrade due to several effects:

– the noise of the receiver on the one hand, and the noise of online amplifiers on the other hand;

– propagation errors which themselves can be divided in linear (intramodal distortion, polarization multimode distortion) and non-linear (self-phase modulation, cross-phase modulation, four wave mixing, stimulated Raman scattering, stimulated Brillouin scattering, modulation jitter) errors.

Some of these errors happen even in a single carrier and their effect can be corrected channel by channel in a WDM system. On the contrary, others (four wave mixing, cross-phase modulation) only happen when several carriers are involved. Non-linear effects [CHR 90, MAR 91] have only become important since the arrival of fiber amplifiers and for two reasons:

– power injected in the fiber has reached high values (several dozens of dBm). These values may seem modest, but we must not forget that the critical parameter is the surface density of power in the fiber, which is equal to the ratio of power transmitted over the useful surface of the mode, i.e. the area on which energy is concentrated. Let us recall that this is typically of a few dozens of mm (a standard fiber is 50 mm^2). Power of 20 dBm (100 mW) corresponds to a density of 2·105 W/cm^2 ;

– as with linear effects, non-linear effects accumulate along an amplified link.

26.5. Multiplexing techniques

It is obviously more cost-effective to transmit a larger quantity of traffic on the same optical fiber than to use different fibers and enabling a lower throughput transmission. That is why multiplexing different traffics at relatively low throughput on the same optical support is so interesting. There are essentially two ways to increase transmission capacity on a fiber. Figure 26.3 presents both multiplexing types used in optical networks.

The first method is based on the increase of transmission throughput with which the nodes inject their traffic on the support. Unfortunately, this requires electronic components which are able to process traffic very quickly. In this perspective, several low throughput traffics are multiplexed in order to form a unique high throughput traffic with the help of the time division multiplex (TDM) method. For example, flows of 64,155 Mb/s can be multiplexed to obtain traffic at 10 Gb/s. The highest successful and commercialized throughput is currently at 10 Gb/s and 40 Gb/s TDM will be available soon. In order to push the TDM technology to surpass these throughputs, research laboratories are developing a new method which makes it possible to multiplex and demultiplex traffic optically. This method is called OTDM (optical time division multiplex).

Even if the time division multiplex technique increases transmission throughput, it is still limited by the transmission speed of the electronic components and therefore we still cannot take advantage of all the bandwidth offered by the optical fiber.

The second method used to increase the capacity of an optical fiber is the WDM (wavelength division multiplex) technology [RAM 02]. WDM is simply multiplexing of traditional frequencies (FDM, frequency division multiplex), a technique which is traditionally adopted in radio transmission. Even though the FDM terminology is the most widely used in radio communications, the term WDM is preferred in optical jargon, maybe because the first one is an innovation from telecommunication people, whereas the latter is an invention from the physics world.

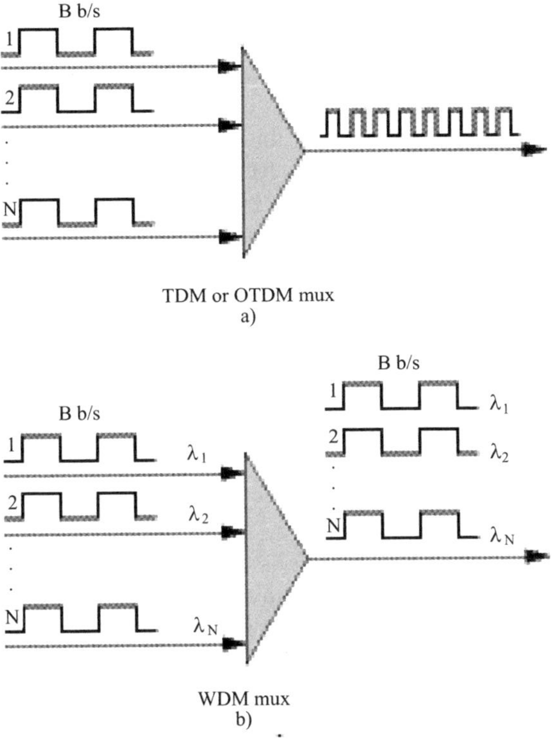

Figure 26.3. *Different multiplexing techniques. (a) Electronic or optical time division multiplex. (b) Wavelength division multiplex*

The basic idea behind WDM is the simultaneous transmission of several traffic flows on a single physical support but on different frequencies or wavelengths.

From this perspective, the wavelengths must be separated by a small frequency space to avoid interference and overlapping between signals. According to the standards of the International Telecommunication Union (ITU), the space between channels is either 50 or 100 GHz. In this way, WDM transforms an optical fiber into a set of parallel channels. WDM is now a very important technology which enables the transmission of all types of voice (voice, video-IP, ATM, etc.) over the same physical support. With the development of WDM, optical networks provide service providers with the means to integrate their different networks and services on the same physical infrastructure. For example, let us take an operator that has an infrastructure simultaneously based on both ATM and SONET technologies. By using the WDM technology, the operator does not have to multiplex the ATM flows until the throughput of SONET to be transmitted over the physical support.

Currently, WDM is favored over time division multiplex technique due to the complexity of the hardware required for TDM. In addition, TDM presents many more technological constraints as the very complex synchronization must be taken into account. On the other hand, OTDM is considered as a long term solution since it is based on components that are still immature. And, WDM is based on simple components already on the market. However, WDM has a very important property which is transparency. WDM channels behave as independent fibers. Once a connection is established between two end nodes through a wavelength, the source and destination have complete power to choose the throughput for the connection, the format for the traffic, the protocol, etc. The transparency property enables the optical network to simultaneously support different services.

WDM and TDM technologies can also be seen as two complementary techniques that can be used together to increase the transmission capacity over optical fiber. For example, we currently execute time division multiplex of traffic over a given wavelength. This combination is very similar to what is currently happening in the GSM norm.

One last comment before ending this section on multiplexing techniques concerns trials in research laboratories in order to use the CDM (code division multiplex) method. This method is feasible but it requires synchronization between the network's components, as with the TDM method. For now, it has not really been adopted by the manufacturers since the technology is still immature.

26.6. Second generation optical networks

Optical transmission has become the preferred solution in high throughput networks, in large part due to the success of the WDM technology. In addition to the simple point-to-point transmission, those involved with optics have recently realized the possibility of developing other intelligent functions in the optical world such as switching or routing. This realization is motivated by the limitations of electronic switching in terms of processing speed. For example, to process a block of data of 53 bytes (ATM cell) at 100 Mb/s at switching, 4.24 μs will be needed, whereas to process the same cell at 10 Gb/s 42.4 ns will be needed. This time in the order of a nanosecond represents a large constraint for the electronic components available today.

Furthermore, in the electronic area, each node must face not only its own traffic, but also the traffic that it must relay to neighboring nodes. This second type of traffic will hereafter be called transit traffic. This event causes a bottleneck at the level of the nodes in the core network. In order to avoid this problem, transit traffic can pass through intermediate nodes in a transparent way, i.e. in the optical domain. Then, each network node will only process local traffic, in other words, traffic to transmit or to receive. In this way, the electronic load over each node will be reduced. This last point is probably one of the major perspectives that has motivated the development of optical switching.

Due to these new capabilities (routing/switching) developed in the optical industry, we can now talk about optical networks. Generally, the switched granularity in optical networks is the wavelength. This type of network, which is largely deployed, is called wavelength switching network. This does not preclude the possibility of even larger switching, for example, a wavelength band or even one whole fiber. However, smaller switching (packet level) is still a subject for research in laboratories.

In fact, switching packets requires optical hardware with very quick processing capacities. These components are unfortunately still immature. However, knowing the importance of packet switching and the flexibility it can bring to the network, this approach remains a very important challenge and the main long term goal to reach by researchers [OMA 01]. Furthermore, the main advantage of packet switching over wavelength switching, which is nothing more than circuit switching, is a more efficient use of resources. We now know that the main concern of operators is the efficient use of their networks.

In order to create a compromise between circuit switching, with its non-optimal resource usage, and packet switching, which is quite complex, a new switching concept has emerged and it is called optical burst switching [BAT 03]. A burst is simply a set of packets, with, for example, the same destination or belonging to the same service class, which will go through the network as one entity. For intermediate nodes, the burst is processed as one block with a unique header. Since the burst is longer than a packet, the switches will have more time to process these blocks. Nevertheless, this solution remains an intermediate short term solution, until we get to the ultimate solution of optical packet switching.

For the remainder of this chapter, we will mainly focus on wavelength switching optical networks.

26.7. Wavelength switching optical networks

Wavelength switching optical networks are mainly composed of wavelength cross connects interconnected by optical fibers. Each cross connect contains a set of input and output ports. The switching decision is taken depending on the input port (fiber) and the signal carrier. Signals which are switched toward the same output port must necessarily be on two separate wavelengths. On the other hand, two signals will not share the same fiber throughout their travels, but can obviously use the same wavelength. This wavelength reuse aspect is very important since it enables the reduction of the number of distinct wavelengths required to construct an extended network.

Depending on the network architecture, a wavelength cross connect can have different capabilities. The switching matrix of the cross connect can be static or possibly dynamic. The cross connect can have the capacity of converting wavelengths. In other words, the cross connect can have the possibility to change the color of an incoming wavelength to another color in the output fiber. This last conversion aspect offers much more flexibility for the network. Let us suppose, for example, that two signals using the same wavelength and coming from two different fibers are heading towards the same output fiber. In the case where the cross connect does not have conversion capabilities, one of the connections must be blocked, whereas in the case where the conversion is executable, one of the two arriving signals will be converted on a new wavelength available on the output fiber. In this way, both connections can continue their route toward their destinations.

The wavelength reuse aspect present in optical circuit switching networks enables this type of network to become the preferred choice of extended network developers. In fact, this perspective enables to fundamentally differentiate this type of network from passive optical networks and broadcast and select networks. Broadcast and select networks were largely deployed in the 1990s, mostly in access networks, but the non-reuse of wavelengths was a problem. Indeed, the number of stations in the network is limited by the number of distinct wavelengths available on the optical fiber.

In wavelength switching optical network topology, a connection between two end nodes can pass through several cross connects. In addition, we avoid the electronic conversion of the optical fiber along its route. Thus, this end-to-end connection behaves like a transparent optical tunnel between end nodes. This tunnel is called lightpath. A lightpath is simply a wavelength concatenation between cross connects in order to go from source to destination. The lightpath must conserve the same wavelength over all the fibers on its route if cross connects do not possess conversion capabilities. This constraint is called continuity constraint. For example, let us examine the network in Figure 26.4. We only have two wavelengths per fiber. If we want to establish a new lightpath between nodes 1 and 3 passing by intermediate node 4, cross connect 4 must have a wavelength converter, or else the connection is rejected.

Two lightpaths on two disjointed routes will not share any fiber along their route, but they can use the same wavelength. We then recall the wavelength reuse property in this type of network. For example, lightpaths 1-3 and 2-4 are assigned in the same wavelength since they do not share a fiber on their routes. In this context, it is possible to establish several lightpaths by using a limited number of wavelengths.

In order to establish a lightpath between two nodes, a source transmitter and a destination receiver must be tuned to the same wavelength as the one used by the lightpath. In fact, a node can use two types of transceivers (transmitter and receiver couple), either fixed or tunable. A fixed transceiver is always tuned to the same wavelength, whereas a tunable transceiver can be tuned to different channels but only to one at a time. In this way, to offer more flexibility to a network node, it can be equipped either by a tunable transceiver or by a fixed transceivers table. Finally, the number of fixed or tunable transceivers determines the number of lightpaths that a network node can establish.

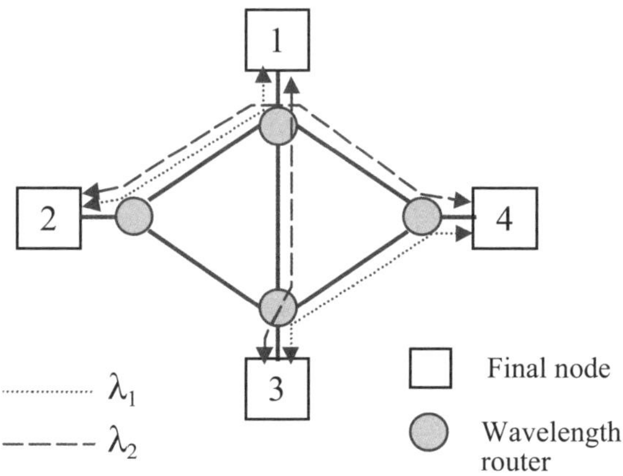

Figure 26.4. *Wavelength switching network*

With this wavelength switching concept, systems based on the use of high capacity optical fibers have become very frequent in transport networks, where multiplexing and concentration make it possible to share the infrastructure among the numerous users. In access networks, their deployment is made more difficult due to the fact that it is absolutely necessary to provide a dedicated line to each subscriber. For national operators with their own network, the deployment of an optical fiber to each subscriber means significant investment costs, which are nevertheless reduced due to the development of new techniques for mass production of optical components as well as new methods of installation of optical fiber cables adapted to access networks. Also, the identification of the most economic solution to support a large bandwidth access network in order to complete the network puzzle and to facilitate the transition toward the "large band" era is strategic for telecommunication operators. A key factor in this process is a solution capable of satisfying both the needs in terms of service types and bandwidth use for residential clients and for companies. It is possible that separate networks are required, but a unique solution could bring a significant advantage in terms of economy of scale and network management. In the remainder of this chapter, we will focus on the introduction of optical networks in what we call the last kilometer or access networks, mostly through the concept of a passive optical network.

26.8. Distribution by optical fiber

Access networks constitute the communication links between the personal area network (PAN) or local and storage area network (LAN and SAN) on the one hand, and the wide area network (WAN) on the other hand. To ensure this connection mission (communication), there are different digital techniques, which are those of cable, optical fiber or radioelectric networks. For digital flows distribution within locals, different cable or radio technique modes are proposed.

All these solutions, which are either in development phase or recently available, should enable companies and individuals to benefit from a wide range of systems and digital services that present each of the original characteristics and performances. Nine major technical families have emerged to ensure this goal of information services distribution in the access network. The normalization of ITU-T authorizes the combination of these different techniques. Technical and economic choices between the different proposed solutions are still being studied.

We now distinguish:

– optical fiber connection;

– HFC technique that introduces a hybrid fiber/coaxial solution;

– ADSL, HDSL, VDSL, etc. digital systems which focus on subscriber telephone line norms in metallic pair whose technical performances are different;

– carrier current techniques;

– distribution by satellite with or without radio return;

– radio connection solutions (MMDS, LMDS, etc.);

– radio distribution by GPRS or UMTS, which are being developed;

– digital terrestrial TV;

– optical in free space.

In this chapter we will focus on the solution based on optical fiber distribution.

26.8.1. *Passive optical network (PON)*

Several types or fiber optical connection (FITL, fiber in the loop) are available today. The increase in fiber capacity, of its range and the use of WDM technique have lead the operators to research components and assemblies which are able to offer lower cost for the highest throughput distribution. The targeted market is first of all the companies with enough computing power for these digital communication performances.

Companies are placed in front of a variety of fiber optical connection solutions associating their local Ethernet network to the metropolitan network or to the long distance network. Compared to other processes using ATM or IP type protocols, these connection types combine simplicity, decrease of transit time and higher reliability. They must make high throughput available over the subscriber's optical network unit (ONU).

The passive optical network enables the distribution of high throughputs. An optical line termination (OLT) is linked by optical fiber to a passive optical splitter (POS), which distributes signals with the use of subscriber optical fibers (see Figure 26.5). As in the preceding connection systems, distribution and interactivity toward information providers must be ensured.

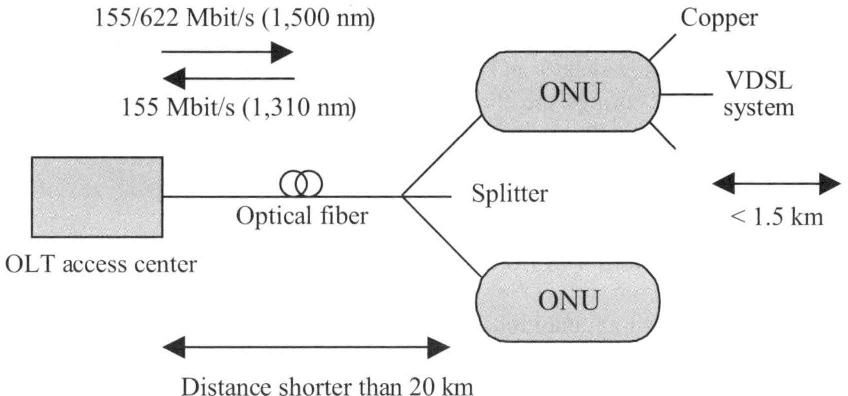

Figure 26.5. *PON optical fiber connection*

It is possible to distribute wavelength multiplexing to broadcast the 12 to 64 overlapping channels over the OLT and to execute distance channel switching as needed with the help of a digital signal as recommended by G.983 of the ITU-T.

Passive distribution reduces investment costs. Synchronization and surveillance of passive optical networks still have management and location problems. The passage to throughputs higher than 1 Gb/s implies more important investments. Each ONU can service 10 VDSL (very high speed digital subscriber line) extensions.

26.8.2. *Super PON*

It is possible to organize high capacity optical trees, which are able to connect, over 200 km, 2,048 ONUs, with a dozen users by ONU. The physical arborescence complements the aggregation function of the ATM. Current developments are also geared toward simple and economic industrial developments involving 8 to 16 wavelengths and flow carriers of 10 to 100 Mb/s, which are switchable with the help of holograms created by polarizable crystals under low tension.

26.8.3. *Ethernet PON*

Ethernet PON (EPON) [KRA 01, KRA 02] constitutes an evolution of the APON technique (ATM PON). The digital communication flow of a company uses a specific wavelength for the transport of 10 or 100 Mb/s, or 1 Gb/s, of the local Ethernet network data over the optical fiber of the access network. The advantage of this solution consists of the continuity of throughput (which experiences no intermediate processing delay) and of the flexibility of the link's upgradeability, since the wavelength brings the flow in all the throughput range. The maximum length of the link currently reached in EPON at 10 Gb/s is s of 69 km on single-mode fiber at 1,550 nm.

26.8.4. *Possible evolution with PONs*

Japan has launched an ambitious program of connecting customers with optical fiber. Almost 20% of subscribers are connected with optical fiber and benefit from a complete optical service at 100 Mb/s. In other countries, connection costs are such that the operators have only connected by optical fiber those subscribers with high traffic. In the USA, 11% of local networks were connected by optical fiber at the end of 2000 and according to KMI Corporation, FTTH (fiber to the home) techniques and FTTC should reach respectively 2.65 and 1.9 million connections in 2006.

Different access network architectures can be found according to the penetration of the optical fiber in these networks. Architectures such as FTTH, FTTB (fiber to the business), FTTCab (fiber to the cabinet), FTTCurb (fiber to the curb) or still FTTEx (fiber to the exchange) deliver enough necessary bandwidth for the new wideband services. Passive optical networks have the characteristic of being able to penetrate each of these architectures.

Passive optical networks can be used to transport enough information to an electric cabinet where the optoelectronic conversion operates in order to develop the last kilometers (ADSL: asymmetric digital subscriber line technology) or the last few hundreds of meters (VDSL technology). The passive optical network complements the xDSL technologies in the FTTEx, FTTCab and FTTCurb architectures. When the passive optical network is deployed all the way to the final subscriber, the architecture is called FTTH or FTTB depending on whether the final subscriber is residential or professional.

26.8.4.1. *FTTEx*

The FTTEx architecture serves approximately a thousand lines. This network configuration reduces temperature or space constraints since line terminations are located inside of a building and not outside as in an FTTCab architecture, for example.

This architecture is attractive for the operator that, with his twisted pair network, wants to offer ADSL services. It constitutes the first step in the progressive migration toward a wideband access network.

A passive optical network will be used as long as the penetration remains low. Indeed, in the case of high penetration, the passive optical network will be limited in bandwidth and will not serve all the connected users. The use of a metropolitan network, of SDH (synchronous digital hierarchy) type, or of a higher throughput bidirectional passive optical network, 622 Mb/s for example, corrects this problem.

26.8.4.2. *FTTCab*

The FTTCab architecture uses the optical fiber until an electric cabinet located at approximately one kilometer from the final client. This cabinet contains VDSL line terminations and the optical fiber is used to provide enough bandwidth for the cabinet. Such a network configuration guarantees data confidentiality since the twisted pair is dedicated to each user.

This architecture works well in access networks with low penetration or presenting an uncertain service request for reasons previously mentioned. The electric cabinet services around 100 homes. Fixed costs remain limited if the cable network is already in place.

26.8.4.3. *FTTCurb/FTTBuilding*

The FTTCurb architecture somewhat increases the penetration of optical fiber in access networks. As with the FTTCab configuration, the optical fiber ends at a dispatcher located less than 300 meters from the final client. This cabinet contains VDSL line terminations which enable a transmission at a higher throughput due to the shorter distance separating them from associated network terminations.

This architecture is a progression toward the FTTH concept but poses a difficult choice for the telecommunications operator. Indeed, he must decide between renovating the twisted pair to support the high transmission throughputs of the VDSL technology or invest directly in an FTTH architecture.

The configuration of the FTTCurb network makes it possible to serve a dozen clients. This topology can also be used for an FTTBuilding (fiber to the building) architecture which consists of placing the dispatcher at the foot of a building and to connect the building residents with the help of the VDSL technology.

26.8.4.4. *FTTH/FTTB*

The FTTH architecture uses the optical fiber up to the final subscriber and consequently is the most apt to deliver a very large bandwidth to the final user.

The advantage of this network configuration is multiple in the sense that the network can be updated at transmission throughput level without limitations due to the transmission media limitation. This update can be done with the help of wavelength multiplexing techniques. The access network is completely insensitive to electromagnetic disturbances. The absence of the active component or metallic cables ensures better network reliability as well as electric power economy. This architecture presents the lowest operation and maintenance costs among all the architectures. The constant decrease of the cost of optical components as well as the normalization of the passive optical network are some of the factors that have revived the interest of telecommunications operators in this type of system.

The FTTH architecture mainly attracts the new users who, not having their own network, must invest in an access network which enables them to deliver a set of services immediately and can adapt to a growing bandwidth demand. Some traditional operators wanting to differentiate their service offering from their direct competitors are studying the possibility of optical fiber deployment to the final client. This means, in a first phase, a deployment in greenfield instead of the replacement of their copper network.

When the passive optical network is used to provide information to an industrial center, its architecture is called FTTB. The type of services offered is very different from the one designed for residential users. Professional users have more symmetric service needs (telecommuting, videoconference, Internet learning, file transfers, LAN interconnection, voice over Internet, etc.).

26.9. Conclusion

The last decade of the 20th century coincided with the arrival of the Internet which has superseded voice traffic in terms of volume of exchanged data. The explosion of the number of users connected to the worldwide network has clearly brought to the forefront the limitations of current networks which are mainly based on the analog telephone network. The network of networks has developed new services requiring high transmission throughputs that the traditional copper pairs will not be able to provide. Optical networks are here to respond to new user requirements. Optical networks have proven their efficiency in the transport network. However, there is still a long way to go with access networks in order to reach users' homes. Finally, optical networks are progressively moving toward packet switching which now presents the biggest challenge for manufacturers in order to develop the compromise between the high bandwidth offered and the efficiency of resource utilization.

26.10. Bibliography

[BAT 03] BATTESTILLI T., PERROS H., "An introduction to optical burst switching", *IEEE Commun. Mag.*, p. 10-15, August 2003.

[CHR 90] CHRAPLYVY A.R., "Limitations on Lightwave Communications Imposed by Optical Fiber non-Linearities", *Journal on Lightwave Technology*, vol. 8, no. 10, November 1990.

[DES 96] DESURVIRE E., *Erbium Doped Fiber Amplifiers: Principles and Applications*, Wiley, 1996.

[KRA 01] KRAMER G., MUKHERJEE B., PESAVENTO G., "Ethernet PON (ePON): Design and Analysis of an Optical Access Network", *Phot. Net. Commun.*, vol. 3, no. 3, p. 307-319, July 2001.

[KRA 02] KRAMER G., PESAVENTO G., "Ethernet Passive Optical Network (EPON): Building a Next-Generation Optical Access Network", *IEEE Commun. Mag.*, p. 66-73, February 2002.

[MAR 91] MARCUSE D., CHRAPLYVY A.R., TKACH R.W., "Effects of Fiber non-Linearity on Long Distance Transmission", *Journal on Lightwave Technology*, vol. 9, no. 1, January 1991.

[OMA 01] O'MAHONY M.J., SIMEONIDOU D., HUNTER D.K., TZANAKAK A., "The Application of Optical Packet Switching in Future Communication Networks", *IEEE Commun. Mag.*, p. 128-135, March 2001.

[RAM 02] RAMASWAMI R., SIVARAJAN K.N., *Optical Networks – A Practical Perspective*, 2nd edition, Morgan Kaufmann Publisher, 2002.

Chapter 27

GMPLS-enabled Optical Networks

27.1. Introduction

Historical networks, predominated by voice services, grew from scratch at a rate of less than 10% each year. However, since 1995, networks have witnessed a drastic change in the overall picture due to the explosive growth in IP-centric data traffic. This type of traffic, which has already surpassed voice traffic, will continue to outpace voice in the years to come while becoming more and more demanding in terms of Quality of Service (QoS). This shift, driven primarily by the proliferation of the Internet as well as Virtual Private Networks (VPNs), has created a demand for capacity that doubles every year. At the same time, and in parallel with the increase in traffic capacity, there is increasingly strong demand from customers to keep the cost of networking down. These recent trends have established a situation where service providers must find solutions to enable them to transport the largest volume of traffic while optimizing transport cost.

In such conditions, technology such as Wavelength Division Multiplexing (WDM) becomes attractive to operators. This, coupled with the efficient data transport, should optimize the cost of multiplexing as data switching does over a large range of traffic volume [BAN 01]. WDM is a cost-efficient multiplexing technique that offers significant technical advantages. This technique increases the bandwidth-carrying capacity of a single optical fiber by effectively multiplexing many non-overlapping wavelength signals (WDM channels) whose carriage capacity is in the order of several gigabits/second (Gb/s) onto the fiber. This provides a multifold increase in bandwidth while leveraging the existing fiber

---

Chapter written by Wissam FAWAZ and Belkacem DAHEB.

infrastructure. Likewise, Optical Cross-Connects (OXC) has emerged as the preferred option for switching multi-gigabit or even terabit data streams in a transparent way, ensuring furthermore cost-effectiveness since electronic per-packet processing is avoided. In fact, at very high data rates, the electronic processing of packets during switching must be avoided since it decreases the maximum throughput and increases the cost of optical switching equipments.

However, it is important to note that the simplification of management and maintenance of network operations are very important criteria in the choice of architectures. Especially since these management and maintenance operations cause recurring costs that rapidly become higher than the amortization of the investment. In this regard, a WDM-based architecture is not an exception. Building on this, efforts have been put in place to implement a control plane for optical networks to simplify management and maintenance tasks. In order to facilitate the implementation of a distributed control feature in optical networks, it is now possible to reuse in WDM networks the same protocols as those used in MPLS (Multi-Protocol Label Switching) but with extensions related to the optical context. The MPLS architecture has been generalized to a new architecture called Generalized MPLS (GMPLS). The particular architecture used to control optical networks is called MPλS. In this way we can develop an IP over WDM network with a unified control plan for IP and WDM parts. The electronic switch that is responsible for the aggregation of IP traffic directly integrates WDM interfaces to connect to the optical cross-connects of the WDM network.

The first part of this chapter introduces the MPLS technology. The second part will present the WDM technology as well as the factors that have contributed to the arrival of a control plane for optical networks using WDM as multiplexing technique. Finally, the MPλS control plane is presented with the corresponding protocol extensions.

27.2. Label switching (MPLS)

27.2.1. *Introduction*

From the introduction of computer networks until now, a wide variety of network technologies have started to emerge. Let us take ATM, for example, or the more recent SONET/SDH. Each of these technologies has been developed to respond to specific needs, without deviating from the main goal which is to transport information via a computer network at increasing speed.

In this regard, the label switching concept remained in the research phase for a few years. Moreover, several manufacturers developed proprietary label (reference) switching scenarios which are not compatible with each other (such as IP-switching proposed by Ipsilon [PUJ 02] or Tag switching proposed by CISCO).

However, the IETF (Internet Engineering Task Force), which is the Internet normalization organization, has standardized the label switching technique. Thus, we now have MPLS, [ROS 01] which has been growing in popularity in the last few years as many operators are using it in their own networks.

27.2.2. *Packet switching networks and MPLS*

At the end of the 1960s, companies and organizations started to develop packet switching networks. Thus, X.25 networks started to make their appearance in the mid-1970s. These networks enable packet switching based on a logical channel number. The MPLS technology uses the same type of reference.

Other technologies also became available, such as frame relay (FR) or ATM (Asynchronous Transfer Mode). All use the virtual circuit notion based on a reference called data link connection ID in FR and virtual path ID/virtual channel ID (VPI/VCI) in ATM.

MPLS is a transfer and reference translation technology. MPLS must integrate, depending on the specifications, the different types of packet transport networks including the one that is taking over the market: the Internet protocol (IP).

We could ask ourselves: why use references? There are several reasons, among them:

– expedited switching;

– scalability;

– simplicity;

– resource consumption;

– route control.

In the following we will explain the MPLS function in more detail.

27.2.3. *MPLS network operation*

27.2.3.1. *MPLS architecture*

In MPLS networks, we do not use network addresses but labels (i.e., references) coded on 32 bits. The difficulty is now doubled: attributing a reference to a client's traffic and associating this reference to the traffic's destination address. In order to do this, MPLS provides reference distribution mechanisms that notify the network's routers of the associations between the references and the network addresses.

A router that supports MPLS is called a label switch router (LSR). The role of LSR is to recover the reference of a packet in order to switch it [BLA 02]. If the reference does not exist, the packet is routed. Figure 27.1 explains how an MPLS-enabled network works. There are three types of nodes:

– at the edge of the MPLS cloud there is an ingress LSR (A in the figure) which recovers user traffic, attributes references and sends them to the following nodes;

– in the middle of the network there are transit LSRs (B, C and D in the figure); the role of these LSRs is to switch the arriving packets toward the next LSR based only on the label value (reference);

– at the other network edge there is an egress LSR (E in the figure).

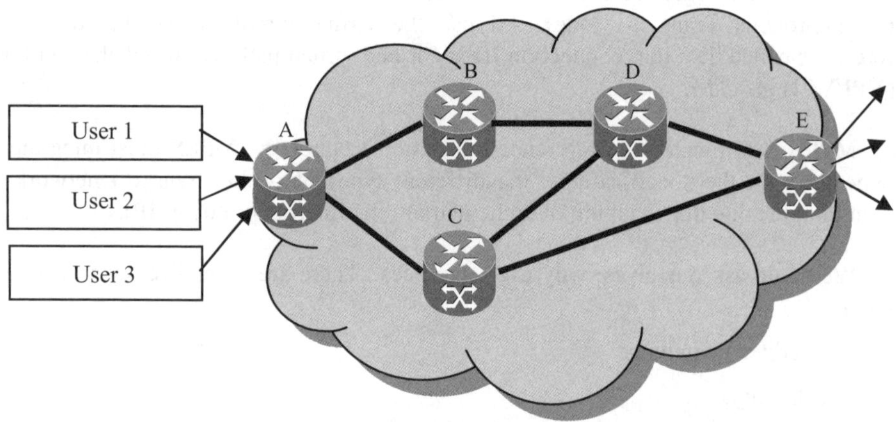

Figure 27.1. *MPLS network architecture*

MPLS eliminates the need for NHRP (Next Hop Resolution Protocol) protocol. Therefore, it eliminates the delay associated with these operations. Figure 27.2 shows how we avoid the routing operation with MPLS in IP architecture over Ethernet.

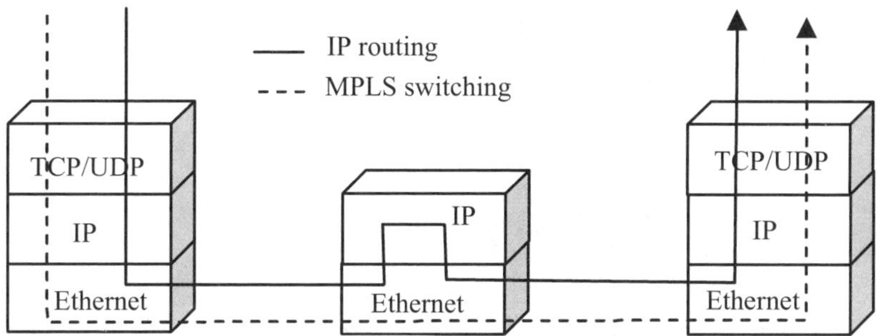

Figure 27.2. *Routing and switching*

We can clearly see that the role of transit LSRs is simplified. On the other hand, the complexity is pushed back to the edge of the MPLS network. It corresponds to the association of a packet (which could not be MPLS to start with) with a reference representing a traffic class at the network's extremities.

MPLS defines the concept of FEC (Forwarding Equivalent Class) which regroups user traffic with common characteristics, for example, the same destination address. Traffic belonging to the same class gets the same treatment within the network. It is therefore simpler for an operator to regroup FEC user traffic to facilitate management in the network. In general, we add QoS considerations to the FEC grouping criteria.

Within the network, the succession of references defines the route followed by the flow packet set. Figure 27.3 illustrates the case of a packet transfer (be it IP or any other) in a switched network. The complete route through the network is called Label Switch Path (LSP). The route is made up of a set of LSRs passed through by the LSP. When traffic is sent over an LSP, it goes through a tunnel which enables it to reach an LSP exit without corruption or modification. The objective is to examine only the label header and no other header.

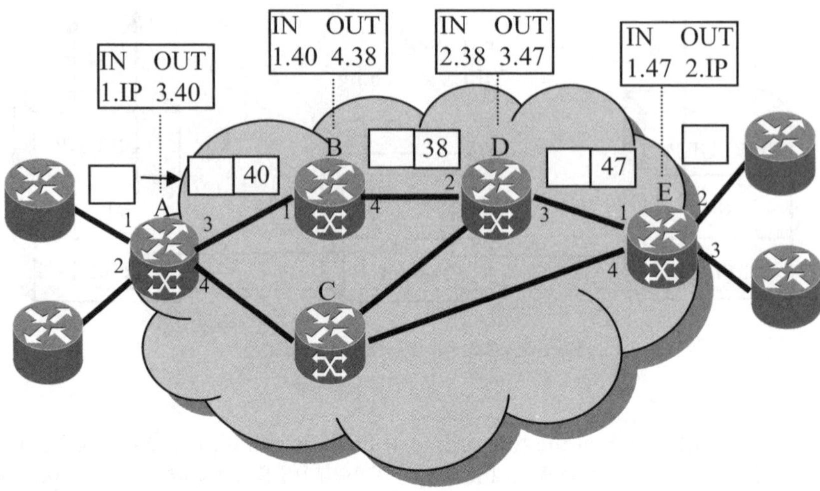

Figure 27.3. *Label switching and permutation*

27.2.3.2. *MPLS header*

If we had to describe where MPLS is situated in the 7 layer stack of the OSI architecture, we would say that MPLS is a level 2.5 protocol. It defines a header that will be located between the level 3 header and the level 2 header.

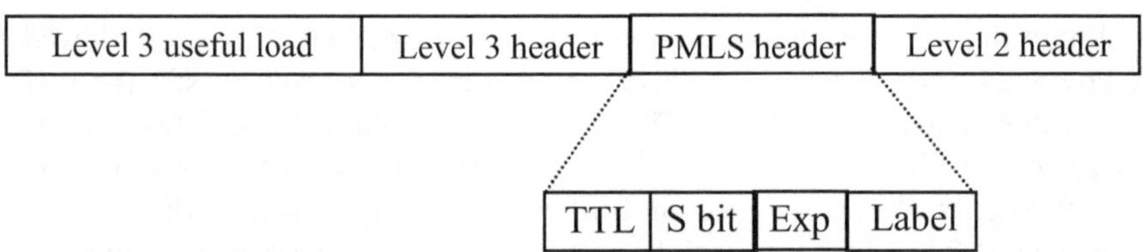

Figure 27.4. *Header format*

Edge routers will use the level 3 header to associate the packet to an FEC and will add an MPLS header to it. Intermediate routers will only use the reference found in the MPLS header to process the packet.

The format of an MPLS header consists of the following fields, as described in Figure 27.4:

– a 20 bit reference which represents the MPLS reference;

– a 3 bit experimental field, for future use such as MPLS over DiffServ;

– a stacking bit, which indicates the existence of a stack of references (see below);

– an 8 bit time to live (TTL), which is responsible for limiting the number of hops the packet executed.

27.2.3.3. *Start of MPLS*

The first process that needs to be executed in an MPLS network is the creation of the Label Information Base (LIB) [BLA 02]. This base determines which reference to use to reach an LSR. It is established by exchanging "hello" messages between the different neighboring LSRs.

The next step is the attribution of references or labels to the different FECs. In other words, once the FEC classes have been defined, which reference should be attributed to each of them in order to reach the expected destination? This reference represents the reference of the start of an LSP thus created. There are two methods for executing this operation, independent and orderly control:

– in independent control, each router assigns a reference to each recognized FEC. In this case, each router chooses the reference that is meant for it. There remains one constraint: that two adjacent LSRs choose different references;

– the orderly control makes it possible to assign references in an orderly way, from the input or the output of an LSP. This forces the other LSRs to use the same FEC as the initial announcer. In addition, this technique enables a network administrator to intervene during LSP creation. He can, for example, give a list of FECs to correlate with an LSP.

There are two ways of distributing references in order to develop LSPs: solicited and non-solicited. In the first case, an association of references is done only if the request to the LSR is made. In the second case, the LSR associates its reference to the ones of the neighbors without it being used by the LSPs.

In order to be able to transfer incoming packets to a neighboring LSR, an LSR needs an association between the network address and the reference. This association is registered on each LSR in a Label Information Base (LIB). It associates each network address with a reference to use to send packets to this LSR. It only focuses on the network addresses of the adjacent nodes.

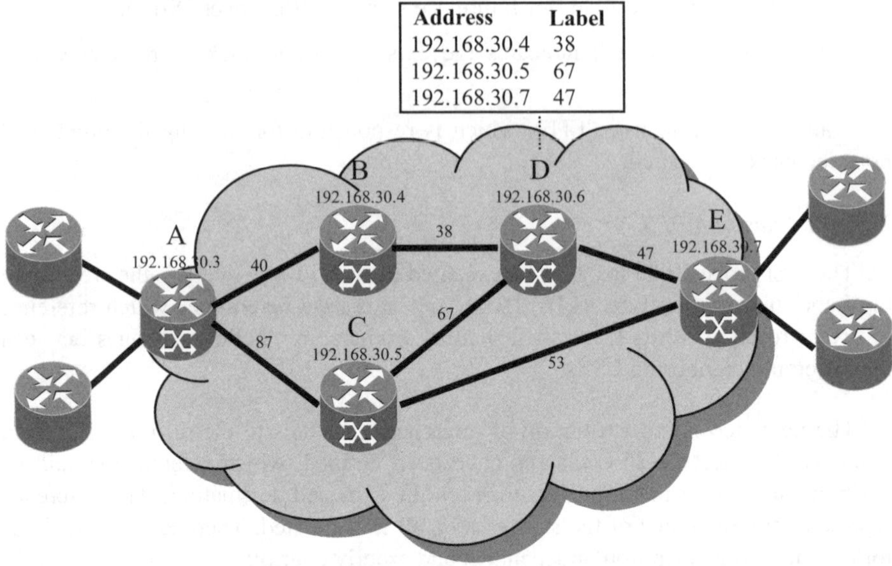

Figure 27.5. *LIB of node D*

An LSR bases itself on a switching table to find out where to forward an incoming packet in output and with which reference. This table is called Label Forwarding Information Base (LFIB). It contains an association between two references and an interface linking two nodes. The first reference is the local reference and the second is the output reference.

27.2.3.4. *Failure recovery in MPLS*

MPLS is a technology planned for deployment over large networks, taking advantage of its scalability. However, the larger the network, the more vulnerable to failures. For this reason, MPLS must control network failures, whether they are link or node failures.

An important step in failure recovery can be detected either by the reference distribution protocol or by the network routing protocol via "hello" messages.

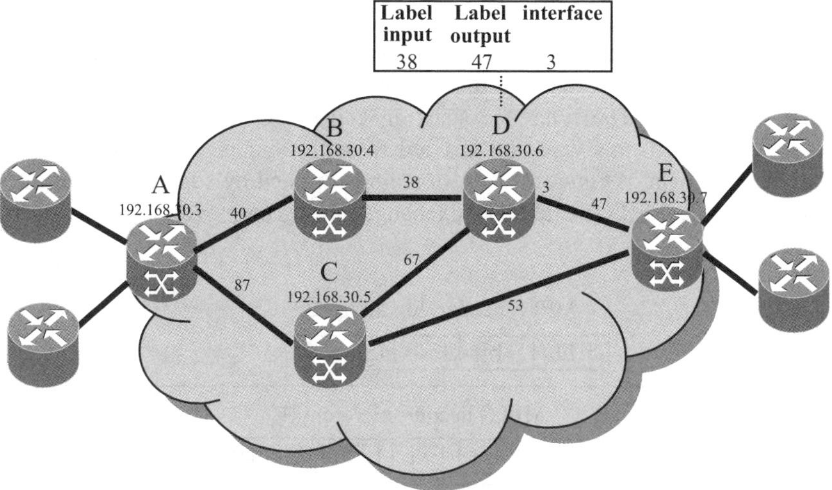

Figure 27.6. *LFIB of node D*

Once an LSR detects failure in a link, it updates its LFIB to announce the failed link. The link is then deleted from the routing table. This activates the routing protocol (for example, OSPF) to choose another route and to place it in the routing table as illustrated in Figure 27.7. It is important to note that there may not be an emergency route.

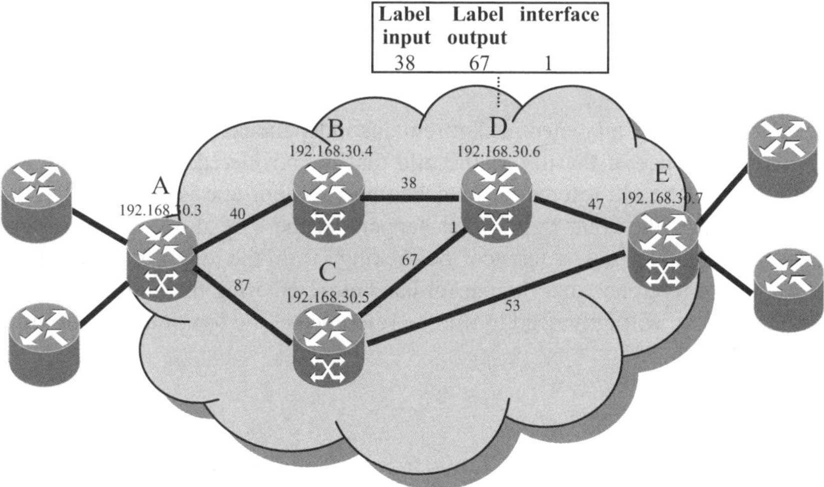

Figure 27.7. *Failure recovery*

27.2.3.5. *The pile of references*

The MPLS architecture defines a concept that will enable it to expand over several MPLS domains, each with its own control, while the MPLS header contains only one reference. The reference stacking concept enables the prioritization of crossed domains. In this way, the last fed through domain will be the last in the hierarchy, as shown in Figure 27.8. A domain is defined by edge LSRs and internal LSRs. The edge LSRs are able to exchange information with LSRs from other domains.

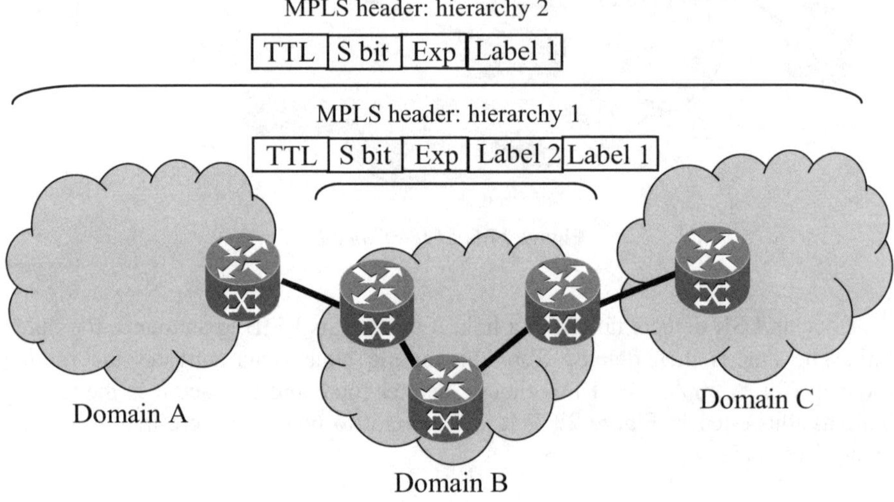

Figure 27.8. *Reference hierarchy*

The processing of a labeled packet remains independent from the hierarchy. It is always the reference at the top of the pile that is considered. When a new MPLS domain arrives, the old reference is stacked and another one is added at the top of the pile, which is the contrary of what happens at the exit of a domain, where the reverse operation happens: reference "destacking" from the top of the pile. Thus, we should find the reference that the packet has before entering the domain. The LSRs inside the domain will only need to use the reference at the top of the pile to transfer the packet.

27.2.3.6. *Aggregation and merging*

In order to transport multiple flows from different routers, we can attribute to each an FEC. If it is possible to have multiple flows through the MPLS network of a single source toward one single destination, it would be helpful to attribute the same reference to these FECs. This procedure is called aggregation. Thus, the aggregation reduces the number of references required in order to manipulate a set of packets.

Merging is receiving several packets with different labels (already belonging to FECs) and attributing one single reference or label. This facilitates the use of MPLS over ATM.

27.2.3.7. *Types of routing*

MPLS uses two methods to choose the route of an LSP. The first solution is hop-by-hop routing where each node decides the next hop. It is the routing used today on the Internet via protocols such as OSPF.

The other method is explicit routing in which no node has the right to choose the next hop. However, one node, which is often the input or output node, specifies the LSRs included in the route. There are two modes:

– strict explicit routing, where the whole route is given;

– loose explicit routing, where only one part of the route is determined.

Explicit routing plays an important part in operations of traffic engineering and QoS introduction in MPLS.

27.2.4. *MPLS extensions*

27.2.4.1. *QoS and label switching*

QoS is the capability of a service provider to support the requirements of usage applications such as bandwidth, delay, jitter and traffic loss as well as security considerations.

Label switching enables the improvement of the delay and jitter of traffic, thus contributing to the guarantee of QoS in an operator network.

27.2.5. *Conclusion*

We have explained in detail the MPLS technology. The goal of this technology is to standardize and group the different switching techniques that have preceded it. This section has helped us understand the different MPLS components and to understand how such networks function.

At the beginning, this technology focused on packet transfer networks. Considerable efforts are provided today to extend the MPLS technique toward other network categories, in particular time division multiplex networks such as SONET/SDH or even circuit switching networks such as WDM.

27.3. Evolution of IP/MPLS signaling for optical WDM networks

27.3.1. *WDM technology*

The increase in network capacity is based on the increase in transmission capacity. Optical fiber is the medium which today helps to increase transmission capacity. The idea behind WDM is based on the well known FDM (Frequency Division Multiplexing) concept [LIU 02]. This technology consists of dividing the total bandwidth offered by the fiber in parallel channels. Each channel represents a wavelength which operates at a smaller rate than the total capacity of the fiber.

To illustrate the WDM concept, let us consider the example of a very simple operator network made up of two network components connected by optical fiber (as illustrated in Figure 27.9).

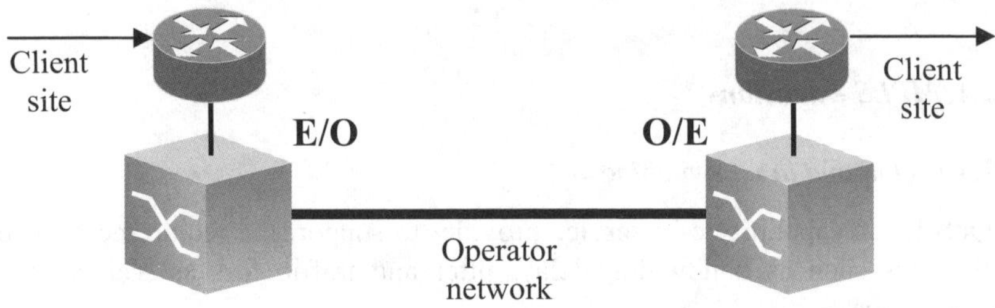

Figure 27.9. *Optical fiber connection between two client sites*

The network equipment starts an electronic process in order to relay information to client sites, so an optoelectronic conversion is necessary to transmit over optical fiber. The network cards are limited in throughput; on the one hand, we find 10 Gb/s cards on the market and, on the other hand, the transmission capacity of an optical fiber is of several terabit/s. The fact that we use optical fiber to transport a 10 Gb/s signal indicates a large under-utilization of the fiber. The WDM technology optimizes the utilization of fibers.

The idea behind WDM is to multiplex several signals at 2.5 or 10 Gb/s in parallel on a single fiber. In order to do this, each flow is modulated with a different color, which enables the destination equipment to separate them with the help of a prism. Upon reception of each individual flow, the flow is converted into an electric signal that can be processed electronically.

27.3.2. *An effective optical layer*

27.3.2.1. *Basic factors*

The data transport service offered by an optical layer depends on two factors:

– the multiplexing technology used;

– the switching equipment used to transmit wavelengths.

As presented in the previous section, WDM as a multiplexing technology has made it possible to optimize the utilization of bandwidth offered by the optical fiber. As for switching equipment, its impact can be seen at client service level.

In fact, the time to establish a connection in the optical network mainly depends on the time necessary to configure the switching equipment used by this connection. In the first generation of WDM networks, we were limited to static point-to-point connections because the time for equipment configuration was relatively long. Therefore, the type of service that the network could deliver was a "rigid leased line" type service.

However, during the second generation of WDM networks, the introduction of optical cross connect such as switching equipment has improved the situation. Network nodes naturally become switching centers where the different wavelengths transported over a fiber are geared toward their respective destination fiber (see Figure 27.10). The equipment configuration time has become shorter. All of a sudden doors opened to the introduction of new more dynamic transport services. However, in order to reach this goal, there needs to be an automatic control plane for this equipment to introduce the network dynamic.

The main objective to reach is to put in place an effective optical layer which is able to take into account new requirements as they arrive [DEL 02]:

– in provisioning, the client wishes to use the bandwidth as soon as possible, on demand or based on a determined calendar;

– increasingly the client wants the network to be transparent, but wants to control end-to-end the transport capacity he needs;

– different qualities of protection or SLAs are necessary to respond to different needs in the best way and in the most economic way possible;

– the operator also needs to control as much as possible his network in order to be competitive, by maintaining wavelength availabilities, by using dynamic and automatic rerouting which enable him to reduce network operational costs. For example, it is possible to dynamically allocate capacity between the different types of optical client networks (IP, SDH/SONET, even Ethernet).

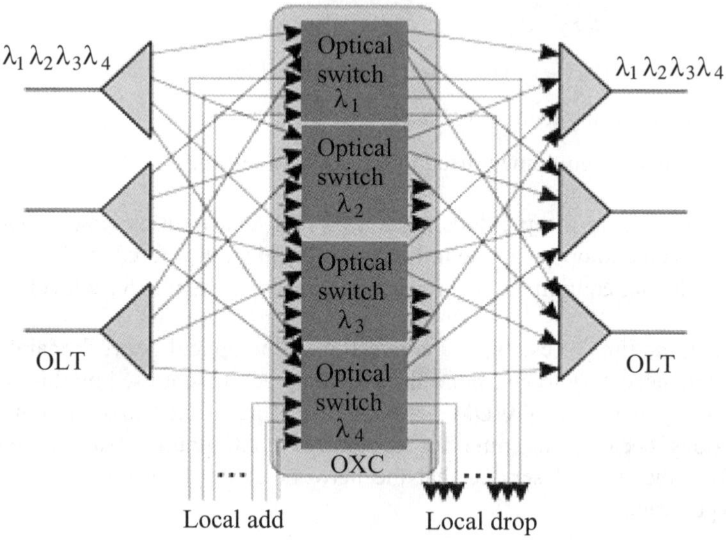

Figure 27.10. *Architecture of an optical cross connect with 4 wavelengths per fiber*

To highlight the value added by cross connect equipment and by a control plane in the optical layer, two types of service examples which can be considered in optical networks are presented below.

27.3.2.2. *Typology of services*

The classification of services in the optical layer can be considered from the protection level [DEL 02]. A first class of service corresponds to the implementation of dedicated protection mechanisms, which are similar to duct protection over SDH rings. Each optical connection is routed over two physically disjointed routes. A second class will rely on restoration by dynamic rerouting via the control plane. In case of failure, optical cross-connect equipment uses the shared backup (i.e., reserve) resource to transmit the affected service class. The lower class service is a non-protected bandwidth service, which we may even call a "preemptible" bandwidth service, i.e. transported over reserve resources that may be mobilized for higher priority needs (of class 2, for example). The definition of these three levels of service in the optical links layer responds to requirements linked to the junction between the optical and customer layers. Ballparks announced are compatible with the 50 ms of APS protection for the first class of service, whereas dynamic rerouting mechanisms require several hundreds of milliseconds to reestablish the bandwidth. The difference between what dedicated protection and dynamic rerouting offer enhances the classic compromise: better optimization of the resource necessary to the protection of the network due to the intelligence of rerouting mechanisms can only be achieved at the cost of a longer reconfiguration time. Non-protected services in the optical layer are based on the protection mechanisms of client layers.

Another possible classification of future optical layer services can be considered at the time of establishment and availability of the bandwidth. We will then distinguish the semi-permanent bandwidth, the fast provisioning bandwidth, etc. The functions of an intelligent control plane characterize the definition of these offerings adapted to the bandwidth provider market.

These two criteria of transport service classifications at wavelength level in an intelligent optical layer emphasize two functional axes of the network:

– wavelength switching capacity by nodes;

– reconfiguration possibility of this function, on the one hand, to protect the network and, on the other hand, to ensure bandwidth control.

To simplify, we can see the two elements constituting the optical layer: cross connect machines data plane corresponding to the reconfigurable switching function and the network control plane providing software elements necessary to the dynamic of the network.

27.3.3. *Centralized control of optical networks*

Historically, the control of transport network elements (cross connects, add multiplexers, frames) was until now integrated to the management and operation of these networks when it was not simply executed manually. Which means that the networks had two major functional planes (as illustrated in Figure 27.11) containing:

– the data plane, where information is transported on wavelengths switched by the switching equipment;

– the management plane, which acts as control plane for the switching equipment.

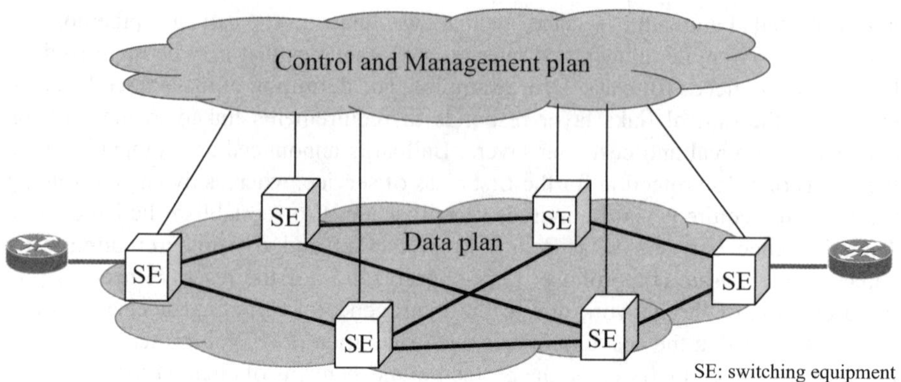

Figure 27.11. *Functional planes of centralized control*

This approach satisfied the telephone networks' transport needs, which are major clients of these networks. Apart from timely and exceptional media events, requirement in transport has a predictable character and is a good match for a centralized and planned control. This control, ever since the 1990s, also applies to a transport plane based on a unique technology, SDH or its US version SONET, which is defined according to the technical characteristics of client telephone networks. The transport networks respond well to the criteria for quality of telephone service, specifically with regards to the transport service protection.

However, three factors are putting this statement in question [DEL 02]:

– weakness of centralized control: the centralized approach has limited responsiveness. Speed of reconfiguration during failure of 50 ms is only obtained through a constrained ring architecture and/or oversizing of resources necessary for the implementation of a dedicated protection mechanism. The incompatibility of equipment management systems coming from different manufacturers has motivated the operators to implement a software integration layer for management tools. This additional complexity is problematic in itself and is liable to hamper the efficiency of the control plane and therefore to limit the flexibility of line operations;

– Internet and data networks: from the time Internet networks have become commercially available, Internet traffic has experienced a very significant growth. This traffic has completely different characteristics from those of the telephone network: asymmetry, absence of geographic location characteristic, very large variability and exponential growth. All this makes predictions of Internet traffic volume and its distribution much more difficult than with telephone networks, which emphasizes the need for adaptive network engineering;

– distributed computing: the growth in processing power of microprocessors unfavorably affects the major centralized computing servers compared to distributed servers. The success of distributed routing on the Internet and the continued growth of server processing power calls into question the validity of a transport architecture based on a set of equipment managed from a centralized command point.

Breaking from our historic concepts, transport networks must take into account a requirement that has lost its deterministic and predictable character, even in a period of strong growth.

This new approach requires the use of an optical, upgradeable, easily configurable and economically competitive transport network with the help of an optimized reserve capacity.

Distributed control mesh networks seem better at efficiently responding to these demands due to their non-restrictive topology, of the flexibility that automatic training of active transport resources brings and of the capabilities of in-car information processing in network components.

In order to face the limitations of a centralized approach in relation to the emerging context, the general trend is toward distributed control of optical networks.

Thus, the new functional network vision also includes a distributed control plane which automatically assures the control function of the switching equipment (see Figure 27.12).

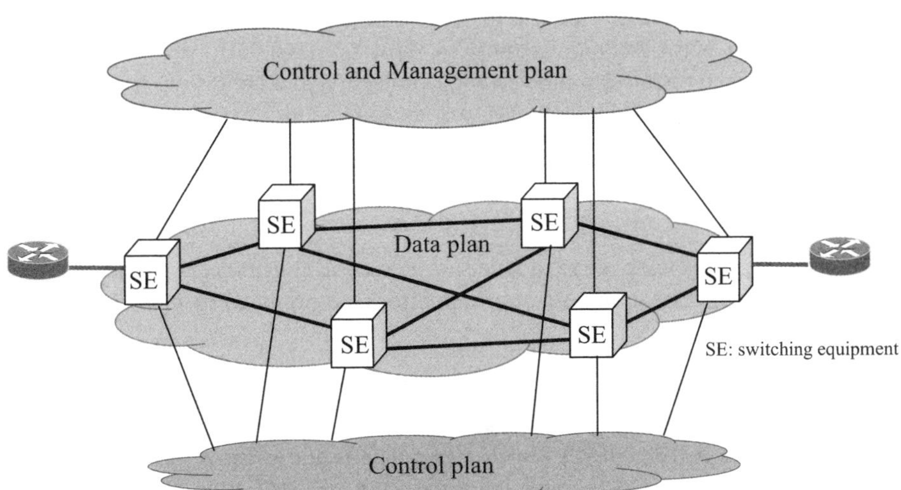

Figure 27.12. *Functional planes for distributed control*

27.3.4. *Distributed control optical networks*

The distributed control plane of an optical cross connect network must execute the following functions:

– discovery, distribution and synchronization of information on installed transport resources, availability, reserved within the optical network;

– routing: routing takes over calculation of bidirectional routes between the access points for the service (optical circuit termination points). The calculated route is optimized for given metrics and in a field of given constraints translating the transport resource usage policy;

– establishment, management and release of optical circuits: real optical circuits bring notable differences from virtual circuits, whether they are ATM or IP/MPLS. The switching function which consists of determining the output interface according to the circuit identifier transported by the frame or cell header is absent in the optical case. Indeed, the interface as well as the output wavelength in the optical context are determined according to those at input based on a configured cross connecting matrix at the optical cross connect level. Therefore, signaling transported by the optical control plane must be able to transport the cross connection matrix configuration.

The choice for control and distributed command techniques must meet the characteristics of dynamic services expected by the optical transport networks. The target control network must then take into account the following points:

– information controlled by the control network and the optical circuit routing must be able to respond to traffic engineering policies that can be considered for optical transport networks, for example, it must be possible to physically define disjointed routes to respond to the needs for protection and optimization of resource usage in the presence of a fine granularity of the bandwidth;

– the control network must enable the establishment of circuits in a short time in the order of a fraction of a second in order to respond to the need for protection, for example;

– the support of a wide range of levels of protection constitutes, with the time for establishing the circuit, an important criterion for QoS provided by optical transport networks.

From the points already discussed, it is clear that the packet mode is the transfer mode that is the most adapted for the support of control networks, although optical transport networks offer a real circuit service. The result is a necessary decoupling between the data plane itself and the control plane. In return, this decoupling requires the definition of specific procedures to enable discovery, learning and maintenance of resource knowledge.

The current retained approach uses routing protocols available in the Internet (OSPF, IS-IS and in a smaller measure BGP) and MPLS signaling, by integrating their extensions for traffic engineering while adapting them to the context of optical transport networks.

The advantages of this approach are summarized below:

– it enables the reuse of protocols and algorithms developed and validated within the context of extended and diversified routing domains. In this way, it enables both the associated research and development efforts;

– it should facilitate the definition of control plane function interfaces between optical transport networks and the Internet, first current data network, especially as the protocols used in the approach already presented come from the Internet world;

– it leaves the door open for the introduction of dynamic transport service in the context of large scale networks;

– it potentially simplifies for ISPs the supervision and operation of main networks by adopting a technology and semantics which are common to optical transport networks and to data networks.

Due to the expected characteristics of the transport network control plane, this orientation requires, in addition to existing protocol adaptation, the definition of a new protocol. That is the case with the LMP protocol which has been defined within the frame of the optical layer in the control plane. LMP is a link management protocol responsible for notifying the control plane about the data or resource plane due to the decoupling of control and data planes.

In conclusion, a generalized version of MPLS (GMPLS) is currently being standardized by the IETF [MAN 03]. The goal of GMPLS is the definition of a common control plane with different transport techniques: transport by virtual circuit (at packet and cell level) or by actual circuit (time division multiplex, wavelength, fiber) mode. The particular control case for this architecture for optical networks is called MPλS.

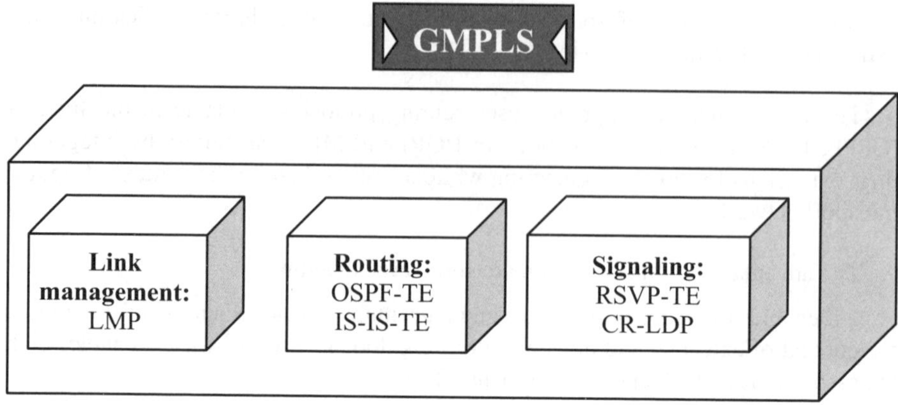

Figure 27.13. *GMPLS functional blocks*

Most of the large manufacturers of telecommunications transport networks are today proposing optical cross connects integrating a controller which implements functional blocks presented in Figure 27.13.

27.3.5. *MPλS*

27.3.5.1. *Overview*

During the establishment of an optical connection, a cross connect must have a view of the installed transport resources, which are available and reserved at the optical link level. Especially as the performance of an optical connection, or in other words a light conduit, is largely correlated with the characteristics of the fed through links by the connection [TOM 02]. For this, it must be possible to restrict the route taken by the light conduit, in order to be able to respect the requirements of customers for whom the optical transport is offered. This need explains the role of routing protocol extensions in the case of MPλS. In fact, the routing protocols are used to represent at each node the actual state of the optical network, such as the occupation state of the wavelengths. In addition, the need for instantiation mechanisms of the route covered by the optical connection justifies the signaling protocol extensions such as RSVP-TE within MPλS. It is through such an extension that the configuration details of an optical cross connect switching matrix are provided.

MPλS's functional blocks that must exist within each optical cross connect are presented in Figure 27.14. A link state routing protocol, like OSPF-TE or IS-IS-TE, is responsible for information distribution concerning the state of the optical network and of the availability of resources. Once collected at each node, this information is stored in a traffic engineering database. Based on this information, a multi-constraint routing algorithm calculates the route covered by the optical LSP throughout the optical network. At a later stage, a signaling protocol, such as RSVP-TE or CR-LDP, is used to put the LSP in place along the route calculated by the routing algorithm. During this phase, the references or even the wavelengths used by the optical connection are chosen.

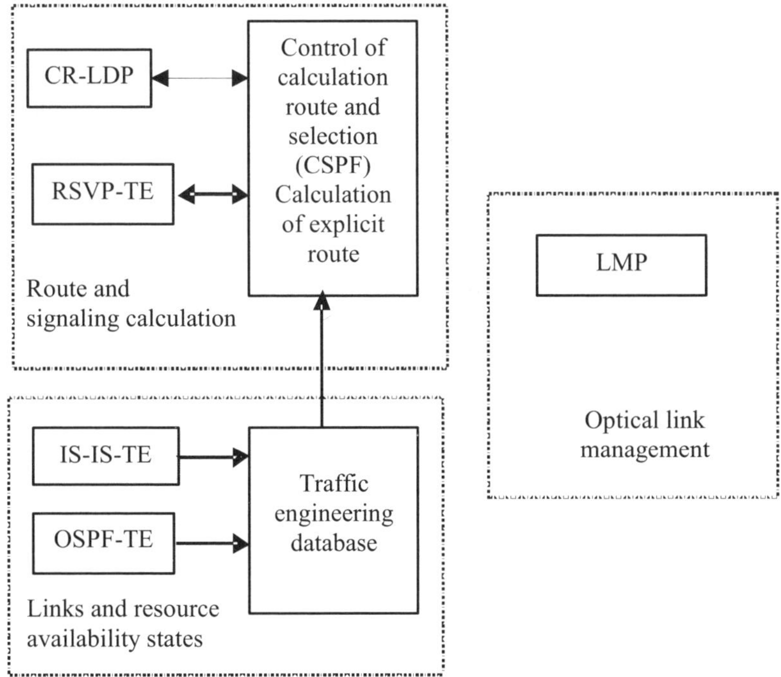

Figure 27.14. *MPλS functional blocks*

27.3.5.2. *Detailed view*

The three main bricks of the optical control plane are:

– routing;

– signaling;

– LMP (Link Management Protocol).

The following sections focus on the protocols and/or evolutions proposed to the IETF.

27.3.5.2.1. Routing

The principle of link state distributed routing is to make sure that each node has, through its link state base, a complete and current knowledge of the complete nodes and links making up the network topology graph. Each node notifies all the other nodes in the routing domain of the state of its links with its direct neighbors as well as their attributes. The role of the routing protocol is to ensure synchronization in the link state base of all the nodes or routers that participate in the routing process. Each router can then calculate, from an algorithm common to all the nodes and for a defined metrics, the positioned routes or outgoing routes by itself and linking to any other destination (sub-network, router, interface, terminal) within the domain. In addition, these routes can be obtained in reference to a set of constraints defining a subset of the graph in order to respond to traffic engineering objectives.

The adaptation to the case of optical cross connected networks from a routing protocol to a link state such as OSPF or IS-IS consists, on the one hand, to extend the attributes attached to the links and the nodes in order to take into account particularities of optical transport networks and, on the other hand, to correlatively redefine the field of constraints taken into account in traffic engineering extensions of these protocols.

Before listing what those attributes can be, it is important to remember that the term link in routing protocols does not cover the optical channel notion but the group (or bundle) of optical channels interconnecting two cross connects and sharing the same attributes. The attributes of the link are [KOM 03]:

– *type of link*: for cross connects it is the point-to-point link;

– *link identifier*: it is the identifier of the remote cross connect for routing;

– *interface identifier*: this identifier is a 32 bit word designating the local interface for the group of the link;

– *link cost*: in the sense of the traffic engineering metrics;

– *multiplexing level*: the levels considered are: packet, time division multiplex, wavelength, frequency band, fiber;

– *maximum bandwidth* (in bytes/s): this is equal to the maximum number of bandwidths of the link group members;

– *total available bandwidth*: it is equal to the sum of non-reserved bandwidths of the link group members;

– *total bandwidth*: this is equal to the sum of bandwidths of the link group members;

– *SRLG:* a shared risk identifier. A single link has several SRLGs;

– *local link protection attribute*:

- "preemptible" link used in protection from other links,

- unprotected link,

- link protected 1 + 1,

- link protected 1 : 1,

- link in shared protection.

In addition, each link must be associated with a control channel making the two end cross connects of the link direct or adjacent neighbors in the control network. This control channel can be shared with the different links interconnecting both cross connects. The association of a group of physical links intended for the transportation of useful data with the control channel defines a link for the MPλS. The notion of link covers more of a logical than a physical notion and matches the information on physical transmission resources with the calculation of the route particular to the routing process.

27.3.5.2.2. Signaling

MPLS signaling defines the process of establishment, management or deletion of MPLS circuits established between two end routers. The information transported by signaling enables each router that is part of the circuit to identify it by the attribution of references with a local significance and logically define the switching process particular to the circuit. The signaling phase takes place after the identification of the route followed by the circuit, which is done by the router initiating the circuit, from its link state base and due to applied constraints; this route is explicitly specified in the signaling messages.

The generalization of MPLS, CR-LDP or RSVP-TE signaling protocols, in the case of transport networks requires the modification of the attributes and the functions of the signaling messages to extend reference signaling, take into account the need for bidirectional circuits or define the manner in which the failures will be treated [BER 03].

Generalized label significance

Transport networks are based on real circuit transfer mode. Used in this context, the reference explicitly identifies the wavelength or the fiber carrying the circuit. The choice of the link in the routing sense, i.e. the group of physical links with the same characteristics interconnecting two cross connects, is controlled by the constrained routing process. Precisely identifying the physical link carrying the circuit is similar to the attribution of a reference logically identifying a virtual circuit between two IP LSRs. The explicit identifying character associated to references on transport networks is also required by the logical decoupling of the control plane and the data plane. The use of a control network distinct from the transfer network means that one single control channel transporting signaling between two cross connects can be shared by several links or data channels. The role of a reference is to point to the link carrying the invoked circuit in order to define the configuration of the cross connect matrix required by the circuit.

As with MPLS, the validation of the circuit is done by the returning signaling message (Resv message with RSVP-TE), since the initial message (PATH/RSVP-TE) was only used to pre-reserve the resources. It is therefore the signaling messages transmitted from the downstream cross connect toward the upstream cross connect that validate the attribution of references and thus the physical links carrying the circuit. However, with GMPLS, the upstream cross connect can suggest the references to the downstream cross connect. In this way, the upstream cross connect can initiate the configuration of its switching matrix and save valuable time in the case of a physical system with a considerable configuration latency. However, this is only an optimization: in case of conflict, the downstream cross connect is the one that is in control of the attribution of the reference through the returning signaling message.

Bidirectional circuits

GMPLS generalizes signaling for bidirectional connections. This constitutes the rule for transport networks. We then talk about "initiating" cross connect instead of upstream cross connect to designate the cross connect initiating signaling. The goal is to establish two counter-directional circuits sharing the same attributes (bandwidth, protection level, delay). The joint establishment of two circuits has several advantages:

– we reduce the possibility of running into a contention problem over resources during the establishment of a circuit or its counter-directional equivalent by coupling resource requests for both circuits, thus increasing the probability of success of the request;

– the time of establishment can be reduced to its minimum, i.e. round trip between end cross connects. In this way we save at least the transit delay from the initiating cross connect to the cross connect terminating the bidirectional connection. This gain is especially significant during the establishment of an emergency circuit during failure.

27.3.5.2.3. *Link management protocol*

The specific configuration of links between cross connects has made necessary the definition of a link management protocol. In general, the interconnection of two cross connects is carried by several fibers, which are themselves carrying several WDM channels, and these can also transport an SDH multiplex. The links (WDM channels, SDH section) of identical characteristics making this interconnection will be advantageously grouped in a single logical link appearing in the link state base controlled by the routing protocol. This grouping in a logical link has the advantage of reducing the size of the link state base and to save addressing space since the physical links are not identified by IP addresses but simply by the cross connect address and a local identifier.

The control channel between two cross connects can be physically disjointed from the links carrying the flow of "useful" data and is in any case necessarily logically separated from these links that do not process information at packet granularity level. Therefore the information exchange particular to the control plane can be carried by a dedicated Ethernet network. Another approach consists of having the control channel between cross connects carried by a dedicated wavelength or "Data Communication Channel or DC", defined from the bytes in the header of the SDH frame.

Contrary to current IP networks, for which a single link layer is shared between information flows linked to the execution of protocols and data flows, the transfer in parallel of information flow inherent to wavelength division multiplexing and the decoupling between control network, network and data network demand the definition of a protocol appropriate for the management of links or "Link Management Protocol", whether it is links carrying the control network or links transporting the data flow. The four basic functions of LMP to manage the links interconnecting two cross connects are [LAN 03]:

– *control channel management*: the management of the control channel consists of establishing and supervising the interconnection link of the cross connect control planes at IP level. The link layer carrying this IP link can be anything (Ethernet, DCC, etc.). However, this link must be terminated at both cross connects. The cross connects will execute an IP peering through their control channel. Consequently, the control network topology will reflect the transfer network topology;

– *correlation of transfer links properties*: this function correlates, for each physical transfer link, the identifiers and properties attributed to this link by each end cross connect. The procedure executed when the link was activated can be repeated at regular intervals. Its aim is to identify possible cabling mistakes to which every human intervention is exposed;

– *connectivity verification*: the goal of connectivity verification is the verification of the physical connectivity of the transfer links and uses for this the transmission of dedicated messages in the band;

– *fault location*: LMP uses the exchange of messages between adjacent cross connects through their control channel to locate faults. The identification of these failures enables in return to generate the protection or restoration process.

27.4. Conclusion

For operators, distributed control optical networks are a promise for economic competitiveness as well as having the capacity to respond to future demands. In this context, the use of a control plan such as MPλS appears as a direction able to respond to the evolutions associated with new information techniques, even if the technique is recent.

27.5. Bibliography

[BAN 01] BANERJEE A., DRAKE J., LANG P. J., TURNER B., KOMPELLA K., REKHTER Y., "GMPLS: An Overview of Routing and Management Enhancements", *IEEE Communication Magazine*, p. 144-150, January 2001.

[BER 03] BERGER L., "GMPLS Signaling Functional Description", *RFC 3471*, January 2003.

[BLA 02] BLACK U., "MPLS and Label Switching Networks", *Advanced Communications Technologies*, 2nd edition, Prentice Hall, 2002.

[DEL 02] DELISLE D., GUILLEMOT C., TILLEROT C., "Commande et Couche Optique", *Les communication optiques du futur*, technical reminder no. 19, France Telecom, June 2002.

[KOM 03] KOMPELLA K., REKHTER Y., "Routing Extensions in support of GMPLS", *internet draft*, draft-ietf-ccamp-gmpls-routing-09.txt, October 2003.

[LAN 03] LANG J., "Link Management Protocol", Internet draft, draft-ietf-ccamp-lmp-10.txt, October 2003.

[LIU 02] LIU H. K., *IP over WDM*, Wiley, 2002.

[MAN 03] MANNIE E., "GMPLS Architecture", Internet draft, draft-ietf-ccamp-gmpls-architecture-07.txt, May 2003.

[PUJ 02] PUJOLLE G., *Les Réseaux*, Eyrolles, 2002.

[ROS 01] ROSEN E., VISWANATANATHAN A., CALLON R., "Multi-Protocol Label Switching Architecture", *RFC 3031*, January 2001.

[TOM 02] TOMSU P., SCHMUTZER C., *Next Generation in Optical Networks*, Prentice Hall Series, 2002.

List of Authors

Nadjib ACHIR
University of Paris-Nord
France

Issam AIB
LIP6
University of Paris 6
France

Denis BEAUTIER
Institut supérieur d'électronique de Paris (ISEP)
Paris
France

Abderrahim BENSLIMANE
Informatics Laboratory
University of Avignon and Pays du Vaucluse
France

Nizar BOUABDALLAH
Alcatel
Marcoussis
France

Benoît CAMPEDEL
Institut supérieur d'électronique de Paris (ISEP)
Paris
France

Hakima CHAOUCHI
National Institute of Telecommunications (INT)
Evry
France

Laurent CIARLETTA
Informatics Laboratory
University of Avignon and Pays du Vaucluse
France

Belkacem DAHEB
LIP6
University of Paris 6
France

Wissam FAWAZ
University of Paris-Nord
Villetaneuse
France

Idir FODIL
6WIND
Montigny le Bretonneux
France

Mauro FONSECA
PUC
Curitiba
Brazil

Yacine GHAMRI-DOUDANE
Institut d'informatique d'entreprise (IIE)
Evry
France

Paulo GONÇALVES
Federal University of Rio de Janeiro
Brazil

Vincent GUYOT
ENST
Paris
France

Stéphane LOHIER
SRC departement
University of Marne-la-Vallée
France

Anelise MUNARETTO
PUC-PR
Curitiba
Brazil

Thi Mai Trang NGUYEN
LIP6/ENST
University of Paris 6
France

Laurent OUAKIL
LIP6
University of Paris 6
France

Guy PUJOLLE
LIP6
University of Paris 6
France

Julien RIDOUX
Sprint
San Francisco
USA

Julien ROTROU
Ucopia
Paris
France

Sidi-Mohammed SENOUCI
LIP6
University of Paris 6
France

Vedat YILMAZ
LIP6
University of Paris 6
France

Index

Q

S

V

W